# Calculus&*Mathematica:*
## Basics, Tutorials and Literacy Sheets

# Calculus&*Mathematica:*
# Basics, Tutorials and Literacy Sheets

Part 1

Preliminary Edition

**Donald P. Brown**
University of Illinois, Urbana-Champaign

**Horacio Porta**
University of Illinois, Urbana-Champaign

**J. Jerry Uhl**
University of Illinois, Urbana-Champaign

with the assistance of
Jonathan Manton

**ADDISON-WESLEY PUBLISHING COMPANY**
*The Advanced Book Program*
Redwood City, California • Menlo Park, California • Reading, Massachusetts
New York • Don Mills, Ontario • Wokingham, United Kingdom • Amsterdam
Bonn • Sydney • Singapore • Tokyo • Madrid • San Juan

The graphic on the cover of Calculus&*Mathematica* illustrates the concept of *osculating circles:*

"When we drive on a winding road we have to adjust our steering wheel to follow the bends. If we kept the steering wheel frozen in a turn, we would continue on a circle (most probably leaving the road), and after completing a full circle, we would come back to the road at the same point where we left it. This is the *osculating circle* for that exit and entry point on that road."

Publisher: *Allan M. Wylde*
Marketing Manager: *Laura Likely*
Production Manager: *Jan V. Benes*
Production Assistant: *Karl Matsumoto*
Mathematics Editor: *Barbara Holland*
Cover Design: *Rick Valdez*

**Library of Congress Cataloging-in-Publication Data**
Brown, Donald P.
    Calculus & mathematica: Part 1, basics, tutorials, and literacy sheets/Donald P. Brown, Horacio Porta, J. Jerry Uhl.—Prelim. ed.
        p. cm.
    1. Calculus—Study and teaching—Data processing. 2. Mathematica (Computer program) I. Porta, Horacio. II. Uhl, J. J. (J. Jerry) III. Title. IV. Title: Calculus and mathematica.
QA303.B8848   1991                    515—dc20                    91-19376
ISBN 0-201-57270-2

This preliminary edition was typeset by the authors, using the *Mathematica* programming language.

0-201-57270-2
2345678910-AL-9594939291

# PREFACE FOR THE INSTRUCTOR

*Calculus as currently taught is, alas, full of inert material....*
    -Peter Lax, 1986 Tulane Calculus Conference

*Unfortunately calculus as presently taught has little in common with the way calculus is used...Many of those who finish (calculus) learn little beyond a series of memorized techniques now more commonly performed by computers.*
    -National Research Council Report, **Everybody Counts**

*The purpose of computing is insight, not numbers.*
    -Richard Hamming, **Numerical Methods for Scientists and Engineers**

This preliminary version of Calculus&*Mathematica* is intended to address the concerns and issues reflected in the above quotes. Through the medium of *Mathematica* notebooks, it emphasizes the art of calculation and measurement by exposing students to real mathematics, real data and real problems. The overall goal of Calculus &*Mathematica* is to produce confident, daring students who can fall back on their calculational experience with calculus to move into new situations with the right mathematical reflexes.

Calculus&*Mathematica* uses calculations and graphing as a dynamic device to explore and discover concepts. The materials are constructed of a sequence of *Mathematica* notebooks. The *Mathematica* notebooks constitute a new medium of communication that combine the advantages of a standard word processor, the advantages of an enormously powerful easy-to-use computer algebra system, and superb graphics capabilities. These notebooks comprise an electronic text in which every calculation and plot is alive and can be reexecuted with different parameters; instead of a single passive example, each has the potential to be infinitely many examples. The reasons for an upcoming calculation are discussed, the calculation is executed, and the meaning of the result is assessed in one single medium. Students experience the enchantment and excitement of a calculation in progress in context. Furthermore, because of *Mathematica's* graphic and calculational power, students in Calculus&*Mathematica* make computations that lead to a mathematical experience far richer than that gained by students taking a "traditional" calculus course.

Calculus&*Mathematica's* goals are to promote:

**1. Active Learning Through Experimentation.** Students are encouraged to explore, conjecture and interpret. They quickly learn to ask "why" instead of "how." They are excited to get answers almost instantaneously and to explain them in their own terms. With a little guidance (found right in their Calculus&*Mathematica* electronic text), they soon announce correct results based on their experience.

**2. Conceptual Emphasis on the Process of Problem-Solving.** Problem-solving is the heart of Calculus&*Mathematica*, and because of *Mathematica's* calculational power students can work with real data. Modern students cannot get excited about calculating how fast a shadow moves or a ladder slides, but they are eager to work on problems that they think are real. Some of the real problems found in the course deal with financial planning for college expenses, radioactive decay, nuclear reactors' cooling, national debt projections, car loans, annuities, mortgages, spread of infection, combat models, pollution elimination, programming a robot, safe trajectories for airplane landings, projectile motion, chemical reactions, safe drug doses, the ripple effect of spending, etc.

**3. Geometric Emphasis.** Students learn to view mathematical concepts geometrically. Calculus&*Mathematica* capitalizes on the eye as a mathematical tool by healthy emphasis on student-produced graphics. Students in Calculus&*Mathematica* visualize the mathematical meaning of the process of differentiation by producing their own graphics depicting the plots of $(f(x+h) - f(x))/h$ approaching the plot of $f(x)$ as $h$ goes to 0. They visualize the trapezoidal rule by producing their own graphics depicting the plots of polygonal approximations approaching the plot of the integrand as the mesh goes to 0. They visualize the meaning of convergence of Taylor polynomials by producing their own graphics depicting the plots of partial sums approaching the plot of the function under expansion as more and more terms are used.

Graphic visualizations play another important role in Calculus&*Mathematica*. Too often the traditional student depends on the "answer in the back of the book" for confirmation of his or her efforts. The answer to many substantial calculus problems can be confirmed with a plot. Calculus&*Mathematica* students are urged to do this as a matter of checking their work. As a result, students usually know whether their answers are right or wrong. When they determine an answer is wrong, they can fix it. Seldom in Calculus&*Mathematica* does one see a completed problem turned in with the wrong answer.

**4. Mathematical Maturity.** Calculus&*Mathematica* students can visualize issues never addressed in traditional calculus – like why the graph of a truncated expansion of a periodic function must eventually pull away from the graph of the base function. They can analyze consumer loans and they can tell you why the U.S. population is not growing exponentially. They know that exponential growth beats power growth and can discuss what the result of interest payments on the national debt will be. They know why the fundamental theorem is fundamental and why the natural logarithm is natural. But most of all, their calculational experience gives them the mathematical maturity seldom seen in traditional students.

## Calculus&*Mathematica* Organization

Calculus&*Mathematica*, Part I, contains 30 lesson units covering the first two terms of calculus, both on disk and in printed form. The format is that of problem solving.

The Calculus&*Mathematica* materials consist of:

**1. The electronic course on floppy disks.** Each disk-based unit is organized into three problem-solving components: *Basics*, *Tutorial*, and *Give It A Try*.

Each notebook opens with *Basic Problems* which introduce many of the new ideas in the material under study.

*Tutorial Problems* then introduce techniques and applications. Full electronically active solutions are provided to each *Basic* and *Tutorial* problem.

Closing each notebook is a section of *Give it a Try* problems. Here no solutions are given, but the student through word-processing and calculation adds his or her solution directly to the notebook and turns in the completed notebook for comments, suggestions and grading.

**2. A manual of problems and solutions.** The Calculus&*Mathematica* manual is a lightly-edited print-out of the *Basic* and *Tutorial* problems of the electronic course. This manual supports the electronic courseware by allowing students to review the course when they are not working on the computer.

The manual also includes a component called *Literacy Sheets*. The problems in the *Literacy Sheets* deal with the ideas and calculations that the student should be able to handle without using the computer. You might think of these questions as forming an oral exam on the material. The *Literacy Sheets* were furnished in response to strong student demand. If a student can complete these problems and discuss their under-lying principles, they clearly understand the material. If a student struggles with the *Literacy Sheets*, they need to return to the courseware and manual to learn more about these calculus concepts. They are not present in the electronic text because questions in the *Literacy Sheets* are not meant to be answered with the help of *Mathematica*.

**3. A binder containing student exercises.** An additional booklet provided in the binder includes printed *Give It A Try* sheets. The binder itself has pockets for storage of the floppy disks. The students should be encouraged to store their printed and electronic work in this binder.

## Other Key Features

**Calculus&*Mathematica* is flexible.** Imagine a text in which the teacher can add or delete material as seen fit. If you can, then you understand the flexibility of Calculus &*Mathematica*. If you don't like the introduction to the integral, then you can replace it with your own. In short, you can revise as much of the text as your time allows. Simi-larly the student can do the same by entering notes wherever the student likes. On color machines, the student can highlight sentences with whatever colors are avail-able. Some students make their own core course notes by adding and editing cuts from the Calculus&*Mathematica* electronic course. The possibilities seem limitless.

**Calculus&*Mathematica* is always lively and can be as lean as the instructor wishes.** One mathematics professor at a workshop asked,"This course is very lively, but is it lean?" The answer is that it can be as lean as desired. It is easy to delete topics in Calculus&*Mathematica*. You just cut them electronically to adjust the course to the desired heft.

**Calculus&*Mathematica* is class tested.** Calculus&*Mathematica* has been under classroom testing at University of Illinois, Ohio State University, Stevens Institute of Technology, Knox College, Westmont College, and Vanderbilt University. An addi-tional dozen or so universities, colleges or high schools have signed on for this year for additional testing. At its home university, Illinois, the course has been revised and repaired in a way reminiscent of the way that race cars are serviced during the Indianapolis 500. Unsuccessful problems were fixed or replaced and unsuccessful explanations were rewritten. Probably no new mathematics course has ever been shaken down and revised in the light of classroom experience to the extent of Calculus&*Mathematica*.

## The Mathematics of Calculus&*Mathematica*

**Many mathematical features of Calculus&*Mathematica* are new.** The danger in the current rush to bring technology into the calculus classroom is that the mathematical foundations and the choice of topics will not get the attention they deserve. Furthermore many of the topics and techniques emphasized in traditional text are either inert or not appropriate in the presence of *Mathematica*. A lot of effort has been expended to make Calculus&*Mathematica* a genuinely new calculus course driven by calculation, plotting and approximation.

Here are a few of the mathematical features of the course:

**a.** Differentiation is introduced via the instantaneous growth rates. Students visualize the process of differentiation by plotting, as functions of $x$, the curves $f'(x)$ and $(f(x = h))/h$ on the same axis. They see the limit materialize when the two plots run together. From this viewpoint, it is clear that a function with a positive derivative is increasing. This notion feeds a concept called the "Race Track Principle" which says that if $f(a) = g(a)$ and $f'(x) \geq g'(x)$ for $x \geq a$, then $f(x) \geq g(x)$ for $x \geq a$. The Race Track Principle, which is the active form of the the the Mean Value Theorem, is used frequently throughout the course. The Mean Value Theorem appears as a rather easy consequence of the Race Track Principle.

**b.** The epsilon-delta definition of the derivative is absent, but the process that the epsilon-delta definitions formalizes is present.

**c.** The natural logarithm and the exponential function are introduced in a calculational way in the second of the 30 calculus lessons. Students get plenty of experience with the exponential function throughout the course. This is as it should be because most of the shining applications of calculus involve the exponential function.

**d.** The differential equation $y' = ay$ and situations modeled on this differential equation and related differential equations are studied early in the course.

**e.** Functions of more than one variable come in right at the start.

**f.** Most traditional courses use continuity to push through the proof of the product rule and make not no further use of the continuity. In Calculus&*Mathematica*, the product rule is obtained directly from the chain rule. This eliminates the need to study continuity in the abstract. But concrete continuity is studied when the students are given a function $f(x)$ and asked to report on how many accurate decimals of $x$ are needed to pin down a set number of accurate decimals of $f(x)$.

**g.** Given a value of $f(x)$ at a point and given $f'(x)$, the students learn how to get good approximations of $f(x)$ by a non-linear version of Euler's method.

**h.** Following the leads of Richard Courant and Emil Artin, the course introduces the integral as a measurement of area. Thus the integral is defined in terms of area and not the opposite. This eliminates the bureaucracy of Riemann sums and allows a quick attack on the fundamental theorem.

**i.** The fundamental theorem is restored to a central role assigned to it by Isaac Newton as the calculational base for measurements made by integration.

**j.** The definite integral plays the predominate role in integration. The indefinite integral is included only for the purpose of calculus literacy.

**k.** Curve fitting via interpolation, approximation, differentiation, integration, least squares and splines is a recurrent theme. Students in this course wonder why old-timers think that the Pade theory of rational approximation is mysterious. Log paper and semi-log paper are studied.

**l.** Approximations are studied throughout the course.

**m.** Euler's approach is followed to obtain expansions in powers of $x$ algebraically and students are led to discover convergence and approximations by plotting. Series of numbers are downplayed but not entirely absent. Convergence at the endpoints is dropped.

**n.** Complex numbers are used as appropriated. The complex exponential comes in a natural way. Students use the complex criteria for convergence of Taylor series.

**p.** Students learn when to use a high degree Taylor approximation and they learn when not to use a specific Taylor polynomial.

**q.** Taylor's formula for the coefficients is introduced only after the student has obtained the basic expansions algebraically and is comfortable with them. L'Hôpital's rule appears as an application of Taylor's formula.

## Supplementary Materials

**1. Guide for the Teacher.** Part I covers the authors' general philosophy about calculus instruction, suggestions for using Calculus&*Mathematica*, and addresses common questions.

Part II contains guides to each Calculus&*Mathematica* electronic session. These include suggestions for creating a "richer" or "leaner" course, main ideas and goals of the session, and "notable" problems.

**2. Guide for the Student.** A booklet written by a former Calculus&*Mathematica* student and lab assistant, to help other students get the most out of the course. *Guide for the Student* answers common questions, and introduces basic techniques and advice for successfully completing assignments.

## Acknowledgements

For financial support, the authors thank the National Science Foundation, Wolfram Research, Apple Computer and Addison-Wesley Publishing.

The long list of those who contributed intellectually to this preliminary edition will be given in the next edition.

We hope that you have as much fun in this course as we are having during its development.

Don Brown, undergraduate student in Physics
Horatio Porta, Professor of Mathematics
Jerry Uhl, Professor of Mathematics
Jonathan Manton, undergraduate students in Mathematics
University of Illinois at Urbana-Champaign

Address for comments:
Calculus&*Mathematica*
Calculus Curriculum Development Center
Department of Mathematics
University of Illinois
1409 West Green Street
Urbana, Illinois 61801

# PREFACE FOR THE STUDENT

Welcome to Calculus&*Mathematica*!

In 1986 Peter Lax, a former president of the American Mathematical Society, echoed the sentiments of many calculus instructors when he stated that "Calculus as currently taught is, alas, full of inert material...." Responding to his and others' call for calculus reform, the National Science Foundation began a large-scale initiative to revitalize calculus instruction. Calculus&*Mathematica* is an outcome of this initiative.

## The first calculus adventure in 300 years....

Arithmetic is the introduction to the science of counting. Calculus is the introduction to the science of measurements - both exact and approximate, of quantities that are often changing or in motion. You might think of calculus as a toolbox of powerful measurement devices. In Calculus&*Mathematica* you will learn what these tools are and how to use them. In this sense, this course is much different than the traditional courses which attempt to teach you calculus by observing others doing it, and which spend most of their time studying the tools but seldom using them for anything interesting. Calculus&*Mathematica* students learn calculus by using calculus to explore concepts and solve real-world problems, with the help of the *Mathematica*, the most important calculational software ever produced.

Calculus&*Mathematica* endeavors to make full use of *Mathematica's* highly developed calculational and graphical abilities to present the tools of calculus and to put you in the position of using them. The medium is the *Mathematica* notebook which allows fully word-processed text to be inserted between active *Mathematica* instructions and calculations. This medium allows students to learn *Mathematica* instructions rather easily because you see them in context. No prior experience with *Mathematica* is necessary. Students in early test groups report that after the first two or three weeks most of them were comfortable with the computer.

## How is Calculus&*Mathematica* set up?

Calculus&*Mathematica* is presented as a course based on problem solving. There are three problem-solving components:

1. *Basic* problems present new ideas,
2. *Tutorial* problems introduce techniques and applications, and
3. *Give It A Try* problems are for you to do.

Your Calculus&*Mathematica* package consists of three parts:

**1. The Calculus&*Mathematica* Electronic Text.**
When you open the course on the computer, you will see fully-electronic text. You can progress through the *Basic* and *Tutorial* problems at your own pace, activating each

calculation as you go. Sometimes you will want to rerun some of the electronic problems with new numbers and functions. This is a good idea because it means that each example has the potential to be infinitely many examples. The *Give It A Try* problems for you to do follow the *Basic* and *Tutorial* problems. Here is where you learn by doing because you solve the assigned *Give It A Try* problems within the electronic text. you can print your solutions or save them electronically for grading and study.

## 2. The Calculus&*Mathematica* Manual.
This consists of a lightly-edited printout of the *Basic* and *Tutorial* problems from the electronic course. This manual supports the electronic course by allowing you to review the *Basic* and *Tutorial* problems when you are away from the computer. For best results, you should not look at a problem in the manual until after you have looked at the corresponding problem in the electronic course on the computer. Reading the manual for reviewing is a good idea. Reading the manual before immersing yourself in the electronic course is not a good idea because in doing so you run the risk of acquiring only a superficial understanding of the material.

The manual also contains important sections called *Literacy Sheets*. These are the problems you should be able to do away from the computer. They also ask you to discuss ideas. If you have any trouble with these, you need to return to the courseware and manual to learn more about the calculus concepts.

## 3. The Calculus&*Mathematica* Binder.
This binder contains all the *Give It A Try* problems in the electronic screen course. You can make best use of this material by looking over the problems before you go to the computer to work on them. As you progress through Calculus&*Mathematica*, you can remove the problem lists from this binder and replace each one with your own worked solutions and anything else you care to add. You can even go so far as to copy important parts of the text, annotate them in any way you like, and store them in this binder. You may end up producing your own version of the text. The possibilities are limitless.

### Student Feedback

This is a preliminary edition. The authors welcome student comment and suggestions. You can write to the address below, or complete the questionaire in the back of the manual. If you suggest something really good, your name will be mentioned in future editions.

We hope that you have as much fun in this course as we are having during its development.

Address for comments:

Calculus&*Mathematica* Laboratory
Calculus Curriculum Development Center
Department of Mathematics
University of Illinois
1409 West Green Street
Urbana, Illinois 61801

# Concise Table of Contents

# Table of Contents

# Integral Calculus

## Series and Approximations

# Numbers and Algebra

## Guide

This introductory lesson has the main goal of letting you get familiar with how to use the *Mathematica* system in calculations with numbers and algebra.

## Basics

### ◼ B.1) Arithmetic and *Mathematica*.

Here are some calculations with simple numbers:

#### B.1.a) Sums

Calculate $23 + 51$

**Answer:**

The sum is:

```
In[1]:=    23 + 51
Out[1]=    74
```

Do not type the *In[1]:=* or *Out[1]=* symbols. The machine does this for you.

#### B.1.b) Differences

Calculate $436 - 117$

**Answer:**

The difference is:

```
In[1]:=    436-117
Out[1]=    319
```

#### B.1.c) Products

Calculate 702 times 38.

**Answer:**

The product is:

```
In[1]:=    702 38
Out[1]=    26676
```

If you prefer you can use

```
In[2]:=    702 * 38
Out[2]=    26676
```

#### B.1.d) Powers and Products

Calculate 15 square times 8.

**Answer:**

The result of multiplying the square of 15 times 8 is

```
In[1]:=    15^2 8
Out[1]=    1800
```

Notice that 15 to the power "2 times 8" is calculated by

```
In[2]:=    15^(2 8)
Out[2]=    6568408355712890625
```

#### B.1.e) All together

Calculate $(561 + 290)^5(-341 + 3^{21}) - 897707$

**Answer:**

The result of this horrible operation is

```
In[1]:=    (561+290)^5 (-341+3^21)-897707
Out[1]=    4668680298612356457068655
```

#### B.1.f) Size

How big is this number?

**Answer:**

Of course this number is pretty big. The best way to describe the size of a number is by getting its value in *scientific notation*, i.e., as a product of a small number times a power of 10. Here is the way to do it:

```
In[1]:=    N[ (561+290)^5 (-341+3^21)-897707]
Out[1]=                  24
           4.66868 10
```

### B.1.g.i) Fractions

Add the fractions 2/3 and 11/17.

---

**Answer:**

Here it is:

```
In[1]:=    2/3 + 11/17
Out[1]=    67
           --
           51
```

That's the exact form of the answer. If you want decimals then you can use:

```
In[2]:=    N[2/3 + 11/17]
Out[2]=    1.31373
```

If you want *more* decimals, then you can use:

```
In[3]:=    N[51/23,20]
Out[3]=    2.217391304347826087
```

or

```
In[4]:=    N[51/23,100]
Out[4]=    2.2173913043478260869565217391304347826
           0869565217391304347826086956521739130
           434782608695652173913O435
```

This decimal expression has 100 accurate places. Of course it repeats by chunks of 22 places, so there is no need to calculate any further decimals in this case.

### B.1.g.ii)

---

Simplify the operation on fractions

(13/17)/((23/11) + (21/8)).

---

**Answer:**

Here it is:

```
In[1]:=    (13/17)/((23/11) + (21/8))
Out[1]=    1144
           ----
           7055
```

That's the exact form of the answer. You may want to keep only the numerator:

```
In[2]:=    Numerator[(13/17)/((23/11) + (21/8))]
Out[2]=    1144
```

or the denominator:

```
In[3]:=    Denominator[(13/17)/((23/11) + (21/8))]
```

```
Out[3]=    7055
```

or you may want to numerical value with 35 decimal places:

```
In[4]:=    N[ (13/17)/((23/11) + (21/8)) , 35 ]
Out[4]=    0.16215450035435861091424521615875266
```

### B.1.h.i) Square Roots

---

Calculate the square root of 1369.

---

**Answer:**

The square root of 1369 is:

```
In[1]:=    Sqrt[1369]
Out[1]=    37
```

### B.1.h.ii)

---

Calculate the square root of 1368.

---

**Answer:**

The square root of 1368 is:

```
In[1]:=    Sqrt[1368]
Out[1]=    6 Sqrt[38]
```

Several things happened here. First, the answer we get is not an integer, and this is because 1368 is not the square of any integer. Second, the answer is a product of an integer and an indicated square root. *Mathematica* considers that this is the simplest form of the answer. If you want to see the decimal value, use:

```
In[2]:=    N[Sqrt[1368]]
Out[2]=    36.9865
```

Most times this is what you will be looking for.

### B.1.i) Other Roots

---

Calculate the fifth root of 6436343

---

**Answer:**

The fifth root of 6436343 is:

```
In[1]:=    (6436343)^(1/5)
Out[1]=    23
```

### B.1.j.i) Complex Numbers

---

Calculate the product of complex numbers $(2 - 3I)(4 + I)$
(Here $I$ is the square root of $-1$.)

**Answer:**

The product is:

```
In[1]:=    (2 + 3 I) (4 + I)
Out[1]=    5 + 14 I
```

Notice that *Mathematica* insists in capitalizing I.

**B.1.j.1)**

Calculate the quotient of complex numbers
$(3 + 5I)/(-6 + 7I)$.

**Answer:**

The quotient is:

```
In[1]:=    (3 + 5 I)/(-6 + 7 I)
Out[1]=    1   3 I
           - - ---
           5   5
```

We can verify that the result is OK by multiplication:

```
In[2]:=    (1/5 - 3 I / 5) (-6 + 7 I)
Out[2]=    3 + 5 I
```

**B.1.k) Special operations: factorials, etc.**

*Mathematica* knows some operations by name. For example:

**B.1.k.i**

Calculate the factorial 10! of 10

**Answer:**

The factorial 10! of 10 is the product 10! = 10 times 9 times 8 times 7 etc. all the way down to 1. Here is a way to do it:

```
In[1]:=    10 9 8 7 6 5 4 3 2 1
Out[1]=    3628800
```

*Mathematica* understands the usual abbreaviation:

```
In[2]:=    10!
Out[2]=    3628800
```

**B.1.k.ii**

Calculate the binomial coefficient $\binom{9}{5}$.

**Answer:**

The binomial $\binom{9}{5}$ has the value

$$\binom{9}{5} = 9!/(5!(9 - 5)!)$$

Here is a way to do it:

```
In[1]:=    9!/(5!(9-5)!)
Out[1]=    126
```

*Mathematica* understands the abbreaviation:

```
In[2]:=    Binomial[9,5]
Out[2]=    126
```

**B.1.k.iii**

Calculate Pi.

**Answer:**

Pi (usually written $\pi$) is the ratio of the circumference to the diameter of any circle. Its value is approximately 3.14 *Mathematica* knows about Pi, so you can use it in expressions:

```
In[1]:=    Pi/6 + 2 Pi
Out[1]=    13 Pi
           -----
             6
```

or calculate its numerical value to any accuracy you want:

```
In[2]:=    N[Pi]
Out[2]=    3.14159
```

```
In[3]:=    N[Pi,20]
Out[3]=    3.1415926535897932385
```

```
In[4]:=    N[Pi,200]
Out[4]=    3.1415926535897932384626433832795028841
           9716939937510582097494459230781640628
           6208998628034825342117067982148086513
           2823066470938446095505822317253594081
           2848111745028410270193852110555964462
           2948954930382
```

Unlike the decimals of ordinary fractions, the decimals of Pi do not repeat. Notice that *Mathematica* insists in capitalizing Pi.

**B.1.k.iv**

Calculate the number of degrees in a radian and the number of radians in a degree.

---

**Answer:**

It takes $2\pi$ radians to make a full turn and it takes 360 degrees to make a full turn. Then it takes $360/2\pi$ degrees to make a radian:

```
In[1]:=    360/(2 Pi)
Out[1]=    180
           ---
           Pi
```

The numerical value is

```
In[2]:=    N[360/(2 Pi)]
Out[2]=    57.2958
```

In the same manner we can get the number of radians in one degree:

```
In[3]:=    N[2 Pi/360]
Out[3]=    0.0174533
```

*Mathematica* knows this value under the name of **Degree**

```
In[4]:=    N[Degree]
Out[4]=    0.0174533
```

This is useful in cases like this: if you have a formula that requires the value of an angle to be given in radians and you want to calculate the value for the angle of $30°$ por example, then you evaluate at "30 **Degree** ". Here is an example:

```
In[5]:=    N[Sin[30 Degree]]
Out[5]=    0.5
```

Notice that Sin[30] is quite a different thing:

```
In[6]:=    N[Sin[30]]
Out[6]=    -0.988032
```

More about this below in the trig section.

■ **B.2) Algebra and *Mathematica*.**

*Mathematica* will do almost any operation with symbols like $x$, $y$, $a$, $b$, $c$, etc. For example:

### B.2.a.i) Products

---

Multiply out $(x - 2)(x - 1)x(x + 1)(x + 2)$.

---

**Answer:**

The product is:

```
In[1]:=    Expand[(x - 2) (x - 1) x (x + 1) (x + 2)]
Out[1]=            3     5
           4 x - 5 x  + x
```

If you don't hang the command **Expand** with the square brackets **[ ]** onto the product, then *Mathematica* will not multiply out the terms because it was not told to do so. Try it.

```
In[2]:=    (x - 2) (x - 1) x (x + 1) (x + 2)
Out[2]=    (-2 + x) (-1 + x) x (1 + x) (2 + x)
```

Get it?

**B.2.a.ii)**

---

Multiply out $(x - 2y)(3x - y)(x + y)$.

---

**Answer:**

The product is:

```
In[1]:=    Expand[(x - 2y) (3x -y) (x + y)]
Out[1]=         3      2          2      3
           3 x  - 4 x  y - 5 x y  + 2 y
```

**B.2.b.i) Factoring**

---

Factor $24 - 50x + 35x^2 - 10x^3 + x^4$.

---

**Answer:**

The factorization is:

```
In[1]:=    Factor[24 - 50 x + 35 x^2 - 10 x^3 + x^4]
Out[1]=    (-4 + x) (-3 + x) (-2 + x) (-1 + x)
```

This simply means that if we multiply out the terms $(-4 + x)$ times $(-3+x)$ times $(-2+x)$ times $(-1+x)$ we will get the polynomial $24 - 50x + 35x^2 - 10x^3 + x^4$. Notice that *Mathematica* orders the symbols inside each parenthesis by putting numbers first, then letters.

**B.2.b.ii)**

---

Factor $4x^4 - 20x^3y + 33x^2y^2 - 20xy^3 + 4y^4$.

---

**Answer:**

The factorization is:

```
In[1]:=    Factor[4 x^4 - 20 x^3 y + 33 x^2 y^2 - 20 x
           y^3 + 4 y^4]
Out[1]=               2          2
           (x - 2 y)  (2 x - y)
```

Check:

```
In[2]:=    Expand[(x - 2 y)^2 (2 x - y)^2]
Out[2]=       4      3        2 2        3
           4 x  - 20 x  y + 33 x y  - 20 x y  +

              4
           4 y
```

On the money.

### B.2.c.i) Together

Combine
$$4/x + 7/(x-1) - 8x^2/(x+2)$$
into a single expression.

### Answer:

The result is:

```
In[1]:=    Together[ 4/x + 7/(x - 1) - 8 x^2/(x + 2)]
Out[1]=              2      3      4
           -8 + 18 x + 11 x + 8 x - 8 x
           --------------------------------
                 (-1 + x) x (2 + x)
```

Don't worry about multiplying out the denominator; the factored form is usually favored because it is simpler and displays more information than the expanded form.

### B.2.c.ii) Numerators and Denominators

Calculate the numerator that results from combining
$$4/x + 7/(x-1) - 8x^2/(x+2)$$
into a single fraction.

### Answer:

The result is:

```
In[1]:=    Numerator[Together[
           4/x + 7/(x - 1) - 8 x^2/(x + 2)]]
Out[1]=              2      3      4
           -8 + 18 x + 11 x + 8 x - 8 x
```

### B.2.d) Simplify

Simplify $(a^6 - 1)/(a^2 - 1)$.

### Answer:

The result is:

```
In[1]:=    Simplify[(a^6 - 1)/(a^2 - 1)]
Out[1]=         2    4
           1 + a  + a
```

That's pretty simple.

### B.2.e) Polynomial division

Give the quotient and remainder that you get after dividing $4x^6 + x - 1$ by $x^2 + 1$.

### Answer:

The quotient is:

```
In[1]:=    PolynomialQuotient[
           4 x^6 +x - 1, x^2 + 1, x]
Out[1]=          2      4
           4 - 4 x  + 4 x
```

The remainder is:

```
In[2]:=    PolynomialRemainder[
           4 x^6 +x - 1, x^2 + 1, x]
Out[2]=    -5 + x
```

This means
$$(4x^6 + x - 1)/(x^2 + 1)$$
$$= (4 - 4x + 4x^4) + (-5 + x)/(x^2 + 1).$$

### B.2.f.i) Solve equations

Solve $x^2 + 2x - 8 = 0$ for $x$.

### Answer:

Solution is given by: (Note the use of the double equal sign within the **Solve** command)

```
In[1]:=    Solve[x^2 + 2 x - 8 == 0,x]
Out[1]=    {{x -> -4}, {x -> 2}}
```

This is in agreement with the factorization:

```
In[2]:=    Factor[x^2 + 2 x - 8]
Out[2]=    (-2 + x) (4 + x)
```

### B.2.f.ii)

Solve $x^2 + bx + c = 0$ for $x$ in terms of $b$ and $c$.

### Answer:

Solution is given by: (Note again the use of the double equal sign within the Solve command)

```
In[1]:=    Solve[x^2 + b x + c == 0,x]
Out[1]=                      2
                   -b + Sqrt[b  - 4 c]
           {{x -> -------------------},
                          2
```

```
          -b - Sqrt[b  - 4 c]
{x -> -------------------}}
              2
```

Look familiar?

Note that a small typing error can throw everything off:

```
In[2]:=    Solve[x^2 + b x + c = 0,x]
```

*Set::write:*

*Symbol Plus is write protected.*

*Solve::eq:*

*0 is not an equation or system of*

*equations.*
```
Out[2]=    Solve[0, x]
```

*Mathematica* is squawking like a duck. The trouble here is that we forgot the double equal sign within the Solve command.

Here is another small typing error that can cause misunderstandings:

```
In[3]:=    Solve[x^2 + bx + c == 0,x]
Out[3]=    {{x -> Sqrt[-bx - c]},

           {x -> -Sqrt[-bx - c]}}
```

The trouble here is that we forgot the space between the $b$ and the $x$. *Mathematica* took the symbol $bx$ as the name of a new quantity instead of taking it as $b$ times $x$. Putting the space between $b$ and $x$ fixes everything:

```
In[4]:=    Solve[x^2 + b x + c == 0,x]
Out[4]=
                     2
          -b + Sqrt[b  - 4 c]
{{x -> -------------------},
              2

                     2
          -b - Sqrt[b  - 4 c]
 {x -> -------------------}}
              2
```

That's better.

### B.2.f.iii)

Solve $x^3 + x - 2 = 0$.

### Answer:

Solution is given by:

```
In[1]:=    Clear[x,b,c]
```

```
In[2]:=    Solve[x^3 + x - 2 == 0,x]
Out[2]=                      -1 + Sqrt[-7]
           {{x -> 1}, {x -> -------------},
```

```
                          2
          -1 - Sqrt[-7]
{x -> --------------}}
              2
```

Note the real root

$x = 1$

and the two complex roots

$x = (-1 + I\sqrt{7})/2$

$x = (-1 - I\sqrt{7})/2$.

### B.2.f.iv)

Solve the system of equations

$3x + 2y - 2 = 0$

$7x + 5y + 4 = 0$

### Answer:

Solution is given by:

```
In[1]:=    Solve[{3 x + 2 y - 2 == 0,
                  7 x + 5 y + 4 == 0},{x,y}]
Out[1]=    {{x -> 18, y -> -26}}
```

## Tutorial

### ■ T.1) Numbers.

### T.1.a)

Express 28640000 in scientific notation.

### Answer:

In scientific notation, 186000 is:

```
In[1]:=    N[28640000]
Out[1]=                 7
           2.864 10
```

### T.1.b)

Express 0.0000000863 in scientific notation.

### Answer:

In scientific notation, 0.0000000863 is:

```
In[1]:=    N[0.0000000863]
Out[1]=                 -8
           8.63 10
```

### T.1.c)

Calculate $\sqrt{3} = 3^{1/2}$ to ten accurate decimals.

**Answer:**

We can try:

```
In[1]:=   N[Sqrt[3],10]
Out[1]=   1.732050808
```

These decimals are all accurate, but there are only *nine* of them, although there are *ten* accurate digits including the 1 out front. To get ten accurate decimals, use:

```
In[2]:=   N[Sqrt[3],11]
Out[2]=   1.7320508076
```

If you want twenty accurate accurate decimals, use:

```
In[3]:=   N[Sqrt[3],21]
Out[3]=   1.73205080756887729353
```

### T.1.d)

Calculate $\sqrt{170} = 170^{1/2}$ to nine accurate decimals.

**Answer:**

We can try:

```
In[1]:=   N[Sqrt[170],10]
Out[1]=   13.03840481
```

These decimals are all accurate, but there are only eight of them, although there are ten accurate digits including the 13 out front. To get ten accurate decimals, use:

```
In[2]:=   N[Sqrt[170],12]
Out[2]=   13.0384048104
```

If you want eighteen accurate accurate decimals, use:

```
In[3]:=   N[Sqrt[170],20]
Out[3]=   13.038404810405297429
```

### T.1.e)

Expand $(1 + I)^4$.

**Answer:**

Here it is:

```
In[1]:=   Expand[(1 + I)^4]
```

```
Out[1]=   -4
```

That's right: the fourth power of the *complex* number $(1 + I)^4$ is the *real* number $-4$ because

$$(1 + I)^4 = 1 + 4I + 6I^2 + 4I^3 + I^4$$

$$= 1 + 4I - 6 - 4I + 1 = -4.$$

## ■ T.2) $\pi$.

### T.2.a.i)

What is the area in square units of a circle of radius 5?

**Answer:**

The exact value is given by:

```
In[1]:=   Pi 5^2
Out[1]=   25 Pi
```

If you want a decimal value, then use:

```
In[2]:=   N[Pi 5^2]
Out[2]=   78.5398
```

If you want 10 accurate decimals, then use:

```
In[3]:=   N[Pi 5^2,12]
Out[3]=   78.5398163397
```

### T.2.a.ii)

Is it true that $\pi = 22/7$?

**Answer:**

The numerical value of $\pi$ to five accurate places is:

```
In[1]:=   N[Pi]
Out[1]=   3.14159
```

The numerical value of 22/7 to five accurate places is:

```
In[2]:=   N[22/7]
Out[2]=   3.14286
```

They are close, but they are not equal. It's a shame, but that's the way nature is. 22/7 was used for centuries as a reasonable approximation for $\pi$. Other reasonable approximations for $\pi$ are 355/113 and 392699/125000.

## ■ T.3) Algebra.

### T.3.a)

Expand $(1 - x)(1 + x + x^2 + x^3 + x^4)$

**Answer:**

Here it is:

```
In[1]:=   Expand[(1 - x) (1 + x + x^2 + x^3 + x^4)]
Out[1]=        5
          1 - x
```

Wow! What is going on here?

### T.3.b)

Divide $x^6 - 1$ by $x - 1$.

**Answer:**

There are many ways of using *Mathematica* to arrive at the answer.

Here is one way:

```
In[1]:=   PolynomialQuotient[(x^6 - 1), (x - 1), x]
Out[1]=               2    3    4    5
          1 + x + x  + x  + x  + x
```

```
In[2]:=   PolynomialRemainder[(x^6 - 1), (x - 1), x]
Out[2]=   0
```

Consequently

$$(x^6 - 1)/(x - 1) = 1 + x + x^2 + x^3 + x^4 + x^5.$$

Here is another way:

```
In[3]:=   Cancel[(x^6 - 1)/(x - 1)]
Out[3]=               2    3    4    5
          1 + x + x  + x  + x  + x
```

Finally we can calculate the quotient *in factored form* by:

```
In[4]:=   Factor[(x^6 - 1)/(x - 1)]
Out[4]=                 2            2
          (1 + x) (1 - x + x ) (1 + x + x )
```

Of course this expands to the correct expression:

```
In[5]:=   Expand[Factor[(x^6 - 1)/(x - 1)]]
Out[5]=               2    3    4    5
          1 + x + x  + x  + x  + x
```

### T.3.c.i)

Reduce
$$(y^2 - xy - 2x^2)/(12x^2 - 3y^2)$$
to lowest terms by canceling common factors.

**Answer:**

Here we go:

```
In[1]:=   Cancel[(y^2 - x y - 2 x^2)/
                 (12 x^2 - 3 y^2)]
Out[1]=    -x - y
          ---------
          6 x + 3 y
```

If you want to get the idea of what happened here, look at the factored versions of the numerator and denominator:

```
In[2]:=   top = Factor[y^2 - x y - 2 x^2]
Out[2]=   -((2 x - y) (x + y))
```

```
In[3]:=   bottom = Factor[12 x^2 - 3 y^2]
Out[3]=   3 (2 x - y) (2 x + y)
```

Now you can cancel the common factor $(2x - y)$ and read off the reduced form:

$$-(x + y)/(3(2x + y)) = (-x - y)/(6x + 3y).$$

The names **top** and **bottom** are quite arbitrary: you can choose any names *Mathematica* does not know anything about. The advantage in using names is that we can use them again and again. For example:

```
In[4]:=   top
Out[4]=   -((2 x - y) (x + y))
```

```
In[5]:=   top/bottom
Out[5]=    -(x + y)
          -----------
          3 (2 x + y)
```

```
In[6]:=   Expand[top^2 + bottom^2]
Out[6]=          4      3          2 2          3
          148 x  + 4 x  y - 75 x  y  - 2 x y  +

                4
          10 y
```

Have fun.

### T.3.c.ii)

Reduce
$$(-a^3 + 2a^2b - 2ab^2 + 4b^3)/(-2a^3 + 4a^2b - ab^2 + 2b^3)$$
to lowest terms by canceling common factors. Then calculate the product of the numerator times the denominator of the resulting fraction.

**Answer**

Easy does it. First we want to see what we have to do. Here is the given fraction :

```
In[1]:=    (-a^3 + 2 a^2 b - 2 a b^2 + 4 b^3)/
           (-2a^3 + 4 a^2 b - a b^2 + 2 b^3)

Out[1]=      3     2       2     3
           -a + 2 a b - 2 a b + 4 b
           ---------------------------
              3     2      2     3
           -2 a + 4 a b - a b + 2 b
```

Let us give it a name that has not been used yet:

```
In[2]:=    givenfraction=
           (-a^3 + 2 a^2 b - 2 a b^2 + 4 b^3)/
           (-2a^3 + 4 a^2 b - a b^2 + 2 b^3)

Out[2]=      3     2       2     3
           -a + 2 a b - 2 a b + 4 b
           ---------------------------
              3     2      2     3
           -2 a + 4 a b - a b + 2 b
```

Good. Now we can **Cancel** the common factors:

```
In[3]:=    Cancel[givenfraction]

Out[3]=      2     2
           -a - 2 b
           ----------
             2     2
           -2 a - b
```

Great! Next we want the product of the numerator and denominator:

```
In[4]:=    top bottom

Out[4]=                2
           -3 (2 x - y) (x + y) (2 x + y)
```

What? This is obviously wrong, since there are no $x$'s or $y$'s in the expression. What happened is that we used **top** and **bottom** for the numerator of the fraction in the previous problem and *Mathematica* has remembered them. This is the main source of conflict when using names. The way to avoid this conflict is by defining again the values with new names (if we want to keep the old expressions alive) or with the same names. A good idea is to **Clear** the old names if we are going to replace them. Here is the complete procedure:

```
In[5]:=    Clear[top,bottom]
```

This erases the values assigned to the names **top** and **bottom** so we can use them again for any expressions we want:

```
In[6]:=    top=Numerator[Cancel[givenfraction]]

Out[6]=      2     2
           -a - 2 b
```

```
In[7]:=    bottom=Denominator[Cancel[givenfraction]]

Out[7]=      2     2
           -2 a - b
```

Now we can calculate the product:

```
In[8]:=    top bottom

Out[8]=       2     2      2     2
           (-a - 2 b ) (-2 a - b )
```

or in expanded form:

```
In[9]:=    Expand[top bottom]

Out[9]=      4     2 2     4
           2 a + 5 a b + 2 b
```

Not bad, eh?

## T.3.d.i)

Is it true that
$$\sqrt{a+b} = \sqrt{a} + \sqrt{b}?$$

Why or why not? Here $\sqrt{x} = x^{1/2}$ is the number whose square is $x$.

**Answer:**

If the formula is right, then it must hold up when we plug in numbers.

Try it:

```
In[1]:=    Sqrt[9 + 16]
Out[1]=    5
```

```
In[2]:=    Sqrt[9] + Sqrt[16]
Out[2]=    7
```

The answers are different, so it cannot be right to say $\sqrt{a+b} = \sqrt{a} + \sqrt{b}$.

## T.3.d.ii)

Combine $1/\sqrt{a+b} - 1/\sqrt{a}$ into an expression with a common denominator.

**Answer:**

```
In[1]:=    Together[1/Sqrt[a + b] - 1/Sqrt[a]]
Out[1]=    Sqrt[a] - Sqrt[a + b]
           ---------------------
             Sqrt[a] Sqrt[a + b]
```

Let's rerun this with a name so we can call it up later:

```
In[2]:=    common =
           Together[1/Sqrt[a + b] - 1/Sqrt[b]]
Out[2]=    Sqrt[b] - Sqrt[a + b]
           ---------------------
             Sqrt[b] Sqrt[a + b]
```

Now we can call this expression at any time by evaluating:

```
In[3]:=     common
Out[3]=     Sqrt[b] - Sqrt[a + b]
            ---------------------
            Sqrt[b] Sqrt[a + b]
```

Good.

**T.3.d.iii)**

Take your answer to part ii) and change its form by rationalizing the numerator.

**Answer:**

Recall from part ii):

```
In[1]:=     common
Out[1]=     Sqrt[b] - Sqrt[a + b]
            ---------------------
            Sqrt[b] Sqrt[a + b]
```

Rationalizing the numerator involves changing this expression by multiplication and division of the same term, namely in this case the sum:

```
In[2]:=     sum=Sqrt[b] + Sqrt[a+b]
Out[2]=     Sqrt[b] + Sqrt[a + b]
```

The result will be:

$$((\sqrt{b} - \sqrt{a+b})(\sqrt{b} + \sqrt{a+b})) / (\sqrt{b}\sqrt{a+b}(\sqrt{b} + \sqrt{a+b}))$$
$$= (b - (a+b)) / (\sqrt{b}\sqrt{a+b}(\sqrt{b} + \sqrt{a+b}))$$
$$= -a / (\sqrt{b}\sqrt{a+b}(\sqrt{b} + \sqrt{a+b})).$$

Here is what we do: split the fraction as the quotient of:

```
In[3]:=     top=Numerator[common]
Out[3]=     Sqrt[b] - Sqrt[a + b]
```

and

```
In[4]:=     bottom=Denominator[common]
Out[4]=     Sqrt[b] Sqrt[a + b]
```

Now calculate

( **common sum** )/ **sum** = (( **top** / **bottom** ) **sum** )/ **sum** =

( **top sum** )/( **bottom sum** )

but we must remember to simplify the product **common sum** , which was the whole reason to do this!. Here we go:

```
In[5]:=     rationalized=
            Simplify[top sum]/(bottom sum)
Out[5]=     -(a / (Sqrt[b] Sqrt[a + b]

                    (Sqrt[b] + Sqrt[a + b])))
```

That's is. Let's check our work:

```
In[6]:=     Simplify[common - rationalized]
Out[6]=     0
```

On the target. Another way of checking is to look at:

```
In[7]:=     Simplify[common/rationalized]
Out[7]=     1
```

Fine.

■ **T.4) Logs and exponentials.**

The idea of the logarithm base $b$ is simple:

We say

$$y = \log[b, x]$$

if

$$b^y = x.$$

Before the age of computers all serious computations were based on logarithms. And with good reason because logarithms change heavy multiplications and divisions into lightweight additions and subtractions thanks to the identities

$$\log[b, xy] = \log[b, x] + \log[b, y]$$
$$\log[b, x/y] = \log[b, x] - \log[b, y]$$
$$\log[b, b^x] = x.$$

Now the days of using logarithms as a numerical calculating tool are well in the past, but their place in the family of important ideas is not in doubt because they help us see what is going on in a myriad of symbolic calculations.

**T.4.a)**

Calculate $\log[2.7, 10]$.

**Answer:**

Look at:

```
In[1]:=     N[Log[2.7,10]]
Out[1]=     2.31823
```

This means that $2.7^{2.31823}$ should be 10. Try it:

```
In[2]:=     N[2.7^2.31823]
Out[2]=     10.
```

It checks out

**T.4.b)**

Find a number $x$ such that $3^x = 20$.

**Answer:**

The right $x$ is:

```
In[1]:=    N[Log[3,20]]
Out[1]=    2.72683
```

Check:

```
In[2]:=    3^N[Log[3,20]]
Out[2]=    20.
```

Got it.

### T.4.c)

Explain why log[2, 1024] = 10.

**Answer:**

Look at:

```
In[1]:=    2^10
Out[1]=    1024
```

This tells us that $2^{10} = 1024$. And this is why log[2, 1024] = 10. Check:

```
In[2]:=    Log[2,1024]
Out[2]=    10
```

Got it.

### T.4.d)

Combine log[12, 24] + log[12, 60] − log[12, 10] into a single number and explain the result.

**Answer:**

Take a look:

```
In[1]:=    N[Log[12, 24] + Log[12, 60] - Log[12, 10]]
Out[1]=    2.
```

The answer is 2. To see why note
$$\log[12, 24] + \log[12, 60] - \log[12, 10]$$
$$= \log[12, (2460/10)] = \log[12, 144] = \log[12, 12^2] = 2.$$

■ **T.5) Trigonometry.**

### T.5.a.i)

Use *Mathematica* to find a number $t$ with sin[$t$] = .902.

**Answer:**

Try:

```
In[1]:=    t = ArcSin[.902]
Out[1]=    1.12438
```

Check:

```
In[2]:=    Sin[1.12438]
Out[2]=    0.902
```

Got it.

### T.5.a.ii)

Use *Mathematica* to find a number $t$ with tan[$t$] = 0.902.

**Answer:**

Try:

```
In[1]:=    t = ArcTan[.902]
Out[1]=    0.733919
```

Check:

```
In[2]:=    Tan[t]
Out[2]=    0.902
```

Got it.

### T.5.a)

Use *Mathematica* to simplify

$(\sin[x])^2 + (\cos[x])^2$ and $(\cos[x])^2 - (\sin[x])^2$.

**Answer:**

Use:

```
In[1]:=    TrigExpand[Sin[x]^2 + Cos[x]^2]
Out[1]=    1
```

This means $\sin[x]^2 + \cos[x]^2 = 1$ no matter what $x$ is. Does this ring a bell?

Next look at:

```
In[2]:=    TrigExpand[Cos[x]^2 - Sin[x]^2]
Out[2]=    Cos[2 x]
```

This means $\cos[x]^2 - \sin[x]^2 = \cos[2x]$ no matter what $x$ is. Does this ring another bell?

## T.5.b)

Expand $(\cos[x] + I\sin[x])^2$. Then express the result in terms of Sines and Cosines of integral multiples of $x$ and assess the result:

**Answer:**

Squaring gives:

```
In[1]:=   square = Expand[(Cos[x] + I Sin[x])^2]
Out[1]=         2
          Cos[x]  + 2 I Cos[x] Sin[x] - Sin[x]
                                              2
```

To express this in terms of Sines and Cosines of integral multiples of $x$, use:

```
In[2]:=   TrigExpand[square]
Out[2]=   Cos[2 x] + I Sin[2 x]
```

Both expressions are different forms of the same thing; so we are forced to decide that

$$\cos[x]^2 - \sin[x]^2 + I(2\sin[x]\cos[x])$$
$$= \cos[2x] + I\sin[2x].$$

This means

$$\cos[2x] = \cos[x]^2 - \sin[x]^2$$

and $\sin[2x] = 2\sin[x]\cos[x]$.

■ **T.6) Solving equations and fishing out the results.**

### T.6.a.i)

To solve for one variable, you need at least one equation. Solve for $x$ if $3x^2 + 7x - 20 = 0$.

**Answer:**

Just in case, let us clear the name of $x$

```
In[1]:=   Clear[x]
```

Generally it is best to give the equation a name.

```
In[2]:=   eqn = 3 x^2 + 7 x - 20 == 0
Out[2]=              2
          -20 + 7 x + 3 x  == 0
```

and also to give a name to the solutions

```
In[3]:=   sols = Solve[eqn,x]
Out[3]=                        5
          {{x -> -4}, {x -> -}}
                             3
```

This gives the answers $x = -4$ and $x = 5/3$. Not much to it. We could have found the numbers without using the name **sols**. Try it:

```
In[4]:=   Solve[eqn,x]
Out[4]=                       5
          {{x -> -4}, {x -> -}}
                            3
```

Naming the solutions sometimes adds convenience because we can call up the individual solutions by activating:

```
In[5]:=   sols[[1,1,2]]
Out[5]=   -4
```

```
In[6]:=   sols[[2,1,2]]
Out[6]=   5
          -
          3
```

If you want to get an idea of what the funny numbers like `[[1,1,2]]` mean, then look at:

```
In[7]:=   sols[[1]]
Out[7]=   {x -> -4}
```

```
In[8]:=   sols[[2]]
Out[8]=          5
          {x -> -}
                3
```

```
In[9]:=   sols[[1,1]]
Out[9]=   x -> -4
```

```
In[10]:=  sols[[2,1]]
Out[10]=         5
          x -> -
               3
```

```
In[11]:=  sols[[1,1,1]]
Out[11]=  x
```

```
In[12]:=  sols[[1,1,2]]
Out[12]=  -4
```

Experiment with these.

### T.6.a.ii)

Solve

$$1 + 2x - 2x^2 - 2x^3 + 2x^4 + x^5 - 2x^6 + x^8 = 0$$

for $x$.

**Answer:**

Try:

```
In[1]:=   Clear[x]
          eqn =
          1 + 2 x - 2 x^2 - 2 x^3 + 2 x^4 +
```

```
            x^5 - 2 x^6 + x^8 == 0
Out[1]=                  2      3     4    5
            1 + 2 x - 2 x  - 2 x  + 2 x  + x  -

                       6    8
                   2 x  + x  == 0
```

```
In[2]:=     Solve[eqn,x]
Out[2]=                            2        3
            {ToRules[Roots[2 x - 2 x  - 2 x  +

                   4    5      6    8
               2 x  + x  - 2 x  + x  == -1, x]]}
```

This is *Mathematica* 's way of telling you that it cannot find exact solutions of this equation. This is not a defect in *Mathematica* because there is no general formula for the exact solution of polynomial equations when the degree is bigger than 4. This was proved by the French mathematician Evariste Galois. In addition to being a mathematical genius, Galois (1811-1832) was an interesting fellow. A chronic "problem student," he was denied admission to the Ecole Polytechnique because he threw an eraser at one of the examiners. He became enmeshed in the Revolution of 1830 and died in a duel at age 21.

Failing to find the exact solutions, we can get decimal approximations by hanging **N[ ]** around the solve command:

```
In[3]:=     N[Solve[eqn,x]]
Out[3]=     {{x -> -1.30678},

               {x -> -0.880513 - 0.57655 I},

               {x -> -0.880513 + 0.57655 I},

               {x -> -0.417648},

               {x -> 0.674284 - 0.952273 I},

               {x -> 0.674284 + 0.952273 I},

               {x -> 1.06844 - 0.270831 I},

               {x -> 1.06844 + 0.270831 I}}
```

There are eight complex solutions: two are real and six are not. Note how the complex roots come up in conjugate pairs $a + Ib$ and $a - Ib$.

### T.6.b)

To solve for two variables, you need at least two equations.

Solve for $x$ and $y$ if the equations are:

$$x^2 + 4x - 20y^2 = 0$$

and

$$x + y = 16.$$

**Answer:**

```
In[1]:=     Clear[x,y]
```

```
In[2]:=     eqn1 = x^2 + 4 x - 20 y^2 == 0
Out[2]=              2         2
            4 x + x  - 20 y  == 0
```

```
In[3]:=     eqn2 = x + y == 16
Out[3]=     x + y == 16
```

Now we can solve the equations (note the squiggily brackets: anytime you want to write a *list* of symbols in *Mathematica* , you need the squiggily brackets):

```
In[4]:=     sols = Solve[{eqn1,eqn2},{x,y}]
Out[4]=              322 - 2 Sqrt[1601]
            {{x -> -------------------,
                           19

                   -36 + 4 Sqrt[1601]
               y -> -------------------},
                           38

               322 + 2 Sqrt[1601]
               {x -> -------------------,
                           19

                   -36 - 4 Sqrt[1601]
               y -> -------------------}}
                           38
```

There they are. The first pair $\{x, y\}$ of solutions is given by:

```
In[5]:=     {sols[[1,1,2]],sols[[1,2,2]]}
Out[5]=     322 - 2 Sqrt[1601]   -36 + 4 Sqrt[1601]
            {-------------------, -------------------}
                    19                   38
```

The second pair $\{x, y\}$ of solutions is given by:

```
In[6]:=     {sols[[2,1,2]],sols[[2,2,2]]}
Out[6]=     322 + 2 Sqrt[1601]   -36 - 4 Sqrt[1601]
            {-------------------, -------------------}
                    19                   38
```

Get the idea?

### T.6.c)

You need a minimum of how many equations to solve for five variables?

**Answer:**

You need a minimum of five equations.

### T.6.d)

What can you try to do when you have, say, three equations and five unknowns?

**Answer:**

You can hope to solve 3 of the variables in terms of the remaining $5 - 3 = 2$ variables. For instance here are three equations in the five variables $h, k, r, x$ and $y$:

```
In[1]:=    Clear[h,k,r,x,y]

In[2]:=    first = r + h + k == 1
Out[2]=    h + k + r == 1

In[3]:=    second = (x - h)^2 + (y - k)^2 == r^2
Out[3]=            2          2     2
           (-h + x) + (-k + y) == r

In[4]:=    third = y == 3 + x
Out[4]=    y == 3 + x
```

These are *three* equations in the *five* unknowns $x, y, h, k$ and $r$.

We cannot hope to solve these for $h, k, r, x$ and $y$. But we can hope to solve for 3 of the variables in terms of the $5 - 3 = 2$ other variables.

Just for the hell of it, we'll solve for $x, y$ and $r$ in terms of $h$ and $k$:

```
In[5]:=    Solve[{first,second,third},{x,y,r}]
Out[5]=    {{r -> 1 - h - k,

            y -> (3 + h + k +

                            2
                 Sqrt[(3 - h - k) -

                   4 (4 + h - 2 k - h k)]) / 2,

            x -> (-3 + h + k +

                            2
                 Sqrt[(3 - h - k) -

                   4 (4 + h - 2 k - h k)]) / 2},

            {r -> 1 - h - k,

            y -> (3 + h + k -

                            2
                 Sqrt[(3 - h - k) -

                   4 (4 + h - 2 k - h k)]) / 2,

            x -> (-3 + h + k -

                            2
                 Sqrt[(3 - h - k) -

                   4 (4 + h - 2 k - h k)]) / 2}}
```

That saved a lot of work.

**T.6.e)**

___

Solve the system

$$x^2 y z = 5$$

$$1/xy = 5$$

for $y$ and $z$ in terms of $x$.

___

**Answer:**

```
In[1]:=    Clear[x,y,z]

In[2]:=    Solve[{x^2 y z == 5, 1/(x y) == 5},{y,z}]
Out[2]=          25          1
           {{z -> --,  y -> ---}}
                  x          5 x
```

Neat.

# Literacy Sheet

*These problems give you an idea of what you should be able to discuss with ease and authority away from the machine. You should feel free to use the machine to help you with whatever you do not understand at first.*

**1.** If a polynomial $p[x]$ factors into

$$5(x-1)(x+4)(3x-7),$$

then what are the solutions of $p[x] = 0$?

.

.

.

**2.** A quadratic polynomial $p[x]$ is known to have the form

$$p[x] = 4x^2 + bx + c,$$

but you are not told what $b$ and $c$ are. If new information comes in that $p[x] = 0$ for $x = -1$ and $x = 3$, then what are $b$ and $c$?

.

.

.

**3.** You want to nail down the values of eight variables. To find their values, you need a minimum of how many equations relating the eight variables?

.

.

.

You have eight variables and six equations relating them. How many of the variables can you hope to solve for in terms of the remaining variables?

.

.

.

**4.** True-False: (Write your answer next to the statement.)

**4.a** $(x^t)^3 = x^{3t}$

.

**4.b** $x^s/x^t = x^{s/t}$

.

**4.c** $x^s x^t = x^{s+t}$

.

**4.d** $(a+b)/(ab) = 1/a + 1/b$

.

**4.e** $ab/(a+b) = b + a$

.

.

**4.f** $(x+y)^2 = x^2 + y^2$

.

.

**4.g** $\sin[x+y] = \sin[x] + \sin[y]$

.

**4.h** $\sin[xy] = \sin[x]\sin[y]$

.

.

**4.i** $\log[xy] = \log[x] + \log[y]$ for $x$ and $y > 0$.

.

.

**4.j** $\log[x^y] = y\log[x]$ provided $x > 0$.

.

.

**4.k** $\log[x/y] = \log[x]/\log[y]$

.

.

**4.l** Regardless of the sign of $x$, you can be sure that $3^x > 0$.

.

**5.** Complete the formulas:

$\sin[0] =$

$\cos[0] =$

$\sin[\pi/2] =$

$\cos[\pi/2] =$

$\sin[\pi] =$

$\cos[\pi] =$

$\tan[\pi/4] =$

$\arctan[1] =$

$\arcsin[0] =$

$\arctan[0] =$

$\arcsin[-1] =$

$\arctan[-1] =$

$b^0 =$

$\log[b, 1] =$

$\sin[x]^2 + \cos[x]^2 =$

Here all angles are measured in radians.

**6.** Draw the familiar "$1 - 2 - \sqrt{3}$" triangle and fill the blanks:

$\sin[\pi/6] =$

$\cos[\pi/3] =$

$\sin[\pi/3] =$

$\cos[\pi/3] =$

$\tan[\pi/3] =$

$\tan[\pi/6] =$

Here all angles are measured in radians.

**7.** If $\log[b, 25] = 2$, then what is $b$?

.

.

**8.** Explain why
$$\log[12, 24] + \log[12, 60] - \log[12, 10] = 2.$$

.

.

.

**9.** You have two fairly large numbers $a$ and $b$ with $a > b$. Which do you expect to be larger: $a^b$ or $b^a$?

.

.

.

You have a number $x$ with $0 < x < 1$. Rank in order of size:
$$x, \sqrt{x}, x^2, x^3, x^{10}.$$

.

.

.

**10.** Is it really true that $\pi = 22/7$? In what kinds of calculations should you feel comfortable in using the value $22/7$ for $\pi$?

.

.

.

# Functions and plotting

## Guide

The exact formula for a function $f[x]$ is handy to have because you can use it to compute values of the function. For instance if you want to study the function $f[x] = 2^{-x} \cos[4x]$ then you can set:

```
In[1]:=   Clear[f,x]
          f[x_] = 2^(-x) Cos[4 x]

Out[1]=   Cos[4 x]
          --------
             x
            2
```

Now you can calculate with $f[x]$ :

```
In[2]:=   N[f[2.5]]
Out[2]=   -0.148328
```

Or we can calculate a table of the values of the function at specified points:

```
In[3]:=   Clear[k]
          Table[{k,N[f[k]]},{k,0,6}]
Out[3]=   {{0, 1.}, {1, -0.326822},

          {2, -0.036375}, {3, 0.105482},

          {4, -0.0598537}, {5, 0.0127526},

          {6, 0.0066278}}
```

But to get the inside scoop on what the function is doing, there is nothing like a plot:

```
In[4]:=   Plot[f[x],{x,0,6},AxesLabel->{"x","y"},
          PlotStyle->RGBColor[0,0,1]]
```

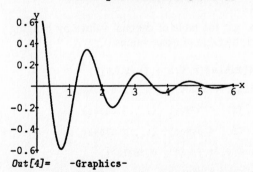

```
Out[4]=   -Graphics-
```

This is a lesson on how you can put *Mathematica* to work for you to find the inner secrets of functions by plotting them.

## Basics

*Mathematica* allows you to set up functions for evaluation and other pleasurable activities.

*Again when you use symbols for functions or variables, it is important to make sure that they are cleared of any values that that someone else may have assigned to them. The way to do this is to use the command* `Clear[f,x]` *or* `Clear[y,a,b]`*, etc.*

**B.1.a.i)**

You want to set up and evaluate the function
$$f[x] = x^2 + 2x + 5.$$
What are some ways of doing this?

**Answer:**

The best way of setting up a function like $f[x] = x^2 + 2x + 5$ is to type and evaluate *(Note the underscore to the right of the variable $x$ and the colon before the equals sign. Also note the square brackets):*

```
In[1]:=   Clear[f,x]
          f[x_] = x^2 + 2 x - 5
Out[1]=              2
          -5 + 2 x + x
```

Now we can evaluate $f[x]$ by plugging in specific numbers in place of $x$ :

```
In[2]:=   f[1]
Out[2]=   -2
```

```
In[3]:=   f[-7]
Out[3]=   30
```

You **do not** get the same effect by typing **without the underscore** :

```
In[4]:=   Clear[f,x]
          f[x] = x^2 + 2 x - 5
Out[4]=              2
          -5 + 2 x + x
```

```
In[5]:=   f[1]
Out[5]=   f[1]
```

```
In[6]:=    f[-7]
Out[6]=    f[-7]
```

This output is useless. Another way to set up functions is to type and evaluate:

```
In[7]:=    Clear[x,y]
           y = x^2 + 2 x - 5
Out[7]=              2
           -5 + 2 x + x
```

But now there is no place to plug in the numbers for $x$.

You can still get the values of $y$ for specific numbers $x$ by typing and activating *(Note the symbols* `/.x->1` *. They tell the machine to replace $x$ by 1. Its advantage is that it assigns the value 1 to $x$ only in the context of this command. The $x$ variable remains clear after this command is activated):*

```
In[8]:=    y/.x->1
Out[8]=    -2
```

```
In[9]:=    y/.x->-7
Out[9]=    30
```

Not bad.

### B.1.a.ii)

What happens when you use normal parentheses instead of square brackets when you set up a function?

### Answer:

Let's try it:

```
In[1]:=    f(x_)=x^2+2 x - 5
```

`Set::write:`

`Symbol Times is write protected.`
```
Out[1]=              2
           -5 + 2 x + x
```

Note the error message. Now try to evaluate $f(2)$ :

```
In[2]:=    f(2)
Out[2]=    2 f
```

The output is 2 times $f$. The reason for this is that *Mathematica* did not understand $f(2)$ the way you meant it. When you evaluated the last statement, *Mathematica* multiplyed 2 times the symbol $f$. The Moral:

**Square brackets are a big deal in *Mathematica*. Unlike pencil and paper calculations, normal parentheses and square brackets are not interchangeable in *Mathematica* calculations. Use normal parentheses for grouping terms, but use square brackets for functions.**

### B.1.b.i)

Take $f[x] = x^7/7^x$. Make a table of the exact values $\{k, f[k]\}$ for $k = 0, 1, 2, 3, ..., 10$. Then make a table of decimal values $\{k, f[k]\}$ for $n = 0, 1, 2, 3, ..., 10$.

### Answer:

It's easy. Just evaluate:

```
In[1]:=    Clear[f,x]
           f[x_] = (x^7)/7^x
Out[1]=     7
           x
           --
            x
           7
```

Here comes the table of exact values:

```
In[2]:=    Clear[k]
           Table[{k,f[k]},{k,1,10}]
Out[2]=        1          128           2187
           {{1, -}, {2, ---}, {3, ----},
               7          49            343

              16384          78125          279936
           {4, -----}, {5, -----}, {6, ------},
              2401           16807          117649

                        2097152          4782969
           {7, 1}, {8, -------}, {9, --------},
                        5764801          40353607

               10000000
           {10, ---------}}
               282475249
```

Here comes the table of decimal values:

```
In[3]:=    Table[{k,N[f[k]]},{k,1,10}]
Out[3]=    {{1, 0.142857}, {2, 2.61224},

           {3, 6.37609}, {4, 6.82382},

           {5, 4.64836}, {6, 2.37942}, {7, 1.},

           {8, 0.363786}, {9, 0.118526},

           {10, 0.0354013}}
```

You can also get the table of decimal values by hanging `N[]` around the table of exact values:

```
In[4]:=    N[Table[{k,f[k]},{k,1,10}]]
Out[4]=    {{1., 0.142857}, {2., 2.61224},

           {3., 6.37609}, {4., 6.82382},

           {5., 4.64836}, {6., 2.37942},

           {7., 1.}, {8., 0.363786},

           {9., 0.118526}, {10., 0.0354013}}
```

Note how the values $f[k]$ go up and down as $k$ goes up.

### B.1.b.i)

Plotting tables of points is easy in *Mathematica* . Take $f[x] = x^7/7^x$. Plot the table of the values $\{k, f[k]\}$ for $k = 0, 1, 2, 3, ..., 10$.

**Answer:**

This is easy too. Just activate:

```
In[1]:=    values = Table[{k,f[k]}, {k,0,10}]
```

$$Out[1]= \{\{0, 0\}, \{1, \tfrac{1}{7}\}, \{2, \tfrac{128}{49}\}, \{3, \tfrac{2187}{343}\},$$

$$\{4, \tfrac{16384}{2401}\}, \{5, \tfrac{78125}{16807}\}, \{6, \tfrac{279936}{117649}\},$$

$$\{7, 1\}, \{8, \tfrac{2097152}{5764801}\}, \{9, \tfrac{4782969}{40353607}\},$$

$$\{10, \tfrac{10000000}{282475249}\}\}$$

Here comes the plot:

```
In[2]:=    ListPlot[values]
```

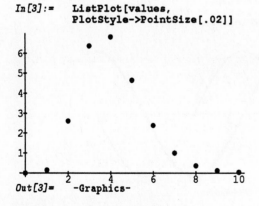

```
Out[2]=    -Graphics-
```

You can make the dots bigger:

```
In[3]:=    ListPlot[values,
           PlotStyle->PointSize[.02]]
```

```
Out[3]=    -Graphics-
```

You can label the $x$ and $y$ axes:

```
In[4]:=    ListPlot[values,AxesLabel->{x,"y"}]
```

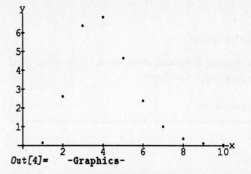

```
Out[4]=    -Graphics-
```

You can do both and if you have a color machine, then you can put the points in color:

```
In[5]:=    ListPlot[values,PlotStyle->
           {PointSize[.02],RGBColor[1,0,0]},
           AxesLabel->{x,"y"}]
```

```
Out[5]=    -Graphics-
```

Nice graphics, *Mathematica!*

### B.1.b.ii)

But why settle for point plots when you can see the whole curve? Take $f[x] = x^7/7^x$ and plot the whole curve $y = f[x]$ for $0 \le x \le 10$.

**Answer:**

This is as easy as apple pie. Just evaluate:

```
In[1]:=    Clear[f,x]
           f[x_] = (x^7)/7^x
           Plot[f[x],{x,0,10},AxesLabel->{"x","y"}]
```

```
Out[1]=    -Graphics-
```

Look at that beauty rise and fall. By the time $x$ gets to 9, the denominator $7^x$ is much bigger than the numerator

$x^7$.

### ■ B.2) Some plotting options.

Browse this section now and refer back to it when you need a little plotting know-how.

### B.2.a) Labeling the axes:

Plot the function $y = 2x - 1$ for $2 \leq x \leq 7$. Label the $x$ and the $y$ axes.

**Answer:**

Here we go:

```
In[1]:=    Clear[x]
           Plot[2 x - 1,{x,2,7},AxesLabel->{"x","y"}]
```

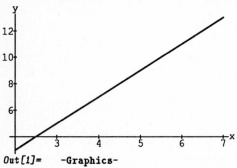

```
Out[1]=    -Graphics-
```

Notice that the "reference axes" shown are not the usual "$x$−axis" and the "$y$−axis", but rather two convenient lines for this function. One is at $y = 4$ ( the $x$−axis), and the other is at $x = 2$.

### B.2.b) True scale plots:

Plot the function $y = 2x - 1$ for $2 \leq x \leq 7$ in true scale.

**Answer:**

To get a true scale plot, add the plotting option

**AspectRatio->Automatic.**

```
In[1]:=    Clear[x]
           Plot[2 x - 1, {x,2,7},
           AxesLabel->{"x","y"},
           AspectRatio->Automatic]
```

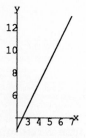

```
Out[1]=    -Graphics-
```

Quite a bit steeper than the plot we saw above.

### B.2.c) Plotting two or more functions on the same axes:

Plot the functions
$$f[x] = \sin[x]$$
and
$$g[x] = \cos[x]$$

on the same axes for $0 \leq x \leq 2\pi$. Describe what you see.

**Answer:**

```
In[1]:=    Clear[x]
           Plot[{Sin[x],Cos[x]},{x,0,2 Pi},
           AxesLabel->{"x","y"}]
```

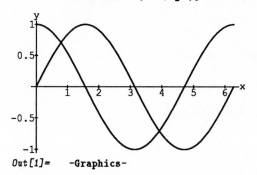

```
Out[1]=    -Graphics-
```

That's $\cos[x]$ starting out on top at the left. Both curves have the same shape. In fact if you can shift either curve onto the other.

If you like, then you can distinguish between the curves by varying the thickness:

The thicker of the two curves is $\cos[x]$:

```
In[2]:=    Clear[x]
           Plot[{Sin[x],Cos[x]},{x,0,2 Pi},
           AxesLabel->{"x","y"},
           PlotStyle->{Thickness[0.005],
           Thickness[0.010]}]
```

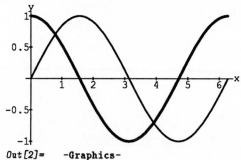

```
Out[2]=    -Graphics-
```

Or if you have a color machine, then you can distingush the curves by varying the colors.

```
In[3]:=    Plot[{Sin[x],Cos[x]},{x,0,2 Pi},
           AxesLabel->{"x","y"},
           PlotStyle->{RGBColor[1,0,0],
           RGBColor[0,0,1]}]
```

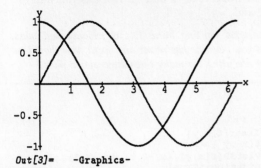

```
Out[3]=    -Graphics-
```

You can do both:

```
In[4]:=    Plot[{Sin[x],Cos[x]},{x,0,2 Pi},
           AxesLabel->{"x","y"},PlotStyle->
           {{Thickness[0.005],RGBColor[1,0,0]},
           {Thickness[0.010],RGBColor[0,0,1]}}]
```

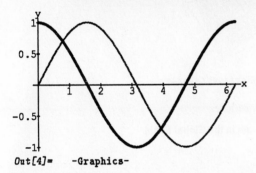

```
Out[4]=    -Graphics-
```

### B.2.d) Showing two plots together:

Give individual plots of $f[x] = \log[2, x]$ for $.1 \le x \le 8$ and $g[x] = 2^x$ for $-4 \le x \le 4$. Then display the plots together in true scale on the same axes. Describe what you see.

**Answer:**

Give names to the individual plots:

```
In[1]:=    Clear[x]
           plot1 = Plot[Log[2,x],{x,.1,8}]
```

```
Out[1]=    -Graphics-
```

```
In[2]:=    plot2 = Plot[2^x,{x,-4,4}]
```

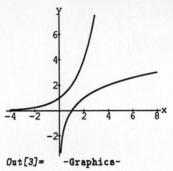

```
Out[2]=    -Graphics-
```

Here they are together in true scale: Note that the **Show** command does not require the squiggilies.

```
In[3]:=    Show[plot1,plot2,AspectRatio->
           Automatic,AxesLabel->{"x","y"}]
```

```
Out[3]=    -Graphics-
```

Mirror images.

### B.2.e) Combining point and curve plots

Take $f[x] = x^6/6^x$. Plot the table of the values $\{k, f[k]\}$ for $k = -1, 0, 1, 2, 3, 4, 5, 6, 7, 8, 9, 10$ and the curve $y = f[x]$ for $-1 \le x \le 10$ on the same axes.

**Answer:**

```
In[1]:=    Clear[x,f]
           f[x_] = (x^6)/6^x
```

```
Out[1]=     6
           x
           --
            x
           6
```

Here are the points:

```
In[2]:=    points = Table[{k,N[f[k]]},{k,-1,10}]
Out[2]=    {{-1, 6.}, {0, 0.}, {1, 0.166667},
             {2, 1.77778}, {3, 3.375},
             {4, 3.16049}, {5, 2.00939}, {6, 1.},
             {7, 0.420271}, {8, 0.156074},
             {9, 0.0527344}, {10, 0.0165382}}
```

Here is the point plot:

```
In[3]:=    pointplot =
           ListPlot[points,
           AxesLabel->{"x","y"},
           PlotStyle->{PointSize[.02],
           RGBColor[1,0,0]}]
```

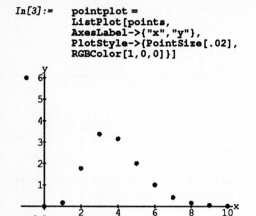

```
Out[3]=    -Graphics-
```

Here is the plot of the curve:

```
In[4]:=    curveplot = Plot[f[x],{x,-1,10},
           AxesLabel->{"x","y"}]
```

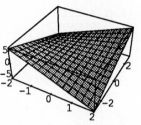

```
Out[4]=    -Graphics-
```

Here they are together in true scale:

```
In[5]:=    Show[pointplot,curveplot,
           AspectRatio->Automatic]
```

```
Out[5]=    -Graphics-
```

As $x$ leaves $-1$ and advances, the numerator, $x^6$, is growing faster than the denominator, $6^x$, but when $x$ gets larger than 4, the denominator begins to grow so fast that the whole fraction $x^6/6^x$ becomes very small.

**B.2.f) Plotting in three dimensions.**

The function
$$f[x,y] = xy$$
measures the product of $x$ and $y$. Plot this function for $-2 \leq x \leq 2$ and $-1 \leq y \leq 1$.

*(Three dimensional plotting burns lots of memory and takes time. Unless you machine has power and speed, you should be careful about executing too many three dimensional plots.)*

**Answer:**

```
In[1]:=    Clear[f,x,y]
           f[x_,y_] = x y
           Plot3D[f[x,y],{x,-2,2},{y,-3,3},
           Lighting->True]
```

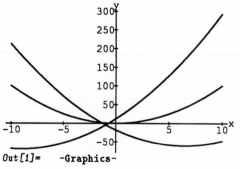

```
Out[1]=    -SurfaceGraphics-
```

Looks like a saddle.

■ **B.3) Polynomials in the Global Scale.**

**B.3.a.i)**

Plot $x^2, x^2 + 18x + 12$ and $x^2 - 13x - 18$ on the same axes for $-10 \leq x \leq 10$. To make sure you see everything, use the plotting option **PlotRange->All**. Describe what you see.

**Answer:**

```
In[1]:=    Clear[x]
           Plot[{x^2, x^2 + 18 x + 12,
           x^2 - 13 x - 18},{x,-10,10},
           AxesLabel->{"x","y"},PlotRange->All]
```

```
Out[1]=    -Graphics-
```

A mess.

**B.3.a.ii)**

Plot the same three functions $x^2, x^2 + 18x + 12$ and $x^2 - 13x - 18$ on the same axes for the huge interval $-1000 \leq x \leq 1000$. Use the plotting option **PlotRange->All**. Describe what you see.

**Answer:**

```
In[1]:=    Clear[x]
           Plot[{x^2, x^2 + 18 x + 12,
           x^2 - 13 x - 18},
           {x,-1000,1000},AxesLabel->
           {"x","y"},PlotRange->All]
```

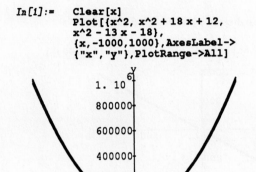

```
Out[1]=    -Graphics-
```

Order out of chaos. In this global scale, they all look the same!

**B.3.b)**

Plot $x^3$, $x^3 - 3x^2 - 61x + 65$, and $x^3 - 18x^2 + 30x - 12$ on the same axes for $-1000 \leq x \leq 1000$. Use the plotting option **PlotRange->All**. Describe what you see.

**Answer:**

```
In[1]:=    Clear[x]
           Plot[{x^3,x^3 -3 x^2 - 61 x + 65,
           x^3 - 18 x^2 + 30 x - 12},
           {x,-1000,1000},AxesLabel->{"x","y"},
           PlotRange->All]
```

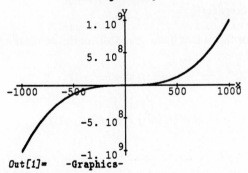

```
Out[1]=    -Graphics-
```

In this global scale, they all look the same!

**B.3.c)**

Plot $x^4$, $x^4 + 8x^3 - 73x^2 - 512x + 571$, and $x^4 - 12x^3 - 73x^2 + 958x - 864$ on the same axes for $-1000 \leq x \leq 1000$. Use the plotting option **PlotRange->All**. Describe what you see.

**Answer:**

```
In[1]:=    Clear[x]
           Plot[{x^4,x^4 + 8 x^3 -
           73 x^2 -512 x + 571,
           x^4 - 12 x^3 -73 x^2 + 958 x - 864 },
           {x,-1000,1000}, AxesLabel->{"x","y"},
           PlotRange->All];
```

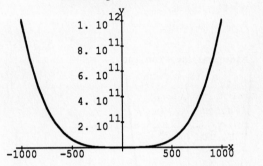

Again-they all look the same in the global scale!

**B.3.d.i)**

What does the global scale plot of

$$x^n + a[n-1]x^{n-1} + a[n-2]x^{n-2} + \ldots + a[2]x^2 + a[1]x + a[0]$$

look like? (Here $a[n-1], a[n-2], ..., a[1]$ and $a[0]$ are numbers whose values we don't know yet.)

**Answer:**

It looks like a plot of $x^n$. Consequently:

$\rightarrow$ If $n$ is odd, then the global scale plot comes up from the lower left and exits the screen at the upper right.

$\rightarrow$ If $n$ is even, then the global scale plot comes down from the upper left and exits the screen at the upper right.

Take a look for various values of $n$ :

```
In[1]:=    Clear[x]
           n = 3
Out[1]=    3


In[2]:=    Plot[x^n, {x,-100,100}]
```

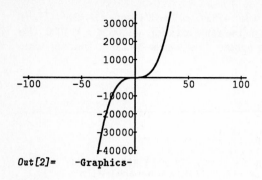

Out[2]=    -Graphics-

In[3]:=    n = 6
Out[3]=    6

In[4]:=    Plot[x^n, {x,-100,100}]

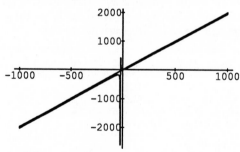

Out[4]=    -Graphics-

### B.3.d.ii)

What **calculational principle** explains why good global scale plots of polynomials look the way they do?

**Answer:**

When $|x|$ is large, then the highest power of $x$ dominates all other powers of $x$.

### B.3.e)

Give good global scale plots of

$$(2x^4 + 5x^2 + 4x + 1)/(x^3 + 6x^2 + 1)$$

and

$$(2x^6 + 5x^3 + 4x + 1)/(x^5 + 23x^4 + 9x^2 + 18)$$

on the same axes. Explain what you see.

**Answer:**

In[1]:=    Clear[x]
           first =
           (2 x^4 + 5 x^2 + 4 x + 1)/
           (x^3 + 6 x^2 + 1)

Out[1]=
$$\frac{1 + 4 x + 5 x^2 + 2 x^4}{1 + 6 x^2 + x^3}$$

In[2]:=    second =
           (2 x^6 + 5 x^3 + 4 x + 1)/
           (x^5 + 23 x^4 + 9x^2 + 18)

Out[2]=
$$\frac{1 + 4 x + 5 x^3 + 2 x^6}{18 + 9 x^2 + 23 x^4 + x^5}$$

Here is a global scale plot of both functions:

In[3]:=    Plot[{first, second}, {x,-1000,1000}]

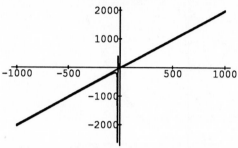

Out[3]=    -Graphics-

The global scale behavior of both functions is the same as the behavior of $2x$ : See what happens when we throw in a plot of $2x$.

In[4]:=    Plot[{first, second, 2x},
           {x,-1000,1000}]

Out[4]=    -Graphics-

The reason for this is that if we ignore all but the dominant terms in

$$(2x^4 + 5x^2 + 4x + 1)/(x^3 + 6x^2 + 1)$$

and

$$(2x^6 + 5x^3 + 4x + 1)/(x^5 + 23x^4 + 9x^2 + 18)$$

then we get

$$2x^4/x^3 = 2x$$

and

$$2x^6/x^5 = 2x.$$

Only the **dominant** terms influence the global scale behavior.

# Tutorial

■ **T.1) Global Scale for polynomials and their quotients.**

### T.1.a.i)

Look at:

```
In[1]:=   Clear[x,f]
          f[x_] =
          x^5 - 1251 x^4 - 28750 x^3 +
          50040 x^2 - 50001 x

          Plot[f[x],{x,-1000,1000},
          AxesLabel->{"x","y"},
          PlotRange->All]
```

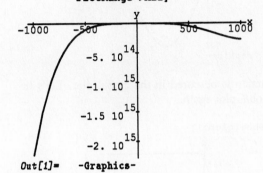

```
Out[1]=   -Graphics-
```

Is this a good global scale plot of

$$f[x] = x^5 - 1251x^4 - 28750x^3 + 50040x^2 - 50001x?$$

Why or why not?

**Answer:**

No, it does not have the characteristic global shape:

```
In[2]:=   Plot[x^5,{x,-100,100},AxesLabel->{"x","y"}]
```

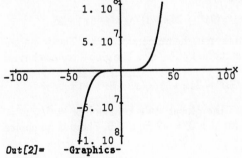

```
Out[2]=   -Graphics-
```

### T.1.a.ii)

Give a good global scale plot of

$$f[x] = x^5 - 1251x^4 - 28750x^3 + 50040x^2 - 50001x.$$

**Answer:**

The plot above looks O.K. on the left; so we'll extend the plotting interval to the right.

```
In[1]:=   Plot[f[x],{x,-1000,2000},
          AxesLabel->{"x","y"},
          PlotRange->All]
```

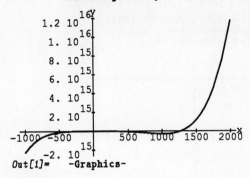

```
Out[1]=   -Graphics-
```

Pretty close

Try again:

```
In[2]:=   Plot[f[x],{x,-2000,3000},
          AxesLabel->{"x","y"},
          PlotRange->All]
```

```
Out[2]=   -Graphics-
```

On the money.

### T.1.b)

Plot

$$f[x] = (7x^4 + 5x^2 + 4x + 1)/(x^4 + 8x^2 - 17)$$

in global scale.

What do you say are the limiting values

$$\lim_{x \to \infty} (7x^4 + 5x^2 + 4x + 1)/(x^4 + 8x^2 - 17)$$

and

$$\lim_{x \to -\infty} (7x^4 + 5x^2 + 4x + 1)/(x^4 + 8x^2 - 17)?$$

**Answer:**

By ignoring all but the dominant terms we see that the global scale behavior of

$$f[x] = (7x^4 + 5x^2 + 4x + 1)/(x^4 + 8x^2 - 17)$$

is the same as the global scale behavior of

$$7x^4/x^4 = 7.$$

Here is a plot:

```
In[1]:=    Clear[x,f]
           f[x_] = (7 x^4 + 5 x^2 + 4 x +1)/
           (x^4 + 8 x^2 - 17)

           Plot[f[x],{x,-1000,1000},
           AxesLabel->{"x","y"},
           PlotRange->{0,8}]
```

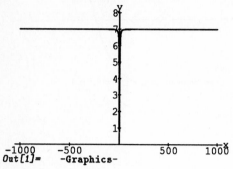

```
Out[1]=    -Graphics-
```

Yep; the global scale behavior is the constant 7.

There is no choice but to say

$$\lim_{x \to \infty} (7x^4 + 5x^2 + 4x + 1)/(x^4 + 8x^2 - 17) = 7$$

and

$$\lim_{x \to -\infty} (7x^4 + 5x^2 + 4x + 1)/(x^4 + 8x^2 - 17) = 7.$$

**T.1.c)**

Plot

$$f[x] = (9x^3 - 24x + 1)/\sqrt{x^6 + 8x^4 + 15}$$

in global scale.

What do you say are the limiting values

$$\lim_{x \to \infty} (9x^3 - 24x + 1)/\sqrt{x^6 + 8x^4 + 15}$$

and

$$\lim_{x \to -\infty} (9x^3 - 24x + 1)/\sqrt{x^6 + 8x^4 + 15}?$$

**Answer:**

Take a look:

```
In[1]:=    Clear[x,f]
           f[x_] = (9 x^3 - 24 x + 1)/
           Sqrt[x^6 + 8 x^4 + 15]

           global =
           Plot[f[x],{x, -1000,1000},
           AxesLabel->{"x","y"}]
```

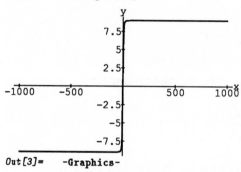

```
Out[1]=    -Graphics-
```

Kowa Bonga!

Just for the hell of it, look at a the local scale plot:

```
In[2]:=    Plot[f[x],{x,-2,2},AxesLabel->{"x","y"}]
```

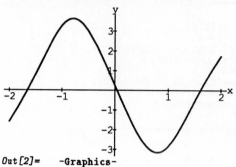

```
Out[2]=    -Graphics-
```

This local action is obscured in the global plot. Let's inspect the global plot again.

```
In[3]:=    Show[global]
```

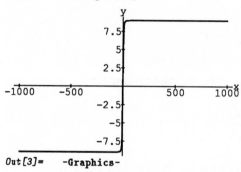

```
Out[3]=    -Graphics-
```

Now look at the dominant terms.

The function is

$$f[x] = (9x^3 - 24x + 1)/\sqrt{x^6 + 8x^4 + 15}$$

The dominant term in the numerator is $9x^3$. The dominant term in the denominator is $\sqrt{x^6}$. The global scale behavior of

$$(9x^3 - 24x + 1)/\sqrt{x^6 + 8x^4 + 15}$$

is the same as the global scale behavior of $9x^3/\sqrt{x^6} = 9x^3/|x|^3 = 9$ for $x > 0; = -9$ for $x < 0$. There is no choice but to say

$$\lim_{x \to \infty} (9x^3 - 24x + 1)/\sqrt{x^6 + 8x^4 + 15} = 9$$

and

$$\lim_{x \to -\infty} (9x^3 - 24x + 1)/\sqrt{x^6 + 8x^4 + 15} = -9.$$

This is in harmony with the plot.

Let's see when $f[x]$ goes into its global scale behavior:

$In[4]:= \quad$ `Plot[{f[x],9,-9},{x,-30,30}]`

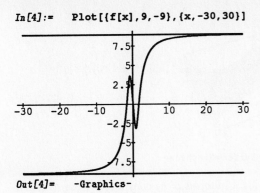

$Out[4]= \quad$ `-Graphics-`

It seems safe to say that $f[x]$ is into its global scale behavior for $|x| > 30$.

### T.1.d)

Plot

$$f[x] = (2x^5 - 4x + 1)/(x^4 + 6x^2 + 12)$$

in global scale.

What simpler function mimicks the global scale behavior of $f[x]$?

### Answer:

Open the door and take a look:

$In[1]:= \quad$ `Clear[x,f]`
`f[x_] = (2 x^5 - 4 x + 1)/`
`(x^4 + 6 x^2 + 12)`

`global =`
`Plot[f[x],{x, -1000,1000},`
`AxesLabel->{"x","y"}]`

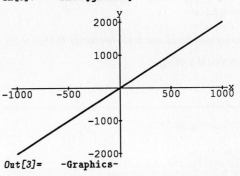

$Out[1]= \quad$ `-Graphics-`

Sure looks like a line in the global scale.

Just for the hell of it, look at a the local scale plot:

$In[2]:= \quad$ `Plot[f[x],{x,-2,2},AxesLabel->{"x","y"}]`

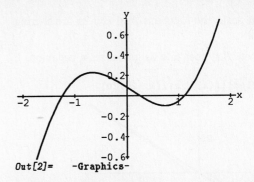

$Out[2]= \quad$ `-Graphics-`

This local action is obscured in the global plot. Let's inspect the global plot again.

$In[3]:= \quad$ `Show[global]`

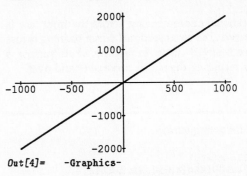

$Out[3]= \quad$ `-Graphics-`

Now look at the dominant terms.

The function is

$$f[x] = (2x^5 - 4x + 1)/(x^4 + 6x^2 + 12).$$

The dominant term in the numerator is $2x^5$.

The dominant term in the denominator is $x^4$.

The global scale behavior of

$$(2x^5 - 4x + 1)/(x^4 + 6x^2 + 12)$$

is the same as the global scale behavior of

$$2x^5/x^4 = 2x.$$

Check:

$In[4]:= \quad$ `Plot[{f[x],2 x},{x,-1000,1000}]`

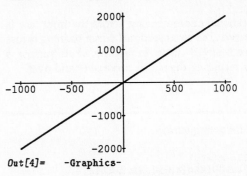

$Out[4]= \quad$ `-Graphics-`

Good.

In this global scale, the functions $f[x]$ and $2x$ are sharing the same ink.

Let's see when $f[x]$ goes into its global scale behavior:

*In[5]:=*　　`Plot[{f[x],2 x},{x,-20,20}]`

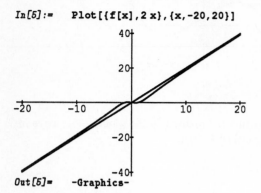

*Out[5]=*　　`-Graphics-`

$f[x]$ seems to be into its global scale behavior for $|x| > 20$.

■ **T.2) Three dimensional plotting.**

**T.2.a)**

Look at the following surface:

*In[1]:=*　　`Clear[x,h]`
　　　　　　`Plot3D[Sin[x],{x,0,2 Pi},`
　　　　　　`{h,0,3},Lighting->True]`

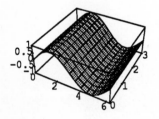

*Out[1]=*　　`-SurfaceGraphics-`

Try to describe it in terms of its constituent cross sectional curves.

**Answer:**

The cross sectional curve running across the front face is the $\sin[x]$ curve. The cross sectional curve running across the back face is $\sin[x]$ curve. In fact the whole surface is made up by repeating the $\sin[x]$ curve over and over.

**T.2.b)**

Look at the following surface:

*In[1]:=*　　`Clear[x,y]`
　　　　　　`Plot3D[Sin[x + y],{x,0,Pi},`
　　　　　　`{y,0,Pi},Lighting->True]`

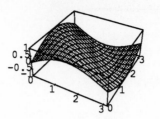

*Out[1]=*　　`-SurfaceGraphics-`

Try to describe it in terms of its constituent cross sectional curves.

**Answer:**

The cross sectional curve running across the front face is the $\sin[x + 0] = \sin[x]$ curve.

The cross sectional curve running across the back face is the $\sin[x + \pi] = -\sin[x]$ curve.

The cross sectional curve running through the middle is the $\sin[x + \pi/2] = \cos[x]$ curve.

In fact the whole surface is made up of constituent cross sectional curves $\sin[x + h]$ with $h$ ranging from 0 to $\pi$.

**T.2.c)**

Look at the following surface:

*In[1]:=*　　`Clear[x,h]`
　　　　　　`Plot3D[Sin[x h],{x,0,2 Pi},{h,1,2},`
　　　　　　`Lighting->True]`

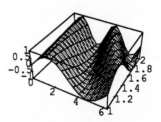

*Out[1]=*　　`-SurfaceGraphics-`

Try to describe it in terms of its constituent cross sectional curves.

**Answer:**

The cross sectional curve running across the front face is the $\sin[x]$ curve.

The cross sectional curve running across the back face is the $\sin[2x]$ curve.

The cross sectional curve running through the middle is the $\sin[(3/2)x]$ curve.

In fact the whole surface is made up of constituent cross
sectional curves sin[$xh$] with $h$ ranging from 1 to 2.

# Literacy Sheet

1. Sketch with a pencil good **global** scale plots of

$f[x] = x$

.

.

$f[x] = x^2 + 3x$

.

.

$f[x] = -x^2 + 3x$

.

$f[x] = x^3 + 5x^2 + 1$

.

.

$f[x] = -x^3 + 5x^2 - x^4$

.

.

2. Explain why a good global scale plot of

$$x + 1/x = (x^2 + 1)/x$$

looks so much like a straight line, but a good global scale plot of

$$x^2 + 1/x = (x^3 + 1)/x$$

looks like a parabola.

What does a good global scale plot of

$$x + 1/x^2 = (x^3 + 1)/x^2$$

look like?

.

.

.

3. Give the values of:

$\lim_{x \to \infty} (2x^2 + 1)/(x^2 + 1)$.

.

.

$\lim_{x \to \infty} (12x^8 + 731x^5 + 645x + 47)/(4x^8 - 61x^3)$.

.

.

$\lim_{x \to \infty} (2x^3 - 849x^2 + 271x + 314)/(\sqrt{4x^6 + 7x^3})$.

.

.

$\lim_{x \to -\infty} (2x^3 - 849x^2 + 271x + 314)/(\sqrt{4x^6 + 7x^3})$.

.

.

$\lim_{x \to \infty} 1 + 1/x^2$.

.

.

4. True or False:

The $y = \cos[x]$ curve is just a shifted version of the $y = \sin[x]$ curve.

.

.

5. Here is a plot of $y = \tan[x]$ for $-3\pi/2 \le x \le 3\pi/2$:

```
In[2]:=   Plot[Tan[x],{x,-3 Pi/2,3 Pi/2},
          AxesLabel->{"x","y"}]
```

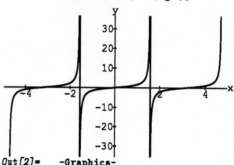

```
Out[2]=   -Graphics-
```

Recall $\tan[x] = \sin[x]/\cos[x]$ and explain the singularities (blow ups) that occur periodically through the plot.

.

.

.

What is the smallest non-zero positive number $t$ such that $\tan[x + t] = \tan[x]$?

.

.

7. If you have a plot of $y = f[x]$ for $-2 \le x \le 2$ on a piece of tracing paper, then how can you use carbon paper to get the plot of $y = f[-x]$ on $[-2, 2]$?

.

.

If you have a plot of $y = f[x]$ on a piece of tracing paper, then how can you use carbon paper to get the plot of $y = f[x + 1]$?

.

.

If you have a plot of $y = f[x]$ on a piece of tracing paper, then how can you use carbon paper to get the plot of $y = f[x - 2]$?

.

.

.

**8.** Look at the following true scale plots:

A: $f[x] = x^2$, $h[x] = x$ and $g[x] = \sqrt{x}$ :

```
In[3]:=    Plot[{x^2,x,Sqrt[x]},{x,0,2},
           AxesLabel->{"x","y"},
           PlotRange->{0,2},
           AspectRatio->Automatic]
```

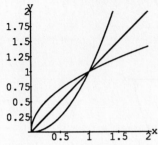

```
Out[3]=    -Graphics-
```

B: $f[x] = x^3$, $h[x] = x$ and $g[x] = x^{1/3}$ :

```
In[4]:=    Plot[{x^3,x,x^(1/3)},{x,0,3},
           AxesLabel->{"x","y"},
           PlotRange->{0,3},
           AspectRatio->Automatic]
```

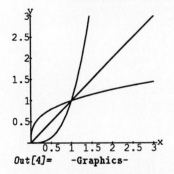

```
Out[4]=    -Graphics-
```

C: $f[x] = 2^x$, $h[x] = x$ and $g[x] = \log[2, x]$ :

```
In[5]:=    Plot[{2^x,x,Log[2,x]},{x,0,16},
           AxesLabel->{"x","y"},
           PlotRange->{0,16},
           AspectRatio->Automatic]
```

```
Plot::notnum:
Log[2, x] does not evaluate to a real
number at x=0..
```

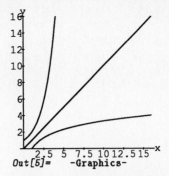

```
Out[5]=    -Graphics-
```

D: $f[x] = \sin[x]$, $h[x] = x$ and $g[x] = \arcsin[x]$ :

```
In[6]:=    Plot[{Sin[x],x,ArcSin[x]},
           {x,0,1.5},AxesLabel->{"x","y"},
           PlotRange->{0,1.5},
           AspectRatio->Automatic]
```

```
Plot::notnum:
ArcSin[x] does not evaluate to a real
number at x=1.0625.
Plot::notnum:
ArcSin[x] does not evaluate to a real
number at x=1.03125.
Plot::notnum:
ArcSin[x] does not evaluate to a real
number at x=1.01563.
General::stop:
Further output of Plot::notnum
will be suppressed during this
calculation.
```

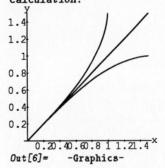

```
Out[6]=    -Graphics-
```

E: $f[x] = \tan[x]$, $h[x] = x$ and $g[x] = \arctan[x]$ :

```
In[7]:=    Plot[{Tan[x],x,ArcTan[x]},
           {x,0,1.5},AxesLabel->{"x","y"},
           PlotRange->{0,1.5},
           AspectRatio->Automatic]
```

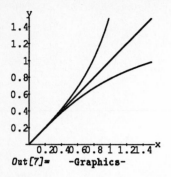

*Out[7]=*    -Graphics-

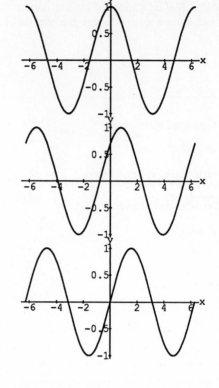

Why do you think these plots all turn out this way?

.

.

.

In each case, if you take a point $\{x, f[x]\}$ on the plot of
$y = f[x]$, then where goes the point $\{f[x], x\}$ land on the
plot?

Go ahead; make some measurements.

If you take a point $\{x, g[x]\}$ on the graph of $y = g[x]$, then
where goes the point $\{g[x], x\}$ land on the plot?

.

.

.

If $f[x]$ and $g[x]$ are any functions that make $f[g[x]] = x$
for all $x's$ in consideration, then how can you take the plot
of $f[x]$ on tracing paper, fold the paper and use carbon
paper and a pencil to trace out a plot of $g[x]$?

.

.

.

**9.** True or False:

a. $\log[3, x] + \log[3, y] = \log[3, xy]$

b. $\log[3, x] - \log[3, y] = \log[3, x/y]$

c. $\log[3, x]\log[3, y] = \log[3, xy]$

d. $\log[3, x^y] = y\log[3, x]$

e. $\log[3, 3^y] = y$

e. $3^{(\log[3, y])} = y$

**10.** Here are three plots. One is is a plot of $\sin[x]$; another
is a plot of $\cos[x]$ and another is the plot of neither Identify
and label the plots of $\sin[x]$ and $\cos[x]$ :

# Linear Growth

## Guide

You might think lines are pretty simple. If so, then you are right. But you can do a heck of a lot with lines and understanding how lines work is a great help in understanding more complicated curves. Jump into this lesson because the seeds of calculus are here.

## Basics

### ■ B.1) Lines and constant growth rates .

#### B.1.a.i)

A new Maytag 17.1 cubic feet refrigerator costs 875 dollars and is projected to cost 71 dollars per year to operate.

Assuming the Maytag people are truthful in saying that their appliances never break down, give a plot showing the total projected costs of buying and operating the Maytag at any time during the first ten years of ownership.

How much has the Maytag owner shelled out for ten years of cold pop and beer?

**Answer:**

The cost in dollars of operating the Maytag for the first $x$ years is

$$f[x] = 875 + 71x.$$

Here is a plot showing the total costs of buying and operating the Maytag at any time during the first ten years of ownership.

```
In[1]:=   Clear[f,x]
          f[x_] = 875 + 71 x
          Plot[f[x],{x,0,10},AxesLabel->{"years","costs"}]
```

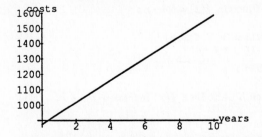

The total cost of buying and operating the Maytag for ten years in dollars is:

```
In[2]:=   f[10]
Out[2]=   1585
```

A lot of bread.

#### B.1.a.ii)

What do you call the plot in part i)?

**Answer:**

A line.

#### B.1.b.i)

Plot $f[x] = 3x + 1$ for $-1 \le x \le 5$ in true scale. $f[x]$ goes up how many times faster than $x$?

**Answer:**

Here is a true scale plot:

```
In[1]:=   Clear[f,x]
          f[x_] = 3 x + 1
          Plot[f[x],{x,-1,5},
          AxesLabel->{"x","y"},
          AspectRatio->Automatic,
          PlotStyle->{Thickness[0.03]}]
```

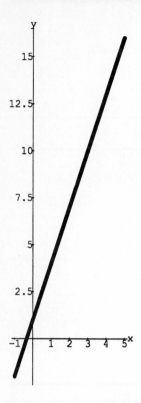

If $x$ increases $h$ units then $f[x]$ increases $f[x + h] - f[x]$ units.

```
In[2]:=    Expand[f[x+ h] - f[x]]
Out[2]=    3 h
```

So $f[x] = 3x + 1$ goes up 3 times faster than $x$. You might say $f[x] = 3x + 1$ has **growth rate** 3.

**B.1.b.ii)**

Plot $f[x] = -2Yx + 8$ for $-1 \leq x \leq 6$ in true scale. $f[x]$ goes up how many times faster than $x$?

**Answer: Here is a true scale plot:**

```
In[1]:=    Clear[f,x]
           f[x_] = -2 x + 8
           Plot[f[x],{x,-1,6},
           AxesLabel->{"x","y"},
           AspectRatio->Automatic,
           PlotStyle->{RGBColor[1,0,0]}]
```

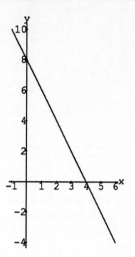

If $x$ increases $h$ units then $f[x]$ increases $f[x + h] - f[x]$ units.

```
In[2]:=    Expand[f[x+ h] - f[x]]
Out[2]=    -2 h
```

So $f[x] = -2x + 8$ goes up $-2$ times faster than $x$ goes up.

(Actually this means $f[x] = -2x + 8$ goes down 2 times faster than $x$ goes up.) You might say $f[x] = -2x + 1$ has **growth rate** $-2$.

**B.1.b.iii)**

What do you call the plots you saw above?

**Answer:**

Lines.

**B.1.c.i)**

What is the growth rate in units on the $y$-axis per unit on the $x$-axis of a line function

$$f[x] = b + rx?$$

**Answer:**

Take a line function $f[x] = b + rx$ :

```
In[1]:=    Clear[x,h,f,r,b]
           f[x_] = b + r x
Out[1]=    b + r x
```

If $x$ increases $h$ units then $f[x]$ increases $f[x + h] - f[x]$ units.

```
In[2]:=    Expand[f[x+ h] - f[x]]
Out[2]=    h r
```

So $f[x] = b + rx$ goes up $r$ times faster than $x$ goes up. Thus $f[x] = b + rx$ has growth rate $r$.

### ■ B.2) Determining formulas of line functions.

#### B.2.a.i)

What do you need to determine the formula for a line function?

**Answer:**

You need the **growth rate** and **one value** of the function. For instance if the growth rate is 3 and $f[4] = 11$, then just put:

```
In[1]:=   Clear[x,f]
          f[x_] = Expand[11 + 3 (x - 4)]
Out[1]=   -1 + 3 x
```

Sure enough the growth rate is 3 and $f[4]$ is:

```
In[2]:=   f[4]
Out[2]=   11
```

Just as we wanted.

#### B.2.a.ii)

Find the line function with growth rate $-1.5$ whose plot passes through $\{1.7, 8.5\}$. Plot this line function for $-2 \le x \le 4$.

**Answer:**

Just put:

```
In[1]:=   Clear[x,f]
          f[x_] = Expand[8.5 + (-1.5) (x - 1.7)]
Out[1]=   11.05 - 1.5 x
```

Sure enough the growth rate is $-1.5$ and $f[1.7]$ is:

```
In[2]:=   f[1.7]
Out[2]=   8.5
```

This shows that plot will go through $\{1.7, 8.5\}$. Let's plot for $-2 \le x \le 4$:

```
In[3]:=   lineplot = Plot[f[x],{x,-2,4}]
```

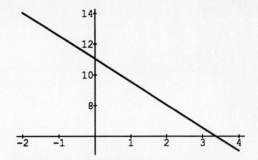

The negative growth rate of $-1.5$ is why the plot goes down. Let's see the point $\{1.7, 8.5\}$ and the line together:

```
In[4]:=   point =
          Graphics[{RGBColor[1,0,0],
          PointSize[0.02],
          Point[{1.7,8.5}]}];

          Show[lineplot,point]
```

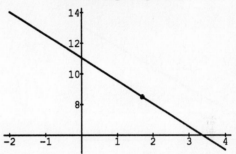

It's not much but it's ours and it's right.

#### B.2.b)

But wait a minute. Most thinking persons determine a line by taking two points $\{a_1, a\}_2$ and $\{b_1, b\}_2$ :

```
In[1]:=   point1 =
          Graphics[{RGBColor[1,0,0],
          PointSize[0.015],
          Point[{1,1}]}]
          label1 =
          Graphics[Text[{"a1","b1"},
          {1,1},{0,-2.5}]]
          point2 =
          Graphics[{RGBColor[1,0,0],
          PointSize[0.015],
          Point[{3,5}]}]
          label2 =
          Graphics[Text[{"a2","b2"},
          {3,5},{0,-2.5}]]

          Show[point1,label1,label2,point2,
          Axes->{0,0},AxesLabel->{"x","y"},
          PlotRange->{{0,5},{-1,10}},Ticks->None]
```

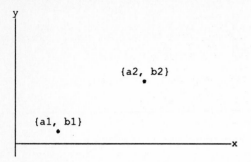

And laying down a ruler and then using a pencil to connect the points:

```
In[2]:=    line = Graphics[Line[{{0,-1},{5,9}}]]

           Show[point1,label1,label2,point2,line,
               Axes->{0,0},AxesLabel->{"x","y"},
               PlotRange->{{0,5},{-1,10}},Ticks-
>None]
```

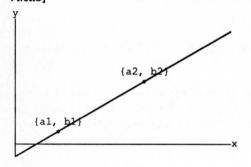

### B.2.b.i)

How can we determine the formula for the line function

$$f[x] = b + rx$$

whose plot goes through the two points $\{a_1, a\}_2$ and $\{b_1, b\}_2$?

**Answer:**

We can do it by the method of undetermined coefficients: Set up a preliminary version of $f[x]$ :

```
In[1]:=    Clear[f,prelimf,x,b,r,a1,a2,b1,b2]
           prelimf[x_] = b + r x

Out[1]=    b + r x
```

Now set up the equations $f[x]$ must satisfy:

```
In[2]:=    eqn1 = (prelimf[a1] == a2)

Out[2]=    b + a1 r == a2
```

```
In[3]:=    eqn2 = (prelimf[b1] == b2)

Out[3]=    b + b1 r == b2
```

Determine the values of $r$ and $b$ by solving the two equations:

```
In[4]:=    randb =Solve[{eqn1,eqn2},{r,b}]

Out[4]=
                  a2          b2
           {{r -> ------- - -------,
                  a1 - b1    a1 - b1

                  a2 b1       a1 b2
           b -> -(-------) + -------}}
                  a1 - b1    a1 - b1
```

Now substitute in these determined values of $r$ and $b$ :

```
In[5]:=    f[x_] = prelimf[x]/.randb[[1]]

Out[5]=
              a2 b1       a1 b2
           -(-------) + ------- +
              a1 - b1    a1 - b1

              a2         b2
           (------- - -------) x
            a1 - b1   a1 - b1
```

A little messy, but we can get the same result with a single *Mathematica* command:

```
In[6]:=    Clear[alternatef]
           alternatef[x_] =
           InterpolatingPolynomial[
           {{a1,a2},{b1,b2}},x]

Out[6]=
                 a1 (-a2 + b2)   (-a2 + b2) x
           a2 - ------------- + -------------
                   -a1 + b1        -a1 + b1
```

To see that both results are the same, look at:

```
In[7]:=    Simplify[f[x] - alternatef[x]]

Out[7]=    0
```

They are the same and we're out of here.

### B.2.b.ii)

How we use common sense to determine the formula for the line function

$$f[x] = b + rx$$

whose plot goes through the two points $\{a_1, a\}_2$ and $\{b_1, b\}_2$?

**Answer:**

If we use our heads a bit we can come up with the same formula with less work. Here's how: Let $f[x] = b + rx$. As $x$ grows from $a_1$ to $b_1$ then $f[x]$ must grow from $a_1$ to $b_2$. This means the growth rate $r$ must be given by

$$r = (b_2 - a_2)/(b_1 - a_1).$$

(Some folks call this number $r$ the **slope** of the line because it is the same as rise/run.)

Because $f[a_1] = a_2$, we see that $f[x]$ is given by:

```
In[1]:=    Clear[f,x,a1,a2,b1,b2]
           f[x_] = a2 + (b2 - a2)/
           (b1 - a1) (x - a1)
```

*Out[1]=*
$$a2 + \frac{(-a2 + b2)\ (-a1 + x)}{-a1 + b1}$$

Compare this to the correct answer we got in the last part:

*In[2]:=*
```
alternatef[x_] =
InterpolatingPolynomial[
{{a1,a2},{b1,b2}},x]
```

*Out[2]=*
$$a2 - \frac{a1\ (-a2 + b2)}{-a1 + b1} + \frac{(-a2 + b2)\ x}{-a1 + b1}$$

To see that they are the same, look at:

*In[3]:=*
```
Simplify[f[x] - alternatef[x]]
```

*Out[3]=*    0

They are the same. Common sense wins again.

### B.2.b.iii)

Determine the formula of the line function

$$f[x] = b + rx$$

whose plot goes through the two points $\{-2.1, 7.6\}$ and $\{5.3, -4.9\}$. Check with a plot.

**Answer:**

Here is the common sense way:

*In[1]:=*
```
Clear[f,x]
f[x_] =
Expand[7.6 + (-4.9 - 7.6)/
(5.3 - (-2.1)) (x - (-2.1))]
```

*Out[1]=*    4.0527 - 1.68919 x

Here is the machine way:

*In[2]:=*
```
InterpolatingPolynomial[
{{-2.1,7.6},{5.3,-4.9}},x]
```

*Out[2]=*    4.0527 - 1.68919 x

Here is the line plot:

*In[3]:=*
```
lineplot =
Plot[f[x],{x,-3,7},
AxesLabel->{"x","y"}]
```

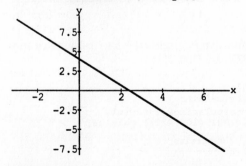

Add the points $\{-2.1, 7.6\}$ and $\{5.3, -4.9\}$ :

*In[4]:=*
```
point1 =
Graphics[{RGBColor[1,0,0],
PointSize[0.02],
Point[{-2.1,7.6}]}]
point2 =
Graphics[{RGBColor[1,0,0],
PointSize[0.02],
Point[{5.3,-4.9}]}]

Show[lineplot,point1,point2]
```

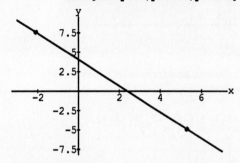

Nailed them both.

### ■ B.3) The logarithm of an exponential function is a line function.

### B.3.a)

Plot the exponential function $f[x] = 6.7(2.8)^x$ on the interval $-2 \le x \le 4$. Then plot $\log[10, f[x]]$ on the same interval. Describe what you see.

**Answer:**

*In[1]:=*
```
Clear[x,f]
f[x_] = 6.7 (2.8)^x

Plot[f[x],{x,-2,4},
AxesLabel->{"x","y"}]
```

Look at that sucker grow. This is characteristic exponential behavior.

Now let's see the plot of $\log[10, f[x]]$ on the same interval:

*In[2]:=*
```
Plot[Log[10,f[x]],{x,-2,4},
AxesLabel->{"x","y"}]
```

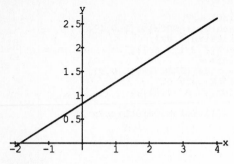

A line!

Maybe this was just an accident; so let's look at $\log[a, f[x]]$ for some other bases $a$:

```
In[3]:=   Plot[Log[2,f[x]],{x,-2,4},
          AxesLabel->{"x","y"}]
```

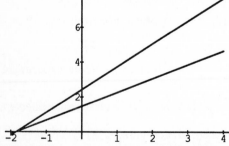

Another line:

Let's plot $\log[a, f[x]]$ for two random numbers $a$ with $2 \leq a \leq 10$:

```
In[4]:=   b1 = Random[Real,{2,10}]
Out[4]=   2.19375
```

```
In[5]:=   b2 = Random[Real,{2,10}]
Out[5]=   3.66598
```

```
In[6]:=   Plot[{Log[b1,f[x]],
          Log[b2,f[x]]},{x,-2,4}]
```

Straight lines again. Reactivate and run again if you like.

If you think that these straight lines are all accidents, then think again.

In mathematics, there are no accidents.

### B.3.b)

Given constants $k$ and $a > 0$ and given $b > 1$, explain why
$$f[x] = \log[b, ka^x]$$
is a line function.

**Answer:**

There's not much to it. Just note that
$$f[x] = \log[b, ka^x] = \log[b, k] + \log[b, a^x] = \log[b, k] + x \log[b, a].$$
This tells us that f[x] is a line function with constant growth rate $\log[b, a]$.

### B.3.c)

What is semi-log paper and why do many scientists like it?

**Answer:**

Semi-log paper has a normal scale on the $x$-axis and a log scale on the $y$-axis. The log scale is fixed up so that when you plot $y = f[x]$ on log paper, then you are actually plotting $y = \log[10, f[x]]$.

This means that if $f[x] = ka^x$ is an exponential function, then the semi-log paper plot of $y = f[x]$ is a straight line.

Neat.

## Tutorial

■ **T.1) Linear models.**

### T.1.a)

Water freezes at 32 degrees Fahrenheit and at 0 degrees Celsius. Water boils at 212 degrees Fahrenheit and at 100 degrees Celsius. Find the line function that takes celsius degrees as input and spits out the corresponding Fahrenheit reading.

Give a plot depicting Celsius readings versus Fahrenheit readings for the range $-30 \leq$ Celsius reading $\leq 50$.

**Answer:**

Use the line function fahr[celsius] = $b + r$ celsius that runs through $\{0, 32\}$ and $\{100, 212\}$:

```
In[1]:=   Clear[fahr,celsius]
          fahr[celsius_] =
          InterpolatingPolynomial[
          {{0,32},{100,212}},celsius]
Out[1]=          9 celsius
          32 + ---------
                   5
```

Here is the plot:

```
In[2]:=   Plot[fahr[celsius],{celsius,-30,50},
          AxesLabel->{Celsius, Fahrennheit}]
```

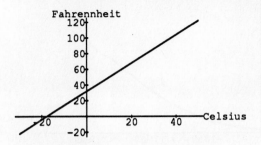

50 degrees Celsius is damn hot.

### T.1.b)

A new John Deere tractor costs $27000. After ten years of heavy use it is predicted to be worth $12000 (in today's dollars).

Give a reasonable estimate of its value in five years.

### Answer:

Run a line through $\{0, 27000\}$ and $\{10, 12000\}$.

```
In[1]:=   Clear[f,t]
          f[t_] = InterpolatingPolynomial[
          {{0,27000},{10,12000}},t]
Out[1]=   27000 - 1500 t
```

Note :

```
In[2]:=   {f[0],f[10]}
Out[2]=   {27000, 12000}
```

So $f[0]$ and $f[10]$ are right on the target. Here is a plot of the linear model for ten years:

```
In[3]:=   Plot[f[t],{t,0,10},AxesLabel->
          {"years","value"}]
```

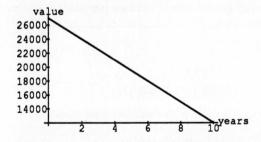

The predicted value in dollars after five years is:

```
In[4]:=   f[5]
Out[4]=   19500
```

Here is another way of doing the same thing: The new tractor costs 27000 dollars and its value falls to 12000 dollars in 10 years. This means its value fell 15000 dollars in 10 years. So it is reasonable to say that its value fell by 1500 dollars per year. This tells us that its value in $t$ years is given by $27000 - 1500t$ dollars. This agrees with the earlier function we produced.

```
In[5]:=   f[t]
Out[5]=   27000 - 1500 t
```

### T.1.c)

Hertz offers you a Ford Taurus for 34 dollars plus 29 cents per mile. Avis offers a Chevrolet Lumina for 29 dollars plus 32 cents per mile. How does the length of your trip tell you which deal is better?

### Answer:

Here is a plot:

```
In[1]:=   Clear[hertz, avis,x]
          hertz[x_] = 34 + .29 x

          avis[x_] = 29 + .32 x

          Plot[{hertz[x],avis[x]},{x,0,300},
          PlotStyle->{{Thickness[0.006]},
          {RGBColor[1,0,0]}}]
```

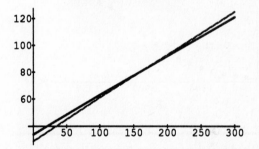

That's the Avis line starting at under the Hertz line and crossing the Hertz line somewhere near $x = 160$.

To find the the exact point where they cross, look at:

```
In[2]:=   Solve[hertz[x] == avis[x],x]
Out[2]=   {{x -> 166.667}}
```

So on trips of fewer than 167 miles, rent the Avis's Lumina; for longer trips go for Hertz's Taurus.

### ■ T.2) Linear interpolation of census figures.

Here are the United States population data in millions $\{t, P[t]\}$ where $t$ corrsponds to the year of each census and $P[t]$ is the official U.S.Census figure for year$t$ (Source: *Statistical Abstracts* ).

```
In[3]:=   USPopData=
          {{1790,4},{1800,5},{1810,7},
```

```
              {1820,10},{1830,13},
              {1840,17},{1850,23},{1860,31},
              {1870,40},{1880,50},
              {1890,63},{1900,76},{1910,92},
              {1920,106},{1930,123},
              {1940,132},{1950,151},{1960,178},
              {1970,203},{1980,227}}
Out[3]=     {{1790, 4}, {1800, 5}, {1810, 7},

              {1820, 10}, {1830, 13}, {1840, 17},

              {1850, 23}, {1860, 31}, {1870, 40},

              {1880, 50}, {1890, 63}, {1900, 76},

              {1910, 92}, {1920, 106}, {1930, 123},

              {1940, 132}, {1950, 151},

              {1960, 178}, {1970, 203}, {1980, 227}}
```

and a plot:

```
In[4]:=    data =
           ListPlot[USPopData,
           AxesLabel->
           {"year","US Population"},
           Ticks->{{1800,1850,1900,1950},
           Automatic},
           PlotStyle->
           {RGBColor[1,0,0],PointSize[0.02]}];
```

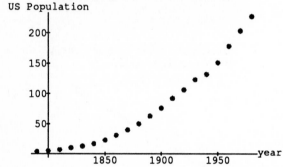

## T.2.a)

Does the United States population increase linearly?

**Answer:**

No, the data points do not come close to all falling on one line.

## T.2.b)

Use linear interpolation of the data to estimate the United States population in 1957.

**Answer:**

Here is a plot that connects the consecutive data points with individual line segments:

```
In[1]:=    ListPlot[USPopData,
           AxesLabel->{"year","US Population"},
```

```
           Ticks->{{1800,1850,1900,1950},
           Automatic},
           PlotJoined->True];
```

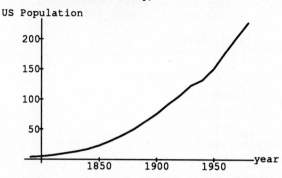

Even though the population over the years did not increase linearly, this plot indicates that it is a good bet that between any two consecutive data points the population did increase approximately linearly. So we estimate population between 1950 and 1960 by running a line through the data points for those years:

```
In[2]:=    Clear[f,x]
           f[x_] =
           InterpolatingPolynomial[
           {{1950,151},{1960,178}},x]
Out[2]=            27 x
           -5114 + ----
                     10
```

Check:

```
In[3]:=    {f[1950],f[1960]}
Out[3]=    {151, 178}
```

Good. Here is the estimated population in 1957 in millions:

```
In[4]:=    N[f[1957]]
Out[4]=    169.9
```

Lots of customers for the hot 1957 Chevy V-8.

## T.2.c)

During the years 1950 − 1960, the U.S. population went up at what average growth rate in millions per year?

**Answer:**

Here are the census figures:

$$\{1950, 151\}, \{1960, 178\}.$$

The average growth rate in millions per year was:

```
In[1]:=    (178 - 151)/(1960 - 1950)
Out[1]=    27
           --
           10
```

Note that this is exactly the growth rate of the interpolating function $f[x]$ in the part immediately above:

```
In[2]:=    f[x]

Out[2]=            27 x
          -5114 + ----
                   10
```

This was not an accident.

### ■ T.3) Data analysis.

Information between the variable $x$ and the function $y$ may consist of nothing more than some measurements indicating the observed value of $y$ for various values of $x$.

A quick plot of the points usually reveals to the alert eye whether the functional relationship of $y$ in terms of $x$ is of the form

$$y = b + rx.$$

If the plot of the points looks like:

```
In[3]:=    data =
           {{-0.8, -0.84},{-0.4, 0.08},
           {-0.2, 0.73},{0.2, 1.46},
           {0.4, 1.92},{0.8, 2.84},{1.4, 4.01},
           {1.8, 5.14},{2.6, 6.98},
           {3.2, 8.36},{3.6, 9.28}}

           dataplot =
           ListPlot[data,AxesLabel->{"x","y"},
           PlotStyle->{RGBColor[1,0,0],
           PointSize[.015]}]
```

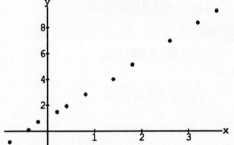

Then we are happy running a line through two of the points with confidence that it will hit (or at least come close to) the others.

Let's try it by taking two data points like $\{0.4, 1.92\}$ and $\{3.2, 8.36\}$, running a line through them and plotting:

```
In[4]:=    Clear[f,x]
           twopoints ={{0.4,1.92},{3.2,8.36}}
           f[x_] =
           InterpolatingPolynomial[twopoints,x]

           lineplot =Plot[f[x],{x,-1,5},
           DisplayFunction->Identity]

           Show[dataplot,lineplot,
           DisplayFunction->$DisplayFunction]
```

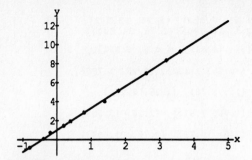

Not bad.

Sometimes the points do not line up in a straight line:

```
In[5]:=    data =
           {{-1., 0.519}, {-0.5, 0.862},
           {-0.25, 1.132},{0.25, 1.795},
           {1., 3.78}, {1.25, 5.204},
           {1.75, 7.912}, {2.25, 13.393},
           {2.5, 16.770}, {2.75, 21.507}}

Out[5]=    {{-1., 0.519}, {-0.5, 0.862},

           {-0.25, 1.132}, {0.25, 1.795},

           {1., 3.78}, {1.25, 5.204},

           {1.75, 7.912}, {2.25, 13.393},

           {2.5, 16.77}, {2.75, 21.507}}
```

```
In[6]:=    dataplot =
           ListPlot[data,AxesLabel->{"x","y"},
           PlotStyle->{RGBColor[1,0,0],
           PointSize[.015]}]
```

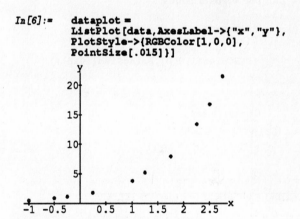

This looks suspiciously exponential.

### T.1.a)

---

How can we better eyeball an exponential relationship?

---

**Answer:**

The points are given in the form $\{x, y\}$. Instead of plotting them in this form, plot them in the form $\{x, \log[a, x]\}$ for any $a > 1$ you like.

Here are the data points again:

```
In[1]:=    data =
           {{-1., 0.519}, {-0.5, 0.862},
           {-0.25, 1.132},{0.25, 1.795},
           {1., 3.78}, {1.25, 5.204},
```

```
                {1.75, 7.912}, {2.25, 13.393},
                {2.5, 16.770},{2.75, 21.507}}
Out[1]=    {{-1., 0.519}, {-0.5, 0.862},

              {-0.25, 1.132}, {0.25, 1.795},

              {1., 3.78}, {1.25, 5.204},

              {1.75, 7.912}, {2.25, 13.393},

              {2.5, 16.77}, {2.75, 21.507}}
```

The following instruction keeps the $x$-coordinates fixed but replaces the corresponding $y$-coordinates by their their base 10 logarithms:

```
In[2]:=    Clear[j]
           logdata =
           Table[{data[[j,1]],
           Log[10,data[[j,2]]]},
           {j,1,Length[data]}]
Out[2]=    {{-1., -0.284833}, {-0.5, -0.0644927},

              {-0.25, 0.0538464}, {0.25, 0.254064},

              {1., 0.577492}, {1.25, 0.716337},

              {1.75, 0.898286}, {2.25, 1.12688},

              {2.5, 1.22453}, {2.75, 1.33258}}
```

Now plot the **logdata** points:

```
In[3]:=    logDataPlot =
           ListPlot[logdata,AxesLabel->
           {"x","y"},PlotStyle->
           {RGBColor[1,0,0],PointSize[.015]}]
```

If you were doing this with pencil and paper, this is what you would see if you plotted the data points on semi-log paper.

They are stretched out in a line. This is a dead give-away that there is an exponential function that runs through or near the original data points.

### T.1.b)

How can we find the exponential function $f[x] = ka^x$ whose plot runs through or near the data points above?

**Answer:**

Look at the logdata points again:

```
In[1]:=    logdata
Out[1]=    {{-1., -0.284833}, {-0.5, -0.0644927},

              {-0.25, 0.0538464}, {0.25, 0.254064},

              {1., 0.577492}, {1.25, 0.716337},

              {1.75, 0.898286}, {2.25, 1.12688},

              {2.5, 1.22453}, {2.75, 1.33258}}
```

Take two of the logdata points, say:

```
In[2]:=    twoPoints =
           {{-0.25, 0.0538464},{2.25,1.12688}}
Out[2]=    {{-0.25, 0.0538464}, {2.25, 1.12688}}
```

And run a line through them:

```
In[3]:=    Clear[logline,x]
           logline[x_] =
           InterpolatingPolynomial[twoPoints,x]
Out[3]=    0.16115 + 0.429213 x
```

If $f[x] = ka^x$ is the exponential function running through the original data points, then taking logarithm base 10 of both sides gives

$$\log[10, f[x]] = \log[10, ka^x]$$
$$= \log[10, k] + \log[10, a^x]$$
$$= \log[10, k] + x\log[10, a].$$

This must agree with

$$0.16115 + 0.429213x.$$

This tells us that

$$\log[10, k] = 0.16115$$

and

$$\log[10, a] = 0.429213.$$

Consequently

$$k = 10^{0.16115}$$

and

$$a = 10^{0.42913}$$

We now know that the exponential $f[x] = ka^x$ through or near the data points is:

```
In[4]:=    Clear[a,k,x]
           f[x_] = k a^x/.{k->10^(0.16115),
           a->10^(0.429213)}
Out[4]=                          x
           1.44927 2.68666
```

Here is its plot:

```
In[5]:=    functionPlot = Plot[f[x],{x,-2,4},
           AxesLabel->{"x","y"}]
```

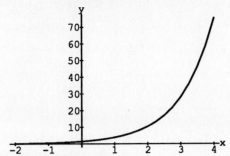

Here is the data plot:

```
In[6]:=    dataplot =
           ListPlot[data,AxesLabel->
           {"x","y"},PlotStyle->
           {RGBColor[1,0,0],PointSize[.015]}]
```

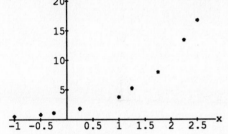

Now let's see whether the plot of the function does go through or near the data points:

```
In[7]:=    Show[dataplot,functionPlot]
```

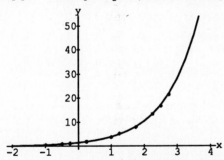

Damn good. The curve goes through most of the points. Chalk up the fact that it misses a couple of them to measurement errors in collecting the original data.

### T.1.c.i)

If there were an entry $\{2.35, y\}$ in the original data table, what would you bet that $y$ would be?

**Answer:**

We could estimate $y$ by linear interpolation but a better bet would be $y = f[2.35]$ where $f[x]$ is the exponential function we found in the last part. So go with the following

value for $y$ :

```
In[1]:=    f[2.35]
Out[1]=    14.7843
```

This is called exponential interpolation because $x = 2.35$ is within the range of the given data and we used an exponential function to arrive at the answer.

### T.1.c.ii)

If there were an entry $\{4.85, y\}$ in the original data table, what would you bet that y would be?

**Answer:**

A reasonable bet would be:

```
In[1]:=    f[4.85]
Out[1]=    174.917
```

This is called exponential extrapolation because $x = 4.85$ is outside the range of the given data. Most experienced data watchers have less confidence in extrapolation than they have with interpolation.

### ■ T.4) Parallel lines.

### T.4.a)

How can you recognize parallel lines?

**Answer:**

It's simple. Two lines are parallel if they have the **same growth rate** . For instance $y = 2x - 1$ and $y = 2x + 3$ are parallel because they both have the same growth rate.

Here's a plot:

```
In[1]:=    Clear[x]
           Plot[{2 x - 1, 2 x + 3},{x,-3,8},
           AxesLabel->{"x","y"}]
```

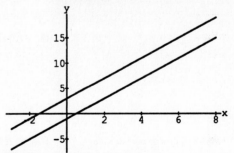

Here is a load of parallel lines:

```
In[2]:=    Clear[k]
           lines = Table[2 x + k, {k,-5,5}]
Out[2]=    {-5 + 2 x, -4 + 2 x, -3 + 2 x,
```

```
        -2 + 2 x, -1 + 2 x, 2 x, 1 + 2 x,
        2 + 2 x, 3 + 2 x, 4 + 2 x, 5 + 2 x}
```

And their plots:

```
In[3]:=    Plot[Release[lines],{x,-8,8},
           AxesLabel->{"x","y"},
           PlotRange->{-8,8}]
```

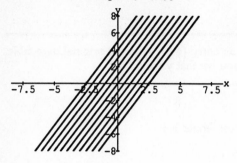

# Literacy Sheet

**1.** What does the plot of a function $f[x] = b + rx$ look like?

**2.** If a line function goes up 4 units on the $y$-axis for every unit on the $x$-axis and the plot goes through $\{0, 2\}$, then what is the formula of the line function?

**3.** How does the sign of the number $r$ tell you whether the line function $f[x] = b + rx$ goes up or down?

**4.** As $x$ advances from left to right, which goes up faster: $f[x] = 2 + 4x$ or $g[x] = 5 + 2x$?

**5.** How can you tell from the formula $f[x] = b + rx$ for a line function whether the plot goes through $\{0, 0\}$?

**6.** If $(f[x+h] - f[x])/h = 3$ for all $x$'s and $h$'s and $f[0] = 6$, then what is the formula for $f[x]$?

**7.** Two lines functions have formulas $f[x] = 12 + r(x - 4)$ and $g[x] = 12 + s(x - 4)$. How do you know $\{4, 12\}$ is on the plot of both functions? If $r > s$, then which line function is on top for $x > 4$ and which line function is on top for $x < 4$?

**8.** What is the formula of the line function whose plot goes through the points $\{-1, -2\}$ and $\{3, 8\}$?

**9.** Are the points $\{2, 3\}$, $\{3, 7\}$, $\{4, 11\}$ and $\{5, 14\}$ on the same line? How about the points $\{2, 3\}$, $\{3, 7\}$, $\{4, 11\}$ and $\{5, 15\}$?

**10.** Given specific positive numbers $a, b$ and $k$ with $b > 1$, then how does $\log[b, ka^x]$ plot out as a function of $x$?

**11.** Solve for $x$ in terms of $a$:

$$2a^x = 7.$$

**12.** Why is it that if you take some data points

$$\{\{x_1, y\}_1, \{x_2, y\}_2, \{x_3, y\}_3, ..., \{x_{12}, y\}\}_{12}$$

and the plot of the points,

$$\{\{x_1, \log[10, y_1]\}, \{x_2, \log[10, y_2]\},$$
$$\{x_3, \log[10, y_3]\}, ..., \{x_{12}, \log[10, y_{12}]\}\}$$

fall onto a straight line, then you know that are constants $k$ and $a$ such that the plot of $f[x] = ka^x$ passes through all the original points?

**13.** Why is it that if you take some data points

$$\{\{x_1, y\}_1, \{x_2, y\}_2, \{x_3, y\}_3, ..., \{x_{12}, y\}\}_{12}$$

and the plot of the points,

$$\{\{\log[10, x_1], \log[10, y_1]\}, \{\log[10, x_2], \log[10, y_2]\},$$
$$\{\log[10, x_3], \log[10, y_3]\}, ..., \{\log[10, x_{12}], \log[10, y_{12}]\}\}$$

fall onto a straight line, then you know that are constants $k$ and $a$ such that the plot of $f[x] = kx^r$ passes through all the original points?

# Instantaneous Growth Rates

## Guide

Almost all mathematical activity in the hard sciences and the social sciences is devoted to finding the laws that govern various changes in the measured quantities under study. The concept of the instantaneous growth rate evolved to do precisely this job.

It is no exaggeration to say that all of calculus is the study of the intantaneous growth rate.

## Basics

■ **B.1) Growth rates of a function.**

### B.1.a.i)

Plot $f[x] = 1 + 2x^3 - x^4$ on the interval $[-1, 2]$.

Measure the net growth of $f[x]$ on the interval $[-1, 2]$.

**Answer:**

```
In[1]:=    f[x_] = 2 x^3 - x^4 + 1
Out[1]=           3   4
           1 + 2 x  - x
```

```
In[2]:=    Plot[f[x],{x,-1,2}]
```

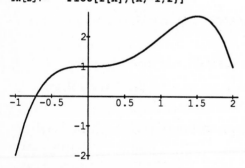

On the interval $[-1, 2]$, the function starts out at:

```
In[3]:=    f[-1]
Out[3]=    -2
```

And it ends up at:

```
In[4]:=    f[2]
Out[4]=    1
```

Its net growth is:

```
In[5]:=    f[2] - f[-1]
Out[5]=    3
```

### B.1.a.ii)

Measure the average growth rate of $f[x] = 1 + 2x^3 - x^4$ on the interval $[-1, 2]$.

Then measure the average growth rate of $f[x] = 1 + 2x^3 - x^4$ on the interval $[1, 3/2]$.

**Answer:**

To measure the average growth rate on the interval $[-1, 2]$, look at:

```
In[1]:=    f[-1]
Out[1]=    -2
```

```
In[2]:=    f[2]
Out[2]=    1
```

So $f[x]$ grows from $f[-1] = -2$ to $f[2] = 1$ as $x$ grows from $-1$ to $2$. Its average growth rate as $x$ grows from $-1$ to $2$ in units on the $y$-axis per unit on the $x$-axis is:

```
In[3]:=    Expand[(f[2] - f[-1])/(2 - (-1))]
Out[3]=    1
```

To measure the average growth rate on the interval $[1, 3/2]$, look at:

```
In[4]:=    f[1]
Out[4]=    2
```

```
In[5]:=    f[3/2]
Out[5]=    43
           --
           16
```

So $f[x]$ grows from $f[1] = 2$ to $f[3/2] = 43/16$ as $x$ grows from $1$ to $3/2$. Its average growth rate as $x$ grows from $1$ to $3/2$ in units on the $y$-axis per unit on the $x$-axis is:

```
In[6]:=    Expand[(f[3/2] - f[1])/((3/2) - (1))]
```

*Out[6]=* $\dfrac{11}{8}$

The average growth rate is larger on $[1, 3/2]$ than on $[-1, 2]$.

### B.1.a.iii)

Measure the average growth rate of $f[x] = 1 + 2x^3 - x^4$ on the interval $[x, x + h]$ for any point $x$.

### Answer:

The average growth rate of this function on the interval $[x, x + h]$ in units on the $y$-axis per unit on the $x$-axis is:

```
In[1]:=   Expand[(f[x + h] - f[x])/h]
```
```
Out[1]=       2    3           2         2
          2 h  - h  + 6 h x - 4 h  x + 6 x  -

               2      3
            6 h x  - 4 x
```

### B.1.a.iv)

Measure the instantanteous growth rate $f'[x]$ of $f[x] = 1 + 2x^3 - x^4$ at a point $x$.

### Answer:

Look at the average growth rate on the interval $[x, x + h]$ in units on the $y$-axis per unit on the $x$-axis:

```
In[1]:=   Expand[(f[x + h] - f[x])/h]
```
```
Out[1]=       2    3           2         2
          2 h  - h  + 6 h x - 4 h  x + 6 x  -

               2      3
            6 h x  - 4 x
```

The instantaneous rate of change at $x$ is the limiting case of the above average rates as $h$ closes in on 0. Evidently this is

$$0 - 0 + 0 - 0 + 6x^2 - 0 - 4x^3 = 6x^2 - 4x^3.$$

So the instantanteous growth rate of $f[x] = 1 + 2x^3 - x^4$ at a point $x$ in units on the $y$-axis per unit on the $x$-axis is given by:

$$f'[x] = 6x^2 - 4x^3.$$

*Mathematica* knows how to compute the instantaneous growth rate $f'[x]$. We simply have to ask for it:

```
In[2]:=   f'[x]
```
```
Out[2]=       2      3
          6 x  - 4 x
```

Nice work *Mathematica* .

### B.1.a.v)

At what point in $[-1, 2]$ is $f[x] = 1 + 2x^3 - x^4$ going up most rapidly?

At what point in $[-1, 2]$ is $f[x] = 1 + 2x^3 - x^4$ going down most rapidly?

At what point in $[-1, 2]$ is $f[x] = 1 + 2x^3 - x^4$ changing least rapidly?

### Answer:

The instantaneous growth rate at $x$ is $f'[x] = 6x^2 - 4x^3$. Let's see where it is the largest and the least:

```
In[1]:=   Plot[f'[x],{x,-1,2},PlotRange->All]
```

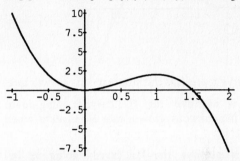

The instantaneous growth rate $f'[x]$ is largest at the left end point $x = -1$; so the function $f[x]$ is **going up most rapidly at** $x = -1$.

The instantaneous growth rate $f'[x]$ is least at the right end point $x = 2$; so the function $f[x]$ is **going down most rapidly at** $x = 2$.

The instantaneous growth rate $f'[x]$ is 0 at $x = 0$ and $x = 1.5$; so the function $f[x]$ is changing the **least rapidly at** $x = 0$ **and** $x = 1.5$.

Let's take a look at the plot of $f[x]$ to make sure that this makes sense:

```
In[2]:=   Plot[f[x],{x,-1,2}]
```

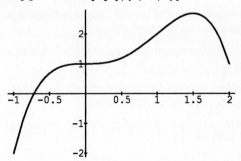

Sure enough: Rapid increase at the left end point; rapid decrease at the right end point and lazy behavior at $x = 0$ and $x = 1.5$.

### B.1.a.vi)

Plot $f[x] = 1 + 2x^3 - x^4$ and its instantaneous growth rate $f'[x]$ on the same axes for $-1 \leq x \leq 2$. Discuss relations between these two plots, paying special attention to what $f[x]$ is doing when $f'[x]$ is negative.

**Answer:**

```
In[1]:=    Plot[{f[x],f'[x]},{x,-1,2},
           PlotStyle->{{Thickness[.008],GrayLevel[0.0]},
           {Thickness[.008],GrayLevel[0.5]}}]
```

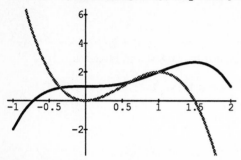

The $f'[x]$ curve represents the instantaneous growth rate.

When $f'[x]$ is **negative**, the $f[x]$ curve is going **down** because the intantaneous growth rate is negative when $f'[x]$ is negative.

When $f'[x]$ is **positive**, the $f[x]$ curve is going **up** because the intantaneous growth rate is positive when $f'[x]$ is positive.

■ **B.2) Instantaneous growth rates in various contexts.**

**Velocity** is the instantaneous growth rate of distance as a function of time.

**Acceleration** is the instantaneous growth rate of velocity as a function of time.

**Jerk** is the instantaneous growth rate of acceleration as a function of time.

**Slope** is the instantaneous growth rate of $f[x]$ as a function of $x$. Thus $f'[x]$ measures the slope of the graph of $y = f[x]$ at the point $(x, f[x])$.

**Density** is the instantaneous growth rate of weight as a function of volume.

**Force** is the instantaneous growth rate of momentum as a function of time.

**Marginal revenue** is the instantaneous growth rate of revenue as a function of level of production.

**Current** is the instantaneous growth rate of electrical charge as a function of time.

# Tutorial

■ **T.1) Instantaneous growth rates for powers of x.**

**T.1.a)**

Measure the instantaneous growth rate $f'[x]$ of $f[x] = x^2$ at any point $x$.

**Answer:**

```
In[1]:=    Clear[x,f]
           f[x_] = x^2
Out[1]=       2
           x
```

Look at the average growth rate on the interval $[x, x + h]$ in units on the $y$-axis per unit on the $x$-axis:

```
In[2]:=    Expand[(f[x + h] - f[x])/h]
Out[2]=    h + 2 x
```

The instantaneous rate of change at $x$ is the limiting case of the above average rates as $h$ closes in on $0$. Evidently this is

$$0 + 2x = 2x$$

So the instantaneous growth rate of $f[x] = x^2$ at a point $x$ in units on the $y$-axis per unit on the $x$-axis is given by:

$$f'[x] = 2x$$

*Mathematica* knows how to compute $f'[x]$:

```
In[3]:=    f'[x]
Out[3]=    2 x
```

Good.

**T.1.b)**

Measure the instantaneous growth rate $f'[x]$ of $f[x] = x^3$ at a point $x$.

**Answer:**

```
In[1]:=    Clear[x,f]
           f[x_] = x^3
Out[1]=       3
           x
```

Look at the average growth rate on the interval $[x, x + h]$ in units on the $y$-axis per unit on the $x$-axis:

```
In[2]:=    Expand[(f[x + h] - f[x])/h]
Out[2]=    2              2
           h  + 3 h x + 3 x
```

The instantaneous rate of change at $x$ is the limiting case of the above average rates as $h$ closes in on $0$.

Evidently this is

$$0 + 0 + 3x^2 = 3x^2$$

So the instantanteous growth rate of $f[x] = x^3$ at a point $x$ in units on the $y$-axis per unit on the $x$-axis is given by:

$$f'[x] = 3x^2$$

Let *Mathematica* compute $f'[x]$:

```
In[3]:=    f'[x]
Out[3]=      2
           3 x
```

Comfy.

**T.1.c)**

Measure the instantaneous growth rate $f'[x]$ of $f[x] = x^4$ at each point $x$.

**Answer:**

```
In[1]:=    Clear[x,f]
           f[x_] = x^4
Out[1]=    4
           x
```

Look at the average growth rate on the interval $[x, x + h]$ in units on the $y$-axis per unit on the $x$-axis:

```
In[2]:=    Expand[(f[x + h] - f[x])/h]
Out[2]=    3       2       2       3
           h  + 4 h  x + 6 h x  + 4 x
```

The instantaneous rate of change at $x$ is the limiting case of the above average rates as $h$ closes in on $0$.

Evidently this is

$$0 + 0 + 0 + 4x^3 = 4x^3$$

So the instantanteous growth rate of $f[x] = x^4$ at a point $x$ in units on the $y$-axis per unit on the $x$-axis is given by:

$$f'[x] = 4x^3$$

Let *Mathematica* compute $f'[x]$:

```
In[3]:=    f'[x]
Out[3]=       3
           4 x
```

Fine.

**T.1.d)**

Measure the instantaneous growth rate $f'[x]$ of $f[x] = x^{12}$ at a point $x$.

**Answer:**

```
In[1]:=    Clear[x, f]
```

```
In[2]:=    f[x_] = x^12
Out[2]=    12
           x
```

Look at the average growth rate on the interval $[x, x + h]$ in units on the $y$-axis per unit on the $x$-axis:

```
In[3]:=    Expand[(f[x + h] - f[x])/h]
Out[3]=    11        10       9  2        8  3
           h  + 12 h   x + 66 h  x + 220 h  x +

             7  4        6  5        5  6
           495 h  x + 792 h  x + 924 h  x +

             4  7        3  8        2  9
           792 h  x + 495 h  x + 220 h  x +

              10        11
           66 h  x  + 12 h
```

Pretty large coefficients, aren't they?

The instantaneous rate of change at $x$ is the limiting case of the above average rates as $h$ closes in on $0$.

Evidently this is

$$0 + 0 + 0 + 0 + 0 + 0 + 0 + 0 + 0 + 0 + 0 + 12x^{11} = 12x^{11}.$$

so the instantaneous growth rate of $f[x] = x^{12}$ at a point $x$ in units on the $y$-axis per unit on the $x$-axis is given by:

$$f'[x] = 12x^{11}$$

Let *Mathematica* compute $f'[x]$:

```
In[4]:=    f'[x]
Out[4]=        11
           12 x
```

This is getting to be old hat.

**T.1.e)**

Measure the instantaneous growth rate $f'[x]$ of $f[x] = 1/x$ at a point $x$.

**Answer:**

```
In[1]:=    Clear[x, f]
           f[x_] = 1/x
Out[1]=    1
           -
           x
```

Look at the average growth rate on the interval $[x, x + h]$ in units on the $y$-axis per unit on the $x$-axis:

```
In[2]:=    Expand[(f[x + h] - f[x])/h]
```

```
Out[2]=        1          1
          -(---)  +  ----------
           h x       h (h + x)
```

The instantaneous rate of change at $x$ is the limiting case of the above average rates as $h$ closes in on 0. But we have a hard time deciding what happens as $h$ closes in on 0, so we try a little algebra by going for a common denominator:

```
In[3]:=    Together[(f[x + h] - f[x])/h]
Out[3]=          1
          -(----------)
            x (h + x)
```

As $h$ closes in on 0, this closes in on

$$-1/(x(0 + x)) = -1/x^2.$$

So the instantanteous growth rate of $f[x] = 1/x$ at a point $x$ in units on the $y$-axis per unit on the $x$-axis is given by:

$$f'[x] = -1/x^2.$$

Check:

```
In[4]:=    f'[x]
Out[4]=        -2
           -x
```

We're getting good at this.

### T.1.f)

Measure the instantaneous growth rate $f'[x]$ of $f[x] = 1/x^2$ at a point $x$.

**Answer:**

```
In[1]:=    Clear[x, f]

In[2]:=    f[x_] = 1/x^2
Out[2]=        -2
           x
```

Look at the average growth rate on the interval $[x, x + h]$ in units on the $y$-axis per unit on the $x$-axis:

```
In[3]:=    Expand[(f[x + h] - f[x])/h]
Out[3]=         1            1
          -(----)  +  ----------
            2         2
           h x       h (h + x)
```

The instantaneous rate of change at $x$ is the limiting case of the above average rates as $h$ closes in on 0. But again, in the form above we have a hard time deciding what happens as $h$ closes in on 0. So we try a little more algebra by going again for a common denominator:

```
In[4]:=    Together[(f[x + h] - f[x])/h]
```

```
Out[4]=        -h - 2 x
           -----------
            2         2
           x  (h + x)
```

As $h$ closes in on 0, this closes in on

$$(0 - 2x)/(x^2(0 + x)^2) = -2x/(x^2x^2) = -2/x^3.$$

So the instantanteous growth rate of $f[x] = 1/x^2$ at a point $x$ in units on the $y$-axis per unit on the $x$-axis is given by:

$$f'[x] = -2/x^3.$$

Check:

```
In[5]:=    f'[x]
Out[5]=        -2
           --
            3
           x
```

Hot Ziggety.

### T.1.g)

Measure the instantaneous growth rate $f'[x]$ of $f[x] = 1/x^3$ at each point $x$.

**Answer:**

```
In[1]:=    Clear[x, f]

In[2]:=    f[x_] = 1/x^3
Out[2]=        -3
           x
```

The average growth rate on the interval $[x, x + h]$ is:

```
In[3]:=    Expand[(f[x + h] - f[x])/h]
Out[3]=         1            1
          -(----)  +  ----------
            3          3
           h x       h (h + x)
```

We go for a common denominator:

```
In[4]:=    Together[(f[x + h] - f[x])/h]
Out[4]=        2                2
           -h  - 3 h x - 3 x
           ------------------
            3         3
           x  (h + x)
```

As $h$ closes in on 0, this closes in on

$$(-0 - 0 - 3x^2)/(x^3(0 + x)^3)$$
$$= -3x^2/(x^3x^3) = -3x^2/x^6 = -3/x^4.$$

Thus the instantanteous growth rate of $f[x] = 1/x^3$ at a point $x$ in units on the $y$-axis per unit on the $x$-axis is given by:

$$f'[x] = -3/x^4.$$

Check:

```
In[5]:=      f'[x]
Out[5]=      -3
             --
              4
             x
```

On the money.

### ■ T.2) Instantaneous growth rate for Sqrt[x].

### T.2.a)

Measure the instananeous growth rate $f'[x]$ of $f[x] = \sqrt{x}$ at a point $x$.

**Answer:**

```
In[1]:=      Clear[x,f]

In[2]:=      f[x_] = Sqrt[x]
Out[2]=      Sqrt[x]
```

Look at the average growth rate on the interval $[x, x + h]$ in units on the $y$-axis per unit on the $x$-axis:

```
In[3]:=      Expand[(f[x + h] - f[x])/h]
Out[3]=        Sqrt[x]      Sqrt[h + x]
             -(-------) + -----------
                h              h
```

The instantaneous rate of change at $x$ is the limiting case of the above average rates as h closes in on 0. But this limiting case is not easy to see from what we have above. It is going to be really hard to spot $f'[x]$ by looking at the last formula.

But here's a thought! Let's take another look:

```
In[4]:=      (f[x + h] - f[x])/h
Out[4]=      -Sqrt[x] + Sqrt[h + x]
             ----------------------
                       h
```

What if we rationalize this? After all,

$$(f[x + h] - f[x])/h$$
$$= ((f[x+h] - f[x])(f[x+h] + f[x])/(f[x+h] + f[x]))(1/h)$$

because the $(f[x+h] + f[x])$ terms in the denominator and numerator cancel. So $(f[x + h] - f[x])/h$ is given by the product of three factors:

```
In[5]:=      factor1=
             Expand[(f[x+h] - f[x])(f[x+h] + f[x])]
Out[5]=      h

In[6]:=      factor2 = 1/(f[x + h] + f[x])
```

```
Out[6]=                 1
             ----------------------
             Sqrt[x] + Sqrt[h + x]

In[7]:=      factor3 = (1/h)
Out[7]=      1
             -
             h
```

Thus an equivalent from of the average growth rate

$$(f[x + h] - f[x])/h$$

is :

```
In[8]:=      factor1 factor2 factor3
Out[8]=                 1
             ----------------------
             Sqrt[x] + Sqrt[h + x]
```

The instantaneous growth rate at $x$ is the limiting case of the above average growth rates as $h$ closes in on 0. Evidently this is

$$1/(\sqrt{x} + \sqrt{0 + x}) = 1/(2\sqrt{x}).$$

So we go with

$$f'[x] = 1/(2\sqrt{x}).$$

This is in agreement with *Mathematica*'s calculation:

```
In[9]:=      f'[x]
Out[9]=          1
             ---------
             2 Sqrt[x]
```

And this makes us very happy.

### ■ T.3) Instantaneous growth rate for Sin[x].

### T.3.a)

Measure the instantaneous growth rate $f'[x]$ of $f[x] = \sin[x]$ at each point $x$.

**Answer:**

```
In[1]:=      Clear[x,f]
             f[x_] = Sin[x]
Out[1]=      Sin[x]
```

The average growth rate on the interval $[x, x + h]$ in units on the $y$-axis per unit on the $x$-axis is:

```
In[2]:=      (f[x + h] - f[x])/h
Out[2]=      -Sin[x] + Sin[h + x]
             --------------------
                      h
```

The limiting case is hard to see from what we have above and it is going to be really difficult to spot $f'[x]$ by looking at this.

But we never give up without a fight.

Maybe we can spot the limiting case of $(f[x+h] - f[x])/h$ by **plotting** $(f[x + h] - f[x])/h$ as a function of $x$ for some small values of $h$.

Let's give this a try:

*In[3]:=*    `plot1 = Plot[(f[x + .1] - f[x])/.1,`
                    `{x, 0, 2 Pi}]`

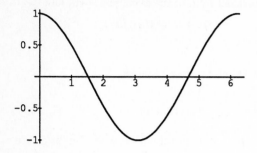

*In[4]:=*    `plot2 = Plot[(f[x + .01] - f[x])/.01,`
                    `{x, 0, 2 Pi}]`

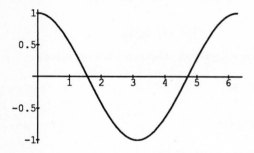

*In[5]:=*    `plot3 = Plot[(f[x + .001] - f[x])/.001,`
                    `{x, 0, 2 Pi}]`

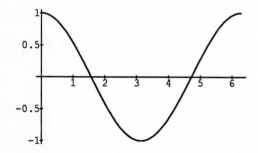

Damned if these curves don't look like cosine curves!

Let's check this out by superimposing a plot of cos[x] on each of the plots above. Here is a plot of cos[x]:

*In[6]:=*    `cosineplot = Plot[Cos[x], {x, 0, 2 Pi}]`

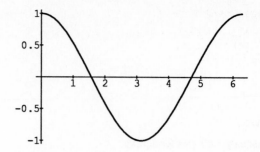

Here is a plot of $(f[x + .1] - f[x])/.1$ and cos[x] on the same axes:

*In[7]:=*    `Show[plot1, cosineplot]`

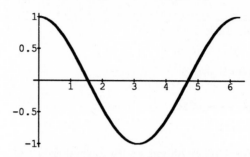

Running pretty close. Here is a plot of $(f[x + .01] - f[x])/.01$ and cos[x] on the same axes:

*In[8]:=*    `Show[plot2, cosineplot]`

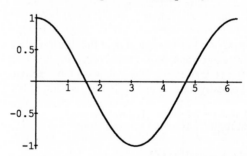

The two plots are almost concurrent. Here is a plot of $(f[x + .001] - f[x])/.001$ and cos[x] on the same axes:

*In[9]:=*    `Show[plot3, cosineplot]`

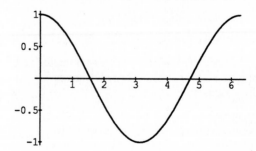

$(f[x + .001] - f[x])/.001$ and cos[x] seem to be plotting out the same. The evidence tells us that the limiting case of the average growth rates

$$(f[x + h] - f[x])/h = (\sin[x + h] - \sin[x])/h$$

as $h$ closes in on 0 is

$$f'[x] = \cos[x].$$

Check:

```
In[10]:=   f'[x]
Out[10]=   Cos[x]
```

How sweet it is. Numerical and graphical prospecting pays off.

### T.3.b)

How well does the average growth rate

$$(\sin[x + .0001] - \sin[x])/.0001$$

approximate the instantaneous growth rate of $\sin[x]$ at a point $x$ for $0 \le x \le 2\pi$?

**Answer:**

Well, the instantaneous growth rate of $f[x] = \sin[x]$ at a point $x$ is given by $f'[x] = \cos[x]$. Let's see how close

$$(\sin[x + .0001] - \sin[x])/.0001$$

is to $f'[x] = \cos[x]$ by plotting their difference for $0 \le x \le 2\pi$:

```
In[1]:=   difference = Cos[x] - ((Sin[x + .0001] -
Sin[x])/.0001)
Out[1]=   Cos[x] - 10000. (-Sin[x] +

          Sin[0.0001 + x])
```

```
In[2]:=   Plot[difference, {x, 0, 2 Pi}]
```

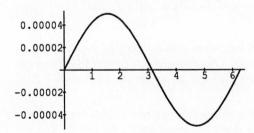

The plot shows $(\sin[x + .0001] - \sin[x])/.0001$ is $f'[x] = \cos[x]$ to four decimals. For most practical purposes, there is no difference between the average growth rate $(\sin[x + .0001] - \sin[x])/.0001$ and the instantaneous growth rate $f'[x] = \cos[x]$.

### T.3.c)

Plot $f[x] = \sin[x]$ and the instantaneous growth rate $f'[x] = \cos[x]$ on the same axes for $0 \le x \le 2\pi$. Discuss the relationships beween the two curves.

**Answer:**

Here is a plot of $\sin[x]$ for $0 \le x \le 2\pi$ :

```
In[1]:=   sineplot = Plot[Sin[x], {x, 0, 2 Pi}]
```

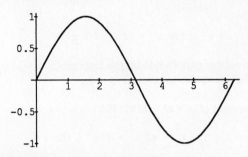

Here is a plot of instantaneous growth rate $f'[x] = \cos[x]$ for $0 \le x \le 2\pi$ :

```
In[2]:=   cosineplot = Plot[Cos[x], {x, 0, 2 Pi}]
```

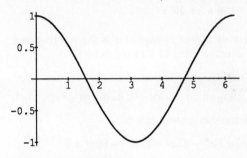

Here they are together:

```
In[3]:=   Show[sineplot, cosineplot]
```

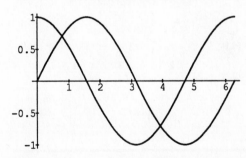

When the instantaneous growth rate, $\cos[x]$, of $\sin[x]$ is positive, then the $\sin[x]$ curve is going up and when the instantaneous growth rate, $\cos[x]$, of $\sin[x]$ is negative, then the $\sin[x]$ curve is going down.

■ **T.4) What the instantaneous growth rate reveals about the function.**

### T.4.a)

Measure the instantaneous growth rate $f'[x]$ of

$$f[x] = 6x^5 - 45x^4 + 110x^3 - 90x^2 + 7$$

at a point $x$.

**Answer:**

```
In[1]:=    Clear[x, f]
           f[x_] = 6 x^5 - 45 x^4 + 110 x^3 -
               90 x^2 + 7
                  2       3       4       5
Out[1]=    7 - 90 x + 110 x - 45 x + 6 x
```

Look at the average growth rate on the interval $[x, x + h]$ in units on the $y$-axis per unit on the $x$-axis:

```
In[2]:=    Expand[(f[x + h] - f[x])/h]
                     2       3      4
Out[2]=    -90 h + 110 h - 45 h + 6 h - 180 x +

                    2        3        2
           330 h x - 180 h x + 30 h x + 330 x -

               2       2 2       3
           270 h x + 60 h x - 180 x +

              3       4
           60 h x + 30 x
```

The instantaneous rate of change at $x$ is the limiting case of the above average rates as $h$ closes in on 0.

Evidently this is

$-0+0-0+0-180x+0-0+0+330x^2-0+0-180x^3+0+30x^4$.

So the instantaneous growth rate of

$$f[x] = 6x^5 - 45x^4 + 110x^3 - 90x^2 + 7$$

at a point $x$ in units on the $y$-axis per unit on the $x$-axis is given by:

$$f'[x] = -180x + 330x^2 - 180x^3 + 30x^4$$

Check:

```
In[3]:=    f'[x]
                      2       3      4
Out[3]=    -180 x + 330 x - 180 x + 30 x
```

Good.

**T.4.b)**

Take the same function

$$f[x] = 6x^5 - 45x^4 + 110x^3 - 90x^2 + 7$$

and plot the instantaneous growth rate $f'[x]$ and the average growth rate $(f[x + .01] - f[x])/.01$ on the same axes for $-1 \le x \le 1$.

Describe what you see.

**Answer:**

```
In[1]:=    Plot[{f'[x],(f[x + .01] - f[x])/.01},
              {x,-1,1}]
```

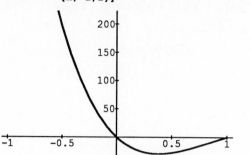

The instantaneous growth rate $f'[x]$ and the average growth rate $(f[x + .01] - f[x])/.01$ are almost the same for this function.

**T.4.c)**

Take the same function $f[x] = 6x^5 - 45x^4 + 110x^3 - 90x^2 + 7$, plot the instantaneous growth rate $f'[x]$ on [-0.5, 3.5] and use it to predict how the plot of $f[x]$ looks. Finally test your prediction by plotting $f[x]$.

**Answer:**

Here is a plot of the instantaneous growth rate $f'[x]$.

```
In[1]:=    growthplot = Plot[f'[x], {x,-0.5,3.5},
              PlotRange->All]
```

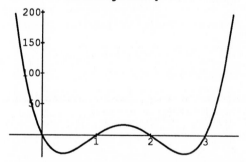

When the instantaneous growth rate $f'[x]$ is positive, then $f[x]$ is going up. Consequently, we predict that the graph of the function $f[x]$ will go up on the intervals $[-0.5, 0]$, $[1, 2]$ and $[3, 3.5]$.

When the instantaneous growth rate $f'[x]$ is negative, then $f[x]$ is going down. Consequently, we predict that the graph of the function $f[x]$ will go down on $[0, 1]$, $[2, 3]$. Let's test the prediction:

```
In[2]:=    functplot = Plot[f[x], {x,-0.5,3.5}]
```

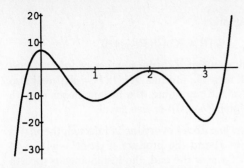

The prediction is right on the money.

Here is a plot of $f[x]$ and $f'[x]$ on the same axes for your enjoyment and good use:

```
In[3]:=    Show[growthplot,functplot]
```

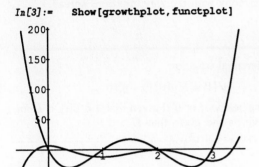

Can you see which is the plot of $f[x]$ and which is the plot of the instantaneous growth rate?

■ **T.5) Going up or down?**

**T.5.a)**

As $x$ leaves $x = 2.6$ and advances a little bit, does $x^3 - 3x^2$ go up or down?

As $x$ leaves $x = 1.7$ and advances a little bit, does $x^3 - 3x^2$ go up or down?

**Answer:**

Find the instantaneous growth rate of $x^3 - 3x^2$ at a point $x$:

```
In[1]:=    Clear[f,x,h]
           f[x_] = x^3 - 3 x^2
Out[1]=         2    3
           -3 x  + x
```

```
In[2]:=    Expand[(f[x+h] - f[x])/h]
Out[2]=          2                      2
           -3 h + h  - 6 x + 3 h x + 3 x
```

The instantaneous growth rate at $x$ is $-0 + 0 - 6x + 0 + 3x^2$.
The instantaneous growth rate at $x = 2.6$ is

```
In[3]:=    (-6 x + 3 x^2)/.x-> 2.6
Out[3]=    4.68
```

This is positive ; so as $x$ leaves $x = 2.6$ and advances a little bit,

$x^3 - 3x^2$ goes UP.

The instantaneous growth rate at $x = 1.7$ is

```
In[4]:=    (-6 x + 3 x^2)/.x-> 1.7
Out[4]=    -1.53
```

This is negative ; so as $x$ leaves $x = 1.7$, $x^3 - 3x^2$ goes **DOWN**.

**T.5.b)**

You are sitting at $x = 1$ and $y = 2$. What happens to $xy + 8/(x^2y^3) + 6x$ as you increase $x$ a little bit? What happens as you increase $y$ a little bit?

**Answer:**

```
In[1]:=    Clear[f,x,y]
           f[x_,y_] = (x y + 8/(x y^3) + 6 x)
Out[1]=              8
           6 x + ---- + x y
                    3
                 x y
```

To see what happens as $x$ increased, look at the average growth rate as $x$ grows from $x$ to $x + h$ :

```
In[2]:=    Together[(f[x+h,y] - f[x,y])/h]
Out[2]=                3      2 3      4     2 4
           -8 + 6 h x y  + 6 x y  + h x y  + x y
           -------------------------------------
                                3
                        x (h + x) y
```

The instantaneous growth rate as $x$ grows is

$$(-8 + 0 + 6x^2y^2 + 0 + x^2y^4)/(xy^3).$$

The instantaneous growth rate at $\{1,2\}$ as $x$ leaves 1 is:

```
In[3]:=    ((-8 + 6 x^2 y^2 + x^2 y^4)/(x y^3))/.
           {x->1,y->2}
Out[3]=    4
```

This is positive. So if we are at $x = 1$ and $y = 2$ and $x$ increases a bit, then $xy + 8/(x^2y^3) + 6x$ goes **UP**.

To see what happens as $y$ increased, look at the average growth rate as $x$ grows from $y$ to $y + h$ :

```
In[4]:=    Together[(f[x ,y + h] - f[x,y])/h]
Out[4]=          2              2      3 2 3
           (-8 h  - 24 h y - 24 y  + h x y  +

                2 2 4       2 5 6     2 6
             3 h x y  + 3 h x y  + x y ) /

                3       3
           (x y  (h + y) )
```

The instantaneous growth rate as $x$ grows is $(-24y^2 + x^2y^6)/(xy^6)$. The instantaneous growth rate at $x = 1$, $y = 2$, as $y$ leaves 2 is:

```
In[5]:=     (- 24 y^2 + x^2 y^6) / (x y^6) /. {x->1,y->2}
Out[5]=        1
            -(-)
               2
```

This is negative. So if we are at $x = 1$ and $y = 2$ and y increases a bit, then $xy + 8/(x^2y^3) + 6x$ goes **DOWN**. We can plot to confirm:

```
In[6]:=     Plot[f[x,2],{x,.5,2}]
```

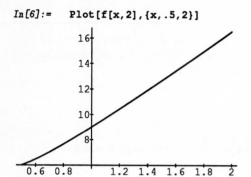

```
In[7]:=     Plot[f[1,y],{y,1,3}]
```

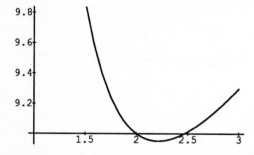

■ **T.6) Differential equations.**

Suppose we have a closed population of $P$ equally suspectible infection-free individuals and that one additional infected individual is introduced. Let $y[t]$ be the number of infected individuals at time $t$.

Scientists studying the spread of the infection often use the logistic differential equation

$$y'[t] = Ky[t](P - y[t])$$

where $y'[t]$ is the instanteous growth rate of the number of infected persons $t$ days after the infection breaks out and $K$ is a positive constant.

**T.6.a)**

Why is this a natural set-up for the study of the spread of the infection?

**Answer:**

Look at the equation

$$y'[t] = Ky[t](P - y[t]).$$

At the beginning when $y[t]$ (the number of infected persons) is small, then the product $Ky[t](P - y[t])$ will also be small. So at the beginning, the instantaneous growth rate $y'[t] = Ky[t](P - y[t])$ is also small.

Similarly when just about everyone is infected, the $y[t]$ is almost equal to $P$ and the product $Ky[t](P - y[t])$ will also be small. So near the end, the instantaneous growth rate $y'[t] = Ky[t](P - y[t])$ is also small.

This means that the infection spreads slowly at first and after it has run its course, it is again spreading slowly.

This is in harmony with the way infections spread.

**T.6.b)**

Use the differential equation

$$y'[t] = Ky[t](P - y[t])$$

to predict the behavior of $y[t]$ given that $0 < y[t] < P$ for small positive $t$'s and given that $K > 0$.

**Answer:**

When $t$ is small, we are given that $0 < y[t] < P$. So for small positive $t$,

$$y'[t] = Ky[t](P - y[t]) > 0$$

because all the factors on the right are positive. This means that for small positive $t'$ s, $y[t]$ goes up. And for larger $t'$ s, $y[t]$ has no choice but to continue to go up until $y[t]$ reaches the value $P$ because there is nothing on the right hand side of

$$y'[t] = Ky[t](P - y[t])$$

that allows the growth rate $y'[t]$ to be negative.

Now ask the question: Can $y[t]$ ever get larger than $P$? The answer is *no* because if $y[t]$ ever does get to $P$, then the right hand side of

$$y'[t] = Ky[t](P - y[t])$$

would force $y'[t]$ to be negative as soon as $y[t]$ exceeds $P$. Thus if $y[t]$ gets beyond $P$, then the differential equation $y'[t] = Ky[t](P - y[t])$ would send $y[t]$ right down to $P$ as $t$ increases.

This means $y[t]$ can get near $P$ but it can never get beyond $P$.

## Literacy sheet

**1.** How do we measure the instantaneous growth rate $f'[x]$ at a point $x$ ?

.

.

.

**2.** What happens to the plots of the $(f[x+h] - f[x])/h$ curves as $h$ closes in on $0$ ?

.

.

.

**3.** For a function $f[x]$, the function $f'[x]$ is another function that measures what quantity?

.

.

.

**4a.** Sketch with a pencil the graph of a function $f[x]$ such that $f'[x]$ is positive for all $x$'s.

.

.

.

**4b.** If $f'[x]$ is positive for all $x$ with $0 \le x \le 1$, then which is the larger: $f[0]$ or $f[1]$?

.

.

.

**5a.** Sketch with a pencil the graph of a function $f[x]$ such that $f'[x]$ is negative for all $x$'s.

.

.

.

**5b.** If $f'[x]$ is negative for all $x$ with $0 \le x \le 1$, then which is the larger: $f[0]$ or $f[1]$?

.

.

.

**6a.** Sketch with a pencil the graph of a function $f[x]$ such that $f'[x]$ is positive for all $x$ with $0 \le x \le 1$, and $f'[x]$ is negative for $1 \le x \le 2$.

.

.

**6b.** If $f[x]$ is a function such that $f'[x]$ is positive for all $x$ with $0 \le x \le 1$, and $f'[x]$ is negative for $1 \le x \le 2$. Then which $x$ in $[0,2]$ makes $f[x]$ the biggest?

.

.

**7.** Sketch with a pencil then graph of a function $f[x]$ such that $f'[x]$ is positive for $0 \le x \le 1$ and $f[x]$ is negative for $0 \le x \le 1$.

.

.

**8.** Sketch with a pencil the graph of a function $f[x]$ such that $f'[x] = 0$ for all $x$'s.

.

.

**9.** Sketch with a pencil the graph of a function $f[x]$ such that $f'[x] = 2$ for all $x$'s.

.

.

**10.** Sketch with a pencil the graph of a function $f[x]$ such that $f'[x] = -2$ for all $x$'s.

.

.

**11.** If $f[x] = \sin[x]$, then what is $f'[x]$?

.

**12.** If $f[x] = x^9$, then what is $f'[x]$?

.

.

**13.** If $f[x] = 7x^9$, then what is $f'[x]$?

.

.

.

**14.** If $f[x] = 7x^9 - x/2$, then what is $f'[x]$?

.

.

**15.** If $f[x] = \cos[x]$, then what is $f'[x]$?

.

.

.

**16.** We measure the instantaneous growth rate $f'[x]$ of $f[x]$ at each point of $[0, 1]$ and in so doing observe that $f'[0] > 0$ but $f'[1] < 0$. Why can we be sure that there is at least one number $s$ with $0 < s < 1$ such that instantaneous growth rate is zero at $s$ : (that is, $f'[s] = 0$ )?

.

.

.

# The Chain Rule Unlocks the Secrets of Logarithms, Exponentials, and Money.

## Guide

For simple functions $f[x]$ we can usually find the instantaneous growth rate by an analysis of

$$(f[x+h] - f[x])/h.$$

But for a complicated function like

$$f[x] = \sin[x^4 + 3\sqrt{x}]$$

this analysis can become terribly complex.

Thankfully, there is an easy mechanical procedure called the chain rule that allows us and the machine to spit out the instantaneous growth rates of many complicated functions quickly and easily:

```
In[1]:=    f[x_] = Sin[x^4 + 3 Sqrt[x]]
                            4
Out[1]=    Sin[3 Sqrt[x] + x ]
```

```
In[2]:=    f'[x]
Out[2]=       3                         4
           (--------- + 4 x ) Cos[3 Sqrt[x] + x ]
            2 Sqrt[x]
```

This is a lesson about the chain rule and how the chain rule can be used to unlock the secrets of calculus.

We'll use the chain rule to help reveal why the usual logarithm base 10 is unsuitable for serious calculation and we'll learn what the best base is.

And if you're interested in money and investments, there is some useful information for you in this lesson.

## Basics

### ■ B.1) Growth rates, derivatives and *Mathematica*

#### B.1.a)

What is the derivative of a function $f[x]$?

**Answer:**

The derivative of $f[x]$ is another function that measures the *instantaneous growth rate* of $f[x]$ at each point $x$.

Thus *"derivative"* is just a short-hand way of saying *"instantaneous growth rate."*

Mathematicians have programmed *Mathematica* to find derivatives of most common functions. Let's see how *Mathematica* handles some simple functions.

To get the derivative of $f[x] = ax^2 + bx + c$, we have several options:

Define:

```
In[1]:=    f[x_] = a x^2 + b x + c
                          2
Out[1]=    c + b x + a x
```

One way to evaluate the derivative is to look at:

```
In[2]:=    Factor[(f[x + h] - f[x])/h]
Out[2]=    b + a h + 2 a x
```

As $h$ closes in on 0, we see that

$$(f[x+h] - f[x])/h$$

closes in on

$$f'[x] = 2ax + b.$$

To get the same result from *Mathematica*, simply type:

```
In[3]:=    f'[x]
Out[3]=    b + 2 a x
```

It works! Or you can type:

```
In[4]:=    D[f[x],x]
Out[4]=    b + 2 a x
```

Chalk up another mark for *Mathematica*. You can also eliminate the need to define $f[x]$:

```
In[5]:=    D[a x^2 + b x + c,x]
Out[5]=    b + 2 a x
```

Another way of doing this is to type:

```
In[6]:=    y = a x^2 + b x + c
                          2
Out[6]=    c + b x + a x
```

The command:

```
In[7]:=    y'
```

```
Out[7]=                    2
              (c + b x + a x )'
```

is not successful. But you can use:

```
In[8]:=    D[y,x]
Out[8]=    b + 2 a x
```

This is reminiscent of the Leibniz notation $dy/dx$ for the derivative of $y$ with respect to $x$.

## B.1.b)

Use the various *Mathematica* options to get the formula for the derivative of $f[x] = a/x$.

**Answer:**

Define:

```
In[1]:=    f[x_] = a/x
Out[1]=    a
           -
           x
```

Look at:

```
In[2]:=    Factor[(f[x + h] - f[x])/h]
Out[2]=          a
           -(---------)
             x (h + x)
```

As $h$ closes in on $0$, we see that

$$(f[x + h] - f[x])/h$$

closes in on

$$f'[x] = -a/x(0 + x) = -a/x^2.$$

To get the same result from *Mathematica*, simply type:

```
In[3]:=    f'[x]
Out[3]=        a
           -(--)
             2
             x
```

Or you can type:

```
In[4]:=    D[f[x],x]
Out[4]=        a
           -(--)
             2
             x
```

Or:

```
In[5]:=    D[a/x,x]
Out[5]=        a
           -(--)
             2
             x
```

You can also define:

```
In[6]:=    y = a/x
Out[6]=    a
           -
           x
```

and type:

```
In[7]:=    D[y,x]
Out[7]=        a
           -(--)
             2
             x
```

## B.1.c)

The **D[ ,x]** notation is particularly useful when the expression you want to differentiate contains parameters or other independent variables. For example, if we are given:

```
In[1]:=    f[x_,z_] = a x^2 + 5 x Sin[z] - 3
Out[1]=             2
           -3 + a x  + 5 x Sin[z]
```

```
In[2]:=    Factor[(f[x + h, z] - f[x,z])/h]
Out[2]=    a h + 2 a x + 5 Sin[z]
```

As h closes in on 0, we see that

$$(f[x + h, z] - f[x, z])/h$$

closes in on on the derivative with respect to $x$

$$a0 + 2ax + 5\sin[z] = 2ax + 5\sin[z].$$

You can use *Mathematica* to find a formula for the derivative with respect to $x$ by looking at:

```
In[3]:=    D[f[x,z],x]
Out[3]=    2 a x + 5 Sin[z]
```

Here we think of $z$ as a second variable and think of $a$ as a parameter. This distinction is not God-given. There is nothing but custom preventing us from thinking of $a$ as another variable and $z$ as a parameter. In fact we could think of both as parameters or we could think of both as new independent variables. Here are two calculations for specific values of $z$ :

```
In[4]:=    D[f[x,Pi/8],x]
Out[4]=                    Pi
           2 a x + 5 Sin[--]
                          8
```

```
In[5]:=    D[f[x,Pi/4],x]
Out[5]=    5 Sqrt[2]
           --------- + 2 a x
               2
```

You can evaluate the derivative at a particular $x$ :

```
In[6]:=    D[f[x,Pi/4],x]/.x->8
```
```
Out[6]=    5 Sqrt[2]
           --------- + 16 a
               2
```

This is not the same as:

```
In[7]:=    D[f[8,Pi/4],x]
```
```
Out[7]=    0
```

Why? Because

```
In[8]:=    f[8,Pi/4]
```
```
Out[8]=    -3 + 20 Sqrt[2] + 64 a
```

is a constant. It does not vary as $x$ varies. Thus its instantaneous growth rate is 0.

### B.1.c.i)

Use *Mathematica* to calculate the derivative of
$$f[x,y] = ax^2 + bxy + cy^2 + dx + gy + h$$
as function of $x$.

**Answer:**

The derivative with respect to $x$ is:

```
In[1]:=    D[a x^2 + b x y + c y^2 + d x + g y + h,x]
```
```
Out[1]=    d + 2 a x + b y
```

### B.1.c.ii)

Use *Mathematica* to calculate the derivative of
$$f[x,y] = ax^2 + bxy + cy^2 + dx + gy + h$$
as function of $y$.

**Answer:**

The derivative with respect to $y$ is:

```
In[1]:=    D[a x^2 + b x y + c y^2 + d x + g y + h,x]
```
```
Out[1]=    d + 2 a x + b y
```

### ■ B.2) The chain rule.

Let's check out the derivative of the composition of two functions: Here is the derivative of $\sin[x^2]$ :

```
In[2]:=    D[Sin[x^2],x]
```
```
Out[2]=              2
           2 x Cos[x ]
```

This is interesting because the derivative of $\sin[x]$ is $\cos[x]$ and the derivative of $x^2$ is $2x$. It seems that the derivative

of $\sin[x^2]$ is manufactured from the derivative of $\sin[x]$ and the derivative of $x^2$.

Here is the derivative of $(x^2 + \sin[x])^8$ :

```
In[3]:=    D[(x^2 + Sin[x])^8,x]
```
```
Out[3]=                         2        7
           8 (2 x + Cos[x]) (x  + Sin[x])
```

This is interesting because the derivative of $x^8$ is $8x^7$ and the derivative of $x^2 + \sin[x]$ is $2x + \cos[x]$. It seems that the derivative of $(x^2 + \sin[x])^8$ is manufactured from the derivative of $x^8$, the derivative of $\sin[x]$ and the derivatrive of $x^2$.

Here is the derivative of $f[g[x]]$ :

```
In[4]:=    D[f[g[x]],x]
```
```
Out[4]=    f'[g[x]] g'[x]
```

Very interesting and of **undeniable importance**. This formula, which says that the derivative of
$$h[x] = f[g[x]]$$
is
$$h'[x] = f'[g[x]]g'[x],$$
is called the **chain rule.**

The chain rule tells us how to build the derivative of $f[g[x]]$ from the derivatives of $f[x]$ and $g[x]$. Here is the chain rule in action:

If
$$h[x] = \sin[x^2],$$
then
$$h'[x] = \cos[x^2]2x$$
in accordance with:

```
In[5]:=    D[Sin[x^2],x]
```
```
Out[5]=              2
           2 x Cos[x ]
```

And if $h[x] = (x^2 + \sin[x])^8$, then
$$h'[x] = 8(x^2 + \sin[x])^7(2x + \cos[x])$$
in accordance with:

```
In[6]:=    D[(x^2 + Sin[x])^8,x]
```
```
Out[6]=                         2        7
           8 (2 x + Cos[x]) (x  + Sin[x])
```

### B.2.a.i)

Give an explanation for why the derivative of $f[g[x]]$ is $f'[g[x]]g'[x]$.

**Answer:**

The explanation is a snap. Recall that $s[t]$ grows $s'[t]$ times as fast as $t$. Accordingly $f[g[x]]$ grows $f'[g[x]]$ times as fast as $g[x]$, and $g[x]$ grows $g'[x]$ times as fast as $x$. As a result $f[g[x]]$ grows $f'[g[x]]g'[x]$ times as fast as $x$.

This explains why the derivative of $f[g[x]]$ is $f'[g[x]]g'[x]$.

### B.2.a.ii)

Give the derivative of $\sin[5x]$. Check with *Mathematica*.

**Answer:**

The derivative of $\sin[5x]$ is $\cos[5x]5 = 5\cos[5x]$. Check:

```
In[1]:=    D[Sin[5 x],x]
Out[1]=    5 Cos[5 x]
```

Got it.

### B.2.a.iii)

Give the derivative of $\sin[x^4]$. Check with *Mathematica*.

**Answer:**

The derivative of $\sin[x^4]$ is $\cos[x^4](4x^3)$. Check:

```
In[1]:=    D[Sin[x^4],x]
Out[1]=       3     4
           4 x   Cos[x ]
```

### B.2.a.iv)

Give the derivative of $(g[x])^2$. Check with *Mathematica*.

**Answer:**

The derivative of $(g[x])^2$ is $2(g[x])g'[x]$. Check:

```
In[1]:=    D[g[x]^2,x]
Out[1]=    2 g[x] g'[x]
```

### B.2.a.v)

Give the derivative of $f[x^3y^2]$ with respect to $y$. Check with *Mathematica*.

**Answer:**

The derivative of $f[x^3y^5]$ with respect to $y$ calculates the instantaneous growth rate of $f[x^3y^5]$ as $y$ changes. So $x$ is regarded as a constant.

The derivative of $f[x^3y^5]$ with respect to $y$ is
$$f'[x^3y^5]5x^3y^4.$$

Check:

```
In[1]:=    D[f[x^3 y^5],y]
Out[1]=       3  4       3  5
           5 x  y  f'[x  y ]
```

Got it.

This means, for example, that the derivative of $\sin[x^3y^5]$ with respect to $y$ is $\cos[x^3y^5]5x^3y^4$.

### B.2.a.vi)

Give the derivative of $(3x^2 - 7x - 5 + 4/x)^7$. Check with *Mathematica*.

**Answer:**

The derivative of $(3x^2 - 7x - 5 + 4/x)^7$ is
$$7(3x^2 - 7x - 5 + 4/x)^6(6x - 7 - 4/x^2).$$

```
In[1]:=    D[ (3 x^2 - 7 x - 5 + 4/x)^2,x]
Out[1]=
                    4             4         2
           2 (-7 - -- + 6 x) (-5 + - - 7 x + 3 x )
                    2             x
                   x
```

Got it.

### ■ B.3) Using the chain rule to unlock the secrets of logs and exponentials.

The idea of the logaritm base $b$ is simple: We say $y = \log[b, x]$ if $b^y = x$.

Before the age of computers all serious computations were based on logarithms. And with good reason because logarithms change heavy multiplications and divisions into lightweight additions and subtractions thanks to the equations
$$\log[b, xy] = \log[b, x] + \log[b, y]$$
$$\log[b, x/y] = \log[b, x] - \log[b, y].$$

(Here $\log[b, x]$ stands for the logarithm with base $b$.)

Now the days of using logarithms for calculating tool are well in the past, but their place in the family of important functions is guaranteed by many scientific considerations.

### B.3.a)

Set $f[x] = \log[b, x]$. Explain why $f'[x]$ must be given by $f'[1]/x$ for $x > 0$.

**Answer:**

We know $f[xy] = f[x] + f[y]$. On the left, we have:

```
In[1]:=    left = f[x y]
```

```
Out[1]=    f[x y]
```

On the right, we have:

```
In[2]:=    right = f[x] + f[y]
Out[2]=    f[x] + f[y]
```

Differentiate both sides with respect to $y$ and set them equal to each other.

```
In[3]:=    derivedequation =
           (D[left,y] == D[right,y])
Out[3]=    x f'[x y] == f'[y]
```

Set $y = 1$, and solve for $f'[x]$:

```
In[4]:=    Solve[(derivedequation/.y->1),f'[x]]
Out[4]=                f'[1]
           {{f'[x] -> -----}}
                        x
```

Done. That wasn't too bad.

### B.3.b.i)

Estimate the derivative of $\log[10, x]$, the logarithm base 10.

### Answer:

Set $f[x] = \log[10, x]$. The last problem says $f'[x] = f'[1]/x$. So we try to estimate $f'[1]$ by estimating the limiting value of $f[1 + h] - f[1])/h$ as $h$ closes in on 0.

```
In[1]:=    f[x_] := Log[10,x]
           Table[{h,N[(f[1 + h]- f[1])/h]}
           /.h->1/10^n, {n,1,7}]
Out[1]=      1                  1
           {{--, 0.413927}, {---, 0.432137},
             10                100

              1                    1
           {----, 0.434077}, {-----, 0.434273},
            1000                10000

               1
           {------, 0.434292},
            100000

                1
           {-------, 0.434294},
            1000000

                 1
           {--------, 0.434294}}
            10000000
```

It seems that $f'[1]$ is about .434294. So the derivative of $\log[10, x]$ is about $.43294/x$.

It is clear that we can estimate the derivative of $\log[b, x]$ for any base $b$ the same way. But there is something clumsy about this because we haven't found an exact value for the derivative of $\log[10, x]$.

### B.3.b.ii)

Check out derivatives of various logarithms by using the *Mathematica* command `N[D[Log[b,x],x]]` to prepare a table of derivatives of $\log[b, x]$ for $b = 2, 3, 4, \ldots, 10$ to machine accuracy.

### Answer:

```
In[1]:=    Table[{"base =",b,"derivative =",
           N[D[Log[b,x],x]]},{b,2,10}]
Out[1]=                                1.4427
           {{base =, 2, derivative =, ------},
                                         x

                                     0.910239
             {base =, 3, derivative =, --------},
                                          x

                                     0.721348
             {base =, 4, derivative =, --------},
                                          x

                                     0.621335
             {base =, 5, derivative =, --------},
                                          x

                                     0.558111
             {base =, 6, derivative =, --------},
                                          x

                                     0.513898
             {base =, 7, derivative =, --------},
                                          x

                                     0.480898
             {base =, 8, derivative =, --------},
                                          x

                                     0.45512
             {base =, 9, derivative =, -------},
                                          x

                                     0.434294
             {base =, 10, derivative =, --------}}
                                           x
```

Look at those weird rounded numbers in the numerators.

### B.3.b.iii)

What serious drawback is present in each of the derivatives that appear in the table prepared for Part b.ii)?

### Answer:

For each of the values of $b$ we have looked at in the table, the derivative $D[\log[b, x], x]$ contains a **weird rounded number** in its numerator.

Whenever we have a $b$ such that $D[\log[b, x], x]$ contains a weird rounded number in its numerator, then precise calculations involving $\log[b, x]$ and its derivative are hard to do.

What we need is a base $b$ such that $D[\log[b, x], x]$ is as clean as possible.

## B.3.b.iv)

What base is natural for clean, accurate calculation?

**Answer:**

The base for the *cleanest* possible derivative for accurate calculations is the base called $e$ such that $\log[e, x]$ has derivative $1/x$.

This number $e$ is the **natural** base for logarithms and $\log[e, x]$ is called the **natural logarithm** function because calculations involving the derivative of $\log[e, x]$ can be done simply and cleanly.

## B.3.b.v)

Find $e$ to three accurate decimals.

**Answer:**

Look at:

```
In[1]:=   Table[{b,N[D[Log[b,x],x]]},{b,2,10}]
```

$$Out[1]= \left\{\left\{2, \frac{1.4427}{x}\right\}, \left\{3, \frac{0.910239}{x}\right\},\right.$$

$$\left\{4, \frac{0.721348}{x}\right\}, \left\{5, \frac{0.621335}{x}\right\},$$

$$\left\{6, \frac{0.558111}{x}\right\}, \left\{7, \frac{0.513898}{x}\right\},$$

$$\left\{8, \frac{0.480898}{x}\right\}, \left\{9, \frac{0.45512}{x}\right\},$$

$$\left.\left\{10, \frac{0.434294}{x}\right\}\right\}$$

The table shows that

$$D[\log[2, x], x] > 1/x > D[\log[3, x], x]$$

Therefore $e$ is between 2 and 3. Now look for $e$ between 2 and 3:

```
In[2]:=   Table[{b,N[D[Log[b,x],x]]},{b,2,3,.1}]
```

$$Out[2]= \left\{\left\{2, \frac{1.4427}{x}\right\}, \left\{2.1, \frac{1.34782}{x}\right\},\right.$$

$$\left\{2.2, \frac{1.2683}{x}\right\}, \left\{2.3, \frac{1.20061}{x}\right\},$$

$$\left\{2.4, \frac{1.14225}{x}\right\}, \left\{2.5, \frac{1.09136}{x}\right\},$$

$$\left\{2.6, \frac{1.04656}{x}\right\}, \left\{2.7, \frac{1.00679}{x}\right\},$$

$$\left\{2.8, \frac{0.971233}{x}\right\}, \left\{2.9, \frac{0.939222}{x}\right\},$$

$$\left.\left\{3., \frac{0.910239}{x}\right\}\right\}$$

This table shows that

$$D[\log[2.7, x], x] > 1/x > D[\log[2.8, x], x].$$

Therefore $e = 2.7$ to one accurate decimal.

Look for $e$ between 2.7 and 2.8:

```
In[3]:=   Table[{b,N[D[Log[b,x],x]]},
          {b,2.7,2.8,.01}]
```

$$Out[3]= \left\{\left\{2.7, \frac{1.00679}{x}\right\}, \left\{2.71, \frac{1.00306}{x}\right\},\right.$$

$$\left\{2.72, \frac{0.999369}{x}\right\}, \left\{2.73, \frac{0.995717}{x}\right\},$$

$$\left\{2.74, \frac{0.992105}{x}\right\}, \left\{2.75, \frac{0.988532}{x}\right\},$$

$$\left\{2.76, \frac{0.984998}{x}\right\}, \left\{2.77, \frac{0.981501}{x}\right\},$$

$$\left\{2.78, \frac{0.978042}{x}\right\}, \left\{2.79, \frac{0.974619}{x}\right\},$$

$$\left.\left\{2.8, \frac{0.971233}{x}\right\}\right\}$$

This table shows that

$$D[\log[2.71, x], x] > 1/x > D[\log[2.72, x], x].$$

Therefore $e = 2.71$ to two accurate decimals. Go to the well again:

```
In[4]:=   Table[{b,N[D[Log[b,x],x]]},
          {b,2.71,2.72,.001}]
```

$$Out[4]= \left\{\left\{2.71, \frac{1.00306}{x}\right\}, \left\{2.711, \frac{1.00269}{x}\right\},\right.$$

$$\left\{2.712, \frac{1.00232}{x}\right\}, \left\{2.713, \frac{1.00195}{x}\right\},$$

$$\left\{2.714, \frac{1.00158}{x}\right\}, \left\{2.715, \frac{1.00121}{x}\right\},$$

$$\left\{2.716, \frac{1.00084}{x}\right\}, \left\{2.717, \frac{1.00047}{x}\right\},$$

$$\left\{2.718, \frac{1.0001}{x}\right\}, \left\{2.719, \frac{0.999736}{x}\right\},$$

$$\left.\left\{2.72, \frac{0.999369}{x}\right\}\right\}$$

This table shows that $e = 2.717$ to three accurate decimals.

*Mathematica* can deliver $e$ to amazing accuracy:

```
In[5]:=    {N[E],N[E,12],N[E,30]}
Out[5]=    {2.71828, 2.71828182846,
           2.7182818284590452353602874713}
```

Impressive. But *Mathematica* has the *nasty habit of insisting that $e$ be capitalized.*

### B.3.c)

Recall $x = \log[b, b^x]$ no matter what base we use. So $x = \log[e^x]$ for all real numbers $x$. Use this fact to explain why $e^x$ must be its own derivative.

**Answer:**

Start with the known fact

$$x = \log[e^x].$$

Imagine that $f[x] = e^x$, but don't tell the machine about this. Among us in the know, we have

$$x = \log[f[x]].$$

On the left, we have:

```
In[1]:=    left = x
Out[1]=    x
```

On the right, we have:

```
In[2]:=    right = Log[f[x]]
Out[2]=    Log[f[x]]
```

Differentiate both sides with respect to $x$ and set them equal to each other.

```
In[3]:=    derivedequation = (D[left,x] ==D[right,x])

Out[3]=
                 f'[x]
           1 == -----
                 f[x]
```

Solve for $f'[x]$:

```
In[4]:=    Solve[derivedequation,f'[x]]
Out[4]=    {{f'[x] -> f[x]}}
```

This means that if $f[x] = e^x$, then $f'[x] = e^x$ as well. This establishes the **most important derivative formula of them all:**

```
In[5]:=    D[E^x,x]

Out[5]=      x
           E
```

### B.3.d)

Just to see that nothing could be easier than differentiating $e^x$ or $\log[x]$, give the derivatives with respect to $x$ of the following functions. Check with *Mathematica* .

### B.3.d.i)

$$e^{f[x]}$$

**Answer:**

By the chain rule, the derivative is $e^{f[x]} f'[x]$. Check:

```
In[1]:=    D[E^f[x],x]

Out[1]=      f[x]
           E       f'[x]
```

Got it.

### B.3.d.ii)

$$e^{-x^2}$$

**Answer:**

Recall the formula:

```
In[1]:=    D[E^f[x],x]

Out[1]=      f[x]
           E       f'[x]
```

The function $e^{-x^2}$ is $e^{f[x]}$ with $f[x] = -x^2$. The derivative is

$$e^{f[x]} f'[x] = e^{-x^2}(-2x).$$

Check:

```
In[2]:=    D[E^(-x^2),x]

Out[2]=    -2 x
           ----
              2
             x
           E
```

Got it.

### B.3.d.iii)

$$e^{\sin[x]}$$

**Answer:**

Recall the formula:

```
In[1]:=    D[E^f[x],x]
Out[1]=    f[x]
         E        f'[x]
```

The function is $e^{\sin[x]}$ The derivative is $e^{\sin[x]}\cos[x]$ Check:

```
In[2]:=    D[E^Sin[x],x]
Out[2]=    Sin[x]
         E        Cos[x]
```

Got it.

**B.3.d.iv)**

---

$$\log[\sin[x]]$$

---

**Answer:**

Recall the formula:

```
In[1]:=    D[Log[f[x]],x]
Out[1]=    f'[x]
         -----
          f[x]
```

The derivative is

$$(1/\sin[x])D[\sin[x],x] =$$
$$(1/\sin[x])\cos[x] = \cos[x]/\sin[x].$$

Check:

```
In[2]:=    D[Log[Sin[x]],x]
Out[2]=    Cos[x]
         ------
          Sin[x]
```

Got it.

**B.3.d.v)**

---

$$\log[(1 + x^2)/(1 - x^2)]$$

---

**Answer:**

Use

$$\log[(1 + x^2)/(1 - x^2)] = \log[1 + x^2] - \log[1 - x^2]$$

Then the derivative is

$$(1/(1 + x^2))(2x) - (1/(1 - x^2))(-2x).$$

Let's put this over a common denominator:

```
In[1]:=    handanswer =
         Together[2 x/(1 + x^2) + 2 x/(1 - x^2)]
```

```
Out[1]=            4 x
         -------------------
                2         2
         (1 - x ) (1 + x )
```

Check:

```
In[2]:=    machineanswer =
         D[Log[(1 + x^2)/(1 - x^2)],x]
```

```
Out[2]=
                                            2
               2     2 x      2 x (1 + x )
         (1 - x ) (------ + -------------)
                       2          2 2
                  1 - x      (1 - x )
         ------------------------------
                      2
                 1 + x
```

We had better simplify.

```
In[3]:=    Together[machineanswer]
Out[3]=         -4 x
         -------------------
                 2         2
         (-1 + x ) (1 + x )
```

Let's see whether these match up.

```
In[4]:=    Together[handanswer - machineanswer]
Out[4]=    0
```

We got it!

# Tutorial

■ **T.1) Derivatives.**

Calculate by hand the derivative with respect to $x$ of the following functions:

**T.1.a)**

---

$$\sin[3x]$$

---

**Answer:**

The derivative is

$$\cos[3x]D[3x, x] = \cos[3x]3 = 3\cos[3x].$$

Check:

```
In[1]:=    D[Sin[3 x],x]
Out[1]=    3 Cos[3 x]
```

Got it.

**T.1.b)**

---

$$\sin[x]^4$$

**Answer:**

The derivative is $4\sin[x]^3\cos[x]$. Check:

```
In[1]:=    D[Sin[x]^5 ,x]
Out[1]=                    4
           5 Cos[x] Sin[x]
```

Got it.

**T.1.c)**

$$\sin[3x]^4$$

**Answer:**

The derivative is

$$4\sin[3x]^3 D[\sin[3x], x] = 4\sin[3x]^3\cos[3x]D[3x, x]$$
$$= 4\sin[3x]^3\cos[3x]3 = 12\sin[3x]^3\cos[3x].$$

Check:

```
In[1]:=    D[Sin[3 x]^4 ,x]
Out[1]=                       3
           12 Cos[3 x] Sin[3 x]
```

Got it.

**T.1.d)**

$$(x + \log[x])^3$$

**Answer:**

The derivative of $x^3$ is $3x^2$. So the derivative of $(x + \log[x])^3$ is

$$3(x + \log[x])^2 D[x + \log[x], x] = 3(x + \log[x])^2(1 + 1/x).$$

Check:

```
In[1]:=    D[ (x + Log[x])^3 ,x]
Out[1]=            1                2
           3 (1 + -) (x + Log[x])
                  x
```

Got it.

**T.1.e)**

$$(1 + x^2)^5$$

**Answer:**

The derivative of $x^5$ is $5x^4$. So the derivative of $(1 + x^2)^5$ is

$$5(1 + x^2)^4 D[1 + x^2, x] = 5(1 + x^2)^4(0 + 2x) = 10x(1 + x^2)^4.$$

Check:

```
In[1]:=    D[ (1 + x^2)^5 ,x]
Out[1]=                 2 4
           10 x (1 + x )
```

Got it.

**T.1.f)**

$$\log[x^2 + 7x]$$

**Answer:**

The derivative is

$$(1/(x^2 + 7x))D[x^2 + 7x, x] = (1/(x^2 + 7x))(2x + 7)$$
$$= (2x + 7)/(x^2 + 7x).$$

Check:

```
In[1]:=    D[Log[x^2 + 7 x] ,x]
Out[1]=    7 + 2 x
           --------
                 2
           7 x + x
```

Got it.

**T.1.g)**

$$e^{-5x}$$

**Answer:**

The derivative is $e^{-5x}(-5) = -5e^{-5x}$

```
In[1]:=    D[E^(-5 x),x]
Out[1]=      -5
           ----
             5 x
           E
```

Got it.

**T.1.h)**

$$\log[x\sin[x]]$$

**Answer:**

Note
$$\log[x\sin[x]] = \log[x] + \log[\sin[x]]$$

So the derivative of $\log[x\sin[x]]$ is
$$1/x + (1/(\sin[x]))D[\sin[x], x] =$$
$$1/x + (1/(\sin[x]))\cos[x] = 1/x + \cos[x]/\sin[x].$$

Check:

```
In[1]:=    D[Log[x Sin[x]],x]
Out[1]=    x Cos[x] + Sin[x]
           ------------------
               x Sin[x]
```

To compare this answer to our hand answer, look at:

```
In[2]:=    Apart[D[Log[x Sin[x]],x]]
Out[2]=    1   Cos[x]
           - + ------
           x   Sin[x]
```

Got it.

### T.1.i)

___

$$\log[x^8(1 + 7x^5)^{-3}]$$

___

**Answer:**

$$\log[x^8(1 + 7x^5)^{-3}] = \log[x^8] + \log[(1 + 7x^5)^{-3}]$$

$$= 8\log[x] - 3\log[1 + 7x^5]$$

So the derivative of $\log[x^8(1 + 7x^5)^{-3}]$ is
$$8/x - 3(1/(1 + 7x^5))(0 + 35x^4) = 8/x - 105x^4/(1 + 7x^5)$$

```
In[1]:=    Apart[D[Log[x^8 (1 + 7 x^5)^(-3)],x]]
Out[1]=              4
           8   105 x
           - - --------
           x        5
               1 + 7 x
```

Got it.

### T.1.j)

___

$$\log[e^{2x}\sin[6x]]$$

___

**Answer:**

$$\log[e^{2x}\sin[6x]] = \log[e^{2x}] + \log[\sin[6x]]$$
$$= 2x + \log[\sin[6x]].$$

So the derivative of $\log[e^{2x}\sin[6x]]$ is

$$2 + (1/\sin[6x])D[\sin[6x], x] = 2 + (1/\sin[6x])\cos[6x]D[6x, x]$$
$$= 2 + (1/\sin[6x])\cos[6x]6 = 2 + 6\cos[6x]/\sin[6x].$$

Check:

```
In[1]:=    Apart[D[Log[E^(2 x) Sin[6 x]],x]]
Out[1]=          6 Cos[6 x]
           2 + -----------
                Sin[6 x]
```

Got it.

### T.1.k)

___

$$aye^{xy} + bx^3y^8$$

___

**Answer:**

The derivative of $aye^{xy} + bx^3y^8$ "with repect to $x$" (that is "as a function of $x$") is

$$aye^{xy}y + b3x^2y^8 = ay^2e^{xy} + 3bx^2y^8$$

Check:

```
In[1]:=    D[a y E^( x y) + b x^3 y^8 ,x]
Out[1]=     x y   2         2 8
           E    a y  + 3 b x  y
```

Got it.

### ■ T.2) e at the bank-compound interest.

### T.2.a)

___

You deposit 500 dollars in a bank account paying 7 percent simple interest per year and then do not touch the account. How much do you have in the account after:

### T.2.a.i)

___

one year?

___

**Answer:**

You've got the 500 dollars plus interest of (.07)500 dollars or:

```
In[1]:=    (500) ( 1 + .07)
Out[1]=    535.
```

### T.2.a.ii)

___

two years?

**Answer:**

At the end of the first year, you have $500(1 + (.07))$. dollars. At the end of the second year, you have

$$500(1 + (.07))(1 + (.07)) = 500(1 + .07)^2$$

dollars.

```
In[1]:=    500 (1 + .07)^2
Out[1]=    572.45
```

### T.2.a.iii)

$n$ years?

**Answer:**

$500(1 + .07)^n$.

### T.2.a.iv)

Give a plot showing the amount in the account at the end of each of the first twenty years.

**Answer:**

```
In[1]:=    ListPlot[Table[ 500 (1 + .07)^n, {n, 1, 20}]];
```

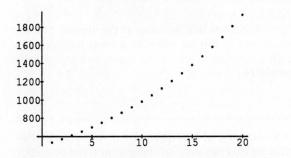

Nice return. And it's growing faster and faster.

### T.2.b)

You deposit $P$ dollars in a bank account paying $100r$ percent simple interest per year and then do not touch the account.

### T.2.b.i)

How much do you have in the account after one year?

**Answer:**

You've got the $P$ dollars plus interest of $rP$ dollars or: $P(1 + r)$ dollars.

### T.2.b.ii)

How much do you have in the account after two years?

**Answer:**

At the end of the first year, you have $P(1 + r)$ dollars. At the end of the second year, you have $P(1 + r)(1 + r) = P(1 + r)^2$ dollars.

### T.2.b.iii)

How much do you have in the account after $n$ years?

**Answer:**

$500(1 + .07)^n$ dollars.

### T.2.c)

You deposit $P$ dollars in a bank account paying $100r$ percent interest compounded twice per year (every 6 months) and then do not touch the account.

### T.2.c.i)

How much do you have in the account after one year?

**Answer:**

Well, after 6 months you've got the $P$ dollars plus half a year's interest or: $P(1 + r/2)$ dollars.

This new amount gathers interest for the second six months; so at the end of the year you have

$$P(1 + r/2)(1 + r/2) = P(1 + r/2)^2$$

dollars.

For $P = 500$ dollars and an interest rate of 7 percent, this amounts to:

```
In[1]:=    500 (1 + .07/2)^2
Out[1]=    535.612
```

### T.2.c.ii)

How much do you have in the account after two years?

**Answer:**

At the end of the first year, you have $P(1 + r/2)^2$ dollars. At the end of the second year, you have

$$P(1 + r/2)^2(1 + r/2)^2 = P(1 + r/2)^4$$

dollars.

For $P = 500$ dollars and an interest rate of 7 percent, this amounts to:

```
In[1]:=    500(1 + .07/2)^4
Out[1]=    573.762
```

## T.2.c.iii)

How much do you have in the account after $n$ years?

**Answer:**

$P(1 + r/2)^{2n}$ dollars.

## T.2.d.i)

You deposit $P$ dollars in a bank account paying $100r$ percent interest compounded 12 times per year and then do not touch the account. How much do you have in the account after one year, two years, $n$ years?

If $r = .07$ , then what simple interest rate $100s$ percent is needed to match this payoff?

The interest rate $100s$ percent is called the **effective interest rate** .

**Answer:**

After one year : $P(1 + r/12)^{12}$ dollars.

After two years: $P(1 + r/12)^{24}$ dollars.

After $n$ years: $P(1 + r/12)^{12n}$ dollars.

Under simple interest $100s$ percent, the payoff after one year would be $P(1 + s)$ dollars.

To find the effective interest rate $100s$ percent corresponding to a compounded rate of 7 percent, we need to find the $s$ that makes $1 + s = (1 + r)^{12}$. This is easy to solve:

```
In[1]:=    Solve[((1 + r/12)^12/.r->.07) == 1 + s, s]
Out[1]=    {{s -> 0.0722901}}
```

The effective interest rate is 7.23 percent.

## T.2.d.ii)

You deposit $P$ dollars in a bank account paying $100r$ percent interest per year compounded $k$ times per year and then do not touch the account. How much do you have in the account after one year, two years, $n$ years?

**Answer:**

After one year : $P(1 + r/k)^{k}$ dollars.

After two years: $P(1 + r/k)^{2k}$ dollars.

After $n$ years: $P(1 + r/k)^{kn}$ dollars.

## T.2.d.iii)

Plot a table of effective interest rates corresponding to an interest rate of 7 percent per year compounded $1, 2, 3, ..., 100$ times.

**Answer:**

For $k$ compounds the effective interest rate $100s$ percent satisfies $(1 + r/k)^k = 1 + s$, or:

```
In[1]:=    s = ((1 + r/k)^k/.r->.07) - 1
Out[1]=              0.07 k
           -1 + (1 + ----)
                       k
```

```
In[2]:=    ListPlot[Table[100 s, {k, 1, 100}],
           AxesLabel->{"compounds", "effective rate"}];
```

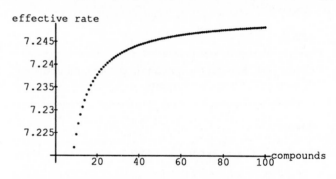

The effective interest rate increases as the number of compounds increases. But the effective interest rate seems to approach a limiting value as the number of compounds becomes large.

## T.2.e.i)

For an interest rate of $100r$ percent per year, the effective interest rate, $100s$ percent , resulting from $k$ compounds is determined by $s = (1 + r/k)^k - 1$. Explain why the limiting value as the number of compounds increases is given by

$$\lim_{k \to \infty} (1 + r/k)^k - 1 = e^r - 1.$$

This is the same as establishing the **basic formula**

$$\lim_{k \to \infty} (1 + r/k)^k = e^r.$$

**Answer:**

Here is a plot showing what it means to say

$$\lim_{k \to \infty} (1 + r/k)^k = e^r$$

in the case that $r = 0.05$. You should play with this plot for other values of r.

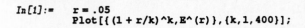

```
In[1]:=    r = .05
           Plot[{(1 + r/k)^k,E^(r)},{k,1,400}];
```

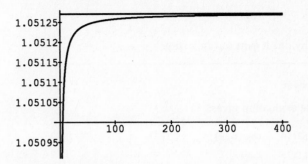

Here is an explanation of why the plot turned out the way it did:

Explaining that $\lim_{k \to \infty}(1 + r/k)^k = e^r$ is the same as:

$$\log[\lim_{k \to \infty}(1 + r/k)^k] = \log[e^r] = r.$$

So this is the same as $\lim_{k \to \infty}\log[(1 + r/k)^k] = r$

which is the same as

$$\lim_{k \to \infty} k \log[1 + r/k] = r,$$

or

$$\lim_{k \to \infty} \log[1 + r/k]/(1/k) = r.$$

Let $h = 1/k$ and note that $h \to 0$ means the same as $k \to \infty$. Then explaining

$$\lim_{k \to \infty} \log[1 + r/k]/(1/k) = r$$

is the same as explaining why the the limiting value of $\log[1 + rh]/h$ as $h \to 0$ is $r$. But because $\log[1] = 0$, this is the same as explaining

why the limiting value of $(\log[1 + rh] - \log[1])/h$ as $h \to 0$ is $r$.

However the limiting value of $(\log[1 + rh] - \log[1])/h$ as $h \to 0$ is the instantaneous growth rate of $(\log[1 + rx]$ as a function of $x$ at the point $x = 0$, that is:

```
In[2]:=    D[Log[1 + r h],h]/.h->0
Out[2]=    0.05
```

This shows that

$$\lim_{k \to \infty}(1 + r/k)^k = e^r.$$

And we're out of here.

### T.2.e.ii)

Plot a table of effective interest rates corresponding to an interest rate of 7 percent per year compounded $1, 2, 3, ..., 500$ times. What is the limiting value of the effective interest rates as the number of compounds grows and grows?

### Answer:

Recall that for $k$ compounds the effective interest rate $100s$ percent satisfies $(1 + r/k)^k = 1 + s$. So define:

```
In[1]:=    s = ((1 + r/k)^k/.r->.07) - 1
Out[1]=              0.07 k
           -1 + (1 + ----)
                       k
```

Here is the plot:

```
In[2]:=    ListPlot[Table[100 s,{k,1,500}],
           AxesLabel->{compounds, effective rate}];
```

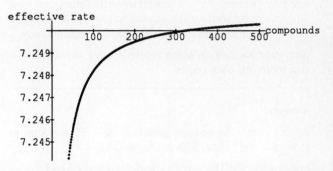

The effective interest rate increases as the number of compounds increases. But the effective interest rate approachs a limiting value as the number of compounds becomes large.

The limiting value of the effective interest rate in percent is:

```
In[3]:=    100(E^(.07) - 1)
Out[3]=    7.25082
```

Lots of compounding can give you more than 0.25 percent more return than simple compounding.

### T.2.e.iii)

Some banks advertise "continuous compounding" at a certain interest rate $100r$ percent per year.

What does this mean?

How do we compute the effective interest rate?

How do we compute the amount in the account after $t$ years?

### Answer:

"Continuous compounding" means that if $P$ dollars are deposited and untouched, then the amount in the account in dollars after one year is

$$\lim_{k \to \infty} P(1 + r/k)^k = Pe^r$$

The effective interest rate is

$$100(e^r - 1)$$

percent.

The amount in the account after $t$ years in dollars is $Pe^{rt}$.

Bet you never expected that the number $e$ could hit you right in the pocket.

### ■ T.3) Financial computations-bank accounts.

### T.3.a

You have $P$ dollars to invest today at a guaranteed interest rate $100r$ percent. Write a short routine that computes how long you must wait for the account to grow to a fixed goal of $g$ dollars if the interest is compounded continuously.

Run your routine on some representative values of $P, r$ and $g$ of your own choice.

### Answer:

After $t$ years, the account grows to $Pe^{rt}$. So we want to solve $g = Pe^{rt}$ for $t$. This is the same as

$$\log[g] = \log[Pe^{rt}] = \log[P] + \log[e^{rt}] = \log[P] + rt.$$

Let us solve for $t$ :

```
In[1]:=    Solve[Log[g] == Log[P] + r t,t]
Out[1]=               -Log[P] + Log[g]
          {{t -> ----------------}}
                         r
```

Here is the routine:

Enter specific values of $P, r$ and $g$. The output is the number of years it will take for the initial deposit of $P$ dollars to grow to g dollars in an account paying interest at interest rate $100r$ percent compounded continuously.

```
In[2]:=    P = 10000
           r = 0.075
           g = 90000
           t = (- Log[P] + Log[g])/r
           N[t]
Out[2]=    29.2963
```

10000 dollars will take almost 29.3 years to grow to 90000 dollars if it is invested in an account paying 7.5 percent interest compounded continuously.

Rerun this with your own data to get a feeling for how your own current or future investments grow.

### ■ T.4) Logs and exponentials for unnatural bases.

### T.4.a)

Look at:

```
In[1]:=    Log[b,x]
Out[1]=    Log[x]
           ------
           Log[b]
```

Why did it turn out this way?

### Answer:

The evaluation gives:

```
In[2]:=    Log[b,x]
Out[2]=    Log[x]
           ------
           Log[b]
```

We know that $y = \log[b, x]$ means $x = b^y$. Take logs base $e$ :

$$\log[x] = \log[b^y] = y \log[b]$$

or $y = \log[x]/\log[b]$. We are done.

Think of it! This means all other logarithm functions can be easily manufactured from the natural logarithm function $\log[x]$.

### T.4.b)

Look at:

```
In[1]:=    D[b^x,x]
Out[1]=     x
           b  Log[b]
```

Why did it turn out this way?

### Answer:

Hmmmm.

Remember $b = e^{\log[b]}$. Therefore $b^x = e^{x \log[b]}$. Consequently the derivative of $b^x$ is

$$e^{x \log[b]} \log[b] = b^x \log[b].$$

# Literacy Sheet

*What you should know away from the computer.*

**1.** Differentiate the following functions with respect to $x$ by hand:

**1.a)** $e^x$

.

.

**1.b)** $e^{-x}$

.

.

**1.c)** $\log[x]$

.

.

**1.d)** $\log[1/x](= -\log[x])$

.

.

**1.e)** $\sin[6x]$

.

.

**1.f)** $\sin[x]^2$

.

.

**1.g)** $\sin[7x]^3$

.

.

**1.h)** $(4x + \log[1-x])^4$

.

.

**1.i)** $(1 - x + x^2)^4$

.

.

**1.j)** $\log[1 - 3x^2]$

.

.

**1.k)** $e^{-2x}$

.

.

**1.l)** $7e^{\{\sin[\pi x]\}}$

.

.

**1.m)** $2\log[x^2 \sin[x^2]^3]$

.

.

**1.n)** $\log[(a + x)/(a - x)]$

.

.

**1.o)** $\log[x^{12}(1 - x^5)^{-8}]/4$

.

.

**1.p)** $\log[\log[x]]$

.

.

**1.q)** $\log[e^x]$

.

.

**1.r)** $2^x$

.

.

**1.s)** $\log[x + y + z]$

.

.

**1.t)** $e^{x+y+z}$

**1.u)** $e^{3x+5y-3z}$

**1.v)** $(e^y)^x$

**1.w)** $3x + f[x-3]$

**1.x)** $f[x] + f[y]$

**1.y)** $f[xy]$

**1.z)** $f[\log[x]]$

**1.zz)** $f[f[x]]$

**2.a)** How are derivatives related to instantaneous growth rates?

**2.b)** What is the chain rule?

**3.a)** Give the value of $e$ to three accurate decimals.

**3.b)** Express $\lim_{x\to\infty}(1 + r/x)^x$ in terms of $e$ and $r$.

**4.a)** Richard Feynman once said:

"The base 10 was used only because we have ten fingers and arithmetic of it is easy, but if we ask for a mathematically natural base (for logarithms), one that has nothing to do with the number of fingers on human beings, we might try to change our scale of logarithms in some convenient and natural manner."

What do you think Feynman was driving at?

**4.b)** The exponential function $a^x$ has the cleanest derivative for what number $a$ ?

**5)** Examine the derivative of $f[x] = x^2 - \log[x]$ to give a reasonably good sketch of the curve $f[x] = x^2 - \log[x]$. What is the minimum value $f[x] = x^2 - \log[x]$ can have?

**6)** Start with the identities $x = \log[e^x]$ and $D[\log[x], x] = 1/x$ and derive the identity $D[e^x, x] = e^x$.

**7)** Examine the derivative of $f[x] = -0.0075x + \log[x]$ to give a reasonably good hand sketch of the curve $y = f[x]$. What is the maximum value $f[x]$ can have?

**8)** Examine the derivative of $f[x] = x - \sin[x]$ to see whether the curve $y = f[x]$ ever goes down.

Is it true that $x \geq \sin[x]$ for $x \geq 0$ ?

**9)** What familar function is approximately equal to $(\log[x + .000001] - \log[x])/.000001$?

**10)** Complete the sentence:

The derivative $f'[x]$ of $f[x]$ is the limiting case of . . .

as . . .

**11)** Are there numbers $x$ that make $e^x$ negative? If so, then what are they?

Is there a number $x$ that makes $e^x = 0$ ? If so then what is it?

Given a positive number $a$ , then what $x$ makes

$e^x = a$ ? Is there more than one such $x$ ? Why?

Given number $a$, then what $x$ makes $\log[x] = a$ ? Is there more than one such $x$ ? Why?

**12)** We differentiate a certain function $f[x]$ and learn

$f'[x] = -(x-2)^2 e^x$.

How do we know that the curve $y = f[x]$ is always going down?

**13)** We differentiate a certain function $f[x]$ and learn

$f'[x] = (x-1)e^{-x}$.

Give a rough sketch of the shape of the curve $y = f[x]$.

**14)** 1000 dollars is put in the bank at an advertised interest rate of 5 percent and left untouched

If the interest is compounded continuously, then what is the effective interest rate and how much is in the account ten years after the deposit?

If the interest is compounded monthly, then what is the effective interest rate and how much is in the account ten years after the deposit?

If the interest is compounded daily, then what is the effective interest rate and how much is in the account ten years after the deposit?

**15)** Given that $\log[2]$ is roughly .72, explain why money invested at $100r$ percent compounded continuously doubles every $72/$ r years.

Some folks call this the "Rule of 72."

**16)** Here are six plots. Three of them are plots of the derivatives of the functions plotted in the other three. Try to match the plots of the functions with the plots of the derivatives.

Function 1:

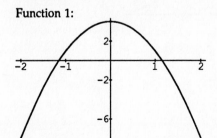

Function 2:

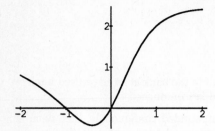

**Function 3:**

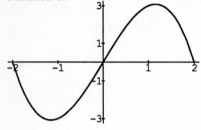

**Function 4:**

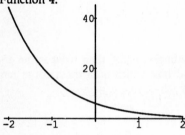

**Function 5:**

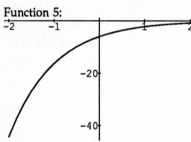

**Function 6:**

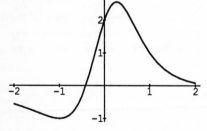

# Powers, products and the trigomometric functions

## Guide

We have one tool, the chain rule, that helps us with some complicated functions. But as powerful as the chain rule is, it does not tell us directly how to differentiate powers, products and quotients.

This is a lesson in how to use the combination of the chain rule and the natural logarithm to differentiate powers, products and quotients.

At the end of it you should be able to differentiate a function like:

```
In[1]:=    f[x_] = (E^x Cos[x] + x^4)^3
Out[1]=     4    x       3
          (x  + E  Cos[x])
```

whose derivative is:

```
In[2]:=    f'[x]
Out[2]=      4    x       2
          3 (x  + E  Cos[x])

               3    x          x
          (4 x  + E  Cos[x] - E  Sin[x])
```

As an added bonus, you will be able to write down the derivative even faster than you can type the function into your computer.

And that's not all! You will also see what the derivatives of all the trigonometric functions are.

To top it off, you will get a glimpse at how differentiation can be used in some concrete situations.

All this in just 12 pages!

## Basics

■ **B.1)  Using the logarithm to calculational advantage: Powers, products and quotients.**

### B.1.a.i)

It may not be clear how to calculate the derivative of
$$f[x] = x^4 \sin[x]^6.$$
But it is very clear how to differentiate the log of $f[x]$ :
$$\log[f[x]] = \log[x^4 \sin[x]^6]$$
$$= 4\log[x] + 6\log[\sin[x]].$$

In fact the chain rule tells us that
$$D[\log[f[x]], x] = 4/x + 6(1/\sin[x])\cos[x]$$
$$= 4/x + 6(\cos[x]/\sin[x]).$$
How does this tell us the way to calculate $f'[x]$?

**Answer:**

Note that on one hand,
$$D[\log[f[x]], x] = f'[x]/f[x].$$
And, as we saw above,
$$D[\log[f[x]], x] = 4/x + 6(\cos[x]/\sin[x]).$$
So
$$f'[x]/f[x] = 4/x + 6(\cos[x]/\sin[x]).$$
Accordingly
$$f'[x] = f[x](4/x + 6(\cos[x]/\sin[x])).$$
Let the machine clean this up:

```
In[1]:=    f[x_] = x^4 Sin[x]^6
Out[1]=     4      6
          x  Sin[x]
```

Our calculation of $f'[x]$ is:

```
In[2]:=    Expand[f[x] ( 4/x + 6 (Cos[x]/Sin[x]))]
Out[2]=       4           5        3       6
          6 x  Cos[x] Sin[x]  + 4 x  Sin[x]
```

Check:

```
In[3]:=    f'[x]
Out[3]=       4           5        3       6
          6 x  Cos[x] Sin[x]  + 4 x  Sin[x]
```

Got it.

### B.1.a.ii): Main idea

Why is $f'[x] = f[x]D[\log[f[x]], x]$ and what calculational advantage does this give us?

**Answer:**

The chain rule tells us that
$$D[\log[f[x]], x] = f'[x]/f[x].$$

Multiply both sides by $f[x]$ to see that

$$f[x]D[\log[f[x]], x] = f[x]f'[x]/f[x] = f'[x].$$

This is a welcome calculational advantage in the cases, as in part i), in which $f[x]$ is hard to differentiate but $\log[f[x]]$ is easy to differentiate. Some old timers call this technique *logarithmic differentiation.* Let's try this out.

If $f[x] = x^{-5}e^x$, then $f[x]$ is hard to differentiate, but

$$\log[f[x]] = -5\log[x] + \log[e^x] = -5\log[x] + x$$

is easy to differentiate. The formula

$$f[x]D[\log[f[x]], x] = f'[x]$$

tells us that $f'[x]$ is given by:

```
In[1]:=    Expand[x^(-5) E^x ((-5/x) + 1)]

Out[1]=         x     x
           -5 E      E
           -----  +  --
             6        5
            x        x
```

Check:

```
In[2]:=    Expand[D[x^(-5) E^x,x]]

Out[2]=         x     x
           -5 E      E
           -----  +  --
             6        5
            x        x
```

Right on the money.

### B.1.a.iii): Powers

By this time most of us believe that if $t$ is any non-zero exponent, then the derivative of

$$f[x] = x^t$$

is

$$f'[x] = tx^{t-1}.$$

Use the identity $f[x]D[\log[f[x]], x] = f'[x]$ to explain this basic formula which some folks call the **power rule.**

### Answer:

Put $f[x] = x^t$. Take the log of both sides:

$$\log[f[x]] = \log[x^t] = t\log[x].$$

Now use

$$f'[x] = f[x]D[\log[f[x]], x] = x^t t/x = tx^{t-1}$$

And that's it! Some may object that this explanation does not cover the case when $x = 0$. This objection is a good one.

*Answer continued: The special case of $f'[0]$*

It is important to realize that $f'[0]$ makes sense only when $t \geq 1$. Here is the reason: $f'[0]$ is the limiting case of

$$(f[0 + h] - f[0])/h = h^t/h = h^{t-1}$$

as $h$ closes in on 0. Clearly this blows up if $t < 1$. But if $t = 1$, then the limiting value is 1 and if $t > 1$, then the limiting value is 0. Thus the formula

$$f'[x] = tx^{t-1}$$

still holds up even when $x = 0$ provided $t \geq 1$.

### B.1.b)

Give the derivatives with respect to $x$ of the following functions. Check with *Mathematica.*

### B.1.b.i)

$$x^{2/3}$$

### Answer:

The derivative is $(2/3)x^{-1/3}$. Check:

```
In[1]:=    D[x^(2/3),x]

Out[1]=        2
           ------
               1/3
           3 x
```

Got it.

### B.1.b.ii)

$$(e^{-7x} - 3e^{2x})^{1/5}$$

### Answer:

The derivative is

$$(1/5)(e^{-7x} - 3e^{2x})^{-4/5}(-7e^{-7x} - 6e^{2x})$$

Check:

```
In[1]:=    D[(E^(-7 x) - 3 E^(2 x))^(1/5),x]

Out[1]=          -7       2 x
               ---- - 6 E
                7 x
               E
           -------------------------
                 -7 x       2 x 4/5
           5 (E      - 3 E     )
```

Got it.

### B.1.c.i)

Use the identity

$$f[x]D[\log[f[x]], x] = f'[x]$$

to calculate the derivative of $f[x] = x^5\log[x]$.

## Answer:

The function is
$$f[x] = x^5 \log[x].$$

We see that $f[x]$ is hard to differentiate, but
$$\log[f[x]] = 5\log[x] + \log[\log[x]]$$

is easy to differentiate. Its derivative is
$$5/x + (1/\log[x])(1/x).$$

The formula $f[x]D[\log[f[x]], x] = f'[x]$ tells us that $f'[x]$ is given by:

```
In[1]:=    Expand[x^5 Log[x]
           ((5/x) + 1/(Log[x]) (1/x))]
Out[1]=     4     4
           x  + 5 x  Log[x]
```

Check:

```
In[2]:=    Expand[D[x^5 Log[x],x]]
Out[2]=     4     4
           x  + 5 x  Log[x]
```

Looks good.

### B.1.c.ii): Products

Use the identity
$$h[x]D[\log[h[x]], x] = h'[x]$$

to explain why the derivative of the product of two functions is the sum of the terms we get by multipying each function times the derivative of the other.

## Answer:

Put $h[x] = f[x]g[x]$. We are trying to see why
$$h'[x] = f'[x]g[x] + f[x]g'[x].$$

We know
$$h'[x] = h[x]D[\log[h[x], x].$$

And we know
$$\log[h[x]] = \log[f[x]g[x]] = \log[f[x]] + \log[h[x]].$$

Differentiate through to get
$$D[\log[h[x], x] = f'[x]/f[x] + g'[x]/g[x]$$

Multiply through by $h[x] = f[x]g[x]$ to get
$$h'[x] = h[x]D[\log[h[x], x]$$

$$= h[x](f'[x]/f[x] + g'[x]/g[x])$$

$$= f[x]g[x](f'[x]/f[x] + g'[x]/g[x])$$

$$= f[x]g[x]f'[x]/f[x] + f[x]g[x]g'[x]/g[x]$$

$$= g[x]f'[x] + f[x]g'[x].$$

And that's it! The derivative of the product of two factors is the sum of the two terms we get by multiplying the derivative of one of the factors times the other factor.

Some may object that this explanation does not cover the case when $f[x] \le 0$ or $g[x] \le 0$. This objection is a good one.

*The case in which $f[x]$ or $g[x]$ have non-positive values.*

If $f[x]$ or $g[x]$ have negative values, then pick any large numbers $C$ and $K$ so that $f[x]+C$ and $g[x]+K$ take only positive values.

So according to what we did above,
$$D[(f[x] + C)(g[x] + K), x]$$
$$= (f[x] + C)D[g[x] + K, x] + D[f[x] + C, x](g[x] + K)$$
$$= (f[x] + C)g'[x] + f'[x](g[x] + K)$$
$$= f[x]g'[x] + Cg'[x] + f'[x]g[x] + Kf'[x].$$

On the other hand if we multiply out and then differentiate, then we get
$$D[(f[x] + C)(g[x] + K), x]$$
$$= D[f[x]g[x] + Kf[x] + Cg[x] + CK, x]$$
$$= D[f[x]g[x], x] + Kf'[x] + Cg'[x] + 0.$$

Equating the two different formulas for $D[(f[x]+C)(g[x]+K), x]$ gives
$$D[f[x]g[x], x] + Kf'[x] + Cg'[x]$$
$$= f[x]g'[x] + Cg'[x] + f'[x]g[x] + Kf'[x].$$

Cancel the common terms on each side:
$$D[f[x]g[x], x] = f[x]g'[x] + f'[x]g[x].$$

This is just what we wanted; so we're done. This was a close one!

### B.2.c.iii)

Calculate the derivative of
$$x^4 e^{3x}$$

Check with *Mathematica*.

## Answer:

The derivative of $x^4$ is $4x^3$; the derivative of $e^{3x}$ is $3e^{3x}$. The derivative of $f[x]g[x]$ is $f[x]g'[x] + f'[x]g[x]$. So the derivative of $x^4e^{3x}$ is
$$x^4 3e^{3x} + 4x^3 e^{3x}.$$

Check:

```
In[1]:=    D[x^4 E^(3 x),x]
```

```
Out[1]=       3 x  3        3 x  4
          4 E   x   + 3 E   x
```

Got it.

### B.2.d): Quotients

We can also handle quotients: Calculate the derivative of
$$\sin[x]/(3 + x^4).$$
Check with *Mathematica.*

**Answer:**

Rewrite the function as $\sin[x](3 + x^4)^{-1}$.

The derivative of $\sin[x]$ is $\cos[x]$. The derivative of $(3 + x^4)^{-1}$ is $-(3 + x^4)^{-2}(0 + 4x^3)$. So the derivative of
$$\sin[x]/(3 + x^4) = \sin[x](3 + x^4)^{-1}$$
is
$$-(3 + x^4)^{-2}(0 + 4x^3)\sin[x] + \cos[x](3 + x^4)^{-1}.$$
Check:

```
In[1]:=    D[Sin[x]/(3 + x^4),x]
Out[1]=            3
          Cos[x]  4 x  Sin[x]
          ------ - -----------
            4          4 2
          3 + x     (3 + x )
```

Got it.

### ■ B.2) The derivatives of other trigonometric functions.

### B.2.a): Cos[x]

The derivative of $\cos[x]$ is $-\sin[x]$. Use the identities
$$\sin[x + \pi/2] = \cos[x]$$
and
$$\cos[x + \pi/2] = -\sin[x]$$
to explain why.

**Answer:**

Differentiating $\cos[x]$ is the same as differentiating $\sin[x + \pi/2]$. So finding the derivative of $\cos[x]$ is purely mechanical.

```
In[1]:=    D[Sin[x + Pi/2],x]
Out[1]=         Pi
          Cos[-- + x]
               2
```

The derivative of $\cos[x]$ is $\cos[x + \pi/2]$ which is the same as $-\sin[x]$.

Let us see:

```
In[2]:=    D[Cos[x],x]
Out[2]=    -Sin[x]
```

The derivative of $\sin[x]$ is $\cos[x]$, but the derivative of $\cos[x]$ is $-\sin[x]$. The extra minus sign is why a lot of calculus people like $\sin[x]$ but regard $\cos[x]$ as a pain in the neck.

### B.2.b)

Explain why the derivative of $\tan[x]$ is
$$1/\cos[x]^2$$
(This is the same as $\sec[x]^2$.)

**Answer:**

$\tan[x] = \sin[x]/\cos[x]$; so the process of getting the derivative of $\tan[x]$ is purely mechanical.

```
In[1]:=    derivoftan = D[Sin[x]/Cos[x],x]
Out[1]=              2
              Sin[x]
          1 + ------
                 2
              Cos[x]
```

Then the derivative of $\tan[x]$ is $1 + \tan^2[x]$. But let's do a little algebra on this:

```
In[2]:=    Together[derivoftan]
Out[2]=          2        2
          Cos[x]  + Sin[x]
          -----------------
                   2
              Cos [x]
```

We recognize this as
$$1/\cos[x]^2 = \sec[x]^2$$
because $\cos[x]^2 + \sin[x]^2 = 1$.

Let's try it.

```
In[3]:=    D[Tan[x],x]
Out[3]=          -2
          Cos[x]
```

Got it.

### B.2.c)

What about the derivatives of the other trigonometric functions?

**Answer:**

You can get by nicely without memorizing these, but in case you're climbing the walls to learn about them here they are:

```
In[1]:=    {D[Cot[x],x],D[Sec[x],x],D[Csc[x],x]}
Out[1]=              1                 Sin[x]
           {-(---------------),       -------,
                  2      2                2
              Cos[x] Tan[x]            Cos[x]

              Cos[x]
            -(------)}
                2
              Sin[x]
```

### B.2.d): ArcTan[x]

Explain why the derivative of arctan$[x]$ is $1/(1 + x^2)$.

***

**Answer:**

arctan$[x]$ is defined for each real number $x$ by

$$\arctan[x] = y$$

if

$$\tan[y] = x$$

and

$$-\pi/2 \leq y \leq \pi/2.$$

This leads directly to the identity

$$\tan[\arctan[x]] = x.$$

This identity is the key to the derivative of arctan$[x]$.

For the moment set $f[x] = \arctan[x]$, but don't tell this to *Mathematica*. This gives

$$\tan[f[x]] = x.$$

Differentiate both sides and solve for $f'[x]$:

```
In[1]:=    Clear[f]
           D[(Sin[f[x]]/Cos[f[x]]),x] == D[x,x]
Out[1]=                    2
                    Sin[f[x]]  f'[x]
           f'[x] + ---------------- == 1
                              2
                       Cos[f[x]]
```

This gives

$$f'[x](1 + \tan[f[x]]^2) = 1.$$

But

$$\tan[f[x]] = \tan[\arctan[x]] = x$$

for all $x$'s. Thus $f'[x](1 + x^2) = 1$. In other words,

$$f'[x] = 1/(1 + x^2).$$

### B.2.e): ArcSin[x]

Explain why the derivative of arcsin$[x]$ is $1/\sqrt{1 - x^2}$.

***

**Answer:**

If $-1 \leq x \leq 1$, then

$$\arcsin[x] = y$$

means

$$\sin[y] = x$$

and

$$-\pi/2 \leq y \leq \pi/2.$$

This definition leads directly to the the identity

$$x = \sin[\arcsin[x]]$$

for $-1 \leq x \leq 1$.

To get a formula for the derivative of arcsin$[x]$, we just differentiate both sides of this identity and then see what happens:

$$1 = D[x, x] = D[\sin[\arcsin[x]]$$

$$= \cos[\arcsin[x]]D[\sin[\arcsin[x], x].$$

Dividing by cos$[\arcsin[x]]$ gives the awkward but correct formula

$$D[\sin[\arcsin[x], x] = 1/\cos[\arcsin[x]].$$

To clean up this formula, recall that

$$\sin[\arcsin[x]]^2 + \cos([\arcsin[x]]^2 = 1$$

and that $\sin[\arcsin[x]] = x$.

Substituting, we find $x^2 + \cos([\arcsin[x]]^2 = 1$. Thus

$$\cos([\arcsin[x]]^2 = 1 - x^2.$$

Because $-\pi/2 \leq \arcsin[x] \leq \pi/2$ and $\cos[y] \geq 0$ for $-\pi/2 \leq y \leq \pi/2$, take the positive square root to get

$$\cos([\arcsin[x]] = \sqrt{1 - x^2}$$

Using

$$D[\sin[\arcsin[x], x] = 1/\cos[\arcsin[x]],$$

we get the clean formula

$$D[\arcsin[x], x] = 1/\sqrt{1 - x^2}.$$

Check:

```
In[1]:=    D[ArcSin[x],x]
Out[1]=          1
           ------------
                    2
           Sqrt[1 - x ]
```

Got it. But this time we had to sweat some.

### B.2.f)

What about the derivatives of the other inverse trigonometric functions?

***

**Answer:**

The other inverse trigonometric functions arccos$[x]$, arccot$[x]$, arcsec$[x]$ and arccsc$[x]$ can be handled in just the same way we handled arctan$[x]$ and arcsin$[x]$. There is not much point in studying these functions in detail because they are rarely important in everyday calculations.

Here are their derivatives:

```
In[1]:=    D[{ArcCos[x], ArcCot[x], ArcSec[x], ArcCsc[x]},{x}][1]=
Out[1]=                1
           {-(-------------), ArcCot'[x],
                          2
                 Sqrt[1 - x ]

                     1                         1
            -----------------, -(-----------------)}
                     -2  2                  -2  2
            Sqrt[1 - x  ] x       Sqrt[1 - x  ] x
```

Not even *Mathematica* seems to know the derivative of arccot [$x$]. You have better things to worry about.

# Tutorial

### ◼ T.1) Derivatives.

Obtain the derivative with respect to $x$ of the following:

### T.1.a)

---

$$x^5 - 5x^{10} + 7x^{-3}$$

---

**Answer:**

The derivative is

$$5x^4 - 50x^9 - 21x^{-4}.$$

Check:

```
In[1]:=    D[x^5 - 5 x^10 + 7 x^(-3),x]
Out[1]=    -21      4      9
           --- + 5 x  - 50 x
            4
            x
```

Got it.

### T.1.b)

---

$$\cos[x]^5(7x^3 - 4/x)^2$$

---

**Answer:**

The derivative is (product rule and chain rule)

$$\cos[x]^5 2(7x^3 - 4/x)(21x^2 + 4/x^2)$$

$$-5\cos[x]^4 \sin[x](7x^3 - 4/x)^2.$$

Check:

```
In[1]:=    D[Cos[x]^5 (7 x^3 - 4/x)^2,x]
```

```
              4        2    -4        3      5
           2 (-- + 21 x ) (-- + 7 x ) Cos[x]  -
              2            x
              x


              -4        3 2      4
           5 (-- + 7 x )  Cos[x]  Sin[x]
              x
```

Got it.

### T.1.c)

---

$$1/(-5x^3 + 7/x^4)^4$$

---

**Answer:**

This function is $(-5x^3 + 7x^{-4})^{-4}$. The derivative is (chain rule)

$$-4(-5x^3 + 7x^{-4})^{-5}D[-5x^3 + 7x^{-4}, x]$$

$$= -4(-5x^3 + 7x^{-4})^{-5}(-15x^2 - 28x^{-5}).$$

Check:

```
In[1]:=    D[1/(-5 x^3 + 7/x^4 )^4,x]
Out[1]=          -28        2
           -4 (--- - 15 x )
                5
                x
           -------------------
               7        3 5
              (-- - 5 x )
               4
               x
```

Got it.

### T.1.d)

---

$$ax^{-3}y + bxy^8$$

---

**Answer:**

The derivative with respect to $x$ is

$$-3ax^{-4}y + by^8$$

Check:

```
In[1]:=    Clear[a,b,x,y]
           D[a (x^(-3)) y + b x y^8 ,x]
Out[1]=    -3 a y       8
           ------ + b y
             4
             x
```

Got it.

### T.1.e)

$$e^{2x} \sin[3x]$$

---

**Answer:**

$$D[e^{2x} \sin[3x], x] = e^{2x} \cos[3x]3 + e^{2x}2\sin[3x]$$

$$= 3e^{2x}\cos[3x] + 2e^{2x}\sin[3x]$$

Check:

```
In[1]:=    D[E^(2 x) Sin[3 x],x]
Out[1]=      2 x            2 x
           3 E    Cos[3 x] + 2 E    Sin[3 x]
```

Got it.

**T.1.f)**

---

$$\sin[5x]\cos[6x]$$

---

**Answer:**

$$D[\sin[5x]\cos[6x], x] = \sin[5x](-6\sin[6x]) + 5\cos[5x]\cos[6x]$$

$$= 5\cos[5x]\cos[6x] - 6\sin[5x]\sin[6x].$$

Check:

```
In[1]:=    D[Sin[5 x] Cos[6 x],x]
Out[1]=    5 Cos[5 x] Cos[6 x] -
             6 Sin[5 x] Sin[6 x]
```

Got it.

**T.1.g)**

---

$$\tan^2[x]$$

---

**Answer:**

$$D[\tan^2[x], x] = 2\tan[x]D[\tan[x], x] = 2\tan[x]/\cos^2[x]$$

Check:

```
In[1]:=    D[Tan[x]^2,x]
```

```
Out[1]=    2 Tan[x]
           --------
              2
           Cos[x]
```

Got it.

**T.1.h)**

---

$$\log[\cos[x]]$$

---

**Answer:**

$$D[\log[\cos[x]], x] = (1/\cos[x])D[\cos[x], x]$$

$$= (1/\cos[x])(-\sin[x]) = -\tan[x]$$

Check:

```
In[1]:=    D[Log[Cos[x]],x]
Out[1]=       Sin[x]
           -(------)
              Cos[x]
```

Got it.

**T.1.i)**

---

$$e^{\arctan[x]}$$

---

**Answer:**

$$D[e^{\arctan[x]}, x] = e^{\arctan[x]}D[\arctan[x], x]$$

$$= e^{\arctan[x]}(1/(1 + x^2)) = e^{\arctan[x]}/(1 + x^2)$$

Check:

```
In[1]:=    D[E^ArcTan[x],x]
Out[1]=     ArcTan[x]
           E
           ----------
                2
            1 + x
```

Got it.

■ **T.2) Using the derivative to help to get a good representative plot.**

**T.2.a)**

---

We want to get good representative plots of functions by giving plots that **exhibit all the dips and the crests of the graph and give a strong flavor of the global scale behavior.**

How can we use the derivative to help to choose the plotting interval?

**Answer:**

The quick way to start this is to make sure your plot includes all points at which the derivative is 0. This way you can be sure that the curve does not change direction as it leaves the screen on the left and on the right.

## T.2.b)

Use the derivative to help you set up a good representative plot of

$$f[x] = e^{-x/10}(48 + 24x + x^2).$$

**Answer:**

```
In[1]:=   f[x_] = (E^(-x/10)) ( 48 + 24 x + x^2)
Out[1]=          2
          48 + 24 x + x
          --------------
               x/10
              E
```

Find where the derivative is 0.

```
In[2]:=   Solve[f'[x] == 0,x]
Out[2]=   {{x -> -16}, {x -> 12}}
```

The plotting interval should include 12 and - 16 .

```
In[3]:=   Plot[f[x],{x,-16 -8,12 + 8}]
```

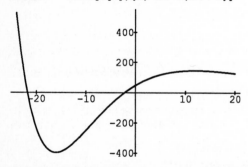

The plot changes direction at $x = -16$ and $x = 12$; it cannot change direction at any other point.

Apparently the $e^{-x/10}$ factor is dominating the quadratic factor on the right.

To get more of the global scale behavior, we should plot more on the right:

```
In[4]:=   Plot[f[x],{x,-16 -8,12 + 18}]
```

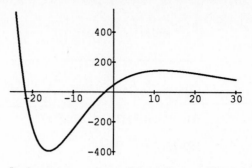

Let's see some more of the action on the right:

```
In[5]:=   Plot[f[x],{x,-16 -8,12 + 38}]
```

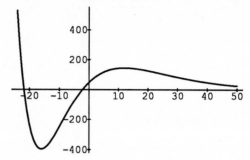

Great plot.

## ■ T.3) Describing the growth of the function by looking at the derivative.

In each case, take the derivative of the given function and use the information contained in the derivative to describe (in words) the behavior of the function. Then confirm your description with a good representative plot of the function.

## T.3.a)

$$xe^x$$

**Answer:**

Factor the derivative:

```
In[1]:=   Factor[D[x E^x,x]]
Out[1]=    x
          E  (1 + x)
```

The instantaneous growth rate (= the derivative) is **positive** for $x > -1$ and it is **negative** for $x < -1$. (Remember $e^x$ is never negative or 0.)

Therefore the function is going **up** for $x > 1$ and going **down** on for $x < -1$. Confirm with a representative plot:

```
In[2]:=   Plot[x E^x,{x,-5,2}]
```

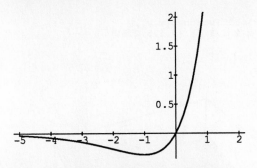

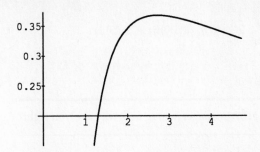

The description is right on the money.

**T.3.b)**

---

$$\log[x]/x$$

---

**Answer:**

Factor the derivative:

```
In[1]:=   fprime[x_] = Factor[D[Log[x]/x,x]]
Out[1]=      -1 + Log[x]
          -(-----------)
                 2
                x
```

To get an idea of what the derivative is doing, plot it:

```
In[2]:=   Plot[fprime[x],{x,2,4}]
```

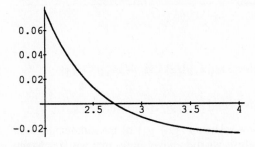

It seems to go from positive to negative somewhere between 2 and 3. Let's take another look at the derivative:

$$(-1 + \log[x])/x^2.$$

The derivative goes from positive to negative when $\log[x] = 1$. But $\log[e] = 1$. Therefore $\log[x]/x$ goes **up** for $x < e$ and $\log[x]/x$ goes **down** for $x > e$.

Confirm with a representative plot:

```
In[3]:=   Plot[Log[x]/x,{x,E - 2, E + 2}]
```

Again we're right on the money because $e$ is approximately equal to 2.7. Let's see more of the action on the right:

```
In[4]:=    Plot[Log[x]/x,{x,E - 2, E + 20}]
```

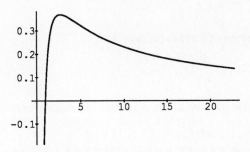

Interesting action. This function will continue to go down as $x$ advances to the right because the derivative is never 0 except at $x = e$. The denominator is in control.

■ **T.4) Efficient design.**

You are working for the MBA Box Co. An order comes in for cardboard boxes with a volume of 6 cubic feet each.

The stupid MBA manager says "That's simple; we'll just make the boxes 3 feet long, 2 feet wide and 1 foot high."

You respond "We can make them cheaper by using different dimensions."

You are right. The shape to choose depends on more information you and the manager have but you know how to use and the manager doesn't. Let

$x$ = length of the box in feet

$y$ = width of the box in feet

$z$ = height of the box in feet.

The material for the box consists of six rectangular pieces.

Two of them (the top and bottom) have area $xy$ each; these cost .40 dollars per square foot.

Another two (the sides) have area $xz$ each; these cost .30 dollars per square foot.

And the other two (the ends) have area $yz$; these cost .35 dollars per square foot

So the cost in dollars of the materials for making each box

is

$$2(.40)xy + 2(.30)xz + 2(.35)yz$$

dollars.

### T.4.a)

---

Why is the MBA manager wrong and why are you right?

---

**Answer:**

We know $xyz = 6$; this means that we can get a formula for $z$ in terms of $x$ and $y$:

```
In[1]:=    xyzeqn = (6 == x y z)
Out[1]=    6 == x y z
```

```
In[2]:=    zsolved = Solve[xyzeqn, z]
Out[2]=
                   6
           {{z -> ---}}
                  x y
```

Also we know the cost of the materials for making each box is :

```
In[3]:=    cost = 2 (.40) x y + 2 (.30) x z + 2 (.35) y z
Out[3]=    0.8 x y + 0.6 x z + 0.7 y z
```

Now replace $z$ by its formula in terms of $x$ and $y$ :

```
In[4]:=    boxcost[x_,y_] = cost/.zsolved[[1]]
Out[4]=
           4.2   3.6
           --- + --- + 0.8 x y
            x     y
```

The MBA boss said that $x = 3$ and $y = 2$. To see why the the stupid MBA boss is full of hot air, look at:

```
In[5]:=    D[boxcost[x,y],x]/.{x->3,y->2}
Out[5]=    1.13333
```

This derivative is positive; so this tells us that by decreasing $x$ a bit from $x = 3$ and holding $y = 2$, we decrease the costs. Thus $x = 3$ was not a good idea.

Also look at:

```
In[6]:=    D[boxcost[x,y],y]/.{x->3,y->2}
Out[6]=    1.5
```

This tells us that by decreasing $y$ from $y = 2$ and holding $x = 3$, we also decrease the costs. Thus $y = 2$ was not a good idea either.

So the MBA manager's quick answer of $x = 3$, $y = 2$ and $z = 1$ would have thrown company money away.

You can see this from the following plot:

```
In[7]:=    Plot3D[boxcost[x,y],
           {x,2.5,3.5},{y,1.5,2.5}]
```

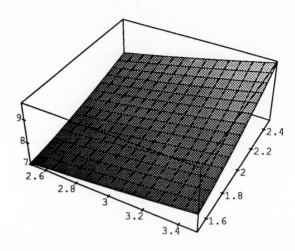

The plot indicates that $x = 2.5$ and $y = 1.5$ is a better choice than $x = 3$ and $y = 2$.

### T.4.b)

---

What dimensions result in least cost?  What is the least cost?  How much more than the least cost does the MBA boss's design cost?

---

**Answer:**

If either

$$D[boxcost[x,y], x]/.\{x \to x_0, y \to y_0\}$$

or

$$D[boxcost[x,y], y]/.\{x \to x_0, y \to y_0\}$$

is not  0 for a selection  $\{x_0, y_0\}$ of measurements, then the same analysis we did above shows that we can change $x$ from  $x_0$ or we can change  $y$ from  $y_0$ to reduce the cost of the box.

Consequently the dimensions  $x$ and  $y$ that result in least cost must be found in the output from:

```
In[1]:=    bestxandy =
           Solve[{D[boxcost[x,y],x] == 0,
           D[boxcost[x,y],y]==0},{x,y}]
Out[1]=    {{x -> 1.82965, y -> 1.56827},

                            (2 I)/3 Pi
              {x -> 1.82965 E         ,

                            (2 I)/3 Pi
               y -> 1.56827 E         },

                            (4 I)/3 Pi
              {x -> 1.82965 E         ,

                            (4 I)/3 Pi
```

y -> 1.56827 E            }}

The only real solution is:

*In[2]:=*    **bestxandy[[1]]**

*Out[2]=*    **{x -> 1.82965, y -> 1.56827}**

The length of the least cost box is:

*In[3]:=*    **bestx= 1.83**
*Out[3]=*    **1.83**

The width of the least cost box is:

*In[4]:=*    **besty = 1.57**

*Out[4]=*    **1.57**

Let's take a look at a plot:

*In[5]:=*    **Plot3D[boxcost[x,y],
{x,1.62,2.02},{y,1.36,1.76}]**

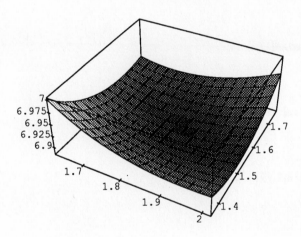

That depression in the center of the plot at $x = 1.83$ and $y = 1.57$ is visual substantiating evidence.

The height $z$ of the box in terms of $x$ and $y$ is $z = 6/(xy)$. So the height of the least cost box is given by:

*In[6]:=*    **bestz = 6/(x y)/.{x->bestx,y->besty}**
*Out[6]=*    **2.08834**

The cost in dollars of the least expensive box is:

*In[7]:=*    **boxcost[bestx,besty]**
*Out[7]=*    **6.88656**

The cost in dollars of the MBA boss's box is:

*In[8]:=*    **boxcost[3.0,2.0]**

*Out[8]=*    **8.**

You have saved the company:

*In[9]:=*    **8.00 - 6.89**
*Out[9]=*    **1.11**

dollars per box.

Time to go to the boss for a pay raise to get enough for a down payment on that red BMW convertible.

# Literacy Sheet.

*What you should know away from the computer.*

**1.** Differentiate the following functions with respect to $x$ by hand and simplify:

$\cos[x]$

.

.

$\arctan[x]$

.

.

$-\log[\cos[x]]$

.

.

$2x^{5/2} - 5x^{2/5}$

.

.

$1/x$

.

.

$x^e$

.

.

$e^x$

.

.

$x^2 \log[x]$

.

.

$x(\log[x] - 1)$

.

.

$e^{-2x} \cos[7x]$

.

.

$xe^{-3x}$

.

.

$(1 + 8x - x^3)/(1 + x)^2$

.

.

$1/\sqrt{x}$

.

.

.

$1/(1 - x)$

.

.

$1/(2(1 - x)^2)$

.

.

$e^x/(1 + e^x)^2$

.

.

$xe^x - e^x$

.

.

$-(2 + 2x + x^2)e^{-x}$

.

.

$-\arctan[1/x]$

.

.

$\arctan[\sqrt{x}]$

.

.

$\arctan[2\sqrt{x}]$

.

.

$(e^{3x} \cos[2x] + x^4)^3$

.

.

**2.** Explain the statement:

Even though the computer has supplanted the logarithm for numerical calculations, the logarithm remains an important calculational tool.

.

.

.

**3.** Examine the derivative of $f[x] = xe^{-x^2}$ to give a reasonably good hand sketch of the curve $y = f[x]$. What is the maximum value $f[x]$ can have?

.

**4.** Examine the derivative of $f[x] = x - \cos[x]$ to see whether the curve $y = f[x]$ ever goes down.

**5.** What familar function is approximately equal to

$$(\arctan[x + .000001] - \arctan[x])/.000001?$$

**6.** Suppose $f[x]$ and $g[x]$ are two functions with $f'[x] > 0$ and $g'[x] > 0$ for all $x$'s. This means both $f[x]$ and $g[x]$ go up as $x$ advances from left to right.

Does it also mean that the product $f[x]g[x]$ also goes up as $x$ advances from left to right? To help you form your opinion, try $f[x] = x - 2$ and $g[x] = x - 3$.

What happens if in addition $f[x] > 0$ and $g[x] > 0$ for all $x$'s?

**7.** If $x$ is measured in radians, then the derivative of $\sin[x]$ with respect to $x$ is $\cos[x]$. Use the formula $\sin[x$ degrees$] = \sin[(2\pi/360)x$ radians$]$ to calculate the derivative of $\sin[x$ degrees$]$ with respect to $x$.

Why does the resulting formula make calculus difficult if we insist on working with degrees instead of radians?

**8.** Find the largest value of $xy$ if $x + y = 4$.

**9.** Differentiate both sides of the identity

$$\sin[2x]/2 = \sin[x]\cos[x]$$

and set the resulting expressions equal to each other. Look familiar?

Hold $y$ constant and differentiate both sides of the identity

$$\sin[x + y] = \sin[x]\cos[y] - \sin[y]\cos[x]$$

with respect to $x$ and set the resulting expressions equal to each other. Look familiar?

Differentiate the four functions, and explain the result.

$2\cos^2[x]$

$-2\sin^2[x]$

$-2\sin^2[x] + 7$

$\cos^2[x] - \sin^2[x]$.

**10.** Is the derivative of the sum given by the sum of the derivatives?

Is the derivative of the product given by the product of the derivatives?

**11.** To say that $y[t]$ is proportional to $f[t]$ means that $y[t] = Kf[t]$ for some constant $K$. To say that $y[t]$ is proportional to both $f[t]$ and $g[t]$ means that $y[t] = Kf[t]g[t]$ for some constant $K$. To say that $y[t]$ is inversely proportional to $f[t]$ means that $y[t] = K/f[t]$ for some constant $K$.

If $y[t]$ is proportional to $t^2$ and $y[1] = 7$, then give a formula for $y[t]$.

·

If $y[t]$ is proportional to $e^{4t}$ and $y[0] = 8.1$, then give a formula for $y[t]$.

·

·

·

If $y[t]$ is proportional to $t^{0.3}$, then what function is $y'[t]$ proportional to? What function is $y'[t]$ inversely proportional to?

·

·

·

If $y[t]$ is proportional to $t^3$ and $y'[1] = 7$, then give a formula for $y[t]$.

·

·

·

If $y[t]$ is proportional to $e^{3t}$, then why is $y'[t]$ proportional to $y[t]$?

·

·

·

If money is invested at interest $100r$ percent compounded continuously and left untouched, then why is the instantaneous growth rate of the account proportional to the amount in the account? Do large accounts grow faster than small accounts?

·

·

·

If $h[t]$ is proportional to $f[t]$, then is $h'[t]$ proportional to $f'[t]$?

·

·

·

If $h[t]$ is proportional to both $f[t]$ and $g[t]$, then is $h'[t]$ proportional to both $f'[t]$ and $g'[t]$?

·

·

·

**12.** Write in mathemtical symbols:

In a certain controlled expansion of a gas, the pressure is

inversely proportional to the the square of the volume.

·

·

·

The number of maggots is proportional to the food supply but is inversely proportional to the temperature.

·

·

The strength of a beam is proportional to both its width and the square root of its depth.

·

·

**13.** Explain why:

The area of a circle is proportional the the square of its diameter.

·

·

If rectangular box with length $x$ inches, width $y$ inches and height $z$ inches is to contain a certain fixed volume $V$ cubic inches, then area of the top is inversely proportional to $z$.

·

·

# The race Track Principle

## Guide

This lesson is built around the following principle:

Suppose $f[a] = g[a]$ and $f'[x] \geq g'[x]$ for $x \geq a$, then $f[x] \geq g[x]$ for $x \geq a$ because both functions start out together and $f[x]$ is growing faster than $g[x]$.

This principle is called the *Race Track Principle*. It can be used for many interesting purposes ranging from comparison of two functions to faking the plot of other functions.

Jump in; the mathematics is fine.

## Basics

■ **B.1) The Race Track Principle.**

**B.1.a.i)**

Plot
$$f[x] = (x - 1)$$
$$g[x] = \sqrt{x} - 1/\sqrt{x}$$
and
$$h[x] = (x - 1)/x$$
on the same axes for $1 \leq x \leq 10$.

**Answer:**

```
In[1]:=    Clear[f,g,h,x]
           f[x_] = (x - 1)

Out[1]=    -1 + x

In[2]:=    g[x_] = Sqrt[x] - 1/Sqrt[x]

Out[2]=        1
           -(-------) + Sqrt[x]
             Sqrt[x]

In[3]:=    h[x_] = (x - 1)/x

Out[3]=    -1 + x
           ------
             x

In[4]:=    Plot[{f[x], g[x], h[x]}, {x,1,10}]
```

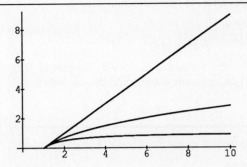

**B.1.a.ii)**

How could we have used derivatives to anticipate what we saw in the plot above?

**Answer:**

First look at:

```
In[1]:=    {f[1],g[1],h[1]}
Out[1]=    {0, 0, 0}
```

All three functions are 0 at $x = 1$.

Next look at the differences of the derivatives:

```
In[2]:=    Factor[{f'[x] - g'[x], g'[x] - h'[x]}]
Out[2]=    (-1 + Sqrt[x]) (1 + Sqrt[x] + 2 x)
           {-----------------------------------,
                          3/2
                       2 x

           (-1 + Sqrt[x]) (2 + Sqrt[x] + x)
           --------------------------------}
                          2
                       2 x
```

Note that if $x \geq 1$, then there is no way the two functions in the list above can be negative.

Therefore if $x \geq 1$, then
$$f'[x] - g'[x] \geq 0$$
and
$$g'[x] - h'[x] \geq 0.$$

Accordingly if $x \geq 1$, then
$$f'[x] \geq g'[x]$$
and
$$g'[x] \geq h'[x];$$

In other words, if $x \geq 1$, then
$$f'[x] \geq g'[x] \geq h'[x].$$
So the functions $f[x]$, $g[x]$ and $h[x]$ all start out together at $x = 1$, but as $x$ advances to the right away from 1,

$f[x]$ shows the greatest rate of increase,

$g[x]$ shows the next greatest, and

$h[x]$ shows the least rate of increase.

As a result for $x \geq 1$,
$$f[x] \geq g[x] \geq h[x].$$
This analysis of the derivative shows why the plot of $f[x]$, $g[x]$ and $h[x]$ on the same axes turned out the way it did.

### B1.b)

Let's take a drive out to the Race Track. Three race horses,

Foolish Pleasure,

Genuine Risk, and

Hill Gail,

leave the starting gate at the same instant. At all times of the race, Genuine Risk runs faster than Foolish Pleasure and Foolish Pleasure runs faster than Hill Gail.

What is the order of finish?

**Answer:**

Win: (1st Place): Genuine Risk

Place: (2ndPlace): Foolish Pleasure

Show: (3rd Place): Hill Gail.

In fact they maintain this order from the starting gate to the finish line.

### B.1.c)

Now we apply what we learned at the race track to using the derivative to predict behavior of functions. Each of the following problems is a manifestation of the Race Track Principle.

### B.1.c.i)

Two functions $f[x]$ and $g[x]$ satisfy
$$f[a] \geq g[a]$$
and
$$f'[x] \geq g'[x]$$
for $x \geq a$. Why does it follow that
$$f[x] \geq g[x]$$
for $x \geq a$?

**Answer:**

For $x = a$ we know $f[a] \geq g[a]$. And for $x > a, g[x]$ is never growing faster than $f[x]$. Therefore if $x \geq a$, we see that $f[x] \geq g[x]$.

### B.1.c.ii)

Two functions $f[x]$ and $g[x]$ satisfy
$$f[a] = g[a]$$
and
$$f'[x] = g'[x]$$
for $x \geq a$. Why does it follow that
$$f[x] = g[x]$$
for $x \geq a$?

**Answer:**

For $x = a$ we know $f[x] = g[x]$. And because for $x \geq a$ we know that $f'[x] = g'[x]$, we see that neither function can ever gain on the other. They grow at exactly the same rate. As a result, neither function can be ahead of the other. Therefore if $x \geq a$, we see that
$$f[x] = g[x].$$
If this were a horse race, then the functions would leave the starting gate at the same time and run the whole race in a dead heat.

### B.1.c.iii)

A function $f[x]$ has derivative $f'[x] = 0$ for all $x$'s in an interval $[a, b]$. Why does it follow that $f[x]$ is a constant function?

How can we find the constant?

**Answer:**

When $x = a$ we know that $f[x] = f[a]$. As $x$ advances from $a$ to $b$, $f'[x]$ remains equal to 0. So as $x$ leaves $a$ and advances to $b$, $f[x]$ cannot change. Therefore
$$f[x] = f[a]$$
for all $x$'s in $[a, b]$.

### B.1.d)

Explain why
$$(1 + x)^{3/2} \geq 1 + (3/2)x$$
for $x \geq 0$. Illustrate with a plot.

**Answer:**

Let $f[x] = (1 + x)^{3/2}$ and let $g[x] = 1 + (3/2)x$. We want to explain why $f[x] \geq g[x]$ for $x \geq 0$.

```
In[1]:=    Clear[f,g,x]
           f[x_] = (1 + x)^(3/2)
```
$$Out[1]=\quad (1 + x)^{3/2}$$

```
In[2]:=    g[x_] = 1 + (3/2)x
```
$$Out[2]=\quad 1 + \frac{3 x}{2}$$

Let's see what the beginning of the race looks like by calculating:

```
In[3]:=    {f[0],g[0]}
Out[3]=    {1, 1}
```

Good; $f[0] = g[0]$; the two functions start out together. Look at the derivatives:

```
In[4]:=    {f'[x],g'[x]}
```
$$Out[4]=\quad \{\frac{3 \,\text{Sqrt}[1 + x]}{2}, \frac{3}{2}\}$$

For $x \geq 0$ we have $\sqrt{1+x} \geq 1$; so for $x \geq 0$, we see
$$f'[x] \geq g'[x].$$

The two functions start out at $x = 0$ together and as $x$ advances away from 0, $f[x]$ grows faster than $g[x]$. The Race Track Principle tells us that $f[x] \geq g[x]$ for $x \geq 0$. Here is a plot:

```
In[5]:=    Plot[{f[x],g[x]},{x,0,24},
           AxesLabel->{"x","y"}]
```

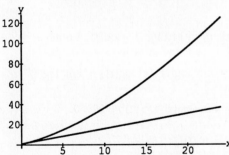

Just as the Race Track Principle predicted.

■ **B.2) Logarithmic and exponential growth.**

We've heard a lot of talk about how powerful exponential growth is and how lethargic logarithmic growth is. Not it's time use the Race Track Principle to deliver a definitive explanation of this.

**B.2.a.i)**

Plot $\log[x]/x$ for $1 \leq x \leq 20$. What do you think is the global scale behavior of $\log[x]/x$?

**Answer:**

```
In[1]:=    Plot[Log[x]/x,{x,1,200},PlotRange->All]
```

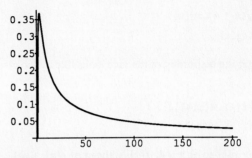

Golly-oskee. It looks like $\lim_{x \to \infty} \log[x]/x = 0$. Let's look again:

```
In[2]:=    Plot[Log[x]/x,{x,1,1000},
           PlotRange->All]
```

And again:

```
In[3]:=    Plot[Log[x]/x,{x,1,10000},
           PlotRange->All]
```

Conclussion: $\log[x]/x$ looks awfully small when $x$ is large.

**B.2.a.ii)**

Use the Race Track Principle to explain why
$$\lim_{x \to \infty} \log[x]/x = 0.$$

**Answer:**

Let $f[x] = \sqrt{x}$ and let $g[x] = \log[x]$ and let's run a race between these two functions starting at $x = 4$.

```
In[1]:=    Clear[f,g,x]
           f[x_] = Sqrt[x]
Out[1]=    Sqrt[x]

In[2]:=    g[x_] = Log[x]
Out[2]=    Log[x]
```

Let's see what the beginning of the race looks like by looking at:

```
In[3]:=    {f[4],N[g[4]]}
Out[3]=    {2, 1.38629}
```

Good: $f[4] \geq g[4]$ so at $x = 4$, $f[x]$ is ahead of $g[x]$. Look at the difference of the derivatives and factor:

```
In[4]:=    Factor[f'[x]- g'[x]]
Out[4]=    -2 + Sqrt[x]
           -------------
               2 x
```

For $x \geq 4$, $\sqrt{x} \geq 2$; so for $x \geq 4$, we see
$$f'[x] - g'[x] \geq 0.$$
This means that for $x \geq 4$,
$$f'[x] \geq g'[x].$$
The two functions start out at $x = 4$ with $f[x]$ ahead of $g[x]$ and as $x$ advances away from 4, $f[x]$ grows faster than $g[x]$. The Race Track Principle tells us that $f[x] \geq g[x]$ for $x \geq 4$. Here is a plot:

```
In[5]:=    Plot[{f[x],g[x]},{x,4,1000}]
```

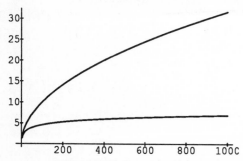

That's $\sqrt{x}$ riding high above $\log[x]$.

You might wonder why all the fuss about $\sqrt{x}$; if so, hang in there: you are going to find out soon enough.

For $x \geq 4$ we know that
$$0 \leq \log[x] \leq \sqrt{x}.$$
Divide through by $x$ to learn
$$0 \leq \log[x]/x \leq \sqrt{x}/x = 1/\sqrt{x}.$$
So
$$0 \leq \lim_{x \to \infty} \log[x]/x \leq \lim_{x \to \infty} 1/\sqrt{x} = 0.$$

This explains why
$$\lim_{x \to \infty} \log[x]/x = 0.$$

### B.2.a.iii) Logarithmic growth vs. power growth

Explain why $\log[x]$ *is dominated by any positive power of $x$* by explaining why
$$\lim_{x \to \infty} \log[x]/x^p = 0$$
for any positive power $p$.

**Answer:**

Look at $\log[x]/x^p$ and change the variable by setting $t = x^p$. This is the same as $t^{1/p} = x$. Note that $t \to \infty$ is the same as $x \to \infty$.

So
$$\lim_{x \to \infty} \log[x]/x^p = \lim_{t \to \infty} \log[t^{1/p}]/t$$

$$= \lim_{t \to \infty} (1/p) \log[t]/t = 0$$

because $\lim_{t \to \infty} \log[t]/t = 0$.

Not a hell of a lot to it.

### B.2.b) Exponential growth vs. power growth

Explain why $e^x$ *dominates any positive power of $x$* by explaining why
$$\lim_{x \to \infty} x^p/e^x = 0$$
for any positive power $p$.

**Answer:**

Change the variable by setting $x = \log[t]$. Recall $e^{\log[t]} = t$. Note that $t \to \infty$ is the same as $x \to \infty$. So
$$\lim_{x \to \infty} x^p/e^x = \lim_{t \to \infty} \log[t]^p/e^{\{\log[t]\}} = \lim_{t \to \infty} \log[t]^p/t$$

$$= \lim_{t \to \infty} (\log[t]/(t^{1/p}))^p = 0^p = 0$$

because $\lim_{t \to \infty} \log[t]/t^{1/p} = 0$.

Watch $x^{30}/e^x$ drop off as $x$ gets large:

```
In[1]:=    Clear[x]
           Plot[x^30/E^x,{x,0,100},PlotRange->All]
```

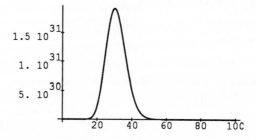

# ■ B.3) The Mean Value Theorem.

### B.3.a)

For each of the following functions $f[x]$ and choices of $a$ and $b$, plot $f'[x]$ for $a \leq x \leq b$.

Use the plot to read off:

the biggest value $M$ of $f'[x]$ for $a \leq x \leq b$, and

the least value $m$ of $f'[x]$ for $a \leq x \leq b$.

Then plot

$$low[x] = f[a] + m(x - a)$$

$$f[x] = f[x]$$

$$high[x] = f[a] + M(x - a)$$

on the same axes for $a \leq x \leq b$. Describe what you see.

### B.3.a.i)

$f[x] = (x^2 + x)e^{-x^2}$, $a = 1$ and $b = 2$.

---

**Answer:**

```
In[1]:=   Clear[f,x]
          f[x_] = (x^2 + x) E^(-x)
```
$$Out[1]= \quad \frac{x^2 + x}{E^x}$$

```
In[2]:=   a = 1
Out[2]=   1
```

```
In[3]:=   b = 2
Out[3]=   2
```

Here is the plot of $f'[x]$ for $a \leq x \leq b$:

```
In[4]:=   Plot[f'[x],{x,a,b},
          PlotRange->All]
```

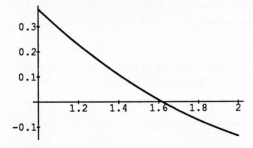

Clearly $f'[x]$ is biggest for $x = a$ and is least for $x = b$; so:

```
In[5]:=   m = f'[b]
```

```
Out[5]=   -2
          ---
          -E
```

```
In[6]:=   M = f'[a]
Out[6]=   1
          -
          E
```

Here is a plot of $y = f[x]$,

$$y = low[x] = f[a] + m(x - a)$$

and

$$y = high[x] = f[a] + M(x - a)$$

on the same axes for $a \leq x \leq b$.

```
In[7]:=   Plot[{f[a] + m(x - a),
          f[x],f[a] + M(x - a)},
          {x,a,b},AxesLabel->{"x","y"}]
```

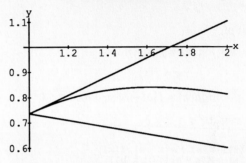

That's $f[x]$ between the two line functions $low[x]$ and $high[x]$.

### B.3.a.ii)

$f[x] = 5 + 129x - 180x^2 + 110x^3 - 30x^4 + 3x^5$, $a = 0.7$ and $b = 2.7$.

---

**Answer:**

```
In[1]:=   Clear[x,f]
          f[x_] =
          5 + 129x - 180x^2 + 110x^3 - 30x^4 + 3x^5
```
$$Out[1]= \quad 5 + 129\,x - 180\,x^2 + 110\,x^3 - 30\,x^4 + 3\,x^5$$

```
In[2]:=   a = 0.7
Out[2]=   0.7
```

```
In[3]:=   b = 2.7
Out[3]=   2.7
```

Here is the plot of $f'[x]$ for $a \leq x \leq b$:

```
In[4]:=   Plot[f'[x],{x,a,b},
          PlotRange->All]
```

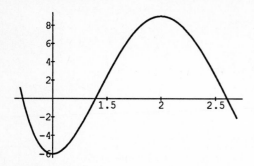

Clearly $f'[x]$ is least for $x = 1$ and is biggest for $x = 2$; so:

```
In[5]:=     m = f'[1]
Out[5]=     -6
```

```
In[6]:=     M = f'[2]
Out[6]=     9
```

Here is a plot of

$$y = low[x] = f[a] + m(x - a),$$
$$y = f[x],$$
$$y = high[x] = f[a] + M(x - a),$$

on the same axes for $a \leq x \leq b$.

```
In[7]:=     Plot[{f[a] + m(x - a),f[x],
            f[a] + M(x - a)},{x,a,b}]
```

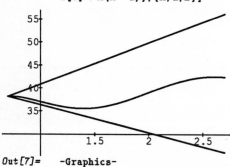

```
Out[7]=     -Graphics-
```

That's $f[x]$ between the two line functions $low[x]$ and $high[x]$.

### B.3.b)

Explain why in each of the above plots we observed that

$$low[x] = f[a]+m(x-a) \leq f[x] \leq f[a]+M(x-a) = high[x]$$

for all the $x$'s in $[a, b]$ and why this will happen for any function we ever want to plot in the future.

### Answer:

Because $m$ is the least value of $f'[x]$ for $a \leq x \leq b$ and because $M$ is the biggest value of $f'[x]$ for $a \leq x \leq b$, we see that

$$low'[x] = m \leq f'[x] \leq M = high'[x]$$

for all $x$'s in $[a, b]$. Thus because

$$low[a] = f[a] = high[a],$$

the Race Track Principle tells us that

$$low[x] \leq f[x] \leq high[x]$$

for all the $x$'s in $[a, b]$.

### B.3.c) The Mean Value Theorem

If $a < b$, then explain why there is a number $z$ between $a$ and $b$ such that

$$f[b] = f[a] + f'[z](b - a).$$

This fact is known as *the Mean Value Theorem*.

### Answer:

From part **b)** above we know that putting $x = b$ gives

$$f[a]+m(b-a) = low[b] \leq f[b] \leq high[b] = f[a]+M(b-a).$$

Thus

$$f[a] + m(b - a) \leq f[b] \leq f[a] + M(b - a).$$

Accordingly

$$f[b] = f[a] + s(b - a)$$

where:

$s$ is a number between $M$ and $m$, ,

$m$ is the least value of $f'[x]$ for $a \leq x \leq b$, and,

$M$ is the greatest value of $f'[x]$ for $a \leq x \leq b$.

Because this number $s$ is between the greatest and least values of $f'[x]$ as $x$ varies in $[a, b]$, there must be a number $z$ in $[a, b]$ with $f'[z] = s$.

For this number $z$ we have $f[a]+s(b-a) = f[a]+f'[z](b-a)$. So for this $z$, we see that

$$f[b] = f[a] + f'[z](b - a)$$

and this is what we wanted to explain!

### B.3.d)

Explain why the Mean Value Theorem is true even if $b < a$.

### Answer:

If $b < a$, then interchanging $a$ and $b$ in the statement of the Mean Value Theorem above gives

$$f[a] = f[b] + f'[z](a - b)$$

for some $z$ between $b$ and $a$.

Multiply left and right by $-1$ to get

$$-f[a] = -f[b] + f'[z](b - a)$$

for the same $z$ between $a$ and $b$.

Move $-f[b]$ to the left side and move $-f[a]$ to the right side to get

$$f[b] = f[a] + f'[z](b - a)$$

for the same $z$ between $a$ and $b$. That's all.

# Tutorial

### ■ T.1) Growth

### T.1.a)

How do you know that as $x$ advances from 1 to 2, then $f[x] = x^2 e^x$ grows at least 8 times faster than $x$ grows but never grows more than 60 times as fast as $x$ grows?

**Answer:**

```
In[1]:=    Clear[f,x]
           f[x_] = x^2 E^x
Out[1]=      x  2
           E   x
```

Look at:

```
In[2]:=    Plot[f'[x],{x,1,2},
           PlotRange->All]
```

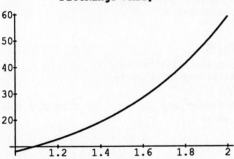

The least growth rate shown on $[1, 2]$ is:

```
In[3]:=    N[f'[1]]
Out[3]=    8.15485
```

The biggest growth rate shown on $[1, 2]$ is:

```
In[4]:=    N[f'[2]]
Out[4]=    59.1124
```

So as $x$ advances from 1 to 2, $f[x] = x^2 e^x$ goes up at least 8.1 times faster than $x$ goes up but never goes up more than 59.2 times as fast as $x$ goes up.

### ■ T.2) Global scale behavior: Limits at infinity.

### T.2.a)

Rank the following functions in order of dominance as $x \to \infty$:

$$\log[x], e^{3x}, x^{100}, e^{2x}/x^{10}, x^{12}e^{-x}, x^{1/100}, 1/x^3, x^x.$$

**Answer:**

For large $x$'s, they line up in the order:

$$x^x, e^{3x}, e^{2x}/x^{10}, x^{100}, x^{1/100}, \log[x], 1/x^3, x^{12}e^{-x}.$$

### T.2.b)

How does

$$(2x^4 + 5x^2 + 4x + 1)/(x^3 + 6x^2 + 1)$$

behave as $x \to \infty$? Illustrate with a plot.

**Answer:**

```
In[1]:=    Clear[x]
           function = (2 x^4 + 5 x^2 + 4 x + 1)/(x^3 + 6
           x^2 + 1)
Out[1]=                      2      4
           1 + 4 x + 5 x  + 2 x
           ---------------------
                      2    3
           1 + 6 x  + x
```

```
In[2]:=    Plot[function,{x,0,1000}]
```

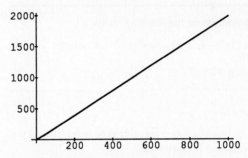

The global scale behavior is the same as the global scale behavior of $2x$. The reason for this is that if we ignore all but the dominant terms in

$$(2x^4 + 5x^2 + 4x + 1)/(x^3 + 6x^2 + 1),$$

then we get

$$2x^4/x^3 = 2x.$$

In other words, only the *dominant* terms influence the global scale behavior.

Here are two plots illustrating just how much

$$(2x^4 + 5x^2 + 4x + 1)/(x^3 + 6x^2 + 1)$$

is like $2x$ as $x \to \infty$:

```
In[3]:=    Plot[{function, 2 x},{x,0,300}]
```

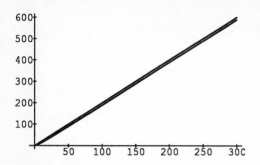

```
In[4]:=    Plot[{function,2 x},{x,1000,3000}]
```

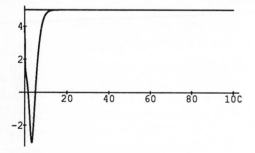

You can't tell them apart without a scorecard.

### T.2.c)

What would you say is the limiting value

$$\lim_{x \to \infty} (5e^x - \cos[x] - 6x^3)/(e^x + 8\sin[x])?$$

Illustrate with a plot

**Answer:**

By ignoring all but the dominant terms as $x$ goes to $\infty$ we see that the limiting behavior of

$$(5e^x - \cos[x] - 6x^3)/(e^x + 8\sin[x])$$

is the same as the limiting behavior of

$$5e^x/e^x = 5.$$

Here is a plot:

```
In[1]:=    Plot[
           {(5 E^x - Cos[x]-6x^3)/(E^x+8 Sin[x]),
           5},{x,0,100}, PlotRange->All]
```

Yep; as $x$ grows from 0 to 100, the plot of

$$(5e^x - \cos[x] - 6x^3)/(e^x + 8\sin[x])$$

cannot stay away from the line $y = 5$.

There is no choice but to say

$$\lim_{x \to \infty} (5e^x - \cos[x] - 6x^3)/(e^x + 8\sin[x]) = 5.$$

### ■ T.3) Using the derivative to guarantee accuracy of calculation.

When we say $x$ is $y$ to $k$ *accurate decimals* we mean $x$ is $y$ *within* $10^{-k}$. This is the same as saying $|x - y| < 10^{-k}$.

### T.3.a)

Find a number $x$ such that $x$ is $\pi$ to ten accurate decimals.

**Answer:**

Any of the following numbers will work for $x$:

```
In[1]:=    N[Pi,11]
Out[1]=    3.1415926536
```

```
In[2]:=    N[Pi,13]
Out[2]=    3.14159265359
```

```
In[3]:=    N[Pi,15]
Out[3]=    3.14159265358979
```

```
In[4]:=    N[Pi,61]
Out[4]=    3.1415926535897932384626433832795028841
           9716939937510582097494945
```

The following number will not work for $x$ because it is not close enough to $\pi$:

```
In[5]:=    N[Pi,10]
Out[5]=    3.141592654
```

The number 3.141592654 is $\pi$ to *nine* accurate decimals but not to *ten* accurate decimals. We are counting the accurate places to the right of the decimal.

### T.3.b.i)

The larger $|f'[x]|$ is, the faster $f[x]$ changes as $x$ changes. So the larger $|f'[x]|$, the more sensitive $f[x]$ is to small changes in $x$. How can the size of the derivative be used to estimate how many accurate decimals of $x$ in an interval $[c, d]$ are needed to compute $k$ accurate decimals of a function $f[x]$?

**Answer:**

The Mean Value Theorem says:

$$f[b] = f[a] + f'[z](b - a)$$

*for some $z$ between $a$ and $b$.*

This is the same as saying

$$f[b] - f[a] = f'[z](b - a)$$

for some $z$ between $a$ and $b$.

Taking $x = b$ and $y = a$ for $x$ and $y$ in $[c, d]$ gives

$$f[x] - f[y] = f'[z](x - y)$$

for some $z$ between $x$ and $y$.

Thus

$$|f[x] - f(y)| = |f'[z]||(x - y)|$$

for some $z$ between $x$ and $y$.

Notice that $z$ is also in $[c, d]$ because it is between $x$ and $y$ and both $x$ and $y$ are in $[c, d]$.

Therefore if we set an integer $p$ such that

$$|f'[z]| < 10^p$$

for all $z$ in $[c, d]$, then we see that

$$|f(x) - f(y)| < 10^p |(x - y)|.$$

Accordingly: if $x$ is $y$ within $10^{-k-p}$, then $|x - y| < 10^{-k-p}$ and

$$|f(x) - f(y)| < 10^p 10^{-k-p} = 10^{-k}.$$

So if $x$ is $y$ within $10^{-k-p}$, then $f[x]$ is $f[y]$ within $10^{-k}$.

Consequently to calculate $k$ accurate decimals of $f[x]$ for $x$ in $[c, d]$:

*Find an integer $p$ with $|f'[z]| < 10^p$ for all $z$ in $[c, d]$ and then feed in $k + p$ accurate decimals of $x$.*

**T.3.b.ii)**

---

Given $f[x] = e^x \sqrt{x^8 + 1}/x^{2/3}$.

If $k$ is a given positive integer, then how many accurate decimals of $x$ in $[x, 2]$ guarantee $k$ accurate decimals of $f[x]$? How many accurate decimals of $x$ in $[2, 6]$ guarantee $k$ accurate decimals of $f[x]$?

---

**Answer:**

```
In[1]:=   Clear[f,x,y,z]
          f[x_] = E^x Sqrt[x^8 + 100] / x^ (2/3)

Out[1]=     x           8
          E  Sqrt[100 + x ]
          -----------------
                 2/3
                x
```

To learn how many accurate decimals of $x$ in $[1, 2]$ guarantee $k$ accurate decimals of $f[x]$, plot $|f'[x]|$ for $1 \leq x \leq 2$ taking care that the plot exhibits all the values of $|f'[x]|$ for $x$ in $[1, 2]$.

```
In[2]:=   Plot[Abs[f'[x]],{x,1,2},PlotRange->All]
```

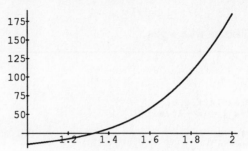

The plot makes it clear that for $x$ in $[1, 2]$,

$$|f'[x]| < 1000 = 10^3$$

To get $k$ accurate decimals of $f[x]$ for $x$ in $[1, 2]$, we can get by with $k + 3$ accurate decimals of $x$.

Try it by looking at:

```
In[3]:=   {f[1.800058],f[1.800057],f[1.800059]}
Out[3]=   {59.2802, 59.2801, 59.2803}
```

The numbers 1.800058, 1.800057 and 1.800059 share 5 accurate decimals; so the numbers $f[1.800058]$, $f[1.800057]$, and $f[1.800059]$ should share at least $5 - 3 = 2$ accurate decimals. They do.

To learn how many accurate decimals of $x$ in $[2, 6]$ we need to calculate $k$ accurate decimals of $f[x]$, plot $|f'[x]|$ for $2 \leq x \leq 6$ taking care that the plot exhibits all the values of $|f'[x]|$ for $x$ in $[2, 6]$.

```
In[4]:=   Plot[Abs[f'[x]],{x,2,6},PlotRange->All]
```

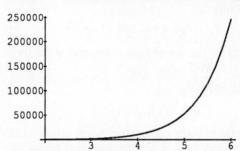

The plot makes it clear that for $x$ in $[2, 6]$,

$$|f'[z]| < 100000 = 10^6.$$

To get $k$ accurate decimals of $f[x]$ for $x$ in $[2, 6]$, we can get by with $k + 6$ accurate decimals of $x$.

■ **T.4) Napier's Inequality.**

Inequalities are the stuff of a large part of mathematics. Here is an inequality originally used to help to compute values of $\log[x]$ by Napier when he invented the natural logarithm. His tables were published in 1614.

**T.4.a)**

---

Use the Race Track Principle to explain why

$$(x-1)/x \le \log[x] \le x-1$$

for $x \ge 1$. Illustrate with a plot.

---

**Answer:**

```
In[1]:=    Clear[x,f,g,h]
           f[x_] = (x - 1)/x
Out[1]=    -1 + x
           ------
             x


In[2]:=    g[x_] = Log[x]
Out[2]=    Log[x]


In[3]:=    h[x_] = x - 1
Out[3]=    -1 + x
```

Look at:

```
In[4]:=    {f[1],g[1],h[1]}
Out[4]=    {0, 0, 0}
```

All three functions start at the same value. Now look at:

```
In[5]:=    Together[{f'[x], g'[x],h'[x]}]
Out[5]=        -2  1
           {x  , -, 1}
                 x
```

For $x \ge 1$, we know that $1 \le x \le x^2$. Consequently for $x \ge 1$, we see that

$$1/x^2 \le 1/x \le 1$$

because the bigger the denominator the smaller the fraction.

Thus for $x \ge 1$, we have

$$f'[x] \le g'[x] \le h'[x].$$

So the functions all start out together when $x = 0$. And as $x$ advances away from 0, the function $h[x]$ shows the greatest growth and $f[x]$ shows the least growth. The Race Track Principle steps in to tell us that for $x \ge 1$, we have

$$f[x] \le g[x] \le h[x].$$

Here comes a plot:

```
In[6]:=    Plot[{f[x],g[x],h[x]},{x,1,10}]
```

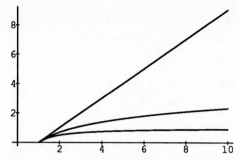

That's $g[x] = \log[x]$ caught between $f[x] = (x-1)/x$ on the bottom and $h[x] = x - 1$ on the top.

■ **T.5) Using the mean value theorem to fake the plot of f[x].**

The mean value theorem tells us that

$$f[x] = f[a] + f'[z](x - a).$$

for some $z$ between $a$ and $x$. The trouble is that it doesn't tell us how to compute the value of $z$ in terms of $x$ and $a$.

### T.5.a.i)

---

In the spirit of fair play, let's see what happens when we split the difference by trying $z = (x + a)/2$ and plotting $f[x]$ and

$$fakef[x] = f[a] + f'[(x + a)/2](x - a)$$

on the same axes for $f[x] = \cos[x]$ and on an interval including $a$ in the case that $a = 0$.

---

**Answer:**

```
In[1]:=    Clear[f,g,x,a]
           f[x_] = Cos[x]
Out[1]=    Cos[x]


In[2]:=    fakef[x_,a_] = f[a] + f'[(x + a)/2] (x - a)
Out[2]=                            a + x
           Cos[a] - (-a + x) Sin[-----]
                                    2


In[3]:=    Plot[{f[x],fakef[x,0]},{x,-2,2}]
```

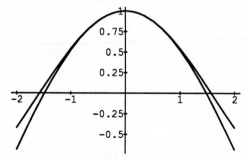

Not a bad fit for $-1 \le x \le 1$. You might say that the plot of

$$fake[x,a] = f[a] + f'[(x + a)/2](x - a)$$

for $a = 0$ and $-1 \leq x \leq 1$ is a pretty good fake of the plot of $f[x]$.

Let's see what happens when we take $a = \pi/2$:

```
In[4]:=    Plot[{f[x],fakef[x,Pi/2]},
           {x,Pi/2 - 1,Pi/2 + 1}]
```

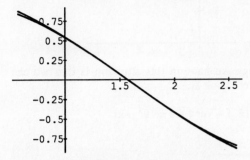

Another masterful forgery. The plot of

$$fake[x, a] = f[a] + f'[(x + a)/2](x - a)$$

for $a = \pi/2$ and $\pi/2 - .5 \leq x \leq \pi/2 + .5$ is a pretty damn good fake of the plot of $f[x]$.

### T.5.a.ii)

---

Again in the spirit of fair play, let's see what happens when we split the difference by trying $z = (x + a)/2$ and plotting $f[x]$ and

$$fakef[x] = f[a] + f'[(x + a)/2](x - a)$$

on the same axes for $f[x] = x^2 e^{-x^2}$ and on an interval including $a$ in the case that $a = 1$.

---

**Answer:**

```
In[1]:=    Clear[f,g,x,a]
           f[x_] = x^2 E^(-x^2)
```
```
Out[1]=       2
             x
            ---
              2
             x
            E
```

```
In[2]:=    fakef[x_,a_] = f[a] + f'[(x + a)/2] (x - a)
```
```
Out[2]=       2
             a                  a + x
            --- + (-a + x) (----------- -
              2                      2
             a              (a + x) /4
            E              E

                  3
               (a + x)
            --------------)
                      2
               (a + x) /4
            4 E
```

```
In[3]:=    Plot[{f[x],fakef[x,1]},{x,0,2}]
```

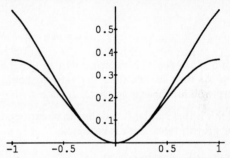

Not a bad fit for $0.6 \leq x \leq 1.4$. The plot of

$$fake[x, a] = f[a] + f'[(x + a)/2](x - a)$$

for $a = 1$ and $0.6 \leq x \leq 1.4$ is a pretty good fake of the plot of $f[x]$.

Let's see what happens when we take $a = 0$ and plot on $[-1, 1]$:

```
In[4]:=    Plot[{f[x],fakef[x,0]},{x,-1,+1}]
```

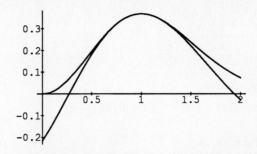

Hot plot. Look at the cohabitation near $a = 0$.

### T.5.b.i)

---

Sometimes you can't get your hands on a formula for $f[x]$ but you do have a formula for $f'[x]$ and you know the value of $f[a]$ at a certain point $a$ and need to get a reasonably accurate plot of $f[x]$ for $x$'s near $a$.

In principle you should be able to deliver the plot because if you are given $f'[x]$ and $f[a]$, then you can find $f[x]$ because you know where $f[x]$ is when $x = a$ and you know how it is growing at all orther points.

### T.5.b.ii)

---

Fake a reasonably accurate plot of the function $f[x]$ with

$$f'[x] = (6 - 4x - x^2)e^{-x}$$

and

$$f[1] = 7/e$$

on an interval containing $x = 1$.

---

**Answer:**

```
In[1]:=    Clear[fprime,a,x,fakef,f]
           fprime[x_] = (6 - 4 x - x^2) E^(-x)
```

Out[1]=
$$\frac{6 - 4\,x - x^2}{E^x}$$

For the faker, we use

$$fakef[x] = f[a] + f'[(x+a)/2](x-a)$$

In[2]:=     ```fakef[x_,a_] =
          f[a] + fprime[(x + a)/2] (x - a)```

Out[2]=
$$\frac{(-a + x)\ (6 - 2\ (a + x) - \dfrac{(a + x)^2}{4})}{E^{(a + x)/2}} +$$

f[a]

In[3]:=     ```f[1] = 7/E```
Out[3]=
$$\frac{7}{E}$$

Now if the plot of $fakef[x, 1]$ is to look anything like the plot of $f[x]$ for $x$'s near 1, then their derivatives should match up pretty well for those $x$'s.

The reason for this is that similar functions can be expected to have similar growth rates. Let's take a look:

In[4]:=     ```fakederiv = D[fakef[x,1],x]
          Plot[{fprime[x],fakederiv},
          {x,1 - 1,1 + 1}]```

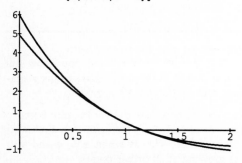

The two plots are sharing the same ink for $.6 \le x \le 1.4$. This tells us that the following plot is a good fake of the plot of $f[x]$ on the interval $[.6, 1.4]$:

In[5]:=     ```Plot[fakef[x,1],{x,.6,1.4},
          PlotRange->All]```

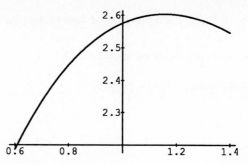

The advantage we have in this situation is that we can check this plot. Look at:

In[6]:=     ```f[x_] = (x^2 + 6 x) E^(-x)```
Out[6]=
$$\frac{6\,x + x^2}{E^x}$$

In[7]:=     ```f[1]```
Out[7]=
$$\frac{7}{E}$$

In[8]:=     ```Together[f'[x]]```
Out[8]=
$$\frac{6 - 4\,x - x^2}{E^x}$$

This shows that the function whose plot we've been faking is

$$f[x] = (x^2 + 6x)e^{-x}.$$

This allows us to check the quality of our fakery.

In[9]:=     ```Plot[{f[x],fakef[x,1]},{x,.6,1.4},
          PlotRange->All]```

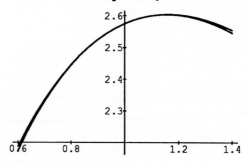

Not bad for beginners.

**T.5.c.i)**

Now it's time to get serious. All you know about a function $f[x]$ is that

$$f'[x] = \sin[x^{3/2}].$$

and $f[1/2] = 1$. Give a reasonably accurate plot of $f[x]$ on an interval containing $a = 1/2$.

### Answer:

Try as we might, we cannot come up with a formula for $f[x]$; so the fakery route is the only possibility open.

```
In[1]:=   Clear[fprime,a,x,fakef,f]
          fprime[x_] = Sin[x^(5/3)]
                  5/3
Out[1]=   Sin[x   ]
```

For the faker, use

$$fakef[x] = f[a] + f'[(x + a)/2](x - a).$$

```
In[2]:=   fakef[x_,a_] =
          f[a] + fprime[(x + a)/2] (x - a)
                            5/3
                     (a + x)
Out[2]=   (-a + x) Sin[----------] + f[a]
                        5/3
                       2
```

```
In[3]:=   f[1/2] = 1
Out[3]=   1
```

Take a look:

```
In[4]:=   fakederiv = D[fakef[x,1/2],x]
          Plot[{fprime[x],fakederiv},
          {x,1/2 - 1,1/2 + 1}]
```

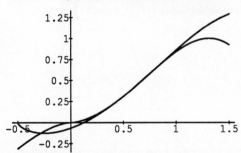

The two plots are sharing the same ink for $0.3 \le x \le 0.8$. This reveals that the following plot is a good fake of the plot of $f[x]$ on the interval $[0.3, 0.8]$:

```
In[5]:=   fake1 = Plot[fakef[x,1/2],{x,0.3,0.8},
          PlotRange->All]
```

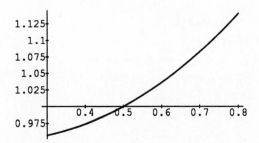

Looks fine.

### T.5.c.ii)

Extend the faked plot to the right beyond $x = 0.8$.

### Answer:

Don't clear the machine. Reset $a = 0.8$. Here is a reasonably good approximation of $f[0.8]$:

```
In[1]:=   f[0.8] = fakef[0.8,1/2]
                         1.54848
Out[1]=   1 + 0.3 Sin[-------]
                        5/3
                       2
```

Here's the new faker:

```
In[2]:=   fakef[x,0.8]
                         1.54848
Out[2]=   1 + 0.3 Sin[-------] +
                        5/3
                       2
                                   5/3
                            (0.8 + x)
          (-0.8 + x) Sin[------------]
                             5/3
                            2
```

To gauge the interval $[0.8, ?]$ on which we can plot $fake[x, 0.8]$ to get a good fake of the plot of $f[x]$, look at:

```
In[3]:=   fakederiv = D[fakef[x,0.8],x]
          Plot[{fprime[x],fakederiv},{x,0.8,1.8}]
```

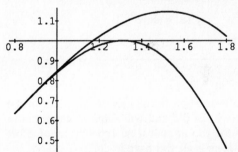

The two plots are sharing the same ink for $0.8 \le x \le 1.0$. So the following plot is a good fake of the plot of $f[x]$ on the interval $[0.8, 1.0]$ :

```
In[4]:=   fake2 = Plot[fakef[x,0.8],{x,0.8,1.0}]
```

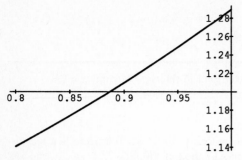

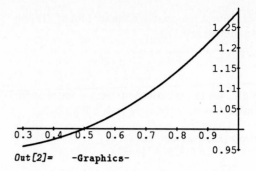

Now put the plots together to get a plot of $f[x]$ on $[0.3, 1.0]$ :

*In[5]:=*     **ourfake = Show[fake1, fake2,**
                  **PlotLabel->"Composite fake plot"]**

           Composite fake plot

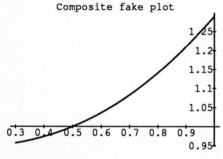

We can use the same technique to extend the plot further to the right of left. Try it.

### T.5.c.iii)

How can *Mathematica* be used to check our faked plot of the above function $f[x]$ with

$$f'[x] = \sin[x^{3/2}].$$

and $f[1/2] = 1$?

**Answer:**

*Mathematica* 's instruction **NIntegrate** operates in the spirit of what we just did and will allow *Mathematica* to do the fakery for you at almost no expenditure of effort on your part. Here is all you have to do:

*In[1]:=*     **machinefakef[x_] :=**
            **NIntegrate[fprime[t],{t,1/2,x}] + f[1/2]**

*The option := is used here because we don't want Mathematica to evaluate the function until we call on it.*

Plotting $machinefakef[x]$ will give an accurate fake plot of $f[x]$ on $[.0.3, 1]$ :

*In[2]:=*     **machinefakeplot =**
            **Plot[machinefakef[x],{x,0.3,1.0}]**

Now let's compare our homemade fake plot with *Mathematica* 's fake plot:

*In[3]:=*     **Show[machinefakeplot,ourfake]**

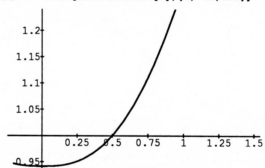

Our fake is certainly close enough for government work. Here is an accurate fake plot of $f[x]$ over an interval larger than $[0.3, 1.0]$:

*In[4]:=*     **Plot[machinefakef[x],{x,-.2,1.5}]**

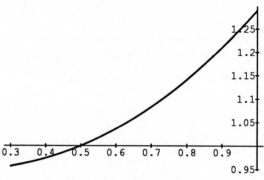

The point is that even though we can't find a formula for $f[x]$ , we can still plot the function $f[x]$. In fact we just did it! If we want an approximate value of, say $f[3.7]$, we can get it from:

*In[5]:=*     **machinefakef[3.7]**

*Out[5]=*     **1.86761**

Cool.

## Literacy sheet

**1.** What is the Race Track Principle and why it is good for establishing inequalities?

.

.

**2.** What is the Mean Value Theorem? How does the mean value theorem help us in faking plots we cannot do otherwise?

.

.

**3.**   Why is exponential growth sometimes called awesome? Which is the most macho:

Logarithmic growth, polynomial growth or exponential growth.

Which is least macho?

.

.

.

**4.**   If the ecomomy holds at a steady inflation rate of 5 percent per year, this means that the price index $t$ years from now, $P[t]$, is given by

$$P[t] = Pe^{.05t}$$

where $P$ is today's price index.

Why should this strike fear into the hearts of all those who are literate in calculus?

.

.

.

**5.** Rank the following function in order of dominance as $x \to \infty$

$$x^2 \log[x], x^3, e^{.001x} \log[x], x, e^{3x}, x^{10}, e^{3x}/x^{40}, e^{2x}.$$

.

.

.

**6.** Give the limits:

$\lim_{x \to \infty} x^6 e^{-x}$

.

.

$\lim_{x \to \infty} e^x/x^{57}$

.

$\lim_{x \to \infty} e^{-x}/x^{57}$

.

.

$\lim_{x \to \infty}(9e^{5x} - 8e^{4x})/(3e^{5x} + 456e^{4x})$

.

.

$\lim_{x \to \infty}(9e^{-5x} - 8e^{-4x})/(3e^{-5x} + 456e^{-4x})$

.

$\lim_{x \to \infty}(17e^{2x} + 6\sin[e^{4x}])/(9x^8 - 8e^{2x})$

.

.

$\lim_{x \to \infty} \log[x]/\sqrt{x}$

.

.

$\lim_{x \to \infty} \log[x]/(x^{1/9}).$

.

.

.

**7.** If $f[a] = f[b] = 0$, then use the Mean Value Theorem to explain why there is a number $z$ between $a$ and $b$ such that $f'[z] = 0$. Sometimes this fact is called *Rolle*'s Theorem.

Remember that the Mean Value Theorem tells you that $f[b] = f[a] + f'[z](b - a)$ for some $z$ between $a$ and $b$.

.

.

.

**8.** If $f'[x] = 0$ at exactly one point in $[0, 1]$ and $f'[x]$ is non-zero at all other points of $[0, 1]$, then how many points $x$ can there be in $[0, 1]$ such that $f[x] = 0$?

Ask yourself: At most how many times can the curve go up and down?

.

.

.

**9.** How many accurate decimals of $x$ guarantee 8 accurate decimals of $\log[x]$ for $x \geq 1$?

.

.

.

**10.** Two functions $f[x]$ and $g[x]$ satisfy

$$f[a] = g[a]$$

and $f'[x]$ and $g'[x]$ are very close for $x's$ near $a$.

Explain why $f[x]$ and $g[x]$ are very close for $x's$ near $a$.

.

.

.

**11.** Two functions $f[x]$ and $g[x]$ satisfy

$$f[a] = g[a]$$

and

$$f'[x] \geq g'[x]$$

for $x \leq a$. Why does it follow that

$$f[x] \leq g[x]$$

for $x \leq a$?

Think of it this way: If the faster horse crosses the finish line in a tie, then which horse was ahead prior to the finish?

.

.

.

**12.** Why does a good representative plot normally depict the function on an interval containing at a minimum all $x's$ such that $f'[x] = 0$?

.

.

.

**13.** Look at:

*In[6]:=*     **Simplify[D[ArcTan[x] + ArcTan[1/x],x]]**

*Out[6]=*     0

This might make us believe that

$$\arctan[x] + \arctan[1/x]$$

is constant. Now look at:

*In[7]:=*     **Plot[ArcTan[x] + ArcTan[1/x],{x,-1,1}]**

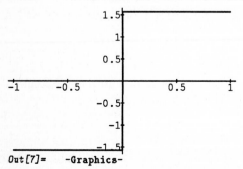

*Out[7]=*     **-Graphics-**

This shows that $\arctan[x] + \arctan[1/x]$ is *not* constant.

Why does this not contradict the statement that: if $f'[x] = 0$ for *all* $x's$, then $f[x]$ is constant?

# The Differential Equation y'[x] = a y[x]

## Guide

Differential equations are the keys to the gates of science and engineering. The biologist or engineer models a process with a differential equation and then turns the differential equation over to the mathematical scientist who tries to provide a solution resulting in a complete description of the orginal process. Differential equations come at all levels of complexity and even today there is active mathematical research in differential equations. In this lesson, we are going to study one of the simplest, yet most important, differential equations.

## Basics

■ **B.1) Natural occurences of the differential equation y'[x] = a y[x].**

### B.1.a) Radioactive decay.

Over a given time interval, each radioactive atom has the same chance for disintegration as any other.

Thus if we have a given quantity $N$ of radioactive atoms of one type, then we can expect twice as much decay if our supply is doubled to $2N$ and three times as much decay if our supply is tripled to $3N$.

This means that the rate of decay $N'[t]$ is proportional to the supply $N[t]$ at time $t$. In other words:

$$N'[t] = aN[t]$$

for some constant of proprotionality $a$.

### B.1.b) Chemical Reactions.

Gaseous nitrogen peroxide decomposes into nitrous oxide and oxygen. The decompostion is caused by collisions of two nitrogen peroxide molecules. So doubling the concentration of the gas doubles the rate of decompostion and tripling the concentration of the gas triples the rate of decompostion.

Thus if $C[t]$ is the concentration of nitrogen peroxide at time $t$, then $C'[t]$ is proportional to $C[t]$. In other words,

$$C'[t] = aC[t]$$

for some constant of proprotionality $a$.

### B.1.c) Continuous compounding.

We know that if $P$ dollars as deposited in an account with advertised interest rate $100r$ percent per year compounded continuously, then the amount $y[t]$ dollars in the account after $t$ years is given by

$$y[t] = Pe^{rt}.$$

Accordingly

$$y'[t] = Pe^{rt}r = ry[t].$$

Thus

$$y'[t] = ry[t].$$

■ **B.2) The differential equation y'[x] = a y[x]: A qualitative analysis.**

### B.2.a)

Fix a positive constant $a$. Describe in general terms how a function $y[x]$ satisfying

$$y'[x] = ay[x]$$

must behave as $x$ gets large.

**Answer:**

If $y[x_0] > 0$ for some $x_0$, then $y'[x_0] = ay[x_0] > 0$; so $y[x]$ is going **up** as $x$ goes through $x_0$. As $x$ pushes further on, the fact that $y[x]$ is going up and the fact that $y'[x] = ay[x]$ means $y[x]$ goes up faster and faster as $x$ advances to $\infty$.

If $y[x_0] < 0$ for some $x_0$, then $y'[x_0] = ay[x_0] < 0$; so $y[x]$ is going **down** as $x$ goes through $x_0$. As $x$ pushes further on, the fact that $y[x]$ is going down and the fact that $y'[x] = ay[x]$ means $y[x]$ goes down faster and faster $x \to \infty$.

In short:

If $y[x]$ is ever positive, then $y[x]$ goes up faster and faster as $x \to \infty$.

If $y[x]$ is ever negative, then $y[x]$ goes down faster and faster as $x \to \infty$.

As a result of these two statements, we see in addition that $y[x]$ cannot change sign as $x$ varies.

■ **B.3) The differential equation y'[x] = a y[x] and how e helps us to get a formula for its solution.**

The differential equation

$$y'[x] = ay[x]$$

says no more or less than the statement:

*The growth rate of $y[x]$ is proportional to $y[x]$ itself.*

Another way to look at it is to divide both sides by $y[x]$ to see that

$$y'[x]/y[x] = a.$$

In other words the *relative instantaneous growth rate of $y[x]$ is constant.*

### B.3.a)

---

Given a fixed number $a$, then how can you use the number $e$ to get a formula for the solution $y[x]$ of the differential equation

$$y'[x] = ay[x]?$$

---

**Answer:**

Set:

```
In[1]:=    y[x_] = K E^(a x)

Out[1]=      a x
           E    K
```

Now look at:

```
In[2]:=    {y'[x], a y[x]}

Out[2]=      a x        a x
           {E    K a, E    K a}
```

They are the same! So we have just seen that for any constant $K$,

$$y[x] = Ke^{ax}$$

solves the differential equation $y'[x] = ay[x]$. The question comes up: Are there any other solutions? The answer is a resounding No! and here is why:

Given that $y'[x] = ay[x]$, we want to see why there is a constant $K$ such that $y[x] = Ke^{ax}$. To do this, we multiply $y[x]$ by $e^{-ax}$ and we hope the result

$$e^{-ax}y[x]$$

is a constant. To check whether it is constant, just diferentiate:

```
In[3]:=    Clear[y]
           D[E^(-a x) y[x],x]

Out[3]=        a y[x]     y'[x]
           -(------) + -----
              a x        a x
             E          E
```

Since $y'[x] = ay[x]$, the numerators are the same so

$$D[e^{-ax}y[x], x] = 0$$

as the following calculation confirms:

```
In[4]:=    D[E^(-a x) y[x],x]/.y'[x]->a y[x]
```

```
Out[4]=    0
```

This tells us that derivative of $e^{-ax}y[x]$ is always 0 no matter what $x$ is. The Race Track Principle steps in to tell us that $e^{-ax}y[x]$ is constant. Call the constant K and note that

$$e^{-ax}y[x] = K;$$

multiply through by $e^{ax}$ to learn

$$y[x] = Ke^{ax}.$$

So our mysterious solution $y[x]$ is nothing other than the solution we already knew about.

This shows that all solutions of $y'[x] = ay[x]$ have the formula $y[x] = Ke^{ax}$. That's all there is to it; bet you were expecting something much harder.

### B.3.b)

---

Now that you know that the solution $y[x]$ of the differential equation $y'[x] = 3y[x]$ is

$$y[x] = Ke^{3x},$$

what data do you need to determine what the constant $K$ is?

---

**Answer:**

You know

$$y[x] = Ke^{3x};$$

so you have one unknown (namely $K$) to determine. You need one data point on $y[x]$ to do this.

For instance if you know that $y[0] = 5$, then you can read off

$$5 = y[0] = Ke^0 = K.$$

This tells you $K = 5$ and gives the fully determined formula

$$y[x] = 5e^{3x}.$$

Or if you happen to know that $y[2] = 7$, then you can read off

$$7 = y[2] = Ke^6.$$

This gives

$$K = 7e^{-6}.$$

The formula for the corresponding solution is

$$y[x] = 7e^{-6}e^{3x} = 7e^{3x-6}.$$

### B.3.c)

---

Sometimes all you know in advance is that $y[x]$ grows at a rate in proportion to itself; in other words

$$y'[x] = ay[x]$$

for some constant of proportionality $a$. If the proportionality constant $a$ is not given in advance, then what data do you need to find a formula for $y[x]$?

**Answer:**

You know $y[x] = Ke^{ax}$, but this time you don't know the actual values of the constants $a$ or $K$. This means that you have two unknowns to determine, so you need two data points on $y[x]$ to do this.

For instance if you know that $y[1.2] = 5.4$ and $y[6.7] = 4.3$, then this leads to

$$5.4 = y[1.2] = Ke^{a1.2}$$

and

$$4.3 = y[6.7] = Ke^{a6.7}.$$

Taking natural logarithms on both sides of both equations gives

$$\log[5.4] = \log[Ke^{a1.2}] = \log[K] + \log[e^{a1.2}]$$
$$= \log[K] + a1.2$$

and

$$\log[4.3] = \log[Ke^{a6.7}] = \log[K] + a6.7.$$

Set these equations:

```
In[1]:=    eqn1 =
           Log[5.4] == Log[K] + a 1.2
Out[1]=    1.6864 == 1.2 a + Log[K]


In[2]:=    eqn2 =
           Log[4.3] == Log[K] + a 6.7
Out[2]=    1.45862 == 6.7 a + Log[K]
```

Solve for $\log[K]$ and $a$ :

```
In[3]:=    Solve[{eqn1,eqn2},{a, Log[K]}]
Out[3]=    {{a -> -0.0414153, Log[K] -> 1.7361}}
```

Nail down $K$ by remembering that $K = e^{\log[K]}$ to get the fully determined formula for $y[x]$ :

```
In[4]:=    y[x_] = K E^(a x)/.
           {a->-0.0414153,K->E^1.7361}
Out[4]=        5.67517
           -------------
           0.0414153 x
           E
```

Here is a plot:

```
In[5]:=    Plot[y[x],{x,0,100}];
```

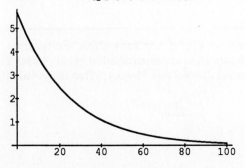

That wasn't very hard.

### ■ B.4) Getting a formula for the solution of the differential equation $y'[x] = ay[x] + b$.

#### B.4.a)

How can you use your knowlege of the formula

$$y[x] = Ke^{ax}$$

for the solution of the differential equation

$$y'[x] = ay[x]$$

to get a formula for the solution of the seemingly more complicated differential equation

$$y'[x] = ay[x] + b$$

where $a$ and $b$ are constants?

**Answer:**

The idea is to change the dependent variable $y[x]$ to reduce this case to the case that we already understand. Here it goes:

Write

$$y'[x] = ay[x] + b = a(y[x] + b/a).$$

Now put $z[x] = y[x] + b/a$ and note that $z'[x] = y'[x]$ because $b/a$ is a constant.

Consequently

$$z'[x] = y'[x] = a(y[x] + b/a) = az[x].$$

Dropping the middle two terms gives $z'[x] = az[x]$. This is great news because we can give a formula for $z[x]$. By what we saw in B.3) above,

$$z[x] = Ke^{ax}$$

for some constant $K$. Recalling that $z[x] = y[x] + b/a$, we see that

$$y[x] + b/a = z[x] = Ke^{ax}.$$

This gives us the formula

$$y[x] = Ke^{ax} - b/a.$$

#### B.4.b)

Given $y[x] = -2y[x] + 5$ and given that $y[1.3] = 7.9$, find a formula for $y[x]$ and plot.

**Answer:**

This is $y[x] = ay[x] + b$ with $a = -2$ and $b = 5$. The first shot at the formula for $y[x]$ is:

```
In[1]:=    prelimy[x_] =
           (K E^(a x) - b/a)/.{a->-2,b->5}
```

```
Out[1]=    5     K
           - + ----
           2    2 x
               E
```

To get the value of $K$, use $y[1.3] = 7.9$ and solve for $K$:

```
In[2]:=    Solve[prelimy[1.3] == 7.9,K]
Out[2]=    {{K -> 72.7042}}
```

This gives us the fully determined formula for $y[x]$:

```
In[3]:=    y[x_] = prelimy[x]/.K->72.7042
Out[3]=    5    72.7042
           - + -------
           2     2 x
                E
```

Here comes a plot:

```
In[4]:=    Plot[y[x],{x,0,4}];
```

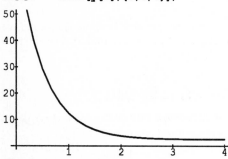

For large $x$, $y[x]$ is close to 5/2.

# Tutorial

### ■ T.1) Quick calculations.

### T.1.a)

Find a formula for $y[x]$ if it is known that the instantaneous growth rate $y'[x]$ is proportional to $y[x]$ and it is known that $y[1] = 2.7$ and $y[3.1] = 5.4$.

Notice that $y[x]$ doubles its value as $x$ advances from 1 to 3.1. Does the value of $y[x]$ double on intervals of length $(3.1 - 1) = 2.1$ years?

**Answer:**

Saying that $y'[x]$ is proportional to $y[x]$ means
$$y'[x] = ay[x]$$
for some constant $a$. According to B.3), this means
$$y[x] = Ke^{ax}$$
where $K$ is another constant. The data tell us that
$$2.7 = Ke^a$$
and
$$5.4 = Ke^{3.1a}.$$

Take logs on both sides these two equations to get
$$\log[2.7] = \log[K] + \log[e^a] = \log[K] + a$$
and
$$\log[5.4] = \log[K] + \log[e^{3.1a}.] = \log[K] + 3.1a.$$

Solve for $\log[K]$ and $a$:

```
In[1]:=    Solve[{Log[2.7] == Log[K] + a,
           Log[5.4] ==
           Log[K] + 3.1 a},{a,Log[K]}]
Out[1]=    {{a -> 0.33007, Log[K] -> 0.663182}}
```

The formula for $y[x]$ is:

```
In[2]:=    y[x_] = K E^(a x)/.
           {a -> 0.33007,K -> E^(0.663182)}
Out[2]=            0.33007 x
           1.94096 E
```

Here comes a plot:

```
In[3]:=    Plot[y[x],{x,0,30},
           AxesLabel->{"x","y"}];
```

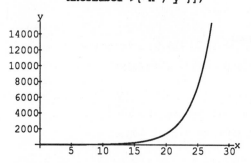

Look at that sucker grow.

To check about the doubling, look at the following table showing the values of $y[x]$ in $x$–increments of 2.1:

```
In[4]:=    Table[{2.1 x ,y[2.1 x]},{x,0,12}]
Out[4]=    {{0, 1.94096}, {2.1, 3.88192}, {4.2, 7.76383},
           {6.3, 15.5277}, {8.4, 31.0553}, {10.5, 62.1106},
           {12.6, 124.221}, {14.7, 248.442}, {16.8, 496.885},
           {18.9, 993.769}, {21., 1987.54}, {23.1, 3975.08},
           {25.2, 7950.15}}
```

Nifty. It sure looks as if $y[x]$ doubles on intervals of length 2.1. Can you see why?

### T.1.b.i)

Find a formula for $y[x]$ if it is known that the instantaneous growth rate $y'[x]$ is proportional to $y[x] - 4$ and it is known that $y[1.2] = 6.7, y[3.9] = 5.4$. What is the value of
$$\lim_{x \to \infty} y[x]?$$

**Answer:**

Saying that $y'[x]$ is proportional to $y[x] - 4$ means

$$y'[x] = a(y[x] - 4) = ay[x] - 4a$$

for some constant $a$.

According to B.4), the solution of the differential equation

$$y'[x] = ay[x] + b$$

is

$$y[x] = Ke^{ax} - b/a.$$

The differential equation we are studying is

$$y'[x] = ay[x] - 4a$$

Make the replacements $a \to a$ and $b \to -4a$ to get the formula for the solution $y[x]$ :

$$y[x] = Ke^{ax} + 4.$$

Now we have two constants $K$ and $a$ to determine. Happily, we have two data points ( $y[1.2] = 6.7, y[3.9] = 5.4$) ; so we can do it.

Rewrite

$$y[x] = Ke^{ax} + 4$$

as

$$y[x] - 4 = Ke^{ax}.$$

Hit both sides with the natural logarithm to get:

$$\log[y[x] - 4] = \log[Ke^{ax}]$$

$$= \log[K] + \log[e^{ax}] = \log[K] + ax.$$

Plug the data points ($y[1.2] = 6.7, y[3.9] = 5.4$) into this and solve for $a$ and $\log[K]$ :

```
In[1]:=    eq1 =
           (Log[y[x] - 4] == Log[K] + a x)/.
           {x->1.2,y[x]->6.7}
Out[1]=    0.993252 == 1.2 a + Log[K]
```

```
In[2]:=    eq2 =
           (Log[y[x] - 4] == Log[K] + a x)/.
           {x->3.9,y[x]->5.4}
Out[2]=    0.336472 == 3.9 a + Log[K]
```

```
In[3]:=    Solve[{eq1,eq2},{a, Log[K]}]
Out[3]=    {{a -> -0.243252, Log[K] -> 1.28515}}
```

This gives the formula for $y[x] = Ke^{ax} + 4$ :

```
In[4]:=    y[x_] = (K E^(a x) + 4)/.
           {a->-0.243252,K->E^(1.28515)}
Out[4]=           3.61521
           4 + -----------
                 0.243252 x
                E
```

Here comes a plot:

```
In[5]:=    Plot[y[x],{x,0,20}];
```

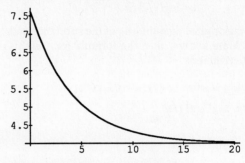

The value of $\lim_{x\to\infty} y[x]$ is 4.

■ **T.2) Radioactive decay and carbon dating.**

Radioactive decay is the process in which a substance disintegrates while its mass is converting to radiation. As pointed out in **B.1.a)** , the rate of decay of mass is proportional to the mass. This means that if $m[t]$ stands for mass of a radioactive substance at time $t$, then

$$m'[t] = am[t]$$

for some proportionality constant $a$. From **B.3)** above, we see that $m[t] = Ke^{at}$.

**T.2.a)**

How is the notion of half-life used to help to determine the constants?

**Answer:**

The constants $K$ and $a$ are determined as follows: Define

```
In[1]:=    m[t_] = K E^(a t)
Out[1]=      a t
           E    K
```

Set $t = 0$ and evaluate $m[0]$ :

```
In[2]:=    m[0]
Out[2]=    K
```

Thus $K = m[0]$ which we will write as $m_0$ and reset $m[t]$ :

```
In[3]:=    m[t_] = m0 E^(a t)
Out[3]=      a t
           E    m0
```

To determine the constant  , the notion of *half-life* comes in. The half-life of a substance is the time required for half the original mass to decay.

If the a substance has a known half-life  $h$, then

$$m_0/2 = m[h] = m_0 e^{ah}$$

and we can solve for  $a$  by cancelling the  $m_0$ 's and taking logs. We get:

$$\log[1/2] = \log[e^{ah}] = ah.$$

This gives

$$a = \log[1/2]/h.$$

Therefore if the original amount $m_0$ of the substance and the half-life $h$ are known, then the formula for $m[t]$ is completely determined:

```
In[4]:=    m[t_] = m0 E^(a t)/.a->Log[1/2]/h
Out[4]=    (t Log[1/2])/h
         E                 m0
```

## T.2.b)

What are the biological principles behind carbon dating?

**Answer:**

The biological principles say that living tissue contains two kinds of carbon. One is radioactive with half-life about 5500 years; the other is not radioactive. In living tissue the ratio of the two is always constant, but when the tissue dies, the radioactive carbon begins to decay while the other carbon remains.

To date a fossil, scientists measure the amount of each kind of carbon present to determine how much of the radioactive carbon has decayed. This gives them what they need to date the fossil.

For example, suppose measurements show that the radioactive carbon has decayed by 50 percent. Since the half-life is 5500 years, we conclude directly that the fossil is 5500 years old.

## T.2.c)

If measurements show that the radioactive carbon in a fossil has decayed by 30 percent, then how old is the fossil?

**Answer:**

Right off the bat (see part a), we know $m[t] = m_0 e^{tLog[1/2]/h}$ where $h = 5500$. Because the fossil has lost 30 percent of its radioactive carbon, we want to solve

$$(.70)m_0 = m[t] = m_0 e^{tLog[.5]/5500}$$

for $t$.

To do this, cancel the $m_0$ 's from both sides to get

$$.70 = e^{tLog[.5]/5500}.$$

Take the natural log of both sides and solve for $t$:

```
In[1]:=    Solve[Log[.70] == t Log[.5]/5500,t]
Out[1]=    {{t -> 2830.15}}
```

If the measurements are accurate, then the fossil is about 2830 years old.

## ■ T.3) Savings Accounts.

You deposit $P$ dollars in a savings account bearing interest rate $100r$ percent per year compounded continuously. This means that if $P[t]$ is the value in dollars of this untouched savings account at time $t$ years after the deposit, then

$$P[t] = Pe^{rt}.$$

## T.3.a)

What differential equation does $P[t]$ satisfy?

**Answer:**

$$P'[t] = Pe^{rt}r = P[t]r = rP[t],$$

so $P[t]$ solves the differential equation

$$P'[t] = rP[t].$$

## T.3.b)

At the time of your original deposit, you arrange with your bank to make a continuous deposit into your account at a rate of $d$ dollars per year. This means that after $t$ years, $td$ additional dollars have been deposited. As the new money is deposited into the account, the new money begins to earn interest at the rate of $100r$ percent compounded continuously.

If $P[t]$ is the amount in the account $t$ years after the arrangement was set up, then what differential equation does $P[t]$ satisfy?

**Answer:**

$$P'[t] = rP[t] + d.$$

(The derivative of the sum is the sum of the derivatives; the $rP[t]$ term comes from interest and the $d$ term comes from your continuous deposit.)

## T.3.c)

Solve this differential equation for $P[t]$. Plot $P[t]$ over the first twenty years for the case in which

$$P = \$2000,$$
$$d = \$480$$
$$100r = 6$$

percent. On the same plot, show how much of $P[t]$ is in the account via interest payments.

**Answer:**

Start with the answer to part b:
$$P'[t] = r P[t] + d.$$

According to B.4) above, the formula for $y[x]$ given
$$y'[x] = a y[x] + b$$
is
$$y[x] = K e^{ax} - b/a.$$

Make the replacements:
$$y[x] \to P[t], a \to r; b \to d$$

to get the formula
$$P[t] = K e^{rt} - d/r$$

where $K$ is a constant to be determined.

```
In[1]:=    PrelimP[t_] = K E^(r t) - (d/r)

Out[1]=      r t     d
          E    K - -
                   r
```

To solve for $K$, remember that the initial deposit = $P[0]$. Plug in $t = 0$ to get $K$:

```
In[2]:=    Ksolved =
           Solve[ Origdeposit == PrelimP[0] ,K]

Out[2]=            d + Origdeposit r
          {{K -> ------------------}}
                         r
```

So the final version of $P[t]$ is:

```
In[3]:=    P[t_] := PrelimP[t]/.Ksolved[[1]]
```

Here it is advantageous to use the  `:=` form for defining `P[t_]`. This allows you to change the parameters at will later on without having to go to the trouble of redefining `P[t_]` each time you change the constants.

The part of $P[t]$ coming from interest is:

```
In[4]:=    Interest[t_] := P[t] - Origdeposit - t d
```

Here comes a plot for the special case of an initial deposit = \$2000, $d$ = \$480 per year (only 40 dollars per month) and a modest interest rate of $100 r = 6$ percent:

```
In[5]:=    Origdeposit = 2000
           d = 480
           r = .06
           Plot[{P[t],Interest[t]},{t,0,30},
           AxesLabel->{years,$}];
```

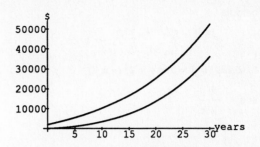

The lower curve is the interest curve. After a feeble start, the accrued interest eventually represents the biggest part of your nest egg. You may be interested in changing the parameters $r, d$ and your original deposit to see what develops in other situations. Do it.

■ **T.4) Restricted growth: Making the differential equation to suit the situation.**

**T.4.a)**

The extension agent tells you that your pond is capable of supporting 500 perch. You stock the lake with 72 perch and after three months the extension agent comes over and estimates that there are 200 perch in the pond. Find a function that gives a reasonable estimate of the number of perch in the pond $t$ months after the original stocking. Plot.

**Answer:**

Let $P[t]$ be the number of perch in the pond $t$ months after stocking it. When $P[t]$ is well beneath the limiting size of 500, it is reasonable to assume that $P[t]$ grows in direct proportion to its size. Thus for $P[t]$ well below 500,
$$P'[t] = a P[t]$$

is a plausible assumption, but it can't be taken too seriously because its solution $P[t] = 72 e^{at}$ allows for an unlimited population explosion.

Back to the drawing board: We want a model in which the growth rate $P'[t]$ is extremely small when $P[t]$ is close to 500 and in which the growth rate is rather generous when $P[t]$ is well below 500.

Here is one such:
$$P'[t] = a(500 - P[t]).$$

This forces the growth rate $P'[t]$ to tend to 0 as $P[t]$ sneaks up on 500. To try out this model, we'll solve and plot.

According to B.4) above, the formula for $y[x]$ given
$$y'[x] = a y[x] + b$$
is
$$y[x] = K e^{ax} - b/a.$$

Make the replacements
$$y[x] \to P[t], a \to -a, b \to 500a$$

to get the formula

$$P[t] = Ke^{-at} + 500.$$

```
In[1]:=    PrelimP[t_] = 500 + K E^(- a t)
Out[1]=         K
           500 + ----
                  a t
                 E
```

Next use the known information to determine the constants K and a. Because we have **two** constants to determine, we need two data points. We know that $P[0] = 72$ and $P[3] = 200$, so we are in luck.

```
In[2]:=    eqn1 = (72 == PrelimP[0])
Out[2]=    72 == 500 + K
```

```
In[3]:=    eqn2 = (200 == PrelimP[3])
Out[3]=                   K
           200 == 500 + ----
                          3 a
                         E
```

Temporarily let $z = 1/e^{3a}$.

```
In[4]:=    neweqn2 = eqn2/. (1/E^(3 a))->z
Out[4]=    200 == 500 + K z
```

```
In[5]:=    Solve[{eqn1,neweqn2},{K, z}]
Out[5]=         75
           {{z -> ---, K -> -428}}
                 107
```

To get $a$, recall $z = 1/e^{3a}$. This is the same as

$$\log[z] = \log[1/e^{3a}] = -3a.$$

Here comes $a$ :

```
In[6]:=    N[Solve[ - 3 a == Log[z]/.z->75/107,a]]
Out[6]=    {{a -> 0.118447}}
```

So $P[t]$ is given by:

```
In[7]:=    P[t_] = PrelimP[t]/.{a->0.118,K->-428}
Out[7]=             428
           500 - --------
                   0.118 t
                  E
```

Here is a plot of the projected fish population over 3 years. The pond capacity of 500 perch is plotted as the line $y = 500$ :

```
In[8]:=    fishplot = Plot[{P[t],500},{t,0,36},
           AxesLabel->{months,perch},
           PlotRange->All];
```

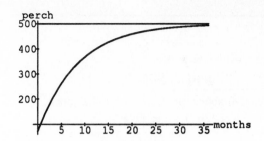

Let's include the original data.

```
In[9]:=    datapoints = ListPlot[{{0,72},{3,200}},
           PlotStyle->{RGBColor[1,0,0],
           PointSize[.02]},
           DisplayFunction->Identity];

           Show[datapoints,fishplot,
           DisplayFunction->$DisplayFunction];
```

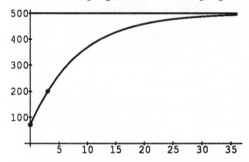

The plot looks pretty good.

A moment of reflection is in order. We have a nice plausible answer to the original problem, but there is no guarantee that this answer is right on the target because another reasonable differential equation that might describe this set up is

$$P'[t] = aP[t](500 - P[t]).$$

We can't handle this differential equation now, but stick around for a while because we'll get to it later.

■ **T.5) Pollution elimination.**

An artesian well supplies new pure water to a 2000 gallon cistern at a rate of 10 gallons per minute. It mixes with the old water and then overflows through a relief tube at a rate of 10 gallons per minute. At a certain instant a complete air-head spills 50 gallons of a polluting liquid into the cistern and the polluting liquid mixes throughly with the water intstantly. As new water flows in and old flows out, the polluting liquid is continually diluted. Let $P[t]$ be the number of gallons of the pollutant in the cistern at time $t$ after the spill.

We want to come up with a formula for $P[t]$. The strategy is to find a differential equation that $P[t]$ satisfies, to make some measurements to set the constants and then come up with a complete formula for $P[t]$. This strategy is very common in making scientific measurements.

**T.5.a)**

Find a differential equation that $P[t]$ satisfies, solve it and obtain a reasonably accurate estimate of the time it will take for only 1/10 gallon of the pollutant to remain in the cistern.

---

**Answer:**

10 gallons of the mixed pollutant and fresh water leaves per minute. If $h$ is small and positive, then from time $t$ to time $t + h$ (in minutes) the amount of pollutant $P[t] - P[t + h]$ leaving satisfies

$$10/2000\,P[t + h]h \le P[t] - P[t + h] \le (10/2000)P[t]h.$$

Therefore

$$(10/2000)P[t + h] \le (P[t] - P[t + h])/h \le (10/2000)P[t].$$

Watching what happens as $h$ approaches 0, we see that

$$P'[t] = \lim_{h \to 0}(P[t + h] - P[t])/h$$

has no choice but to be given by

$$P'[t] = (-10/2000)P[t].$$

This differential equation has solution

$$P[t] = Ke^{-10t/2000}$$

where $K$ is a constant.

To nail down $K$, recall that

$$50 = P[0] = Ke^0 = K.$$

Thus

$$P[t] = 50e^{-10t/2000}.$$

To find the $t$ for which $P[t] = 1/10$, just write $1/10 = 50e^{-10t/2000}$, take the natural log of both sides

$$\log[.1] = \log[50e^{-10t/2000}]$$

$$= \log[50] + \log[e^{-10t/2000}] = \log[50] - 10t/2000$$

and solve for $t$ :

```
In[1]:=    N[Solve[Log[.1] ==
           Log[50] - 10 t/2000,t]]

Out[1]=    {{t -> 1242.92}}
```

It will take about 1243 minutes or

```
In[2]:=    N[1243/60]
Out[2]=    20.7167
```

hours for the pollutant to be reduced to 1/10 gallon in solution.

## Literacy Sheet

**1.** You know that $y'[x] = ay[x]$ for all $x$'s but do not know the value of a. How many data points of the form $\{x, y[x]\}$ do you need to determine a formula for $y[x]$?

.

.

.

**2.** You know that $y'[x] = ay[x]$ for all $x$'s and you do know the value of $a$. How many data points of the form $\{x, y[x]\}$ do you need to determine a formula for $y[x]$?

.

.

.

**3.** You know that $y'[x] = -y[x]$ for all $x$'s and you know that a is negative but $y[x]$ is always positive. Does $y[x]$ go up or down as $x$ advances from left to right?

.

.

.

**4.** You know that $y'[x] = ay[x]$ for all $x$'s and you know that $y[0] = 5$. Can there be an $x$ such that $y[x]$ is negative?

.

.

.

**5.** You know that $y'[x] = ay[x]$ for all $x$'s and you know that $a$ is positive. What is the value of

$$\lim_{x \to \infty} y[x]/x^{10}?$$

.

.

.

**6.** You know that $y'[x] = ay[x]$ for all $x$'s and you know that $y[x]$ is positive for all $x's$. How does the $\log[y[x]]$ curve plot out?

.

.

.

**7.** You know that $y'[x] = ay[x]$ for all $x's$ and you know that $y[10] = 0$. Why does this tell you that $y[x] = 0$ for all $x's$?

.

.

.

**8.** You know that $y'[x] = ay[x] + b$ for all $x's$ but do not know the value of either a or b. How many data points of the form $\{x, y[x]\}$ do you need to determine a formula for $y[x]$?

.

.

.

**9.** You know that $y'[t]$ is proportional to $y[t]$ and you know that $y[0] = 5$ and $y[1] = 8$. Give a formula for $y[t]$.

.

.

.

**10.** You know that $y'[t] = ay[t] + b$ and $y[0] = 6.5$. Outline a procedure for finding a formula for $y[t]$.

.

.

**11.** You know that $y[t] = 83e^{.017t}$. What differential equation does $y[t]$ satisfy?

.

.

.

**12.** You know that your untouched bank account is accruing interest compounded continuously. Why does this tell you that the intantaneous growth rate of the account is proportional to the amount in the account?

.

.

**13.** Suppose $y[x] = Ke^{ax}$ and both $K$ and are positive. Find $L$ such that $y[x]$ doubles on every interval of length $L$.

.

.

**14.** Glucose in the blood is converted and excreted at a rate proportional to the present concentration of the glucose in the blood. If $C[t]$ is the concentration of glucose in the blood at time $t$, what differential equation does $C[t]$ satisfy? Give a formula for $C[t]$. What data do you need to set the constants in your formula for $C[t]$?

If in addition, the concentration of glucose in the blood is increased at a constant rate, then what differential equation does $C[t]$ satisfy? Give a formula for $C[t]$. What data do you need to set the constants in your formula for $C[t]$?

.

**15.** Here are five plots:

They are plots of:

(a) a solution of $y'[t] = ay[t]$ with $a > 0$.

(b) a solution of $y'[t] = ay[t]$ with $a < 0$.

(c) a solution of $y'[x] = ay[x] + b$.

(d) the logarithm of the solution to one of the above.

(e) none of the above.

Which is which?

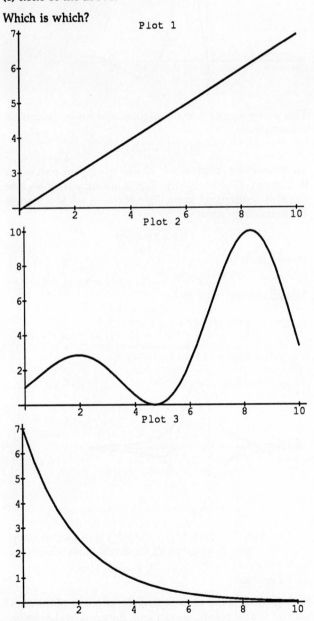

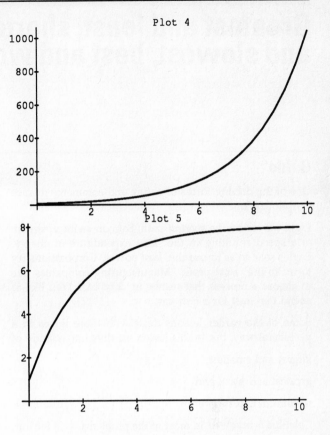

# Greatest and least, shortest and tallest, fastest and slowest, best and worst, . . .

## Guide

One of the driving forces of nature and commerce finance is optimization.

Light moves on the *shortest* path; Salmon swim upstream at a speed resulting in the *least* expenditure of energy. Eagles soar so as to use the *least* power. Corporations try to make the *most* profit. Manufacturing companies like to choose a process that results in *least* cost. You like to spend the *least* for a new car.

Some of the earlier lessons dealt with these issues in a peripheral way, but in this lesson we dive into a study of

largest and smallest,

greatest and least, and

maximum and minimum.

Calculus is involved in most of the problems we'll look at, but some can be done with just common sense.

## Basics

There are no new basic ideas in this lesson; most of the lesson is devoted to polishing ideas that have already emerged. For this reason, there are no Basic problems presented in this lesson. As compensation, the Tutorial section is full of beef.

## Tutorial

■ **T.1) Highest points on the graph.**

**T.1.a)**

Find the *highest* and *lowest* points on the graph of the function
$$y = \sin[x]/(3 - 2x + x^2).$$

**Answer:**

Size up the situation:

```
In[1]:=    plot = Plot[Sin[x]/(3 - 2 x + x^2),
           {x,-6,8},PlotRange->All];
```

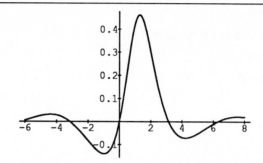

This plot suggests that the highest and lowest points on the graph of
$$\sin[x]/(3 - 2x + x^2)$$

are somewhere between $x = -2$ and $x = 2$. The plot seems to dampen off quite rapidly. This is not surprising because $\sin[x]$ is condemned to oscillate between $-1$ and $1$ while the remaining factor
$$1/(3 - 2x + x^2)$$

shrinks to 0.

Let's look at $1/(3-2x+x^2)$, $-1/(3-2x+x^2)$ and $\sin[x]/(3-2x+x^2)$ on the same plot :

```
In[2]:=    Plot[{1/(3 - 2 x + x^2),
           Sin[x]/(3 - 2 x + x^2),-1/(3 - 2 x + x^2)},
           {x,-8,10},AxesLabel->{"x","y"},
           Ticks->{{-5,5},{-0.2,0.2}}];
```

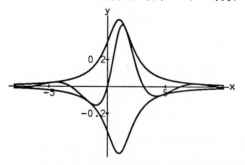

Sure enough. The factor $1/(3-2x+x^2)$ is in control except near the origin. It squashes all the action in the $\sin[x]/(3-2x + x^2)$ curve except near the origin. This confirms our initial reaction.

Let's go back to the plot. To find the locations of the highest and lowest points, we can look for zeros of the derivative:

```
In[3]:=    deriv =
           Together[D[Sin[x]/(3 - 2 x + x^2),x]]
```

118

*Out[3]=*
$$(3 \text{ Cos}[x] - 2 \text{ x Cos}[x] + x^2 \text{ Cos}[x] +$$
$$2 \text{ Sin}[x] - 2 \text{ x Sin}[x]) \, /$$
$$(3 - 2 \text{ x} + x^2)^2$$

For this to be zero we must have a zero numerator:

*In[4]:=*   `numderiv = Numerator[deriv]`

*Out[4]=*
$$3 \text{ Cos}[x] - 2 \text{ x Cos}[x] + x^2 \text{ Cos}[x] +$$
$$2 \text{ Sin}[x] - 2 \text{ x Sin}[x]$$

The plot tells us that the highest point is somehere close to $x = 1.2$.

*In[5]:=*   `highx = FindRoot[numderiv == 0, {x,1.2}]`

*Out[5]=*   `{x -> 1.29516}`

The highest point is:

*In[6]:=*   `{1.29516, Sin[x]/(3 - 2 x + x^2)`
`/.x->1.29516}`

*Out[6]=*   `{1.29516, 0.461043}`

The plot tells us that the lowest point is somewhere close to $x = -1$.

*In[7]:=*   `lowx = FindRoot[numderiv == 0 , {x,-1}]`

*Out[7]=*   `{x -> -0.981371}`

The lowest point is:

*In[8]:=*   `{-0.981371, Sin[x]/(3 - 2 x + x^2)`
`/.x->-0.981371}`

*Out[8]=*   `{-0.981371, -0.140277}`

Not too bad.

### T.1.b)

Find the *highest* point on the graph of

$$y = (10x - 3)\big/(3x^2 - 5x + 8).$$

### Answer:

The global scale behavior of

$$(10x - 3)\big/(3x^2 - 5x + 8).$$

is that of $10/(3x)$. The graph should quickly begin to hug the $x$−axis.

*In[1]:=*   `y = (10 x - 3) / (3 x^2 - 5 x + 8)`

*Out[1]=*
$$\frac{-3 + 10 \text{ x}}{8 - 5 \text{ x} + 3 \text{ x}^2}$$

*In[2]:=*   `Plot[y,{x,-8,8}];`

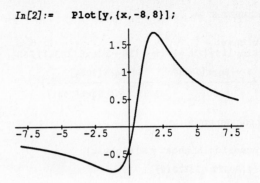

Let's see more:

*In[3]:=*   `Plot[y,{x,-50,50}];`

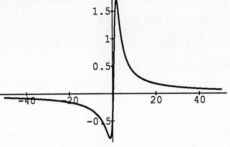

The global scale behavior has kicked in. From the first plot, we see that the highest point is somewhere to the right of $x = 2.5$ and the lowest point is just to the left of $x = 0$.

Because we are dealing with a quotient of low degree polynomials, the **Solve** instruction will tell us all we want to know:

*In[4]:=*   `points = Solve[D[y,x]==0,x]`

*Out[4]=*
$$\left\{\left\{x \to \frac{18 + 2 \text{ Sqrt}[2031]}{60}\right\},\right.$$
$$\left.\left\{x \to \frac{18 - 2 \text{ Sqrt}[2031]}{60}\right\}\right\}$$

Great! Wild looking numbers, but we've got them both.

The smaller must be the lowest point:

*In[5]:=*   `lowest =`
`Simplify[{x,y}/.x->(18-2 Sqrt[2031])/60]`

*Out[5]=*
$$\left\{\frac{18 - 2 \text{ Sqrt}[2031]}{60},\right.$$
$$\left.\frac{-50 \text{ Sqrt}[2031]}{2031 + 16 \text{ Sqrt}[2031]}\right\}$$

Or in more convenient form:

*In[6]:=*   `numericallowest = N[lowest]`

*Out[6]=*   `{-1.20222, -0.818778}`

The highest point is:

```
In[7]:=   highest =
          Simplify[{x,y}/.x->(18 + 2 Sqrt[2031])/60]

Out[7]=     9 + Sqrt[2031]      50 Sqrt[2031]
          {--------------, ---------------------}
                30            2031 - 16 Sqrt[2031]
```

Or in more convenient form:

```
In[8]:=   numericalhighest = N[highest]

Out[8]=   {1.80222, 1.72019}
```

Another problem bites the dust.

### T.1.c)

---

Find the *highest* point on the graph of

$$f[x] = -577 + 736x - 324x^2 + 60x^3 - 4x^4.$$

Is there a *lowest* point on this graph?

---

### Answer:

```
In[1]:=   f[x_] = -577 + 736 x - 324 x^2 + 60 x^3 - 4 x^4

Out[1]=                        2        3      4
          -577 + 736 x - 324 x  + 60 x  - 4 x
```

The global scale behavior is negative because the dominant fourth degree term carries a negative coefficient and is in control when the absolute value of $x$ is large. For this reason, there is no lowest point on the graph.

To find the highest point, take a look:

```
In[2]:=   Plot[f[x],{x,-8,8}];
```

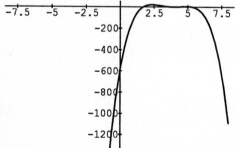

We had better look at what's happening between 1 and 6:

```
In[3]:=   Plot[f[x],{x,1,6},PlotRange->{-1,25}];
```

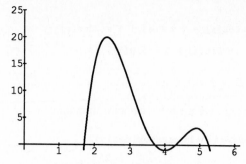

The highest point sits near $x = 2.3$.

```
In[4]:=   xhigh = FindRoot[f'[x],{x,2.3}]

Out[4]=   {x -> 2.34413}
```

The highest point is:

```
In[5]:=   {x,f[x]}/.x->2.34413

Out[5]=   {2.34413, 19.9916}
```

Because $f'[x]$ is a third degree polynomial, the exact answer is available to us via:

```
In[6]:=   exact = Solve[f'[x] == 0,x]

Out[6]=                            29 + Sqrt[105]
          {{x -> 4}, {x -> ---------------},
                                  8

                  29 - Sqrt[105]
            {x -> ---------------}}
                        8
```

To see which solution we want, take numerical values:

```
In[7]:=   nexact = N[exact]

Out[7]=   {{x -> 4.}, {x -> 4.90587},

            {x -> 2.34413}}
```

The exact coordinates of the highest point are:

```
In[8]:=   highexact =
          {(29 - Sqrt[105])/8, f[(29 - Sqrt[105])/8]}

Out[8]=    29 - Sqrt[105]
          {---------------,
                  8

            -577 + 92 (29 - Sqrt[105]) -

                                 2
             81 (29 - Sqrt[105])
             -------------------- +
                     16

                                 3
             15 (29 - Sqrt[105])
             -------------------- -
                     128

                               4
             (29 - Sqrt[105])
             -----------------}
                   1024
```

Or in more tractable form:

*In[9]:=*    **Expand[highexact]**

*Out[9]=*    {$\frac{29}{8}$ - $\frac{\text{Sqrt}[105]}{8}$, $\frac{1483}{128}$ + $\frac{105\ \text{Sqrt}[105]}{128}$}

The exact specification is pleasing, but from a practical point of view, both solutions are equally good.

### T.1.d)

Find the *highest and lowest* points on the graph of

$$f[x] = e^{-x^2}(x^8 + 8x^5 + 16x + 2)$$

### Answer:

The decay of $e^{-x^2}$ is so devastatingly rapid that it will quickly squash any action that $x^8 + 8x^5 + 16x + 2$ might have in mind.

*In[1]:=*    **f[x_] = E^(-x^2) ( x^8 + 8 x^5 + 16 x + 2)**
          **Plot[f[x],{x,-7,7},PlotRange->All];**

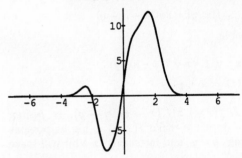

Interesting plot. The $e^{-x^2}$ factor has completely wiped out all trace of the polynomial factor $x^8 + 8x^5 + 16x + 2$ to the right of $x = 4$ and to the left of $x = -4$.

The highest point sits just to the left of $x = 2$.

*In[2]:=*    **xhigh = FindRoot[f'[x],{x,2}]**

*Out[2]=*    **{x -> 1.59545}**

The highest point is:

*In[3]:=*    **{1.59545, f[1.59545]}**

*Out[3]=*    **{1.59545, 11.9388}**

The lowest point sits just to the left of $x = -1$.

*In[4]:=*    **xlow = FindRoot[f'[x],{x,-1}]**

*Out[4]=*    **{x -> -1.11425}**

The lowest point is:

*In[5]:=*    **{-1.11425, f[-1.11425]}**

*Out[5]=*    **{-1.11425, -7.85685}**

That's all folks.

■ **T.2) The largest number in an infinite list.**

### T.2.a)

Find the positive integer $n$ for which the expression

$$(8n^2 - 5n + 1)/(n^4 + n^2 + 1)$$

is the *largest*.   Is there an integer $n$ for which the value of this expression is the *smallest* ?

### Answer:

*In[1]:=*    **f[n_] = (8 n^2 - 5 n - 2)/(n^4 + n^2 + 1)**

*Out[1]=*    $\dfrac{-2 - 5\ n + 8\ n^2}{1 + n^2 + n^4}$

Look at some of the values:

*In[2]:=*    **values = Table[{n,f[n]},{n,1,20}]**

*Out[2]=*    {{1, $\frac{1}{3}$}, {2, $\frac{20}{21}$}, {3, $\frac{55}{91}$}, {4, $\frac{106}{273}$},

          {5, $\frac{173}{651}$}, {6, $\frac{256}{1333}$}, {7, $\frac{355}{2451}$},

          {8, $\frac{470}{4161}$}, {9, $\frac{601}{6643}$}, {10, $\frac{748}{10101}$},

          {11, $\frac{911}{14763}$}, {12, $\frac{1090}{20881}$},

          {13, $\frac{1285}{28731}$}, {14, $\frac{1496}{38613}$},

          {15, $\frac{1723}{50851}$}, {16, $\frac{1966}{65793}$},

          {17, $\frac{2225}{83811}$}, {18, $\frac{2500}{105301}$},

          {19, $\frac{2791}{130683}$}, {20, $\frac{3098}{160401}$}}

Plot:

*In[3]:=*    **plot = ListPlot[values,**
          **PlotStyle->**
          **{PointSize[.02],RGBColor[0,0,1]},**
          **Axes->{0,0},**
          **Ticks->{Range[2,18,4],Automatic}];**

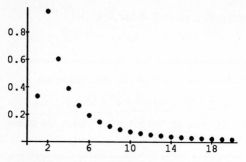

It appears that the largest value is attained at $n = 2$ and there seems to be no smallest value since we see that $f[n]$ continues to decrease towards 0 without ever reaching 0. To verify that $f[n]$ decreases as suspected, look at the derivative:

```
In[4]:=    deriv = Together[f'[n]]
Out[4]=
                       2      3      4
           (-5 + 20 n + 5 n  + 8 n  + 15 n  -

              5       2    4 2
           16 n ) / (1 + n  + n )
```

The denominator is always positive, so to see how $f[n]$ behaves, we look at the numerator:

```
In[5]:=    num[n_] = Numerator[deriv]
Out[5]=
                     2      3      4       5
           -5 + 20 n + 5 n  + 8 n  + 15 n  - 16 n
```

The global scale version of the numerator as $n \to \infty$ is negative.

Let's see whether the global scale has kicked in for $n = 20$:

```
In[6]:=    num[20]
Out[6]=    -48733605
```

```
In[7]:=    Apart[num[n]/(n - 20)]
Out[7]=
                       48733605
           -2436680 - --------- - 121835 n -
                       -20 + n

                 2        3        4
           6092 n  - 305 n  - 16 n
```

All minuses. This tells us that the negative global scale of $f'[n]$ has kicked in for $n \geq 20$.

Thus $f[n] < f[20]$ for $n > 20$. So the largest possible $f[n]$ shows up in the plot. The largest entry is clearly:

```
In[8]:=    f[2]
Out[8]=    20
           --
           21
```

There is no smallest entry because as n advances from 20 to $\infty$, the values $f[n]$ decrease towards the eventual limit:

```
In[9]:=    Limit[f[n],n->Infinity]
```

```
Out[9]=    0
```

■ **T.3) The largest product.**

**T.3.a)**

Write the number 11 as the sum of three positive **numbers** and multiply them together. What is the *largest* product you can get?

---

**Answer:**

We have
$$x + y + z = 11$$
and we want to make
$$xyz$$
as large as we can.

To this end, note that
$$z = 11 - x - y.$$

So making $xyz$ as large as we can is the same as making
$$f[x, y] = xy(11 - x - y)$$
as big as possible.

```
In[1]:=    f[x_,y_] = x y (11 - x - y)
Out[1]=    x (11 - x - y) y
```

Note that if $D[f[x,y], x]/.\{x \to x_0, y \to y\}_0$ is *positive* then $\{x_0, y\}_0$ does not make $f[x, y]$ as big as possible because holding $y = y_0$ and increasing $x$ a bit will make $f[x, y]$ go *up*.

Also note that if $D[f[x,y], x]/.\{x \to x_0, y \to y\}_0$ is *negative* then $\{x_0, y\}_0$ does not make $f[x, y]$ as big as possible because holding $y = y_0$ and decreasing $x$ a bit will make $f[x, y]$ go *up*.

Therefore if $\{x_0, y\}_0$ makes $f[x, y]$ as large as possible, then
$$D[f[x, y], x]/.\{x \to x_0, y \to y\}_0 = 0$$
and
$$D[f[x, y], y]/.\{x \to x_0, y \to y\}_0 = 0.$$

Consequently the $x_0$ and the $y_0$ we are looking for are contained in the output from:

```
In[2]:=    bestxandy = Solve[{D[f[x,y],x] == 0,
           D[f[x,y],y] == 0},{x,y}]
Out[2]=
                     11        11
           {{x -> --,  y -> --}, {x -> 0, y -> 11},
                     3         3

             {x -> 11, y -> 0}, {x -> 0, y -> 0}}
```

Try them out:

```
In[3]:=    {f[11/3,11/3],f[0,11],f[11,0],f[0,0]}
```

```
Out[3]=    1331
       {----, 0, 0, 0}
         27
```

The largest is:

```
In[4]:=    f[11/3,11/3]
Out[4]=    1331
           ----
            27
```

This is the largest value of $xyz$ provided $x + y + z = 11$. The $x, y$ and $z$ that do the job are:

$$x = 11/3, y = 11/3,$$

and $z$ is :

```
In[5]:=    (11 - x - y)/.{x->11/3,y->11/3}
Out[5]=    11
           --
           3
```

### T.3.b)

Write the number 11 as the sum of three positive **integers** and multiply them together. What is the *largest* product you can get?

### Answer:

Let's dip our big toe into the water. For

$$11 = 2 + 4 + 5.$$

the product is

```
In[1]:=    2 4 5
Out[1]=    40
```

The question is whether there is a larger product. The main observations are:

*(i)* If $11 = x + y + z$, then $z$ must be be given by $z = 11 - x - y$.

*(ii)* The number $x$ cannot be bigger than 9 because if $x$ is larger than 9, then there would be no room for $y$ and $z$. Hence we simply calculate

$$xyz = xy(11 - x - y)$$

for all choices of $x$ and $y$ in the appropriate ranges.

Also note that once $x$ is nailed down, then $y$ can't be more than 10 - x.

```
In[2]:=    products =
           Table[{x,y,x y (11 -x - y)},
           {x,1,9},{y,1,10 -x}]
Out[2]=    {{{1, 1, 9}, {1, 2, 16}, {1, 3, 21},
             {1, 4, 24}, {1, 5, 25}, {1, 6, 24},
             {1, 7, 21}, {1, 8, 16}, {1, 9, 9}},
```

```
{{2, 1, 16}, {2, 2, 28}, {2, 3, 36},
 {2, 4, 40}, {2, 5, 40}, {2, 6, 36},
 {2, 7, 28}, {2, 8, 16}},
{{3, 1, 21}, {3, 2, 36}, {3, 3, 45},
 {3, 4, 48}, {3, 5, 45}, {3, 6, 36},
 {3, 7, 21}}, {{4, 1, 24},
 {4, 2, 40}, {4, 3, 48}, {4, 4, 48},
 {4, 5, 40}, {4, 6, 24}},
{{5, 1, 25}, {5, 2, 40}, {5, 3, 45},
 {5, 4, 40}, {5, 5, 25}},
{{6, 1, 24}, {6, 2, 36}, {6, 3, 36},
 {6, 4, 24}}, {{7, 1, 21},
 {7, 2, 28}, {7, 3, 21}},
{{8, 1, 16}, {8, 2, 16}}, {{9, 1, 9}}}
```

The largest product is 48. It results from the choice $x = 3$, $y = 4$, $z = 11 - x - y = 4$ or any shuffling of these values ($x = 4, y = 3, z = 4$ or $x = 4, y = 4, z = 3$).

For lists like this, we can use **Max** to extract the largest element without reading the whole list:

```
In[3]:=    Max[Table[x y (11-x-y),
           {x,1,9},{y,1,10-x}]]
Out[3]=    48
```

### ■ T.4) Fish gotta swim: The least energy.

This problem was adapted from E. Batschelet, *Introduction to Mathematics for Life Scientists,* Springer, New York, 1979. This book is highly recommended for supplementary reading.

A river near the Puget Sound is flowing at a constant speed $v_r$ mph relative to the river bank. A salmon swims upstream at a constant speed $v_f$ mph relative to the water, intent on reaching a spawning point $s$ miles up the river.

### T.4.a.i)

Given that $v_f > v_r$, what is the time $t$ needed for the salmon to complete the journey?

### Answer:

The time $t$ is given by $t = s/(v_f - v_r)$

### T.4.a.ii)

The energy the salmon must expend to maintain its journey is determined by friction in the water and by the time $t$ necessary to reach the spawning point. Fish biologists measuring this energy have come to the conclusion that this energy $E$ is jointly proportional to the time $t$ needed to complete the journey and to $(v_f)^k$ for some empirical constant $k > 2$.

Find an expression for $E$ in terms of $t$ and $v_f$.

---

**Answer:**

Because $E$ is jointly proportional to the time $t$ needed to complete the journey and to $(v_f)^k$ for some empirical constant $k$, we see that $E = cv_f^k t$ where $c$ is a positive constant of proportionality.

### T.4.a.iii)

Calculate the speed $v_f$ that makes the salmon's trip possible with the least energy $E$.

---

**Answer:**

```
In[1]:=    energy = c vf^k t
Out[1]=          k
            c t vf
```

We know (part i) that $t = s/(v_f - v_r)$. So in terms of $v_f$, we see that:

```
In[2]:=    energy = energy/.t-> (s/(vf - vr))
Out[2]=          k
            c s vf
            -------
            vf - vr
```

Now differentiate with respect to $v_f$ and factor:

```
In[3]:=    Factor[D[energy,vf]]
Out[3]=          -1 + k
            c s vf        (-vf + k vf - k vr)
            ---------------------------------
                                2
                        (vf - vr)
```

Remembering that $c$ and $s$ are positive and noticing that the denominator cannot be negative, we see that the sign of the derivative is the same as the sign of:

$(-v_f + kv_f - kv_r) = (vf(k-1) - kv_r) = (k-1)(v_f - (k/(k-1))v_r)$.

Accordingly because $k > 2$, $D[\text{energy}, v_f]$ is *negative* for $v_f < (k/(k-1))v_r$ and $D[\text{energy}, v_f]$ is *positive* for $v_f > (k/(k-1))v_r$.

This means that as a function of $v_f$, the energy *decreases* as $v_f$ advances from $v_r$ to $(k/(k-1))v_r$ and the energy *increases* as $v_f$ advances from $(k/(k-1))v_r$ to $\infty$.

Therefore the *least enegy* velocity is $v_f = (k/(k-1))v_r$.

Note that the optimal speed $v_f$ of the fish does not depend on the distance $s$.

■ **T.5) Largest and smallest area.**

If a one quart can of fruit punch has least possible surface area, then what is the ratio of the its height to the diameter of its base?

---

**Answer:**

Let

$V$ = volume of a one quart can

$h$ = its height

$r$ = radius of its base.

Then $V = \pi r^2 h$. Also: area of the top = area of the bottom $= \pi r^2$. The area of the cylindrical side is the height times the circumference of the top. Thus:

```
In[1]:=    SurfArea = 2 Pi r^2 + 2 Pi r h
Out[1]=                                    2
            2 Pi h r + 2 Pi r
```

Next solve $V = \pi r^2 h$ for $h$.

```
In[2]:=    hsolved = Solve[V == Pi r^2 h,h]
Out[2]=               V
            {{h -> -----}}
                       2
                   Pi r
```

Substiute into the expression $V = \pi r^2 h$ immediately above. The result is surface area as a function of $r$.

```
In[3]:=    SurfArear = SurfArea/.h->V/(Pi r^2)
Out[3]=     2 V        2
            --- + 2 Pi r
             r
```

The physical set-up tells us that there is a least surface area and that this must happen when the derivative is 0.

```
In[4]:=    optimalr = Solve[D[SurfArear,r] == 0,r]
Out[4]=                     1/3
                         V
            {{r -> ----------},
                    1/3   1/3
                   2    Pi

                  (2 I)/3 Pi   1/3
                 E           V
             {r -> ----------------},
                      1/3   1/3
                     2    Pi

                  (4 I)/3 Pi   1/3
                 E           V
             {r -> ----------------}}
                      1/3   1/3
                     2    Pi
```

The best radius is:

```
In[5]:=    bestr = optimalr[[1,1,2]]
```

*Out[5]=*
$$\frac{V^{1/3}}{2^{1/3} \, Pi^{1/3}}$$

The corresponding height is:

*In[6]:=*     `besth = V/(Pi r^2)/.r->bestr`

*Out[6]=*
$$\frac{2^{2/3} \, V^{1/3}}{Pi^{1/3}}$$

and the ratio of its height to the diameter of its base is:

*In[7]:=*     `Simplify[besth/(2 bestr)]`

*Out[7]=*     1

For the least surface area, *the height equals the diameter of the base.*

### T.5.b)

Find the rectangle with *largest* area among those rectangles with a given diagonal length $d$.

### Answer:

*In[1]:=*     `area = base height`

*Out[1]=*     `base height`

The Pythagorean theorem says that
$$d = \sqrt{base^2 + height^2}.$$

*In[2]:=*     `pythageqn = d^2 == base^2 + height^2`

*Out[2]=*     $d^2 == base^2 + height^2$

Solve for height :

*In[3]:=*     `heightsolved = Solve[pythageqn,height]`

*Out[3]=*
$$\{\{height \rightarrow Sqrt[-base^2 + d^2]\},$$
$$\{height \rightarrow -Sqrt[-base^2 + d^2]\}\}$$

Pick the positive solution. Substitute this into the area expression to find area as a function of the base:

*In[4]:=*     `areafunctofbase =`
            `area/.height->Sqrt[d^2 - base^2]`

*Out[4]=*     $base \; Sqrt[-base^2 + d^2]$

It is not surprising that area $= 0$ when base $= 0$ or when base $= d$ . At the true optimum value, the curve changes from increasing to decreasing, so the derivative

`D[areafunctofbase, base]`

must be 0 when the base is at its optimum value.

*In[5]:=*     `bestbase =`
            `Solve[D[areafunctofbase,base]==0,base]`

*Out[5]=*
$$\{\{base \rightarrow \frac{d}{Sqrt[2]}\},$$
$$\{base \rightarrow -(\frac{d}{Sqrt[2]})\}\}$$

Take the positive root. The base of the largest rectangle is: $d/\sqrt{2}$. The height of the largest rectangle is:

*In[6]:=*     `Sqrt[d^2 - base^2]/.base->d/Sqrt[2]`

*Out[6]=*
$$\frac{d}{Sqrt[2]}$$

A square! Some folks will argue that a square is not a rectangle. But those of us in the know say that a square is just a funny shaped rectangle.

■ **T.6) The closest approximation of ' by a fraction.**

### T.6.a.i)

Of all fractions $p/q$, where $p$ and $q$ are positive integers and $q$ is no more than 10, find the one that is *closest* to $\pi$.

### Answer:

Looking at all fractions $p/q$ with $1 \le q \le 10$ would be a monumental task because there are infinitely many of them. We can make the job a lot simpler by looking at some decimals of $\pi$.

*In[1]:=*     `N[Pi,10]`

*Out[1]=*     `3.141592654`

We try to look at fractions $p/q$ such that

$3.141592 < p/q < 3.141593$.

This means

$3.141592q < p < 3.141593q$.

All these fractions $p/q$ satisfy

**Floor[3.141592 q]** $< p <$ **Ceiling[3.141593 q]**

where **Floor[x]** is the largest integer smaller than $x$ and **Ceiling[x]** is the smallest integer bigger than $x$ .

Here is a table computing $|\pi - p/q|$ in the first slot and $p/q$ in the second slot for $q = 1, ..., 10$ and for $p$ in the range:

**Floor[3.141592 q]** $< p <$ **Ceiling[3.141593 q]]**

*In[2]:=*     `bestshots =`

```
Flatten[Table[{N[Abs[Pi-p/q]],p/q},{q,1,10},
{p,Floor[3.141592 q],Ceiling[3.141593 q]}],1]
```

*Out[2]=*     {{0.141593, 3}, {0.858407, 4},

$$\{0.141593, 3\}, \{0.358407, \frac{7}{2}\},$$

$$\{0.141593, 3\}, \{0.191741, \frac{10}{3}\},$$

$$\{0.141593, 3\}, \{0.108407, \frac{13}{4}\},$$

$$\{0.141593, 3\}, \{0.0584073, \frac{16}{5}\},$$

$$\{0.141593, 3\}, \{0.025074, \frac{19}{6}\},$$

$$\{0.141593, 3\}, \{0.00126449, \frac{22}{7}\},$$

$$\{0.0165927, \frac{25}{8}\}, \{0.108407, \frac{13}{4}\},$$

$$\{0.0304815, \frac{28}{9}\}, \{0.0806296, \frac{29}{9}\},$$

$$\{0.0415927, \frac{31}{10}\}, \{0.0584073, \frac{16}{5}\}\}$$

Now sort the list in order of increasing first slots:

*In[3]:=*     `Sort[bestshots]`

*Out[3]=*
$$\{\{0.00126449, \frac{22}{7}\}, \{0.0165927, \frac{25}{8}\},$$

$$\{0.025074, \frac{19}{6}\}, \{0.0304815, \frac{28}{9}\},$$

$$\{0.0415927, \frac{31}{10}\}, \{0.0584073, \frac{16}{5}\},$$

$$\{0.0584073, \frac{16}{5}\}, \{0.0806296, \frac{29}{9}\},$$

$$\{0.108407, \frac{13}{4}\}, \{0.108407, \frac{13}{4}\},$$

$$\{0.141593, 3\}, \{0.141593, 3\},$$

$$\{0.141593, 3\}, \{0.141593, 3\},$$

$$\{0.141593, 3\}, \{0.141593, 3\},$$

$$\{0.141593, 3\}, \{0.191741, \frac{10}{3}\},$$

$$\{0.358407, \frac{7}{2}\}, \{0.858407, 4\}\}$$

The winner is the familiar 22/7; it differs from $\pi$ by only 0.00126449 .

We could have picked this one out with the instruction:

*In[4]:=*     `Sort[bestshots][[1]]`

*Out[4]=*
$$\{0.00126449, \frac{22}{7}\}$$

It's nice to see our old school friend, 22/7, show up here. Golly, some of our school chums actually believe $\pi = 22/7$, but we know better.

**T.6.a.ii)**

Of all fractions $p/q$, where $p$ and $q$ are positive integers and $q$ is no more than 50, find the one that is closest to $\pi$.

**Answer:**

We can use the same table, but this time let $q$ range from 1 to 50. Also because the output will be so big, we'll place a semi-colon at the end of the instruction. This way the computer will know the table, but we won't have to look at it.

*In[1]:=*
```
bestshots =
Flatten[Table[{N[Abs[Pi-p/q]],p/q},{q,1,50},
{p,Floor[3.141592 q],Ceiling[3.141593 q]}],1];
```

Just for curiosity, look at a few terms:

*In[2]:=*     `Short[bestshots,3]`

*Out[2]=*     {{0.141593, 3}, {0.858407, 4},

$$\{0.141593, 3\}, <<96>>,$$

$$\{0.0184073, \frac{79}{25}\}\}$$

Now sort the list in order of increasing first slots and pick the winner:

*In[3]:=*     `Sort[bestshots][[1]]`

*Out[3]=*
$$\{0.00126449, \frac{22}{7}\}$$

The winner and still the champion is the our friend 22/7. Even if we allow denominators as large as 50 we cannot do better than our old standby.

**T.6.a.iii)**

Of all fractions $p/q$, where $p$ and $q$ are positive integers and $q$ is no more than 100, find the one that is closest to $\pi$.

Greatest and least, shortest and tallest, fastest and slowest, best and worst, . . .

**127**

**Answer:**

Use the same table, but this time let $q$ range from 1 to 100:

```
In[1]:=    bestshots =
           Flatten[Table[{N[Abs[Pi- p/q]],p/q},
           {q,1,100},{p,Floor[3.141592 q],
           Ceiling[3.141593 q]}],1];
```

and pick the winner:

```
In[2]:=    Sort[bestshots][[1]]
                         311
Out[2]=     {0.000178512, ---}
                          99
```

A new champion: 311/99 beats out 22/7.

Let's check:

```
In[3]:=    {Abs[N[Pi - 311/99]],Abs[N[Pi - 22/7]]}
Out[3]=    {0.000178512, 0.00126449}
```

The fraction 311/99 wins in a knock-out.

### T.6.a.iv)

Of all fractions $p/q$, where $p$ and $q$ are positive integers and $q$ is no more than 200, find the one that is closest to $\pi$.

**Answer:**

Use the same table, but this time let $q$ range from 1 to 200:

```
In[1]:=    bestshots =
           Flatten[Table[{N[Abs[Pi- p/q]],p/q},
           {q,1,200},{p,Floor[3.141592 q],
           Ceiling[3.141593 q]}],1];
```

And pick the winner:

```
In[2]:=    Sort[bestshots][[1]]
                     -7   355
Out[2]=     {2.66764 10  , ---}
                         113
```

A new champion: 355/113 beats out 311/99.

Let's check:

```
In[3]:=    {Abs[N[Pi - 355/113]],Abs[N[Pi - 311/99]],
            Abs[N[Pi - 22/7]]}
                     -7
Out[3]=     {2.66764 10  , 0.000178512, 0.00126449}
```

The fraction 355/113 wins in a BIG way.

Let's look at all past and current champions to eleven accurate decimals:

```
In[4]:=    N[22/7,12]
Out[4]=    3.14285714286
```

```
In[5]:=    N[311/99,12]
Out[5]=    3.14141414141
```

```
In[6]:=    N[355/113,12]
Out[6]=    3.14159292035
```

```
In[7]:=    N[Pi,12]
Out[7]=    3.14159265359
```

The fraction 355/113 is a spectacularly good approximation of $\pi$. Can you find a better fraction?

### ■ T.7) Relative contribution to the minimum cost.

#### T.7.a)

The Calculus& *Mathematica* Company is going to manufacture $x$ widgets and $y$ gizmos. The cost of the raw material for each for each is proportional to the number made. Thus

widget cost = $ax$

and

gizmo cost = $by$

where $a$ and $b$ are positive constants. The overhead cost is inversely proportional to the square of the number of gizmos made and is inversely proportional to the number of widgets made. Thus

overhead cost = $c/(x^2 y)$.

What should the ratios

widget cost : gizmo cost : overhead cost

be if the overall cost

$$ax + by + c/(x^2 y)$$

is at a minimum? What are the corresponding optimal production levels $x$ and $y$?

**Answer:**

```
In[1]:=    widget[x_] = a x
Out[1]=    a x
```

```
In[2]:=    gizmo[y_] = b y
Out[2]=    b y
```

```
In[3]:=    overhead[x_,y_] = c/(x^2 y)
Out[3]=       c
           ----
             2
            x  y
```

```
In[4]:=    cost[x_,y_] =
           widget[x] + gizmo[y] + overhead[x,y]
```

$Out[4]=$

$$a \, x + \frac{c}{x^2 \, y} + b \, y$$

Let's find the $x$ and $y$ that make the cost as small as possible.

Note that if $\{x_0, y\}_0$ is the optimal pair, then we must have $(D[cost[x, y], x]$ and $(D[cost[x, y], y]$ both equal to zero at $x = x_0$, $y = y_0$, and you know why. Set up the equations:

$In[5]:=$    `eqn1 = (D[cost[x, y], x]/.{x->x0,y->y0})`
`== 0`

$Out[5]=$

$$a - \frac{2 \, c}{x0^3 \, y0} == 0$$

$In[6]:=$    `eqn2 = (D[cost[x, y], y]/.{x->x0,y->y0})`
`== 0`

$Out[6]=$

$$b - \frac{c}{x0^2 \, y0^2} == 0$$

$In[7]:=$    `solutions = Solve[{eqn1,eqn2},{x0,y0}]`

$Out[7]=$

$$\left\{\left\{ y0 \rightarrow \frac{4^{1/4} \, \mathrm{Sqrt}[a] \, c^{1/4}}{2 \, b^{3/4}}, \right.\right.$$

$$\left. x0 \rightarrow \frac{4^{1/4} \, b^{1/4} \, c^{1/4}}{\mathrm{Sqrt}[a]} \right\},$$

$$\left\{ y0 \rightarrow \frac{-\frac{I}{2} \, 4^{1/4} \, \mathrm{Sqrt}[a] \, c^{1/4}}{b^{3/4}}, \right.$$

$$\left. x0 \rightarrow \frac{I \, 4^{1/4} \, b^{1/4} \, c^{1/4}}{\mathrm{Sqrt}[a]} \right\},$$

$$\left\{ y0 \rightarrow \frac{-(4^{1/4} \, \mathrm{Sqrt}[a] \, c^{1/4})}{2 \, b^{3/4}}, \right.$$

$$\left. x0 \rightarrow -\left(\frac{4^{1/4} \, b^{1/4} \, c^{1/4}}{\mathrm{Sqrt}[a]}\right) \right\},$$

$$\left\{ y0 \rightarrow \frac{-\frac{I}{2} \, 4^{1/4} \, \mathrm{Sqrt}[a] \, c^{1/4}}{b^{3/4}}, \right.$$

$$\left.\left. x0 \rightarrow \frac{-I \, 4^{1/4} \, b^{1/4} \, c^{1/4}}{\mathrm{Sqrt}[a]} \right\}\right\}$$

Only the real answers make any sense here; so the optimal $x_0$ and $y_0$ are given by:

$In[8]:=$    `optimals = solutions[[1]]`

$Out[8]=$

$$\left\{ y0 \rightarrow \frac{4^{1/4} \, \mathrm{Sqrt}[a] \, c^{1/4}}{2 \, b^{3/4}}, \right.$$

$$\left. x0 \rightarrow \frac{4^{1/4} \, b^{1/4} \, c^{1/4}}{\mathrm{Sqrt}[a]} \right\}$$

The minimum cost is:

$In[9]:=$    `cost[x0,y0]/.optimals`

$Out[9]=$

$$\frac{\mathrm{Sqrt}[a] \, b^{1/4} \, c^{1/4}}{4^{1/4}} +$$

$$\frac{3 \, 4^{1/4} \, \mathrm{Sqrt}[a] \, b^{1/4} \, c^{1/4}}{2}$$

Lets look at the ratios:

$In[10]:=$    `(widget[x0]/gizmo[y0])/.optimals`

$Out[10]=$    2

$In[11]:=$    `(gizmo[y0]/overhead[x0,y0])/.optimals`

$Out[11]=$    1

So the ratios

widget cost : gizmo cost : overhead cost

are $2:1:1$ .

Because

overall cost = widget cost + gizmo cost + overhead cost

this means that to minimize costs, we spend

one half our capital on widget production,

one fourth on gizmos production and

one fourth on overhead.

No matter what the actual values of $a, b$ and $c$ are.

Nice management tool. And a nice use of *Mathematica* .

■ **T.8) Sampling to find the greatest possible weight.**

**T.8.a).**

You grab a handful of identical bolts and dump them onto a scale. The handful weighs 30.465 grams. You grab three more handfuls and get weights of 46.713, 64.992 and 58.899 grams. You never bothered to count the number of bolts in any of the handfuls.

Greatest and least, shortest and tallest, fastest and slowest, best and worst, . . .

**129**

What do these samplings tell you about the greatest possible weight of an individual bolt?

**Answer:**

The measurements you made were:

```
In[1]:=    w[1]=30.465;w[2]=46.713;
           w[3]=64.992;w[4]=58.899;
```

Sort the weights in increasing order:

```
In[2]:=    weights = Sort[Table[w[k],{k,1,4}]]
Out[2]=    {30.465, 46.713, 58.899, 64.992}
```

Say that an individual bolt weighs $g$ grams. Our job is to estimate the largest possible value for $g$.

Let's agree that the $c[k]$ bolts weigh $w[k]$ grams;

thus $w[k] = c[k]g$ for each $j = 1, 2, 3, 4$.

We have no idea of what the values of the $c[k]'s$ are, but this does not stop us. The main point is that each $c[k]$ is a **positive integer** and each $w[k]$ is an integral multiple of $g$.

As a result, all the differences

$$w[k] - w[k-1] = (c[k] - c[k-1])g$$

are integral multiples of $g$. Let's look at these differences sorted in increasing order:

```
In[3]:=    firstdiffs =
           Sort[Table[
           weights[[k]] - weights[[k-1]],{k,2,4}]]
Out[3]=    {6.093, 12.186, 16.248}
```

Because each of these numbers is an integral multiple of $g$, we see at this stage that

$$g \leq 6.093.$$

Further, the differences of these differences are integral multiples of g. Here are these second differences:

```
In[4]:=    seconddiffs =
           Sort[Table[firstdiffs[[k]]
           -firstdiffs[[k-1]],{k,2,3}]]
Out[4]=    {4.062, 6.093}
```

More information. Because each of these numbers is an integral multiple of $g$, we see at this stage that

$$g \leq 4.062.$$

Further, the differences of these differences are integral multiples of $g$. Here are these third differences:

```
In[5]:=    thirddiffs =
           Sort[Table[seconddiffs[[k]]
           -seconddiffs[[k-1]],{k,2,2}]]
```

```
Out[5]=    {2.031}
```

Our last piece of information: An individual bolt cannot weigh more than 2.031 grams. Let's see whether this checks out with the data:

```
In[6]:=    weights/2.031
Out[6]=    {15., 23., 29., 32.}
```

These numbers appear to be integers, so we conclude that *from the information we have*, an individual bolt cannot weigh more than 2.031 grams and that if this is the actual weight of an individual bolt, then there were 15, 23, 29 and 32 individual bolts in each of the various handfuls.

Of course this still allows the possibility that an individual bolt could could actually weigh say (2.031)/2 grams and then all quantities would check out, with twice as many in the count of each handful.

■ **T.9) Greatest and least constrained values.**

**T.9.a)**

Find the greatest and least vaues of

$$(x^7 - 4x^2 + 1)/(2x^6 + 11)$$

for $-4 \leq x \leq 4$.

**Answer:**

Plot the function for $-4 \leq x \leq 4$:

```
In[1]:=    function = (x^7 - 4 x^2 + 1)/(2 x^6 + 11)
Out[1]=            2      7
             1 - 4 x  + x
             -------------
                     6
               11 + 2 x
```

```
In[2]:=    Plot[function,{x,-4,4},PlotRange->All];
```

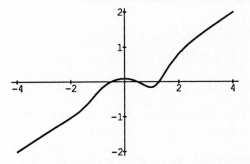

It is clear that the function takes its least value for $-4 \leq x \leq 4$ at $x = -4$ and takes its greatest value at $x = 4$.

Its least value is:

```
In[3]:=    function/.x->-4
Out[3]=       16447
            -(-----)
```

8203

The greatest value is:

```
In[4]:=    function/.x->4
Out[4]=    16321
           -----
           8203
```

Just a breeze.

**T.9.b)**

---

Find the greatest and least vaues of

$$x^{2/3} - 0.9 \sin[2x^2]$$

for $-0.251 \leq x \leq 1.042$.

---

**Answer:**

To see what's happening, plot:

```
In[1]:=    y = x^(2/3) - .9 Sin[2 x^2]
           plot = Plot[y,{x,-0.251,1.042}];
```

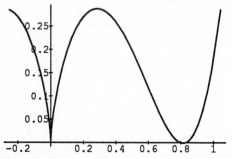

Neat plot! Note the sharp corner at $x = 0$. Some folks call that corner a "cusp."

There are three contenders for the greatest value. Two of the contenders are at the endpoints. The other is the point near $x = 0.3$ at which the derivative is zero.

```
In[2]:=    interiorzero =
           FindRoot[D[y,x]== 0, {x,0.3}]
Out[2]=    {x -> 0.285121}
```

Compare the values:

```
In[3]:=    {y/.x->-0.251,y/.x->0.285,y/.x->1.042}
Out[3]=    {0.284806, 0.287512, 0.285378}
```

It's a tight competition, but the greatest value is:

```
In[4]:=    y/.x->0.285
Out[4]=    0.287512
```

Now let's see about the least value. There are two contenders for the least value. One of them is 0. The other is the point near $x = 0.8$ at which the derivative is 0:

```
In[5]:=    Interiorzero =
           FindRoot[D[y,x]== 0, {x,0.8}]
Out[5]=    {x -> 0.813815}
```

Compare the values:

```
In[6]:=    {y/.x->0,y/.x->0.814}
Out[6]=    {0, -0.0011927}
```

It's another tight competition, but the least value is:

```
In[7]:=    y/.x->0.814
Out[7]=    -0.0011927
```

## Literacy Sheet

**1.** Show that the biggest rectangle with a fixed perimeter is the square.

.

.

.

**2.** R.M. Thrall and his University of Michigan colleagues (Report No. 40241-R-7, University of Michigan(1967)) gave the following crisp description of auto-catalytic reaction of one substance into a new substance:

Auto-catalytic reaction progresses in such a way that the first substance catalyzes its own formation.

In the Thrall model, the reaction rate(with respect to time) is proportional to the amount $x$ of the new substance at time $t$ and the reaction rate is also proportional $(a - x)$ where a is the original amount of the first substance.

Why does this mean that

$$dx/dt = Kx(a - x)$$

for some positive constant K?

.

.

.

Differentiate with respect to $x$ your expression for $dx/dt$ and analyze it to determine the value of $x$ for which the reaction rate is fastest.

For what $x's$ is the reaction rate the slowest? Interpret.

.

.

.

**3.** What are the largest and smallest values of

$$2x^3 - 9x^2 + 12x - 1$$

for $0 \leq x \leq 2$ ?

.

.

.

What are the largest and smallest values of

$$2x^3 - 9x^2 + 12x - 1$$

for $1 \leq x \leq 3$ ?

.

.

.

**4.** Engineering studies have shown that the turning effect of a ship's rudder is proportional to $\cos[d]\sin^2[d]$ where $d$ ( $0 \leq d < \pi/2$ ) is the angle the rudder makes with

the keel. Approximately which deflection $d$ of the rudder makes the rudder most effective?

.

.

.

**5.** Why are $22/7$ and $355/113$ both rather spectacular fractions?

.

.

**6.** Suppose $A$ and $B$ are positive constants and suppose $f[x] = Ax + B/x$. Calculate the positive number $x_0$ such that $f[x_0] \leq f[x]$ for all other positive $x's$ . What is the ratio $Ax_0/(B/x_0)$ ?

.

.

Suppose $A$ and $B$ are positive constants and suppose $f[x] = Ax^2 + B/x$. Calculate the positive number $x_0$ such that $f[x_0] \leq f[x]$ for all other positive $x's$ . What is the ratio $A(x_0)^2/(B/x_0)$ ?

.

.

Suppose $A$ and $B$ are positive constants and suppose $f[x] = Ax^3 + B/x$. Calculate the positive number $x_0$ such that $f[x_0] \leq f[x]$ for all other positive $x's$ . What is the ratio $A(x_0)^3/(B/x_0)$ ?

.

.

.

**7.** New York, Denver, Madrid, Istanbul and Beijing are all at latitude $40°$ North. Consequently, if you are going from New York to Beijing, the due east route will take you over Madrid and Istanbul, but the due west route will take you over Denver.

Tables reveal that the distances (in miles) by air are:

|  | Beijing | Denver | Istanbul | Madrid | NY |
|---|---|---|---|---|---|
| Beijing |  | 6350 | 4380 | 5730 | 6830 |
| Denver | 6350 |  | 6150 | 5010 | 1630 |
| Istanbul | 4380 | 6150 |  | 1700 | 5010 |
| Madrid | 5730 | 5010 | 1700 |  | 3580 |
| NY | 6830 | 1630 | 5010 | 3580 |  |

Is Denver on the shortest route from New York to Beijing? How about Madrid or Istanbul?

.

.

.

**Can the due east route or the the due west route from New York to Beijing be the shortest route?**

.

.

.

# Tangent Lines

## Guide

Tangent lines to circles are easy to find: just take the lines perpendicular to the radius through the point. But how about other shapes?

We want to be able to describe effectively the tangent lines to all sorts of graphs: ellipses, parabolas, sine curves, exponential curves and many others. The clear geometric meaning of this concept has an equally clear calculus counterpart: The *derivative* measures the slope of the tangent line to any graph and provides a splendid calculating device in many geometric situations including root finding, mirrors, radar signals, acoustics, and antennas.

## Basics

■ **B.1) Tangent lines.**

**B.1.a.i)**

What is the equation of the line tangent to the graph of $y = f[x]$ through the point $\{x_0, f[x_0]\}$ ?

**Answer:**

The tangent line must go through $\{x_0, f[x_0]\}$. So the equation of the tangent line has the form

$$y = m(x - x_0) + f[x_0].$$

We still have to determine what the growth rate $m$ is.

To see what $m$ is, recall that the derivative measures the instantaneous growth rate of $y = f[x]$ at all points $x$.

At the point $\{x_0, f[x_0]\}$ of tangency, the line and the function should have the same instantaneous growth rate. So:

$$m = f'[x_0].$$

As a result, the equation of the tangent line to the graph of $y = f[x]$ through the point $\{x_0, f[x_0]\}$ is

$$y = f'[x_0](x - x_0) + f[x_0].$$

**B.1.a.ii)**

Find the tangent line to the curve $y = \sqrt{x}$ through $\{3, \sqrt{3}\}$ and plot it and the curve $y = \sqrt{x}$ on the same axes for $0 \le x \le 10$.

**Answer:**

```
In[1]:=   f[x_] = Sqrt[x]
Out[1]=   Sqrt[x]
```

The equation of the tangent line through $\{3, \sqrt{3}\}$ is:

```
In[2]:=   ytan = f'[3] (x - 3) + f[3]
Out[2]=                  -3 + x
          Sqrt[3]  +  ----------
                       2 Sqrt[3]
```

Here is the plot:

```
In[3]:=   Plot[{f[x],ytan},{x,0,10},
            PlotRange->{0,3}];
```

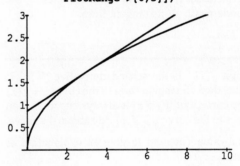

Looks good.

**B.1.a.iii)**

Plot the curve $y = \sqrt{x}$ and its tangent lines through $\{x_0, \sqrt{x_0}\}$ for $x_0 = 0.5, 1, 2, 4$ and $8$ on the same axes for $0 \le x \le 10$. Discuss where the tangent lines lie in relation to the curve.

How does the sign of the second derivative of $\sqrt{x}$ tell you where the tangent lines ride?

**Answer:**

```
In[1]:=   f[x_] = Sqrt[x]
          Plot[{f[x],
          f'[0.5](x -.5) + f[.5],
          f'[ 1](x - 1) + f[1],
          f'[ 2](x - 2) + f[2],
          f'[ 4](x - 4) + f[4],
          f'[ 8](x - 8) + f[8]},{x,0,10},
          PlotRange->{0,3},AxesLabel->{"x","y"}];
```

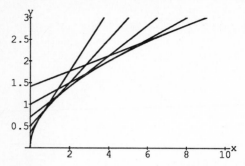

Those are the tangent lines riding on top of the square root curve.

We can see why by looking at the second derivative:

```
In[2]:=    f''[x]
Out[2]=      -1
          ------
            3/2
         4 x
```

For $x \geq 0$, $f''[x]$ is always negative. This means that the growth rates $f'[x]$ of the tangent lines through $\{x, f[x]\}$ decrease as $x$ increases from left to right. As a result, the curve stays underneath all its tangent lines.

### B.1.a.iv)

Take a function $f[x]$. If the second derivative $f''[x]$ is always positive, then all tangent lines to the curve $y = f[x]$ ride below the curve, but if $f''[x]$ is always negative, then all tangent lines to the curve $y = f[x]$ ride above the curve.

Use the Mean Value Theorem to give an explanation of why this is true.

**Answer:**

The equation of the tangent line to the curve $y = f[x]$ through the point $\{a, f[a]\}$ is

$$y_{tan} = f[a] + f'[a](x - a)$$

Fix an $x$. The Mean Value Theorem applied to $f[x]$ tells us

$$f[x] = f[a] + f'[z_1](x - a)$$

for some $z_1$ between $x$ and $a$.

The Mean Value Theorem applied to $f'[z_1]$ tells us

$$f'[z_1] = f'[a] + f''[z_2](z_1 - a)$$

for some $z_2$ between $z_1$ and $a$.

So

$$f[x] = f[a] + f'[z_1](x - a)$$

$$= f[a] + (f'[a] + f''[z_2](z_1 - a))(x - a)$$

$$= f[a] + f'[a](x - a) + f''[z_2](z_1 - a))(x - a)$$

$$= y_{tan} + f''[z_2](z_1 - a))(x - a).$$

Now here's the key point: Because $z_1$ is between $x$ and $a$, the point $z_1$ is on the same side of $a$ that $x$ is on. Therefore $(z_1 - a))(x - a) \geq 0$.

If $f''[x]$ is always positive this means that $f''[z_2](z_1 - a))(x - a) \geq 0$.

Therefore

$$f[x] = y_{tan} + f''[z_2](z_1 - a))(x - a)$$

means $f[x] = y_{tan} +$ a nonnegative number.

This tells us that $y_{tan}$ is **below** $f[x]$ and explains why the tangent lines ride under the curve in the case that $f''[x]$ is always positive.

A similar analysis explains why the tangent lines ride above the curve in the case in which $f''[x]$ is always negative.

### B.1.b)

There is another way to visualize why the derivative is the growth rate or slope of the tangent line: The derivative $f'[x]$ is the limiting case of the average growth rates

$$(f[x + h] - f[x])/h$$

as $h \to 0$. The quantities $(f[x + h] - f[x])/h$ are slopes of the secant lines running through the points $\{x, f[x]\}$ and $\{x + h, f[x + h]\}$.

The limiting case of the slopes of these secants is the slope of the tangent line through $\{x, f[x]\}$. In other words:

$f'[x]$ **is the instantaneous growth rate or slope of the line tangent to the graph of the curve at the point** $\{x, f[x]\}$.

### ■ B.2) Newton's method.

Newton's method is an old workhorse of numerical mathematics. Its chief use is to find approximate solutions of $f[x] = 0$ when symbolic exact methods fail. In fact, the *Mathematica* instruction **FindRoot** is nothing more or less than an automatic implementation of Newton's Method.

### B.2.a)

What is Newton's Method and the idea behind it?

**Answer:**

Here is the idea:

Suppose we start with a rather complicated function $f[x]$:

```
In[1]:=    f[x_] = -8 E^(-x) + Sum[((2 x)^k)/k!, {k,1,6}]
Out[1]=
                              3      4      5
          -8            2   4 x    2 x    4 x
          -- + 2 x + 2 x  + ---- + ---- + ---- +
           x                 3      3      15
          E
```

$$\frac{4 x^6}{45}$$

We may try to find an exact solution of $f[x] = 0$:

```
In[2]:=    Solve[f[x] == 0,x]
```

Solve::ifun:

Warning:  inverse functions are being

used by Solve, so some solutions may

not be found.

Solve::tdep:

The equations appear to involve

transcendental functions of the

variables in an essentially

nonalgebraic way.

```
Out[2]=
           Solve[-8/x + 2 x + 2 x  + 4 x/3 + 2 x/3 +
              E
```
$$\text{Solve}\left[\frac{-8}{E^x} + 2x + 2x^2 + \frac{4x^3}{3} + \frac{2x^4}{3} + \frac{4x^5}{15} + \frac{4x^6}{45} == 0,\ x\right]$$

The function is too much for *Mathematica*'s **Solve** instruction! So we plot it hoping to spot a solution of $f[x] = 0$.

```
In[3]:=    Plot[f[x],{x,0,3}];
```

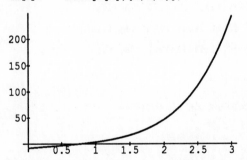

Eyeball a solution of $f[x] = 0$ near $x = 0.8 = a[1]$. This is our first value so we call it $a[1]$ :

```
In[4]:=    a[1] = 0.8
Out[4]=    0.8
```

Next plot the function and its tangent line through $\{a[1], f[a[1]]\}$ :

```
In[5]:=    tanline[x_,1] = f[a[1]] + f'[a[1]] (x - a[1])
Out[5]=    0.351785 + 13.4409 (-0.8 + x)
```

```
In[6]:=    Plot[{f[x],tanline[x,1]},{x, 0.6,1.0}];
```

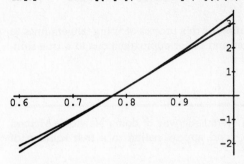

This tangent line crosses the $x$-axis at a number $a[2]$ very close to the true solution of $f[x] = 0$. This point $a[2]$ can be found by the **Solve** command:

```
In[7]:=    a2solved = Solve[tanline[x,1] == 0,x]
Out[7]=    {{x -> 0.773827}}
```

```
In[8]:=    a[2] = x/.a2solved[[1]]
Out[8]=    0.773827
```

Then set up the tangent line through $\{a[2], f[a[2]]\}$ and plot:

```
In[9]:=    tanline[x_,2] = f[a[2]] + f'[a[2]] (x - a[2])
Out[9]=    0.00527523 + 13.0422 (-0.773827 + x)
```

```
In[10]:=   Plot[{f[x],tanline[x,2]},{x, 0.76,0.8}];
```

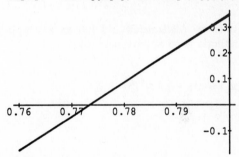

The tangent line crosses the $x$-axis at a number $a[3]$ very very close to the true solution of $f[x] = 0$.

```
In[11]:=   a3solved = Solve[tanline[x,2] == 0,x]
Out[11]=   {{x -> 0.773423}}
```

```
In[12]:=   a[3] = x/.a3solved[[1]]
Out[12]=   0.773423
```

This value $a[3]$ is not bad at all because even though:

```
In[13]:=   f[a[3]]
Out[13]=   0.00000120355
```

is not 0, it is pretty close to 0 .

So 0.773423 is a pretty good approximation to a true solution of $f[x] = 0$.

**Newton's method** is this process of using tangent lines to
generate better and better approximations to a true solution of $f[x] = 0$.

### B.2.b)

How about a systematic way of doing Newton's Method
for getting a good approximation to a true solution of
$f[x] = 0$?

**Answer:**

Start with a reasonable guess $a[1]$ for $x$. The tangent line
through $\{a[1], f[a[1]]\}$ crosses the $x-$ axis at a number
$a[2]$. The equation of this tangent line is

$$y - f[a[1]] = f'[a[1]](x - a[1]).$$

It crosses the $x-$ axis at a point $a[2]$ satisfying

$$0 - f[a[1]] = f'[a[1]](a[2] - a[1]).$$

Solve this for $a[2]$, getting

$$a[2] = a[1] - f[a[1]]/f'[a[1]].$$

Iterate this idea via the update formula

$$a[n] = a[n-1] - f[a[n-1]]/f'[a[n-1]]$$

which we get by making the replacements $2 \to n$ and
$1 \to (n-1)$ in the formula expressing $a[2]$ in terms of
$a[1]$.

Let's turn it over to *Mathematica* and run on a sample
problem of finding a good approximate solution of $e^x - x - 2 = 0$:

```
In[1]:=    f[x_] = E^x - x - 2
Out[1]=         x
           -2 + E  - x
```

Plot:

```
In[2]:=    Plot[f[x],{x,0,3}];
```

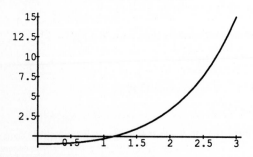

A reasonable choice for $a[1]$ is:

```
In[3]:=    Clear[a]
           a[1] = 1
```

```
Out[3]=    1
```

The update formula is:

```
In[4]:=    a[n_] := a[n] = a[n-1] - f[a[n-1]]/f'[a[n-1]]
```

Let us take a look:

```
In[5]:=    a[1]
Out[5]=    1
```

```
In[6]:=    a[2]
Out[6]=          -3 + E
           1 - ------
                -1 + E
```

```
In[7]:=    a[3]
Out[7]=          -3 + E              -3 + E
           1 - ------ - (-2 - (1 - ------) +
                -1 + E              -1 + E

                1 - (-3 + E)/(-1 + E)
           E                        ) /

                     1 - (-3 + E)/(-1 + E)
           (-1 + E                        )
```

Here are successive decimal approximations $approx[1]$,
$approx[2]$, $approx[3]$, $approx[4]$, . . . of a solution of
$f[x] = 0$:

```
In[8]:=    Clear[approx]
           approx[n_]:=approx[n]=N[a[n],10]
```

```
In[9]:=    first6=Table[{n,approx[n]},{n,1,6}]
Out[9]=    {{1, 1.}, {2, 1.163953414},

            {3, 1.146421185}, {4, 1.146193259},

            {5, 1.146193221}, {6, 1.146193221}}
```

Here's a better look:

```
In[10]:=   ColumnForm[first6]
Out[10]=   {1, 1.}
           {2, 1.163953414}
           {3, 1.146421185}
           {4, 1.146193259}
           {5, 1.146193221}
           {6, 1.146193221}
```

Note how they cluster.

Let's look at the successive approximations and the values
of $f[x]$ at these successive approximations.

```
In[11]:=   ColumnForm[Table[
           N[{n,approx[n],f[approx[n]]},10],
           {n,1,6}]]
Out[11]=   {1., 1., -0.2817181715}
           {2., 1.163953414, 0.0386159498}
           {3., 1.146421185, 0.0004893374545}
```

-8

{4., 1.146193259, 8.173544504 10 }
{5., 1.146193221, 2.282028733 10$^{-15}$ }
{6., 1.146193221, 2.385244779 10$^{-18}$ }

These are damn good results. For example:

```
In[12]:=  N[approx[5],10]
Out[12]=  1.146193221
```

is a great approximation of a solution of $f[x] = 0$ because:

```
In[13]:=  N[f[approx[5]],10]
Out[13]=           -15
          2.282028733 10
```

is so small.

### B.2.c)

How is the *Mathematica* instruction **FindRoot** related to Newton's method?

**Answer:**

**FindRoot[f[x] == 0,x,a]** starts Newton's method with first guess $a$ and continues to run Newton's method until the successive approximations $a[1]$, $a[2]$, $a[3]$, . . . , $a[n]$ begin to cluster.

Let's run on the function above:

```
In[1]:=   f[x_] = E^x - x - 2
          FindRoot[f[x] == 0,{x,1}]
Out[1]=   {x -> 1.14619}
```

This is in agreement with:

```
In[2]:=   approx[5]
Out[2]=   1.146193221
```

which we produced above.

Let's see what happens when we start at other initial guesses:

```
In[3]:=   FindRoot[f[x] == 0,{x,1.20000}]
Out[3]=   {x -> 1.14619}
```

```
In[4]:=   FindRoot[f[x] == 0,{x,2.5000}]
Out[4]=   {x -> 1.14619}
```

If our initial guess really sucks, then **FindRoot** balks:

```
In[5]:=   FindRoot[f[x] == 0,{x,18}]
```

*FindRoot::convNewt:*

*Newton's method failed to converge to*

*the prescribed accuracy after 15 iterations.*

```
Out[5]=   {x -> 3.1574}
```

Garbage in; garbage out.

### B.2.d)

The Newton iteration formula for finding a solution of $f[x] = 0$ is

$$a[n] = a[n-1] - f[a[n-1]]/f'[a[n-1]].$$

If the solution root $x_{root}$ of has the property that $f'[x_{root}] = 0$, then this one thing can make Newton's method less than reliable.

Look at:

```
In[1]:=   FindRoot[x == 0,{x,.2}]
Out[1]=   {x -> 0.}
```

Good.

```
In[2]:=   FindRoot[x^2 == 0,{x,.2}]
Out[2]=            -22
          {x -> -1.05879 10   }
```

Not bad.

```
In[3]:=   FindRoot[x^4 == 0,{x,.2}]
Out[3]=   {x -> 0.0177979}
```

Suspicious, but at the true solution the derivative is 0.

```
In[4]:=   FindRoot[x^8 == 0,{x,.2}]
Out[4]=   {x -> 0.15}
```

Not so good.

```
In[5]:=   FindRoot[x^60 == 0,{x,.2}]
Out[5]=   {x -> 0.2}
```

Awful.

The trouble you are seeing here in not with *Mathematica*: it is with *Newton's method itself* when applied under unfavorable conditions.

■ **B.3) Angle of incidence and angle of reflection problems.**

### B.3.a)

Two summer camps are to be supplied with water from a nearby straight river. The same pumphouse on the river bank is to supply both camps. Explain why the total length of level pipe needed to carry water to the two camps is least if the pumphouse is placed so that the two pipe lines make equal angles with the river bank.

**Answer:**

We need a picture.

```
In[1]:=   campAcoord={0,3}
          campBcoord={5,4}
          pumpcoord ={2,0}
          pipe =
          Graphics[Line[{campAcoord,pumpcoord,
          campBcoord}]]
          riverbank =
          Graphics[Line[{{-2,0},{7,0}}]]
          river =
          Graphics[{RGBColor[0,1,1],
          Polygon[{{-2,0},{7,0},{7,-1},
          {-2,-1}}]}]
          rivername =
          Graphics[Text["river",{4,-0.5}]]
          camplabels =
          Graphics[{RGBColor[0,1,0],
          PointSize[.04],
          Point[campAcoord],Point[campBcoord],
          Text["Camp A",campAcoord,{0,-2}],
          Text["Camp B",campBcoord,{0,-2}]}]
          pumphouse =
          Graphics[{PointSize[.03],
          Point[pumpcoord],Text["Pump",{2,0.6}]}]
          Show[pipe,river,riverbank,camplabels,
          pumphouse,rivername,
          PlotRange->{{-1.5,6.5},{-2,6}},
          AspectRatio->Automatic];
```

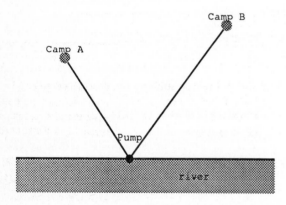

Now let's put in some labels and coordinates. The $x$–axis is the riverbank and the $y$–axis is the vertical on the left:

```
In[2]:=   leftvertical =
          Graphics[Line[{{0,0},campAcoord}]]
          rightvertical =
          Graphics[Line[{{5,0},campBcoord}]]
          campcoords =
          Graphics[Text["{0,ya}",
          campAcoord,{-1.5,0}],
          Text["{xb,yb}",campBcoord,{1.5,0}]];
          pumpcoord =
          Graphics[Text["{x,0}",{2,-.3}]]
          distances =
          Graphics[Text["ya",{-.2,1.6}],
          Text["yb",{5.2,2.6}],
          Text["x",{1,-.2}],
          Text["xb-x",{3.5,-.2}]]
          anglelabels =
          Graphics[Text["s",{1.7,.2}],
          Text["t",{2.3,.2}]]
```

```
          Show[pipe,riverbank,camplabels,
          pumphouse,leftvertical,
          rightvertical,campcoords,
          pumpcoord,distances,
          anglelabels,
          PlotRange->{{-1.5,6.5},{-2,6}},
          AspectRatio->Automatic];
```

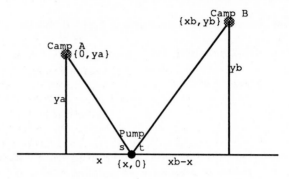

The problem is how to set $x$ to make the combined length of the pipe as small as possible.

The combined length of the pipe is :

```
In[3]:=   Clear[x,ya,xb,yb]
          length =
          Sqrt[x^2 + ya^2] + Sqrt[(xb - x)^2 + (yb)^2]
```

$$Out[3]= \quad Sqrt[x^2 + ya^2] + Sqrt[(-x + xb)^2 + yb^2]$$

Differentiate the combined length of the pipe with respect to $x$:

```
In[4]:=   deriv = D[length,x]
```

$$Out[4]= \quad \frac{x}{Sqrt[x^2 + ya^2]} - \frac{-x + xb}{Sqrt[(-x + xb)^2 + yb^2]}$$

If $x$ is set to minimize the length, we *cannot* have **D[length,x]** $< 0$ because if **D[length,x]** $< 0$, then increasing $x$ slightly will reduce the length. Similarly if $x$ is set to minimize the length, we cannot have **D[length,x]** $> 0$ because if **D[length,x]** $> 0$, then decreasing $x$ slightly will reduce the length.

Therefore if $x$ is set to minimize the length, we must have **D[length,x] = 0** . This says that the following quantity is $0$ :

```
In[5]:=   deriv
```

$$Out[5]= \quad \frac{x}{Sqrt[x^2 + ya^2]} - \frac{-x + xb}{Sqrt[(-x + xb)^2 + yb^2]}$$

So if $x$ is set to minimize the length, we must have

$$x/\sqrt{x^2 + y_a^2} = (x_b - x)/\sqrt{(x_b - x)^2 + y_b^2}$$

Glance at the picture. Evidently the minimizing condition

$$x/\sqrt{x^2 + y_a^2} = (x_b - x)/\sqrt{(x_b - x)^2 + y_b^2}$$

is the same as

$$x/AP = (x_b - x)/BP$$

with $AP$ =distance from Camp A to the Pump, $BP$ =distance from Camp B to the Pump. This is the same as $\cos[s]$ = $\cos[t]$ when $x$ is set to minimize the combined length of the pipes.

But then both pipes make the same angle with the river-bank.

And we're out of here.

### B.3.b)

A number of apparently different problems have the same setup and solution as the pumphouse problem above. A good example is a law of optics that says that when light bounces off a mirror, then the angle of incidence is the angle of reflection.

The basic physical principle is Fermat's principle which says that light always takes the path of minimim time which in this case is the shortest path.

That the angle of incidence equals the angle of reflection is established with an argument identical to the solution in part i). We just change the interpretation of the picture by thinking of the canal bank as the mirror and the pipe to be the path of light. Light originating at A will bounce off the mirror at P and hit B only if the P is situated so that the length AP + PB is a minimum. The analysis in the solution of the pumphouse problem shows that the angle of incidence is the angle of reflection.

## Tutorial

■ **T.1) Tangents.**

### T.1.a)

Find two straight lines through $\{2,6\}$ that are both tangent to the parabola

$$y = 4x - x^2.$$

Confirm with a plot of the parabola and the two tangent lines on the same axes.

**Answer:**

Here is a view of the parabola and the point $\{2,6\}$:

```
In[1]:=    Clear[f,x]
           f[x_] = 4 x - x^2
           parabola = Plot[f[x],{x,-3,7},
           DisplayFunction->Identity]
           point =
           Graphics[{RGBColor[1,0,0],PointSize[.02],
           Point[{2,6}]}]
           Show[parabola,point,AxesLabel->{"x","y"},
           DisplayFunction->$DisplayFunction];
```

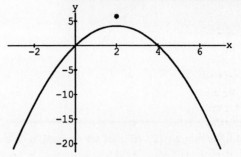

A line through $\{2,6\}$ has the form

$$y - 6 = m(x - 2).$$

If $\{t, f[t]\}$ is a point of tangency, then

$$m = f'[t].$$

```
In[2]:=    tanlineequation =
           (y - 6 == f'[t] (x - 2))
Out[2]=    -6 + y == (4 - 2 t) (-2 + x)
```

All we have to do now is to determine $t$. This is not hard because the point of tangency is $\{t, f[t]\}$ and hence $\{t, f[t]\}$ is on the tangent line. So to get $t$, just plug this point into the tangent line equation and then solve for $t$:

```
In[3]:=    tanpoints =
           Solve[tanlineequation/.{x->t,y->f[t]},t]
Out[3]=          4 + 2 Sqrt[2]
           {{t -> -------------},
                       2

                 4 - 2 Sqrt[2]
            {t -> -------------}}
                       2
```

```
In[4]:=    ytan1 =
           (6 + f'[t] (x - 2))/.t->(4 + 2 Sqrt[2])/2
Out[4]=    6 + (4 - (4 + Sqrt[2])) (-2 + x)
```

The other is:

```
In[5]:=    ytan2 =
           (6 + f'[t] (x - 2))/.t->(4 - 2 Sqrt[2])/2
Out[5]=    6 + (4 - (4 - Sqrt[2])) (-2 + x)
```

Here is the plot:

```
In[6]:=    confirmplot =
           Plot[{f[x],ytan1,ytan2},{x, 2-4,2+4},
           AxesLabel->{"x","y"},
```

```
DisplayFunction->Identity];
Show[point,confirmplot,
DisplayFunction->$DisplayFunction];
```

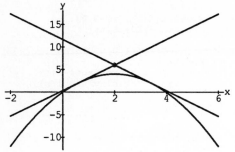

Groovy.

### T.1.b)

---

Are there any lines though {2, −6˝ that are tangent to this parabola?

---

**Answer:**

Look at:

```
In[1]:=    Clear[f,x]
           f[x_] = 4 x - x^2
           parabola = Plot[f[x],{x,-3,7},
           DisplayFunction->Identity]
           newpoint =
           Graphics[{RGBColor[1,0,0],PointSize[.02],
           Point[{2,-6}]}]
           Show[parabola,newpoint,
           AxesLabel->{"x","y"},
           DisplayFunction->$DisplayFunction];
```

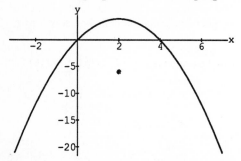

A glance at the above plot indicates a negative answer. To confirm what we see, we'll just copy, paste and edit what we did above and see how the algebra goes.

A line through {2, −6} has the form

$$y + 6 = m(x - 2).$$

If {t, f[t]} is a point of tangency, then

$$m = f'[t].$$

```
In[2]:=    tanlineequation =
           (y + 6 == f'[t] (x - 2))
```

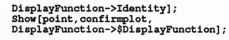

```
Out[2]=    6 + y == (4 - 2 t) (-2 + x)
```

All we have to do now is to determine $t$. This is not hard because the point of tangency is {t, f[t]} and hence {t, f[t]} is on the tangent line. So to get $t$, just plug this point into the tangent line equation and then solve for $t$:

```
In[3]:=    tanpoints =
           Solve[tanlineequation/.{x->t,y->f[t]},t]
```

```
Out[3]=          4 + 2 Sqrt[-10]
           {{t -> ---------------},
                        2

                 4 - 2 Sqrt[-10]
            {t -> ---------------}}
                        2
```

The solutions are not real numbers (because of the negative 10 in the square roots). These complex numbers tells us that no tangent can do the required job.

### ■ T.2) Back-to-back functions: smooth splines.

A spline is a long flexible strip of plastic or the like that is used in drawing smooth curves.

Suppose we have two functions $f[x]$ and $g[x]$ and we have a number $b$. Create a new function $h[x]$, called a **spline** of $f[x]$ and $g[x]$, by putting

$h[x] = f[x]$ for $x \le b$ and

$h[x] = g[x]$ for $x > b$.

Clearly $h'[x] = f[x]$ for $x < b$ and $h'[x] = g'[x]$ for $x > b$. Our interest here is the value of $h'[x]$ for $x = b$ and the tangent line to the curve $y = h[x]$ through the point {b, h[b]}.

### T.2.a.i)

---

Plot $h[x]$ for $b - 1 \le x \le b + 1$ for

$$f[x] = \sin[x],$$

$$g[x] = x \cos[x]^{1/3}$$

and $b = 0$. Try to determine the value of $h'[x]$ for $x = b$.

---

**Answer:**

```
In[1]:=    f[x_] = Sin[x]
           g[x_] = x Cos[x]^(1/3)
           b = 0
           h[x_] := g[x]/;x > b
           h[x_] := f[x] /;x <= b
```

*To get the full effect, use the option* := *in defining* h[x].

```
In[2]:=    spline = Plot[h[x],{x,b - 1,b + 1 }];
```

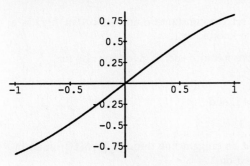

Even though we used different formulas to the left and right of $x = b = 0$, the graph looks really smooth near $x = b = 0$. To see why the graph looks so smooth near $x = 1 = b$, look at the tangent lines to the $y = f[x]$ and $y = g[x]$ curves for $x = b = 0$.

The tangent to the $y = f[x]$ curve through $\{b, f[b]\}$ is:

```
In[3]:=    yf = f'[b] (x - b) + f[b]
Out[3]=    x
```

The tangent to the $y = g[x]$ curve through $\{b, g[b]\}$ is:

```
In[4]:=    yg = g'[b] (x - b) + g[b]
Out[4]=    x
```

Both tangent lines are the same; this is the reason that different formulas to the left and right gave a smooth curve.

The common tangent line to the $y = f[x]$ curve and the $y = g[x]$ curve through $\{b, f[b]\} = \{b g[b]\}$ gives us the tangent line to the spline curve $y = h[x]$ through $\{b, h[b]\}$:

```
In[5]:=    tanline = Plot[yf, {x,b-1,b+1},
           DisplayFunction->Identity];
```

```
In[6]:=    Show[spline,tanline];
```

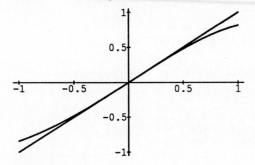

Looks good.

The value $h'[b]$ of the derivative of $h[x]$ at $b = 0$ is given by the common value:

```
In[7]:=    {f'[b],g'[b]}
```

So $h'[b] = h'[0] = 1$.

**T.2.a.ii)**

Plot $h[x]$ for $b - 1 \le x \le b + 1$ for

$$f[x] = x^3,$$
$$g[x] = \sqrt{x}$$

and $b = 1$. Try to determine the value of $h'[x]$ for $x = b$.

---

**Answer:**

```
In[1]:=    f[x_] = Sqrt[x]
           g[x_] = x^6
           b = 1
           h[x_] := g[x]/;x > b
           h[x_] := f[x] /;x <= b
```

```
In[2]:=    spline =
           Plot[h[x], {x,b - 1,b + 1 },
           PlotRange->{0,10}];
```

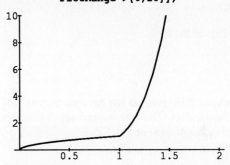

Note the hitch (or corner) at $\{1,1\}$.

The tangent to the $y = f[x]$ curve through $\{1, f[1]\}$ is:

```
In[3]:=    yf = Expand[f'[b] (x - b) + f[b]]
Out[3]=    1   x
           - + -
           2   2
```

The tangent to the $y = g[x]$ curve through $\{1, g[1]\}$ is:

```
In[4]:=    yg = Expand[g'[b] (x - b) + g[b]]
Out[4]=    -5 + 6 x
```

The tangent lines are different; this is the reason that different formulas to the left and right forced a corner at $\{1,1\}$.

```
In[5]:=    tanlines =
           Plot[{yf,yg},{x,b-1,b+1},
           PlotStyle->
           {RGBColor[0,1,0],RGBColor[0,1,1]},
           DisplayFunction->Identity]
           Show[tanlines,spline,
           DisplayFunction->$DisplayFunction];
```

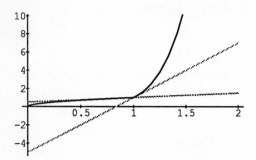

A mess. The tangent lines make the same angles that the curves make and there is no way to determine a tangent line line to $y = h[x]$ through the point $\{b, h[b]\}$.

The value $h'[b]$ of the derivative should be given by the common value:

```
In[6]:=    {f'[b],g'[b]}
Out[6]=     1
           {-, 6}
            2
```

But these values are different! So we have no way to determine the value of $h'[b]$. The best we can say is that the "left hand" derivative of $h[x]$ at $x = b$ is:

```
In[7]:=    f'[b]
Out[7]=    1
           -
           2
```

And the "right hand" derivative of $h[x]$ at $x = b$ is:

```
In[8]:=    g'[b]
Out[8]=    6
```

### T.2.a.iii)

Under what general circumstances does $h'[b]$ make sense? In the case that $h'[b]$ makes sense, some folks say that $h[x]$ is a **smooth spline knotted at** $b$.

**Answer:**

There is no problem provided $f[b] = g[b]$ and $f'[b] = g'[b]$. In this case we have no trouble in agreeing that $h'[b]$ is given by the common value of $f'[b]$ and $g'[b]$.

### T.2.b)

Put
$$h[x] = c(x^2 + 1) + k$$

for $x \leq 1$ and
$$h[x] = x^4$$

for $x > 1$.

Say how to set the constants $c$ and $k$ so that $h[x]$ is a smooth spline knotted at 1.

Once you have found $c$ and $k$, give formulas for

$h'[x]$ for $x < 1$,

$h'[x]$ for $x > 1$, and

$h'[x]$ for $x = 1$.

Plot $h[x]$ and its tangent line through $\{1, h[1]\}$ on an interval centered on 1.

**Answer:**

OK. Let us set a preliminary version of the function:

```
In[1]:=    prelimf[x_] := c (x^2 + 1) + k
```

```
In[2]:=    g[x_] = x^4
Out[2]=     4
           x
```

We have to find $c$ and $k$ to make the following equations hold:

```
In[3]:=    eqn1 = prelimf[1] == g[1]
Out[3]=    2 c + k == 1
```

```
In[4]:=    eqn2 = prelimf'[1] == g'[1]
Out[4]=    2 c == 4
```

Now solve:

```
In[5]:=    sols = Solve[{eqn1,eqn2},{c,k}]
Out[5]=    {{c -> 2, k -> -3}}
```

Now put:

```
In[6]:=    f[x_] = prelimf[x]/.sols[[1]]
Out[6]=                  2
           -3 + 2 (1 + x )
```

and set:

```
In[7]:=    h[x_] := g[x]/;x > 1
           h[x_] := f[x] /;x <= 1
```

Here comes the plot of the spline $h[x]$ on an interval centered on 1:

```
In[8]:=    spline =
           Plot[h[x],{x,0,2},PlotRange->{-3,5}];
```

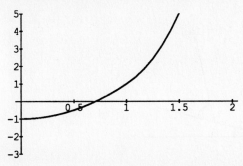

Nicely smooth through $\{1, h[1]\}$. Let's throw in the tangent line:

The tangent to the $y = f[x]$ curve through $\{1, f[1]\}$ is:

```
In[9]:=    yf = Expand[f'[1](x - 1) + f[1]]
Out[9]=    -3 + 4 x
```

The tangent to the $y = g[x]$ curve through $\{1, g[1]\}$ is:

```
In[10]:=   yg = Expand[g'[1](x - 1) + g[1]]
Out[10]=   -3 + 4 x
```

Both the same. We are rolling.

```
In[11]:=   tanlines =
           Plot[{yf,yg},{x,0,2},
           DisplayFunction-> Identity];
           Show[spline,tanlines,PlotRange->{-3,5}];
```

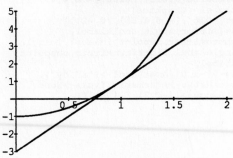

Just one tangent line: we are home free.

The value of $h'[1]$ is given by:

```
In[12]:=   {f'[1],g'[1]}
Out[12]=   {4, 4}
```

Thus $h'[1] = 4$.

■ **T.3) Square roots by Newton's method.**

**T.3.a)**

Newton's method gives a simple method for extracting square roots on a very cheap calculator.

Here is the idea: Given a number $A$, we try to find a root of $f[x] = x^2 - A$ by Newton's method. Say what the update formula is and then show how to punch the approprate numbers into the calculator.

**Answer:**

Set the function:

```
In[1]:=    f[x_] = x^2 - A
                      2
Out[1]=    -A + x
```

Take the formula

$$a[n] = a[n - 1] - f[a[n - 1]]/f'[a[n - 1]]$$

from B.2.b) above. So $a[n]$ is given by:

```
In[2]:=    a[n] = a[n-1] - f[a[n-1]]/f'[a[n-1]]
Out[2]=                                2
                          -A + a[-1 + n]
           a[-1 + n]  -  ----------------
                            2 a[-1 + n]
```

Let's put this in a more agreeable form:

```
In[3]:=    updateformula = Apart[a[n]]
Out[3]=          A            a[-1 + n]
           -----------  +  ----------
           2 a[-1 + n]         2
```

So the update formula is:

$$a[n] = (1/2)(A/a[n - 1] + a[n - 1]).$$

Now to compute $\sqrt{3}$, we just punch the following numbers into the calculator:

We'll take 1.5 for the first guess which leads to the first approximation:

```
In[4]:=    (1/2)(3/1.5 + 1.5)
Out[4]=    1.75
```

The second approximation is:

```
In[5]:=    (1/2)(3/1.75 + 1.75)
Out[5]=    1.73214
```

The third approximation is:

```
In[6]:=    (1/2)(3/1.73214 + 1.73214)
Out[6]=    1.73205
```

Here is $\sqrt{3}$ to machine accuracy:

```
In[7]:=    N[Sqrt[3]]
Out[7]=    1.73205
```

On the calculator, we would have gotten this in three steps.

To compute $\sqrt{10}$, we just punch the following numbers into the calculator:

We'll take 3.0 for the first guess which leads to the first approximation:

```
In[8]:=    (1/2) (10/3.0 + 3.0)
Out[8]=    3.16667
```

The second approximation is:

```
In[9]:=    (1/2) (10/3.16667 + 3.16667)
Out[9]=    3.16228
```

The third approximation is:

```
In[10]:=   (1/2) (10/3.16228 + 3.16228)
Out[10]=   3.16228
```

Here is $\sqrt{10}$ to machine accuracy:

```
In[11]:=   N[Sqrt[10]]
Out[11]=   3.16228
```

On the calculator, we would have gotten this in two steps.

■ **T.4) Reflecting Mirrors.**

A mirror is inclined to make an angle $t$ with the $x$-axis:

```
In[12]:=   mirror =
           Graphics[Line[{{-3,-0.75},{6,1.5}}]]
           angle = Graphics[
           Circle[{0,0},2,{0,ArcTan[(1.5)/6]}]]
           anglelabel =
           Graphics[Text["t",{1.6,.13}]]
           Show[mirror,angle,anglelabel,
           Axes->{-1,0},AxesLabel->{"x","y"},
           AspectRatio->Automatic,Ticks->None,
           PlotRange->{{-2,7},{-1,6}}];
```

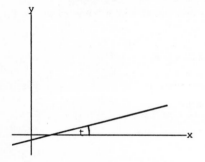

A light ray of slope *sincom* originating at the indicated point hits the mirror at angle $z$:

```
In[13]:=   lighttomirror=Graphics[
           {RGBColor[1,0,0],Line[{{4,5},{3,.75}}]}]
           inpoint=Graphics[
           {PointSize[0.015],Point[{4,5}]}]
           inlabel=
           Graphics[Text["In",{4,5},{0,-2}]]
           inangle=
           Graphics[
           Circle[{3,.75},.9,
           {ArcTan[.25],ArcTan[4.25]}]];
           inanglelabel =
           Graphics[Text["z",{3.5,1.25}]]
           Show[mirror,angle,anglelabel,
           lighttomirror,
           inpoint,inlabel,inangle,inanglelabel,
           Axes->{-1,0},AxesLabel->{"x","y"},
           AspectRatio->Automatic,Ticks->None,
           PlotRange->{{-2,7},{-1,6}}];
```

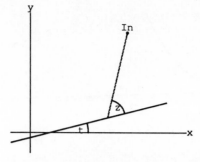

Because the angle of incidence is the angle of reflection (see B.3.b), the light ray bounces off the mirror making an equal angle $z$ with the mirror on the left.

**T.4.a.i)**

Given the slope, *sincom*, of the incoming ray and the slope, *smirror*, of the mirror what is the slope, *soutgo*, of the outgoing ray?

**Answer:**

The outgoing ray goes through the indicated point:

```
In[1]:=    lightfrommirror = Graphics[
           {RGBColor[1,0,0],
           Line[{{-.6,4.85},{3,.75}}]}]
           outpoint = Graphics[
           {PointSize[0.015],Point[{-.6,4.85}]}];
           outlabel = Graphics[
           Text["Out",{-.6,4.85},{0,-2}]]
           Show[mirror,angle,anglelabel,
           lighttomirror,inpoint,inlabel,inangle,
           inanglelabel,lightfrommirror,outpoint,
           outlabel,
           Axes->{-1,0},AxesLabel->{"x","y"},
           AspectRatio->Automatic,Ticks->None,
           PlotRange->{{-2,7},{-1,6}}];
```

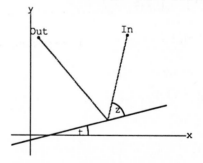

Let's put in some more labels:

The outgoing ray makes a counterclockwise angle

$$z + (\pi - 2z) = \pi - z$$

with the mirror. So the outgoing ray makes a counterclockwise angle

$$t + \pi - z$$

with the $x$- axis, while the incoming ray makes a counterclockwise angle

$$t + z$$

with the $x$-axis.

Thus the slopes of the incoming ray (*sincom*), of the mirror (*smirror*) and of the reflected ray (*soutgo*) are:

$$smirror = \tan[t]$$

$$sincom = \tan[t + z] = (\tan[t] + \tan[z])/(1 - \tan[t]\tan[z])$$

and

$$soutgo = \tan[t - z + \pi] = \tan[t - z]$$

$$= (\tan[t] - \tan[z])/(1 + \tan[t]\tan[z]).$$

($\tan[x]$ repeats itself on intervals of length $\pi$.)

We'll solve the first two equations for $\tan[t]$ and $\tan[z]$ in terms of *smirror* and *sincom*:

```
In[2]:=   Clear[smirror,sincom,soutgo,
          tant,tanz, u, v]
          tanssolved =
          Solve[{smirror == tant,
          sincom == (tant + tanz)/(1 - tant tanz)},
          {tant,tanz}]
Out[2]=              sincom - smirror
          {{tanz -> -------------------,
                     1 + sincom smirror

             tant -> smirror}}
```

So

$$soutgo = \tan[t - z] = (\tan[t] - \tan[z])/(1 + \tan[t]\tan[z])$$

is given by:

```
In[3]:=   soutgo =
          Simplify[(tant -tanz)/(1 + tant tanz)/.
          tanssolved[[1]]]
Out[3]=                                       2
          sincom - 2 smirror - sincom smirror
          ------------------------------------
                                            2
              -1 - 2 sincom smirror + smirror
```

Thank God for *Mathematica*.

### T.4.a.ii)

A flat mirror is aligned with the line $y = x/2 - 1$. A light ray comes in on the line $y = 2.5x - 2$ and bounces off the mirror.

Determine the equation of the line on which the outgoing light ray moves and plot in true scale.

### Answer:

First find where the ray hits the mirror:

```
In[1]:=   Solve[x/2 - 1 == 2.5 x - 2,x]
Out[1]=   {{x -> 0.5}}
```

The light ray bounces off the mirror at the point:

```
In[2]:=   hit = {x,2.5 x - 2}/.x->.5
```

```
Out[2]=   {0.5, -0.75}
```

Here is a picture of the light coming in:

```
In[3]:=   incomplot =
          Plot[2.5 x - 2,{x,.5,2},
          PlotStyle->RGBColor[1,0,0],
          DisplayFunction->Identity];
          mirrorplot =
          Plot[x/2 - 1,{x,-3,3},
          DisplayFunction->Identity];
          Show[incomplot,mirrorplot,Axes->{0,0},
          AxesLabel->{"x","y"},
          AspectRatio->Automatic,
          DisplayFunction->$DisplayFunction];
```

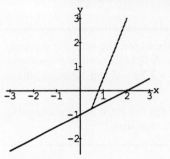

Now find the line on which the relected light moves using the formula developed in part a.i):

Recall from above:

```
In[4]:=   soutgo =
          (sincom - 2smirror - sincom smirror^2)/
          (-1 - 2 sincom smirror + smirror^2)
Out[4]=                                         2
          sincom - 2 smirror - sincom smirror
          ------------------------------------
                                             2
             -1 - 2 sincom smirror + smirror
```

In this problem, $smirror = 1/2$ and $sincom = 2.5$; so the slope of the outgoing light ray is:

```
In[5]:=   soutgo/.{smirror->1/2,sincom->2.5}
Out[5]=   -0.269231
```

The light bounces off the mirror at

```
In[6]:=   hit
Out[6]=   {0.5, -0.75}
```

and the outgoing light moves on the line

$$y = (-.269231)(x - .5) + (-.75).$$

Here is a plot:

```
In[7]:=   outplot =
          Plot[ (-.269231)( x - .5) + (-.75),
          {x,-3,.5},PlotStyle->RGBColor[1,0,0],
          DisplayFunction->Identity]
          Show[incomplot,mirrorplot,outplot,
          Axes->{0,0},
          AspectRatio->Automatic,
          PlotLabel->"Path of light ray",
          DisplayFunction->$DisplayFunction];
```

Path of light ray

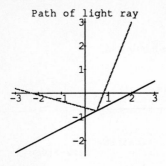

Not bad.

### T.4.a.iii)

A flat mirror is bent to be in alignment with the parabola $y = x^2 - 1$ with $0 \leq x \leq 1$. A light ray comes from above the parabola on the line $y = 2.5x - 2$ and bounces off the mirror.

Determine the equation of the line on which the outgoing light ray moves and plot in true scale.

### Answer:

Take a look:

```
In[1]:=    parabola = x^2 - 1
           lightin = 2.5 x - 2
           Plot[{parabola,lightin},{x,0,1},
           AspectRatio->Automatic,
           PlotRange->{-1,0.5}];
```

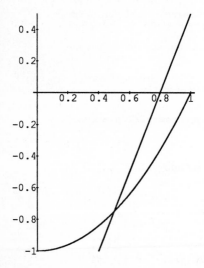

First find where the light ray meets the parabolic mirror :

```
In[2]:=    Solve[parabola == lightin,x]
Out[2]=    {{x -> 2.}, {x -> 0.5}}
```

The light ray bounces off the mirror at the point:

```
In[3]:=    hit = {x,parabola}/.x->.5
```

```
Out[3]=    {0.5, -0.75}
```

Here is a better true scale plot:

```
In[4]:=    parabplot =
           Plot[parabola,{x,0,1},
           DisplayFunction->Identity]
           lightinplot =
           Plot[lightin,{x,.5,1},
           PlotStyle->RGBColor[1,0,0],
           DisplayFunction->Identity]
           Show[parabplot,lightinplot,
           AspectRatio->Automatic,
           PlotRange->{-1,0.5},
           DisplayFunction->$DisplayFunction];
```

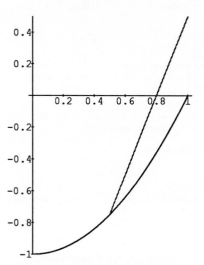

Now let's find the line on which the relected light moves.

The optical principle is that the light bounces off the parabola as if it were bouncing off the **tangent line to the parabola at the point at which the light hits the parabola.** This means we can use:

```
In[5]:=    smirror = D[parabola,x]/.x->.5
Out[5]=    1.
```

The incoming light is moving on the line $2.5x - 2$; so we use:

```
In[6]:=    sincom = 2.5
Out[6]=    2.5
```

Recall from above:

```
In[7]:=    soutgo =
           (sincom - 2smirror - sincom smirror^2)/
           (-1 - 2 sincom smirror + smirror^2)
Out[7]=    0.4
```

The relected light moves with slope given by:

```
In[8]:=    soutgo
Out[8]=    0.4
```

The light hits the parabola at:

```
In[9]:=    hit
Out[9]=    {0.5, -0.75}
```

The outgoing light moves on the line

$$y = .4(x - .5) - .75$$

Here is a plot:

```
In[10]:=   outplot =
           Plot[ .4 (x - .5) -0.75,{x,-.5,.5},
>RGBColor[1,0,0],
           DisplayFunction->Identity];
           Show[parabplot,lightinplot,outplot,
           AxesLabel->{"x","y"},
           AspectRatio->Automatic,
           PlotRange->{-1,0.5},
           DisplayFunction->$DisplayFunction];
```

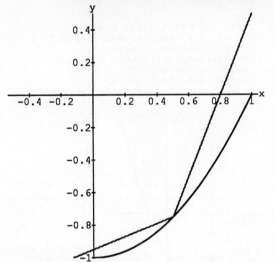

Let's include a plot with the tangent line at the point at which the light hits the parabola. The slope of the tangent line at {0.5, −0.75} is:

```
In[11]:=   D[x^2 - 1,x]/.x->0.5
Out[11]=   1.
```

The tangent line equation is:

```
In[12]:=   ytan = (x - .5) - 0.75
Out[12]=   -1.25 + x
```

```
In[13]:=   tanplot =
           Plot[ytan,{x,0,1},
           PlotStyle->RGBColor[0,0,1],
           DisplayFunction->Identity]
           Show[parabplot,lightinplot,outplot,
           tanplot,
           AspectRatio->Automatic,
           PlotRange->{-1,0.5},
           DisplayFunction->$DisplayFunction];
```

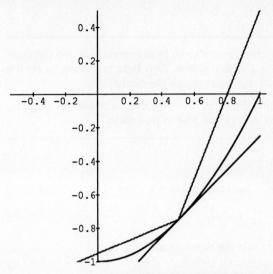

You can see why the the light bounces off the curve the same way it bounces off the tangent line.

**T.4.b.i)**

---

The formula

$$soutgo$$
$$= (sincom - 2smirror - sincomsmirror^2)/$$
$$(-1 - 2sincomsmirror + smirror^2)$$

is defective in the case that the light comes in on a vertical line because in this case $sincom = \infty$.

What is the formula in the case that the light comes in on a vertical line?

---

**Answer:**

Look at the formula

$$soutgo =$$
$$(sincom - 2smirror - sincomsmirror^2)/$$
$$(-1 - 2sincomsmirror + smirror^2)$$

Take the limit as $sincom \to \infty$ by dividing the dominant terms as $sincom \to \infty$ and cancelling:

The quotient of dominant terms is

$$(sincom - sincomsmirror^2)/(-2sincomsmirror)$$
$$= (1 - smirror^2)/(-2smirror)$$
$$= (smirror^2 - 1)/(2smirror).$$

Thus in the case that the light comes in on a vertical line, then

$$soutgo = (smirror^2 - 1)/(2smirror)$$

**T.4.b.ii)**

A flat mirror is bent to be in alignment with the parabola $y = x^2 - 2$ with $0 \leq x \leq 2$. A light ray comes in on the line $x = 1$ and bounces off the mirror.

Determine the equation of the line on which the outgoing light ray moves and plot in true scale.

**Answer:**

```
In[1]:=   parabola = x^2 - 2
Out[1]=        2
          -2 + x
```

The light hits the parabola at the point:

```
In[2]:=   hit ={x, parabola}/.x->1
Out[2]=   {1, -1}
```

Take a look:

```
In[3]:=   lightin =
          Graphics[{RGBColor[1,0,0],
          Line[{{1,-1},{1,2}}]}]
          parabplot = Plot[x^2 - 2, {x,0,2},
          DisplayFunction->Identity]
          Show[parabplot,lightin,
          AspectRatio->Automatic,
          DisplayFunction->$DisplayFunction];
```

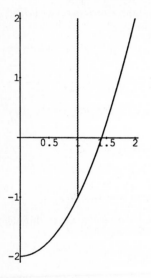

Use the formula

$$soutgo = (smirror^2 - 1)/(2smirror)$$

in part b.i) above.

```
In[4]:=   soutgo = (smirror^2 - 1)/(2 smirror)
Out[4]=            2
          -1 + smirror
          -------------
            2 smirror
```

In this problem *smirror* is:

```
In[5]:=   D[parabola,x]/.x->1
Out[5]=   2
```

*soutgo* is given by:

```
In[6]:=   soutgo/.smirror->2
Out[6]=   3
          -
          4
```

The light ray bounces off the mirror at the point:

```
In[7]:=   hit
Out[7]=   {1, -1}
```

So the reflected light moves on the line

$$y = (3/4)(x - 1) - 1.$$

Here comes the plot:

```
In[8]:=   lightout = Plot[(3/4) (x - 1) - 1,
          {x,-2,1},PlotStyle->RGBColor[1,0,0],
          DisplayFunction->Identity]
          Show[parabplot,lightin,lightout,
          AxesLabel->{"x","y"},
          AspectRatio->Automatic,
          DisplayFunction->$DisplayFunction];
```

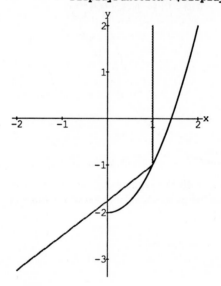

Spiffy.

# Literacy Sheet

**1.** What is the equation of the line tangent to the curve $y = f[x]$ through the point $\{x_0, f[x_0]\}$ ?

In the vicinity of $x_0$ , how close to the curve is this tangent line?

**2.** If the tangent line to the graph of $y = f[x]$ at $x = 1$ is $y = 2x - 3$ and the tangent line to same graph at $x = 4$ is $y = x + 2$, where is $y = f[x]$ going up faster: At $x = 1$ or at $x = 2$ ?

**3.** Find a constant $c$ such that the parabola $y = x^2 + c$ is tangent to the line $y = x$ . Give a hand sketch of the parabola and the line on the same axes.

**4.** The functions $y = f[x]$ and $y = g[x]$ satisfy $f[x_0] = g[x_0]$ and $f'[x_0] = g'[x_0]$ for some $x_0$ . What can you report about the tangent lines to their graphs at $\{x_0, f[x_0]\} = \{x_0 g[x_0]\}$ ? Why are these functions ripe for smoothly splining?

**5.** Four functions $f[x], g[x], h[x]$ and $k[x]$ satisfy

$f[1] = 3$ and $f'[1] = -2$ ,

$g[1] = -2$ and $g'[1] = -2$ ,

$h[1] = 3$ and $h'[1] = -1$ ,

$k[1] = 3$ and $k'[1] = -2$ .

Which pair or pairs of functions would you pick to define a smooth spline(s) knotted at $x = 1$ ?

**6.** How does the sign of the second derivative tell us whether the tangents to the curve $y = f[x]$ are below or above the curve?

**7.** A convex function is a function $f[x]$ with the property that all the tangent lines to the graph of $y = f[x]$ ride below the curve. Suppose $f[x]$ is a convex function and suppose $c = (a + b)/2$. Explain why

(i) $f[a] \geq f[c] + f'[c](a - c) = f[c] + f'[c](a - b)/2$

(ii) $f[b] \geq f[c] + f'[c](b - c) = f[c] + f'[c](b - a)/2$

(iii) $f[b] + f[a] \geq 2f[c]$

(iv) $(f[b] + f[a])/2 \geq f[(a + b)/2]$.

**8.** Two functions $f[x]$ and $g[x]$ are even; so their graphs are symmetrical with respect to the $y-$ axis. What conditions do you need to define a smooth spline knotted at $0$ with $f[x]$ on the left of $x = 0$ and $g[x]$ on the right of $x = 0$ ?

**9.** Here is the graph of a certain function f[x]. With a pencil draw a sketch of the steps Newton's method takes in the following graph for each of the three starts:

$a_1 = -1$,

$a_1 = 1$,

$a_1 = 3$.

Which roots do you approximate in each case?

Mark all possible starts that will approximate the root near $x = 0$ .

```
In[9]:=    graph=Plot[0.3 (x-1)^4 + 3 Sin[3 x],
           {x,-2,4},PlotRange->{-3,7},
           Ticks->{Automatic,None},
           DisplayFunction->Identity]
           starts=Graphics[{PointSize[.02],
           Point[{-1,0}],Point[{1,0}],Point[{3,0}]}]
           Show[graph,starts,
           DisplayFunction->$DisplayFunction]
```

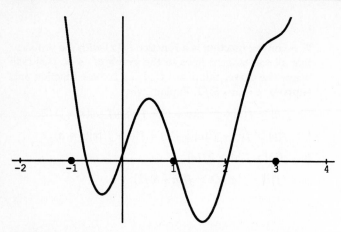

**10.** The formula $a_{n+1} = (a_n + A/a_n)/2$ defines successive approximations to the a quantity related to $A$. Which quantity?

.

.

.

**11.** How does light bounce off a flat surface?

.

.

.

How does light bounce off a curved surface?

.

.

.

**12.** Three flat mirrors are placed to box in a $30° - 60° - 90°$ triangle. With a pencil, start a light ray at the vertex with the right angle and bounce it five times by the usual reflection laws. Does it seem to go to any special place? How does the initial angle affect the outcome?

.

.

.

**13.** What is the underlying principle of the parabolic dish antennas used for TV reception?

.

.

.

# Parameters

## Guide

Here is how the Earth's orbit about the sun looks when it is charted from Mars. Activate and enjoy:

```
In[1]:=    {xearth[t_],yearth[t_]} =
           {Cos[2 Pi t],Sin[2 Pi t]}

           {xmars[t_],ymars[t_]} =
           {1.52 Cos[2 Pi t/2],
           1.52 Sin[2 Pi t/2]}

           earthorbit =
           ParametricPlot[
           {xearth[t],yearth[t]} -
           {xmars[t],ymars[t]},{t,0,2},
           AspectRatio-> Automatic,
           DisplayFunction->Identity]

           mars =
           Graphics[RGBColor[1,0,0],
           PointSize[0.1],Point[{0,0}]]

           Show[earthorbit,mars,
           AspectRatio-> Automatic,
           Ticks->{{1,2},{1,2}},
           DisplayFunction->$DisplayFunction];
```

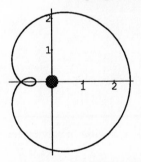

That's the red planet Mars at the origin. Before we can see why generating this plot was easy, let's take a step back. The graph of any function is a curve. There are other curves (the path of Earth relative to Mars above is one such) and we need to have a good way of studying them. We can describe a curve as the path of a moving point. In this interpretation, we just have to say where the point is at any given time.

An example: Say that for any $t$ (= time) between 3 and 6 the point is at $\{x[t], y[t]\}$ where:

```
In[2]:=    x[t_] = t + Sin[3t]

Out[2]=    t + Sin[3 t]
```

```
In[3]:=    y[t_] = (t^2)/5 + Cos[5t]
```

```
Out[3]=     2
           t
           -- + Cos[5 t]
           5
```

The path of the point looks like this:

```
In[4]:=    path =
           ParametricPlot[{x[t],y[t]},
           {t,3,6},Axes->{0,0},
           AspectRatio->Automatic,
           PlotRange->{{-1,10},{-1,9}}];
```

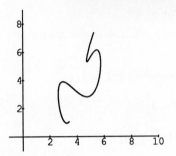

It begins at

```
In[5]:=    start = N[{x[3],y[3]}]

Out[5]=    {3.41212, 1.04031}
```

and ends at

```
In[6]:=    finish = N[{x[6],y[6]}]

Out[6]=    {5.24901, 7.35425}
```

Here is the path (= curve) with its endpoints:

```
In[7]:=    ends =
           Graphics[RGBColor[0,0,1],
           PointSize[.04],Point[start],
           Point[finish]];

           labels=
           Graphics[Text["start",start,{-1.5,0}],
           Text["finish",finish,{0,-2.2}]]

           Show[path,ends,labels,
           AspectRatio->Automatic,
           PlotRange->{{-1,10},{-1,9}}];
```

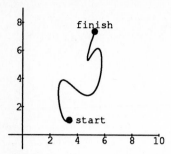

Parametric curves can be fascinating and they have lots of utility.

Jump in and enjoy.

# Basics

■ **B.1) Parametric Plots.**

One way of plotting the circle

$$x^2 + y^2 = 9$$

is to solve for $y$ in terms of $x$ and then to plot the top and the bottom of the circle:

```
In[8]:=    ysolved = Solve[x^2 + y^2 == 9,y]
```
```
Out[8]=
             2
          {{y -> Sqrt[9 - x ]},

                        2
           {y -> -Sqrt[9 - x ]}}
```

```
In[9]:=    top =ysolved[[1,1,2]]
```
```
Out[9]=
              2
          Sqrt[9 - x ]
```

```
In[10]:=   bottom = ysolved[[2,1,2]]
```
```
Out[10]=
               2
          -Sqrt[9 - x ]
```

```
In[11]:=   Plot[{top,bottom},{x,-3,3},
           AspectRatio->Automatic,
           PlotRange->{{-4,4},{-4,4}}];
```

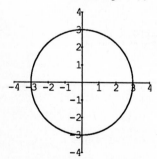

Another way of plotting the circle $x^2 + y^2 = 9$ is to write $x[t] = 3\cos[t]$ and $y[t] = 3\sin[t]$ and then to plot the points

$$\{x[t], y[t]\} = \{3\cos[t], 3\sin[t]\}$$

as $t$ advances from 0 to $2\pi$.

```
In[12]:=   ParametricPlot[
           {3 Cos[t],3 Sin[t]},{t,0,2 Pi},
           AspectRatio->Automatic,
           PlotRange->{{-4,4},{-4,4}}];
```

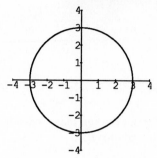

In this manner, the original variables $x$ and $y$ are plotted by means or a third variable ($t$ in this case) called **parameter**. A parameter is an auxillary variable that plays a back-room role. Parameters sometimes give us plotting freedom normal plotting does not allow.

Of course we can always use:

```
In[13]:=   Show[Graphics[Circle[{0,0},3]],
           Axes->{0,0},
           AspectRatio->Automatic,
           PlotRange->{{-4,4},{-4,4}}];
```

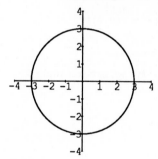

## B.1.a)

Use parametric plotting to plot $y^2 = x$ without solving for $y$ in advance.

**Answer:**

Put:

```
In[1]:=    x[t_] = t^2
```
```
Out[1]=
             2
          t
```

```
In[2]:=    y[t_] = t
```
```
Out[2]=   t
```

This makes

$$y^2 = t^2 = x$$

and we can plot the parabola:

```
In[3]:=   ParametricPlot[
          {x[t],y[t]},{t,-3,3},
          AxesLabel->{"x","y"},
          AspectRatio->Automatic];
```

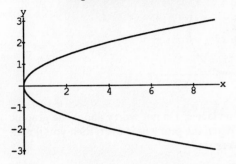

Using more $t$'s gives more of the curve $y^2 = x$ :

```
In[4]:=   ParametricPlot[
          {x[t],y[t]},{t,-5,5},
          AxesLabel->{"x","y"},
          AspectRatio->Automatic];
```

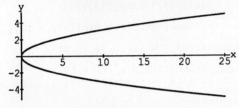

The top part corresponds to the solution $y = \sqrt{x}$ of $y^2 = x$; it comes from the positive $t$'s.

```
In[5]:=   ParametricPlot[
          {x[t],y[t]},{t,0,3},
          AxesLabel->{"x","y"},
          AspectRatio->Automatic];
```

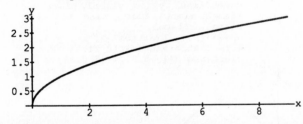

The bottom part corresponds to the solution $y = -\sqrt{x}$ of $y^2 = x$; it comes from the negative $t$'s.

```
In[6]:=   ParametricPlot[
          {x[t],y[t]},{t,-3,0},
          AxesLabel->{"x","y"},
          AspectRatio->Automatic];
```

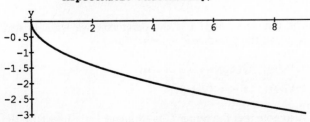

**B.1.b): Spiral.**

Some curves are best described via parametric equations and are more difficult to describe in the usual $y = f[x]$ terms. To see what this means, plot the spiral:

$x = x[t] = t \sin[t]$

$y = y[t] = t \cos[t]$.

**Answer:**

```
In[1]:=   x[t_] = t Cos[t]
Out[1]=   t Cos[t]

In[2]:=   y[t_] = t Sin[t]
Out[2]=   t Sin[t]

In[3]:=   ParametricPlot[
          {x[t],y[t]},{t,0,8Pi},
          AspectRatio->Automatic,
          PlotStyle->
          {{Thickness[0.01],RGBColor[1,0,0]}}];
```

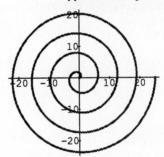

Extend the values of the parameter $t$ to see more of the spiral:

```
In[4]:=   ParametricPlot[
          {x[t],y[t]},{t,0,16 Pi},
          AspectRatio->Automatic,
          PlotStyle->
          {{Thickness[0.01],RGBColor[1,0,0]}}];
```

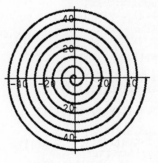

Bull's Eye.

**B.1.c): Cycloid.**

Sometimes parametric equations are the only viable way of setting up a precise description of a curve. Here is a case: A circle of radius $r$ rolls along the $x$-axis. The path traced out by a point $P$ on the circle is called a *cycloid*. Give parametric equations for the cycloid.

Plot in the case that $r = 1$.

**Answer:**

We place the coordinate system so that at the initial point the set-up looks like this:

```
In[1]:=   origin =
          Graphics[{RGBColor[1,0,0],PointSize[0.03],
          Point[{0,0}]}]

          label =
          Graphics[Text["P",{0,0},{2,2}]]

          circle =
          Graphics[Circle[{0,1},1]]

          center =
          Graphics[{PointSize[0.03],Point[{0,1}]}]

          Show[origin,label,circle,center,
          Axes->{0,0},AxesLabel->{"x","y"},
          AspectRatio->Automatic,Ticks->None,
          PlotRange->{{-1.5,1.5},{-0.5,2.5}}];
```

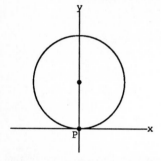

After the circle has rolled to the right a bit, the set-up looks like this:

```
In[2]:=   point = Graphics[
          {RGBColor[1,0,0],PointSize[0.03],
          Point[{1.2-Sin[1.2],1-Cos[1.2]}]}]

          pointlabel = Graphics[Text[
          "P",{1.2 - Sin[1.2],1 - Cos[1.2]},
          {0,-2}]]

          circle =
          Graphics[Circle[{1.2,1},1]]

          center = Graphics[
          {PointSize[0.03],Point[{1.2,1}]}]
          centerlabel =
          Graphics[Text["C",{1.2,1},{0,-2}]]

          radiusline = Graphics[Line[
          {{1.2-Sin[1.2],1-Cos[1.2]},{1.2,1}}]]
          rlabel = Graphics[Text["r",
          {1.2 -Sin[1.2]/2,1.35 - Cos[1/2]/2}]]

          Show[point,pointlabel,circle,center,
          radiusline,centerlabel,rlabel,
          Axes->{0,0},AxesLabel->{"x","y"},
          AspectRatio->Automatic,Ticks->None,
          PlotRange->{{-.5,3},{-.5,2.5}}];
```

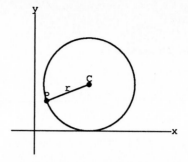

Now put in some labels: Do not worry about the graphic specifications. When the problem is over, then you'll realize how they were set.

```
In[3]:=   yline =
          Graphics[Line[{{0,1 - Cos[1.2]},
          {1.2 - Sin[1.2], 1 - Cos[1.2]}}]]
          ylabel =
          Graphics[Text["y",{-.1,1 - Cos[1.2]}]]
          xline =
          Graphics[Line[{{1.2 - Sin[1.2],0},
          {1.2 - Sin[1.2],1 - Cos[1.2]}}]]
          xlabel =
          Graphics[Text["x",{1.2 - Sin[1.2],-.1}]]
          mlabel =
          Graphics[Text["M",{1.2,0},{0,2}]]
          tline =
          Graphics[Line[{{1.2,0},{1.2,1},
          {1.2 - Sin[1.2],1 - Cos[1.2]}}]]
          tarc =
          Graphics[
          Circle[{1.2,1},.2,{3 Pi/2 -1.2,3 Pi/2}]]
          tlabel = Graphics[Text["t",{1.05,.75}]]
          rrlabel = Graphics[Text["r",{1.25,.5}]];
```

Here is the diagram:

```
In[4]:=   diagram =
          Show[point, pointlabel, circle, center,
          centerlabel, yline, xline,
          xlabel, mlabel, tline, tarc, tlabel,
          rlabel, rrlabel,
          Axes->{0,0},AxesLabel->{"x","y"},
          AspectRatio->Automatic,Ticks->None,
          PlotRange->{{-.5,3},{-.5,2.5}}];
```

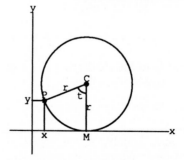

The indicated angle $t$ is the parameter we want to use. Read off the parametric equation for $y$:

```
In[5]:=   y[t_] = r - r Cos[t]

Out[5]=   r - r Cos[t]
```

Also note that the center $C$ is sitting at $\{rt, r\}$ because the length of the arc on the circle from the point of contact $M$ to the mark at $P$ is $rt$. This allows us to read off the parametric equation for $x$ :

```
In[6]:=    x[t_] = r t - r Sin[t]
Out[6]=    r t - r Sin[t]
```

Here comes the plot of the first roll in the case that

```
In[7]:=    r = 1
           firstroll =
           ParametricPlot[
           {x[t],y[t]},{t,0,2 Pi},
           AspectRatio->Automatic,
           Ticks->{Automatic,{1,2}},
           PlotStyle->
           {{RGBColor[1,0,0],Thickness[.008]}}];
```

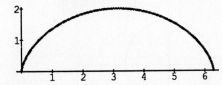

Two rolls:

```
In[8]:=    ParametricPlot[
           {x[t],y[t]},{t,0,4 Pi},
           AspectRatio->Automatic,
           Ticks->{Automatic,{1}},
           PlotStyle->
           {{RGBColor[1,0,0],Thickness[.008]}}];
```

Here is the diagram above with the actual plot of the first roll:

```
In[9]:=    Show[diagram,firstroll];
```

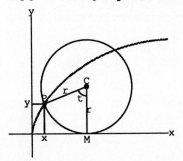

### ■ B.2) Parameterizing the circle and other curves.

There is no set way of taking a curve specified in $xy$-coordinates and finding a parameter $t$ for parametric equations $x = x[t]$, $y = y[t]$. But there are approaches that often pay off.

Here are a three ways of finding parametric equations for the circle $x^2 + y^2 = 1$.

### B.2.a.i)

Use the slope $t$ of the line from $\{-1,0\}$ to a point $\{x,y\}$ on

$x^2 + y^2 = 1$ to generate parametric equations of the circle

$$x^2 + y^2 = 1.$$

**Answer:**

Take a look:

```
In[1]:=    circle =
           Graphics[
           {RGBColor[0,0,1],Thickness[.008],
           Circle[{0,0},1]}]

           points =
           Graphics[
           {RGBColor[1,0,0],PointSize[0.03],
           Point[{1/2,Sqrt[3/4]}],
           Point[{-1,0}]}]

           pointlabel =
           Graphics[Text[{"x","y"},
           {1/2,Sqrt[3/4]},{-1.5,0}]]

           line =
           Graphics[Line[{{-1.2,
           (Sqrt[3]/3)(-1.2+1)},
           {1,(Sqrt[3]/3)(1+1)}}]]

           Show[circle,points,pointlabel,line,
           Axes->{0,0},AxesLabel->{"x","y"},
           AspectRatio->Automatic,
           Ticks->{{1},{1}},
           PlotRange->{{-3,2},{-1.5,1.5}}];
```

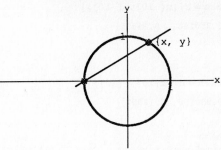

The parameter $t$ is the slope (= growth rate) of the indicated line.

Now to find parametric equations for $x$ and $y$ on the circle, just:

```
In[2]:=    Clear[x,y,t]
           Solve[{x^2 + y^2 == 1,
           y == t (x + 1)},{x,y}]
```

$$Out[2]= \left\{\left\{x \to -1 + \frac{2}{1 + t^2},\ y \to \frac{2t}{1 + t^2}\right\},\right.$$

$$\left.\{x \to -1,\ y \to 0\}\right\}$$

The parametric equations are:

```
In[3]:=    x[t_] =
           Together[-1 + 2/(1+t^2)]
```

$$Out[3]= \frac{1 - t^2}{1 + t^2}$$

```
In[4]:=    y[t_] = 2t/(1+t^2)

Out[4]=     2 t
          ------
               2
          1 + t
```

The parametric equations for *t* ranging form -10 to 10 plot out as:

```
In[5]:=    circ =
           ParametricPlot[
           {x[t],y[t]},{t,-10,10},
           AxesLabel->{"x","y"},
           AspectRatio->Automatic,
           Ticks->{{1},{1}},
           PlotRange->{{-3,2},{-1.5,1.5}}];
```

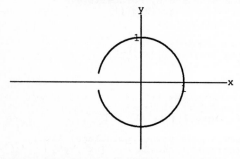

This path covers most of the circle $x^2 + y^2 = 1$. It starts at:

```
In[6]:=    start = N[{x[-10],y[-10]}]

Out[6]=    {-0.980198, -0.19802}
```

and ends at:

```
In[7]:=    finish = N[{x[10],y[10]}]

Out[7]=    {-0.980198, 0.19802}
```

```
In[8]:=    finish

Out[8]=    {-0.980198, 0.19802}
```

Here is the path with start and finish marked:

```
In[9]:=    Show[circ,Graphics[PointSize[.03],
           Point[start],Point[finish],
           Text["start",start,{1.5,0}],
           Text["finish",finish,{1.5,0}]],
           Ticks->{{1},{1}},
           PlotRange->{{-3,2},{-1.5,1.5}}];
```

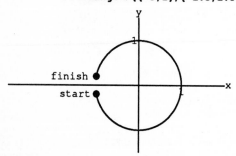

To get even more of the circle, use more *t's*:

```
In[10]:=   start = N[{x[-20],y[-20]}]
```

```
Out[10]=   {-0.995012, -0.0997506}
```

and ends at:

```
In[11]:=   finish = N[{x[20],y[20]}]

Out[11]=   {-0.995012, 0.0997506}
```

```
In[12]:=   Show[circ,Graphics[PointSize[.03],
           Point[start],Point[finish],
           Text["start",start,{1.5,.5}],
           Text["finish",finish,{1.5,-.5}]],
           Ticks->{{1},{1}},
           PlotRange->{{-3,2},{-1.5,1.5}}];
```

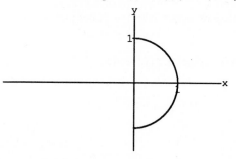

To get right hand side *half* of the circle all we have to do is run the parameter *t* from $-1$ to $+1$ :

```
In[13]:=   ParametricPlot[{x[t],y[t]},{t,-1,1},
           AxesLabel->{"x","y"}, AspectRatio->Automatic,
           Ticks->{{1},{1}},
           PlotRange->{{-3,2},{-1.5,1.5}}];
```

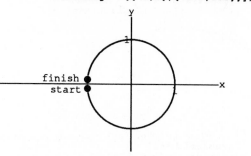

Neat. You may want to add the "start" and "finish", dots, etc.

### B.2.a.ii)

Use the slope *t* of the line from $\{0,0\}$ to a point $\{x, y\}$ on $x^2 + y^2 = 1$ to generate parametric equations of the circle $x^2 + y^2 = 1$.

**Answer:**

Take a look:

```
In[1]:=    circle =
           Graphics[RGBColor[0,0,1],
           Thickness[.008],
           Circle[{0,0},1]]

           point =
           Graphics[{RGBColor[1,0,0],
```

```
                 PointSize[0.03],
                 Point[{1/2,Sqrt[3/4]}]}]
                 pointlabel =
                 Graphics[Text[{"x","y"},{1/2,Sqrt[3/4]},{-
1.5,0}]]

                 line =
                 Graphics[Line[{{0,0},1.2{1/2,Sqrt[3/4]}}]]

                 Show[circle,point,pointlabel,line,
                 Axes->{0,0},
                 Ticks->{{1},{1}},
                 PlotRange->{{-3,2},{-1.5,1.5}}];
```

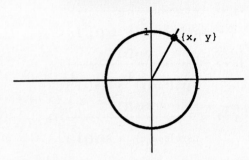

The parameter $t$ is the slope (= growth rate) of the indicated line.

Now to find parametric equations for $x$ and $y$ on

the circle, just:

```
In[2]:=   Clear[x,y,t]
          Solve[{x^2 + y^2 == 1,
          y == t x}, {x,y}]
```

$$Out[2]= \left\{\left\{x \to \frac{\mathrm{Sqrt}\left[1 - \frac{1}{1+t^2}\right]}{t},\right.\right.$$

$$\left.y \to \mathrm{Sqrt}\left[1 - \frac{1}{1+t^2}\right]\right\},$$

$$\left\{x \to -\left(\frac{\mathrm{Sqrt}\left[1 - \frac{1}{1+t^2}\right]}{t}\right),\right.$$

$$\left.\left.y \to -\mathrm{Sqrt}\left[1 - \frac{1}{1+t^2}\right]\right\}\right\}$$

The equations:

```
In[3]:=   x[t_] =
          Simplify[Sqrt[1 - (1/(1 + t^2))]/t]
```

$$Out[3]= \frac{1}{\mathrm{Sqrt}[1 + t^2]}$$

```
In[4]:=   y[t_] =
          Simplify[Sqrt[1 - (1/(1 + t^2))]]
```

$$Out[4]= \frac{t}{\mathrm{Sqrt}[1 + t^2]}$$

plot out as:

```
In[5]:=   right =
          ParametricPlot[{x[t],y[t]},
          {t,-10,10},
          AxesLabel->{"x","y"},
          AspectRatio->Automatic,
          Axes->{0,0},
          Ticks->{{1},{1}},
          PlotRange->{{-3,2},{-1.5,1.5}}];
```

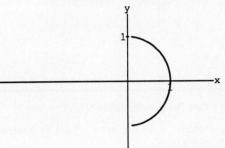

This gives us almost all of the right hand side of the circle.

To get the left half, use the other solutions:

```
In[6]:=   Clear[xx,yy,t]
          xx[t_] =
          Simplify[-Sqrt[1 - (1/(1 + t^2))]/t]
```

$$Out[6]= -\left(\frac{1}{\mathrm{Sqrt}[1 + t^2]}\right)$$

```
In[7]:=   yy[t_] =
          Simplify[-Sqrt[1 - (1/(1 + t^2))]]
```

$$Out[7]= -\left(\frac{t}{\mathrm{Sqrt}[1 + t^2]}\right)$$

```
In[8]:=   left =
          ParametricPlot[
          {xx[t],yy[t]},{t,-10,10},
          PlotStyle->{{RGBColor[1,0,0],
          Thickness[.008]}},
          AxesLabel->{"x","y"},
          AspectRatio->Automatic,
          Axes->{0,0},
          Ticks->{{1},{1}},
          PlotRange->{{-3,2},{-1.5,1.5}}];
```

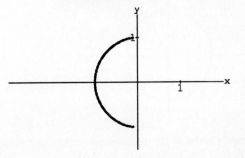

We can get the whole plot: The thicker of the two parts comes from the second set of solutions.

```
In[9]:=    Show[left,right];
```

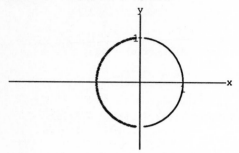

Cool.

### B.2.a.iii)

---

Use the angle $t$ that the $x$-axis makes with the line

from $\{0,0\}$ to a point $\{x,y\}$ on $x^2 + y^2 = 1$ to generate parametric equations of

$$x^2 + y^2 = 1.$$

---

**Answer:**

Take a look:

```
In[1]:=    circle =
           Graphics[RGBColor[0,0,1],
           Circle[{0,0},1]]
           point =
           Graphics[{RGBColor[1,0,0],PointSize[0.015],
           Point[{1/2,Sqrt[3/4]}]}]
           pointlabel = Graphics[Text[{"x","y"},
           {1/2,Sqrt[3/4]},{-1,-1}]]
           line=Graphics[Line[{{0,0},{1/2,Sqrt[3/4]}}]]
           angle =
           Graphics[Circle[{0,0},3/4,
           {0,ArcTan[Sqrt[3/4]/(1/2)]}]];
```

```
In[2]:=    Show[circle,point,pointlabel,line,
           angle,Axes->{0,0},
           AxesLabel->{"x","y"},
           AspectRatio->Automatic,
           Ticks->{{1},{1}},
           PlotRange->{{-3,2},{-1.5,1.5}}];
```

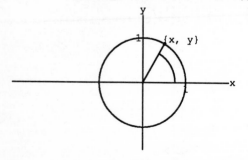

The parameter $t$ is the angle (measured counterclockwise) that the indicated line makes with the $x$-axis.

Now to find parametric equations for $x$ and $y$ on the circle, we just:

```
In[3]:=    params =
           Solve[{x^2 + y^2 == 1,
           y == (Sin[t]/Cos[t]) x}, {x,y}]
```

$$Out[3]= \left\{\left\{x \to \frac{\cos[t]}{\sqrt{\cos[t]^2 + \sin[t]^2}},\right.\right.$$

$$\left.y \to \frac{\sin[t]}{\sqrt{\cos[t]^2 + \sin[t]^2}}\right\},$$

$$\left\{x \to -\left(\frac{\cos[t]}{\sqrt{\cos[t]^2 + \sin[t]^2}}\right),\right.$$

$$\left.\left.y \to -\left(\frac{\sin[t]}{\sqrt{\cos[t]^2 + \sin[t]^2}}\right)\right\}\right\}$$

That's a mess, but we recognize that $\cos[t]^2 + \sin[t]^2 = 1$. So we incorporate this fact:

```
In[4]:=    params/. (Cos[t]^2 + Sin[t]^2)->1
Out[4]=    {{x -> Cos[t], y -> Sin[t]},
           {x -> -Cos[t], y -> -Sin[t]}}
```

Consequently as $t$ advances from 0, the points:

```
In[5]:=    x[t_] = Cos[t]
Out[5]=    Cos[t]
```

```
In[6]:=    y[t_] = Sin[t]
Out[6]=    Sin[t]
```

advance in a counterclockwise fashion around the circle $x^2 + y^2 = 1$.

```
In[7]:=    ParametricPlot[
           {x[t],y[t]},{t,0,2 Pi},
           AxesLabel->{"x","y"},
           AspectRatio->Automatic];
```

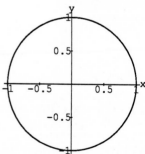

Right on the money. The other solutions $x = -\cos[t]$ and $y = -\sin[t]$ will work equally well. But they carry the fear of minus signs; so we ignore them.

**B.2.b)**

The *folium of Descartes* is described by

$$x^3 + y^3 = 6xy.$$

But usual $x - y$ plotting of this is disgusting because when we solve for $y$ in terms of $x$, then we get a mess:

```
In[1]:=    Solve[x^3 + y^3 == 6 x y, y]
```

```
Out[1]=                      2 x
           {{y -> --------------------------- +
                    3                6
                  -x           3   x    1/3
                 (--- + Sqrt[-8 x  + --])
                   2                4

                    3                6
                  -x           3   x    1/3
                 (--- + Sqrt[-8 x  + --])    },
                   2                4

           {y ->

             (Sqrt[-3] (

                              -2 x
                    ---------------------------
                      3                6
                    -x           3   x    1/3
                   (--- + Sqrt[-8 x  + --])
                     2                4

                      3                6
                    -x           3   x
                 + (--- + Sqrt[-8 x  + --])
                     2                4

                    1/3
                       )) / 2 -

                              2 x
                  (---------------------------  +
                     3                6
                   -x           3   x    1/3
                  (--- + Sqrt[-8 x  + --])
                    2                4

                     3                6
                   -x           3   x    1/3
                  (--- + Sqrt[-8 x  + --])    )
                    2                4

                / 2}, {y ->

             -(Sqrt[-3] (

                              -2 x
                    ---------------------------
                      3                6
                    -x           3   x    1/3
                   (--- + Sqrt[-8 x  + --])
                     2                4

                       3
                     -x
                  + (--- +
                      2

                                   6
                             3   x    1/3
                     Sqrt[-8 x  + --])  )) /
                             4

                 2 - (

                              2 x
```

```
6
```

```
                    -x           3   x    1/3
                   (--- + Sqrt[-8 x  + --])
                     2                4

                    3                6
                  -x           3   x    1/3
                + (--- + Sqrt[-8 x  + --])
                    2                4

           ) / 2}}
```

Yuck. Find parametric equations for the *folium of Descartes* $x^3 + y^3 = 6xy$. Use your parametrizations to plot the curve.

**Answer:**

We'll use the slope $t$ of the line from $\{0,0\}$ to a point $\{x,y\}$ on $x^3 + y^3 = 6xy$ to generate our parametric equations.

```
In[2]:=    Clear[x,y,t]
           Solve[{x^3 + y^3 == 6 x y, y == t x}, {x,y}]
```

```
Out[2]=                 2
                  6 t         6 t
           {{x -> ------, y -> ------},
                      3            3
                  1 + t        1 + t

           {x -> 0, y -> 0}, {x -> 0, y -> 0}}
```

Nice, easy, understandable formulas:

```
In[3]:=    x[t_] = 6 t / (1 + t^3)
```

```
Out[3]=     6 t
           ------
                3
           1 + t
```

```
In[4]:=    y[t_] = (6 t^2) / (1 + t^3)
```

```
Out[4]=        2
            6 t
           ------
                3
           1 + t
```

Here comes the plot:

```
In[5]:=    ParametricPlot[{x[t],y[t]},{t,-40,40},
             AspectRatio->Automatic,
             PlotLabel->"Folium of Descartes"];
```

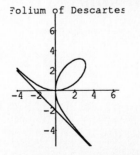

To see which $t$'s give rise to the various parts of this plot,

look at:

```
In[6]:=    loop =
           ParametricPlot[{x[t],y[t]},
           {t,0,40},
           AspectRatio->Automatic];
```

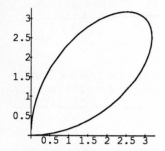

```
In[7]:=    back =
           ParametricPlot[{x[t],y[t]},
           {t,-.5,0},
           AspectRatio->Automatic];
```

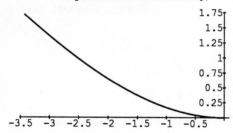

```
In[8]:=    front =
           ParametricPlot[{x[t],y[t]},
           {t,-40,-2}];
```

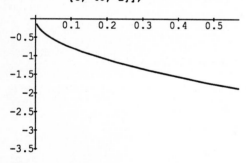

Put the parts together:

```
In[9]:=    Show[loop,back,front,
           PlotRange->{{-8,8},{-8,8}}];
```

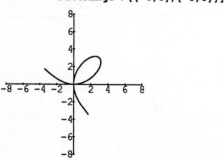

## B.2.c)

Use the parametric equations

$$x = \cos[t]$$
$$y = \sin[t]$$

of the circle $x^2 + y^2 = 1$ to obtain parametric equations of the ellipse

$$(x/5)^2 + (y/3)^2 = 1.$$

**Answer:**

Why not try $x/5 = \cos[t]$ and $y/3 = \sin[t]$? This gives:

```
In[1]:=    Clear[x,y,t]
           x[t_] = 5 Cos[t]
Out[1]=    5 Cos[t]
```

```
In[2]:=    y[t_] = 3 Sin[t]
Out[2]=    3 Sin[t]
```

To see whether this works, look at:

```
In[3]:=    (x[t]/5)^2 + (y[t]/3)^2
Out[3]=         2           2
           Cos[t]  + Sin[t]
```

This is just a disguised way of saying $(x/4)^2 + (y/3)^2 = 1$. Accordingly our choice of $x$ and $y$ will land us on the ellipse

$(x/4)^2 + (y/3)^2 = 1$. To see whether the whole ellipse is covered, plot:

```
In[4]:=    ellipse=
           ParametricPlot[{x[t],y[t]},
           {t,0,2 Pi},
           AxesLabel->{"x","y"},
           AspectRatio->Automatic];
```

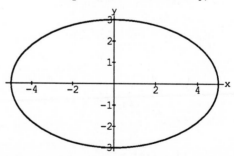

Nice plot. It is worth noting that the parameter $t$ used above does not have the geometric interpretation it had when we parameterized the circle. However it is not hard to see that $t$ describes the angle shown in the following diagram:

■ **B.3) Derivatives for curves given parametrically.**

**B.3.a.i)**

Sometimes a curve, like the cycloid above, comes via parametric equations $x = x[t]$ and $y = y[t]$ but it is clear that the curve is the graph of a function $y = f[x]$. This means that $y[t] = f[x[t]]$ for all $t$'s. Finding the explicit form of $f[x]$ may be impossible or too much trouble. It might seem that in this situation we are out of luck in computing the derivative $D[f[x], x] = f'[x]$, but there is a way to do it. How?

**Answer:**

If $y = f[x]$, then clearly $y[t] = f[x[t]]$. So by the chain rule, $y'[t] = f'[x[t]]x'[t]$. Consequently

$$f'[x[t]] = y'[t]/x'[t].$$

We can do this with *Mathematica* :

```
In[1]:=    Clear[x,y,t,f]
           first = (D[y[t] == f[x[t]],t])
Out[1]=    y'[t] == f'[x[t]] x'[t]
```

```
In[2]:=    Solve[first,f'[x[t]]]
Out[2]=                 y'[t]
           {{f'[x[t]] -> -----}}
                         x'[t]
```

**B.3.b)**

How is the formula

$$f'[x[t]] = y'[t]/x'[t]$$

for the derivative of $f[x]$ with respect to $x$ used?

**Answer:**

Given $x_0$, find $t_0$ with $x[t_0] = x_0$. Then calculate

$$f'[x_0] = f[x[t_0]] = y'[t_0]/x'[t_0].$$

**B.3.c.i)**

Plot the curve

$$x = x[t] = 3t/(1 + t^3)$$

$$y = y[t] = 3t^2/(1 + t^3)$$

for $-5 \le t \le -1$. What is the derivative of $y$ with respect to $x$ at $x = 2$?

**Answer:**

```
In[1]:=    Clear[x,y,t]
           x[t_] = 3 t/(1 + t^3)
Out[1]=        3 t
           ------
                3
           1 + t
```

```
In[2]:=    y[t_] = 3 t^2/(1 + t^3)
Out[2]=         2
           3 t
           ------
                3
           1 + t
```

Here is the plot:

```
In[3]:=    ParametricPlot[{x[t],y[t]},
           {t,-5,-1},
           PlotStyle->
           {{Thickness[0.015],RGBColor[0,0,1]}}];
```

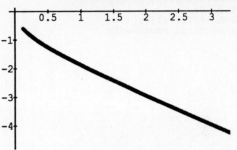

Next find what $t$ makes $x[t] = 2$ :

```
In[4]:=    N[Solve[x[t] == 2,t]]
Out[4]=    {{t -> -1.47569},

           {t -> 0.737843 + 0.365018 I},

           {t -> 0.737843 - 0.365018 I}}
```

Only the first is real, so we use it: The derivative of $y$ with respect to $x$ at $x = 2$ is:

```
In[5]:=    y'[t]/x'[t]/.t->-1.47569
Out[5]=    -1.03588
```

Not too hard.

**B.3.c.ii)**

Plot the curve

$$x = x[t] = 3t/(1 + t^3)$$

$$y = y[t] = 3t^2/(1 + t^3)$$

for $0 \le t \le 10$. If $y$ is regarded to be a function $f[x]$ of $x$, then at which $x's$ do we have $f'[x] = 0$? At what points $\{x, y\}$ on this curve do the tangent lines cut the $x$-axis at an angle of $\pi/4$ radians? Plot the tangent lines and the curves on the same axes.

**Answer:**

```
In[1]:=    Clear[x,y,t]
           x[t_] = 3 t/(1 + t^3)
```
```
Out[1]=      3 t
           ------
                3
           1 + t
```

```
In[2]:=    y[t_] = 3 t^2/(1 + t^3)
```
```
Out[2]=        2
             3 t
           ------
                3
           1 + t
```

Here is the plot:

```
In[3]:=    frootloop =
           ParametricPlot[{x[t],y[t]},
           {t,0,10},
           Ticks->{{1},{1}},
           AspectRatio->Automatic];
```

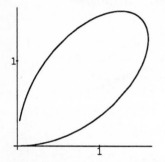

We can spot one point $\{x,y\}$ at which $f'[x] = 0$ at the top of the plot. Maybe there is another one at the lower left.

Because $f'[x] = y'[t]/x'[t]$, we can solve $f'[x] = 0$ by solving $y'[t] = 0$ for $t$.

```
In[4]:=    N[Solve[y'[t] == 0,t]]
```
```
Out[4]=    {{t -> 1.25992},

             {t -> -0.629961 + 1.09112 I},

             {t -> -0.629961 - 1.09112 I},

             {t -> 0.}}
```

The values of $x$ at which $y'[x] = 0$ are:

```
In[5]:=    x[1.25992]
```
```
Out[5]=    1.25992
```

and

```
In[6]:=    x[0.]
```
```
Out[6]=    0.
```

Interesting answers because we have $x[t] = t$ at both $t = 1.25922$ and $t = 0$. The corresponding points on the curve

are

```
In[7]:=    top = {x[t],y[t]}/.t->1.25992
```
```
Out[7]=    {1.25992, 1.5874}
```

```
In[8]:=    origin = {x[0],y[0]}
```
```
Out[8]=    {0, 0}
```

To find the points $\{x,y\}$ on this curve at which the tangent lines cut the $x$-axis at an angle of $\pi/4$ radians, just remember

$$f'[x] = y'[t]/x'[t]$$

and solve:

```
In[9]:=    N[Solve[y'[t]/x'[t] == Tan[Pi/4],t]]
```
```
Out[9]=    {{t -> 2.29663}, {t -> 0.435421},

             {t -> -0.366025 + 0.930605 I},

             {t -> -0.366025 - 0.930605 I}}
```

The two answers are:

```
In[10]:=   point1 = {x[t],y[t]}/.t->2.29663
```
```
Out[10]=   {0.5254, 1.20665}
```

and

```
In[11]:=   point2 = {x[t],y[t]}/.t->0.435421
```
```
Out[11]=   {1.20665, 0.525401}
```

Check:

```
In[12]:=   {{y'[t]/x'[t]}/.t->2.29663,
            {y'[t]/x'[t]}/.t->0.435421}
```
```
Out[12]=   {{1.}, {1.}}
```

Good. The two tangent lines have equations:

```
In[13]:=   Clear[x]
           ytan1 = 1.20665 + (x - 0.5254)
```
```
Out[13]=   0.68125 + x
```

```
In[14]:=   ytan2 = 0.525401 + (x - 1.20665)
```
```
Out[14]=   -0.681249 + x
```

Here comes the plot:

```
In[15]:=   tanplots =
           Plot[{ytan1,ytan2},
           {x,-0.7,1.75},
           DisplayFunction->Identity]
           Show[tanplots,frootloop,
           AspectRatio->Automatic,
           PlotRange->{0,1.8},
           DisplayFunction->$DisplayFunction];
```

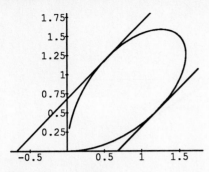

The angle is correct.

# Tutorial

## ■ T.1) Parametric equations of lines: Tangents to parametric curves.

A line can be nailed down by two specifications:

i) A point $\{x_0, y_0\}$ that the line goes through and

ii) A direction stick from $\{0, 0\}$ to $\{a, b\}$ to establish a direction for the line. Most folks use the term "direction vector" for the direction stick from $\{0, 0\}$ to $\{a, b\}$.

### T.1.a.i)

Give parametric equations of the line through $\{3, 1\}$ that moves in the direction set by the stick (or vector) from $\{0, 0\}$ to $\{4, 7\}$. Plot the line parametrically. Then give the usual $xy$-equation of this line.

**Answer:**

The poimt amd the direction are:

```
In[1]:=    point = {3,1}
Out[1]=    {3, 1}
```

```
In[2]:=    direction ={4,7}-{0,0}
Out[2]=    {4, 7}
```

The parametric equations for this line are given by:

```
In[3]:=    {x[t_],y[t_]} =
           point + t direction
Out[3]=    {3 + 4 t, 1 + 7 t}
```

Here is the plot:

```
In[4]:=    lineplot =
           ParametricPlot[{x[t],y[t]},{t,-2,3},
           AxesLabel->{"x","y"}];
```

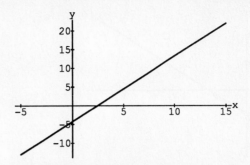

Here is another plot showing the point $\{3, 1\}$ and the stick (or vector) from $\{0, 0\}$ to $\{4, 7\}$.

```
In[5]:=    pointplot =
           Graphics[RGBColor[1,0,0],
           PointSize[0.03],Point[point]]
           vectorplot =
           Graphics[RGBColor[0,0,1],
           Line[{{0,0},direction}]]

           Show[lineplot,vectorplot,pointplot];
```

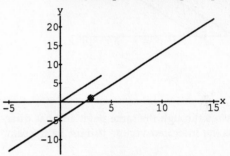

Note how the vector from $\{0, 0\}$ to $\{4, 7\}$ gives the line its direction and the point $\{3, 1\}$ gives the line a definite position. Let's go after the $x - y$ equation of this line:

```
In[6]:=    xyeqn =
           Eliminate[{x == x[t],y == y[t]},t]
Out[6]=    4 y == -17 + 7 x
```

Let us solve for $y$:

```
In[7]:=    xyline[x_]=y/.Solve[xyeqn,y][[1]]
Out[7]=    -(17 - 7 x)
           -----------
                4
```

So the usual $xy$-equation of this line is $y = (7/4)x - 17/4$. Note how the direction vector $\{4, 7\}$ is still visible. Just to see that no deception is underway, let's plot this:

```
In[8]:=    xylineplot =
           Plot[xyline[x],{x,-4,10}];
```

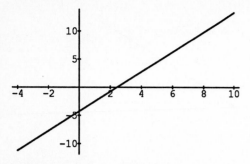

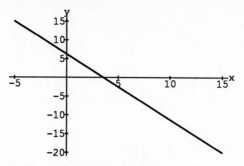

Looks like the same line. Let's put in the point and the direction vector:

```
In[9]:=    Show[xylineplot,
           vectorplot,pointplot,
           AxesLabel->{"x","y"}];
```

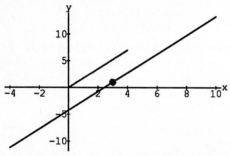

Yep. It still passes through the same point and has direction parallel to the indicated vector. It must be the same line.

### T.1.a.ii)

---

Give parametric equations of the line through $\{3,1\}$ that moves in the direction set by the vector from $\{0,0\}$ to $\{4,-7\}$. Plot the line parametrically. Then give the usual $xy$-equation of this line.

---

**Answer:**

The parametric equations for this line are given by:

```
In[1]:=    point = {3,1}
Out[1]=    {3, 1}
```

```
In[2]:=    direction ={4,-7}-{0,0}
Out[2]=    {4, -7}
```

```
In[3]:=    {x[t_],y[t_]} = point + t direction
Out[3]=    {3 + 4 t, 1 - 7 t}
```

Here is the plot:

```
In[4]:=    lineplot =
           ParametricPlot[{x[t],y[t]},{t,-2,3},
           AxesLabel->{"x","y"}];
```

Here is another plot showing the point $\{3,1\}$ and the stick (or vector) from $\{0,0\}$ to $\{4,7\}$.

```
In[5]:=    pointplot =
           Graphics[RGBColor[1,0,0],
           PointSize[0.015],Point[point]]

           vectorplot =
           Graphics[RGBColor[0,0,1],
           Line[{{0,0},direction}]]

           Show[lineplot,vectorplot,pointplot];
```

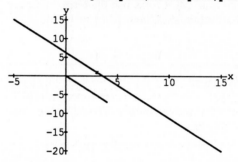

Note how the vector from $\{0,0\}$ to $\{4,-7\}$ gives the line its direction and the point $\{3,1\}$ gives the line a definite position. Let's go after the $xy$-equation of this line:

```
In[6]:=    xyeqn =
           Eliminate[{x == x[t],y == y[t]},t]
Out[6]=    4 y == 25 - 7 x
```

```
In[7]:=    xyline[x_]=y/.Solve[xyeqn,y][[1]]
Out[7]=    -(-25 + 7 x)
           ------------
                4
```

So the usual $xy$-equation of this line is $y = -(7/4)x + 25/4$. Note how the direction vector $\{4,-7\}$ is still visible. Just to see that no deception is underway, let's plot this:

```
In[8]:=    xylineplot =
           Plot[-(7/4) x + 25/4, {x,-4,10}];
```

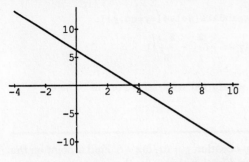

Looks like the same line. Let's put in the point and the direction vector:

```
In[9]:=    Show[xylineplot,vectorplot,pointplot,
           AxesLabel->{"x","y"}];
```

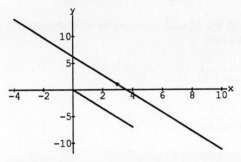

Good. It still passes through the same point and has direction parallel to the indicated vector. It must be the same line.

### T.1.a.iii)

Give parametric equations of the line through $\{c,d\}$ that moves in the direction set by the stick (or vector) from $\{0,0\}$ to $\{a,b\}$. Then give the usual $xy$-equation of this line. How can you read the slope( growth rate) of the line directly off a direction vector from $\{0,0\}$ to $\{a,b\}$?

**Answer:**

The parametric equations for this line are given by:

```
In[1]:=    Clear[a,b,c,d]
           point = {c,d}
Out[1]=    {c, d}

In[2]:=    direction ={a,b} -{0,0}
Out[2]=    {a, b}

In[3]:=    Clear[x,y,t]
           {x[t_],y[t_]} = point + t direction
Out[3]=    {c + a t, d + b t}
```

Let's go after the $xy$-equation of this line:

```
In[4]:=    xyeqn =
           Eliminate[{x == x[t],y == y[t]},t]
```

```
Out[4]=           b c   b x
        a != 0 && d == --- - --- + y ||
                        a     a

        b != 0 && a == 0 && c == x ||

        a == 0 && c == x && b == 0 && d == y
```

```
In[5]:=    Solve[d == (b c)/a - (b x)/a + y, y]
Out[5]=          -(b c) + a d + b x
           {{y -> -------------------}}
                          a
```

So the usual $x-y$ equation of this line is $y = (b/a)x + d - (bc)/a$. Note how the direction vector from $\{0,0\}$ to $\{a,b\}$ reflects itself by telling you that the slope (growth rate) of this line is $b/a$.

### T.1.b.i)

A line has $xy$-equation $y = (9/8)x - (3/2)$. Find a point on the line and find a direction vector for the line and write parametric equations for the line. Plot the line using usual $xy$-plotting and show the direction vector and the point you picked on the same axes.

Check yourself by converting your parametric equations back to the $xy$-equation.

**Answer:**

Here is a plot:

```
In[1]:=    Clear[x]
           lineplot =
           Plot[(9/8) x - 3/2,{x,-1,8},
           AspectRatio->Automatic];
```

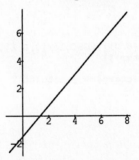

Setting $x = 0$ in the equation $y = (9/8)x - 3/2$ reveals that $\{0, -3/2\}$ is on this line.

```
In[2]:=    point = {0,-3/2}
Out[2]=          3
           {0, -(-)}
                2
```

A direction vector is:

```
In[3]:=    direction = {8, 9} -{0,0}
Out[3]=    {8, 9}
```

Parametric equations are given by:

```
In[4]:=    Clear[x,y,t]
           {x[t_],y[t_]} = point + t direction
```
```
Out[4]=
                  3
           {8 t, -(-) + 9 t}
                  2
```

Here is the plot: The thicker line segment indicates the direction vector.

```
In[5]:=    pointplot =
           Graphics[RGBColor[1,0,0],
           PointSize[0.04],Point[point]]

           vectorplot =
           Graphics[Thickness[0.01],RGBColor[0,0,1],
           Line[{{0,0},direction}]]

           Show[lineplot,vectorplot,pointplot];
```

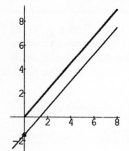

Looks good. If we like, then we can use a smaller direction vector:

```
In[6]:=    newvector = (1/4) direction
```
```
Out[6]=
               9
           {2, -}
               4
```

```
In[7]:=    newvectorplot =
           Graphics[Thickness[0.01],RGBColor[0,0,1],
           Line[{{0,0},newvector}]]

           Show[lineplot,newvectorplot,pointplot];
```

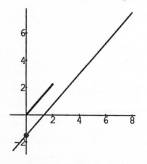

Looks even better. For an additional check, let's convert the our parametric equations back to the $xy$-equations:

```
In[8]:=    xyeqn =
           Eliminate[{x == x[t],y == y[t]},t]
```
```
Out[8]=    8 y == -12 + 9 x
```

```
In[9]:=    ExpandAll[Solve[xyeqn,y]]
```
```
Out[9]=
                  3     9 x
           {{y -> -(-) + ---}}
                  2      8
```

This is what we started with; so we're out of here.

### T.1.b.ii)

A line has $xy$-equation $y = (a/b)x + c$. Find a point on the line and find a direction vector for the line and write parametric equations for the line. Check yourself by converting your parametric equations back to the $xy$-equation.

### Answer:

Setting $x = 0$ in the equation $y = (a/b)x + c)$ reveals that $\{0, c\}$ is on this line.

```
In[1]:=    Clear[a,b,c]
           point = {0,c}
```
```
Out[1]=    {0, c}
```

A direction vector is:

```
In[2]:=    direction = {b, a}
```
```
Out[2]=    {b, a}
```

Parametric equations are given by:

```
In[3]:=    Clear[x,y,t]
           {x[t_],y[t_]} = point + t direction
```
```
Out[3]=    {b t, c + a t}
```

Now convert the our parametric equations back to the $x-y$ equations:

```
In[4]:=    xyeqn = Eliminate[{x == x[t],y == y[t]},t]
```
```
Out[4]=
                                 a x
           b != 0 && c == -(---) + y ||
                                  b

              a != 0 && b == 0 && x == 0 ||

              b == 0 && x == 0 && a == 0 && c == y
```

```
In[5]:=    Solve[c == -(a/b) x + y, y]
```
```
Out[5]=
                   b c + a x
           {{y -> ---------}}
                      b
```

Good; this what we started with.

### T.1.c.i)

Plot the curve

$$x = x[t] = t^2 \sin[t]$$

$$y = y[t] = 10(1 + t^5)^{1/5}$$

for $0 \le t \le 10$. Then find parametric equations of the tangent line through the point $\{x[8], y[8]\}$ and show the tangent line with your plot.

---

**Answer:**

```
In[1]:=    Clear[x,y,t]
           x[t_] = t^2 Sin[t]
Out[1]=    2
           t  Sin[t]
```

```
In[2]:=    y[t_] = 10 (t^5 + 1)^(1/5)
Out[2]=              5 1/5
           10 (1 + t )
```

Here is the plot:

```
In[3]:=    curveplot =
           ParametricPlot[{x[t],y[t]},{t,0,10},
           PlotStyle->
           {{Thickness[0.005],RGBColor[0,0,1]}},
           AspectRatio->Automatic];
```

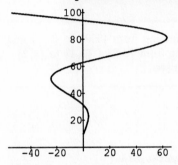

Now let's go after parametric equations of the tangent line through $\{x[8], y[8]\}$. We've got a built in point:

```
In[4]:=    point = N[{x[8],y[8]}]
Out[4]=    {63.3189, 80.0005}
```

The $xy$-equation for the tangent line is

$$y = y[8] + (y'[8]/x'[8])(x - x[8]).$$

This allows us to read off a direction vector:

```
In[5]:=    direction = N[{x'[8],y'[8]}] - {0,0}
Out[5]=    {6.51773, 9.99976}
```

Parametric equations are given by:

```
In[6]:=    Clear[xtan,ytan,t]
           {xtan[t_],ytan[t_]} = point + t direction
Out[6]=    {63.3189 + 6.51773 t,

           80.0005 + 9.99976 t}
```

Here is the plot: The thicker line segment at the bottom indicates the direction vector of the tangent line.

```
In[7]:=    tanplot = ParametricPlot[
           {xtan[t],ytan[t]},{t,-8,2},
```

```
DisplayFunction->Identity]

pointplot = Graphics[
{RGBColor[1,0,0],PointSize[0.015],Point[point]}]
vectorplot =
Graphics[
{Thickness[0.01],Line[{{0,0},direction}]}]

Show[curveplot,tanplot,vectorplot,
pointplot,
AspectRatio->Automatic,
DisplayFunction->$DisplayFunction];
```

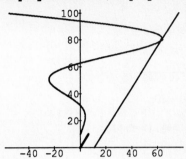

Looks great.

**T.1.c.ii)**

---

Given the parametric curve $x = x[t]$, $y = y[t]$, write down parametric equations of the tangent line through $\{x[a], y[a]\}$.

---

**Answer:**

$\{xtan[t], ytan[t]\} = \{x[a], y[a]\} + t\{x'[a], y'[a]\}$. This is the same as $xtan[t] = x[a] + tx'[a]$ and $ytan[t] = y[a] + ty'[a]$.

**T.1.c.iii)**

---

Again plot the curve

$$x = x[t] = t^2 \sin[t]$$

$$y = y[t] = 10(1 + t^5)^{1/5}$$

for $0 \le t \le 10$. Then find parametric equations of the tangent line through the point at which $y = 30$ and show the tangent line with your plot.

---

**Answer:**

```
In[1]:=    Clear[x,y,t]
           x[t_] = t^2 Sin[t]
Out[1]=    2
           t  Sin[t]
```

```
In[2]:=    y[t_] = 10 (t^5 + 1)^(1/5)
Out[2]=              5 1/5
           10 (1 + t )
```

Here is the plot:

```
In[3]:=  curveplot =
         ParametricPlot[{x[t],y[t]},{t,0,10},
         PlotStyle->
         {{Thickness[0.005],RGBColor[0,0,1]}},
         AspectRatio->Automatic];
```

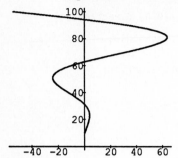

Find the $t$ that makes $y = 30$ :

```
In[4]:=  Plot[{y[t],30},{t,0,6}];
```

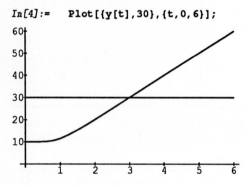

It is somewhere near $t = 3$ :

```
In[5]:=  root=t/.FindRoot[y[t] == 30,{t,3}]
Out[5]=  2.99753
```

The tangent line we are after is the tangent line through $\{x[.99753], y[2.99753]\}$.

```
In[6]:=  point = N[{x[t],y[t]}/.t->root]
Out[6]=  {1.28998, 30.}
```

```
In[7]:=  direction =
         N[{x'[t],y'[t]}/.t->root]-{0,0}
Out[7]=  {-8.03139, 9.96706}
```

Parametric equations are given by:

```
In[8]:=  Clear[xtan,ytan,t]
         {xtan[t_],ytan[t_]} =
         point + t direction
Out[8]=  {1.28998 - 8.03139 t, 30. + 9.96706 t}
```

Here is the plot: The thicker line segment at the bottom indicates the direction vector of the tangent line.

```
In[9]:=  tanplot =
         ParametricPlot[{xtan[t],ytan[t]},{t,-3,5},
         DisplayFunction->Identity]
         pointplot =
         Graphics[RGBColor[1,0,0],
```

```
         PointSize[0.015],Point[point]]
         vectorplot =
         Graphics[Thickness[0.01],
         Line[{{0,0},direction}]]
```

```
         Show[curveplot,tanplot,
         vectorplot,pointplot,
         AspectRatio->Automatic,
         DisplayFunction->$DisplayFunction];
```

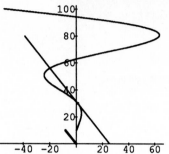

Looks great.

### ■ T.2) Tangent vectors.

#### T.2.a)

Plot the stick (=vector) running from $\{0,0\}$ to $\{3,2\}$ and the stick (=vector) running from $\{0,0\} + \{2,1\}$ to $\{3,2\} + \{2,1\}$. Describe what you see.

**Answer:**

Look at:

```
In[1]:=  vector1 = Graphics[Line[
         {{0,0},{3,2}}]];
```

```
         vector2 = Graphics[Line[
         {{0,0}+ {2,1},{3,2}+{2,1}}]];
```

```
         Show[vector1,vector2,
         Axes->{0,0},AxesLabel->{"x","y"}];
```

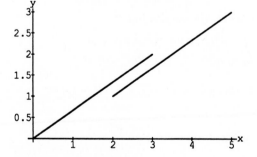

These sticks are parallel.

#### T.2.b)

Given a parametric curve $x = x[t]$, $y = y[t]$ and a point $t = t_0$ then what does the vector from $\{x[t_0], y[t_0]\}$ to $\{x[t_0], y[t_0]\} + \{x'[t_0], y'[t_0]\}$ represent? Try out your answer on some sample parametric curves.

**Answer:**

Well, the stick from $\{0,0\}$ to $\{x'[t_0], y'[t_0]\}$ is the direction vector for the tangent line through $\{x[t_0], y[t_0]\}$. The stick (= vector) from $\{x[t_0], y[t_0]\}$ to

$$\{x[t_0], y[t_0]\} + \{x'[t_0], y'[t_0]\}$$

is parallel to the stick from $\{0,0\}$ to $\{x'[t_0], y'[t_0]\}$ and starts at the point $\{x[t_0], y[t_0]\}$. Consequently, the stick from $\{x[t_0], y[t_0]\}$ to $\{x[t_0], y[t_0]\} + \{x'[t_0], y'[t_0]\}$ runs concurrently with the tangent line through $\{x[t_0], y[t_0]\}$. That's why some folks call this a **"tangent vector"**. Let's try this out: Here is a familiar parametric curve:

```
In[1]:=   Clear[x,y,t]
          x[t_] = Cos[t]
          y[t_] = Sin[t]

          curveplot =
          ParametricPlot[{x[t],y[t]},{t,0,2 Pi},
          AspectRatio->Automatic];
```

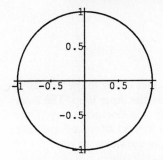

Now set up the plot of the vector from $\{x[t_0], y[t_0]\}$ to $\{x[t_0], y[t_0]\} + \{x'[t_0], y'[t_0]\}$ for $t_0 = \pi/4$.

```
In[2]:=   Clear[tanvector]
          tanvector[t_] = Graphics[
          Line[{{x[t],y[t]},
          {x[t],y[t]}+{x'[t],y'[t]}}]]

          Show[curveplot,tanvector[Pi/4],
          AspectRatio->Automatic];
```

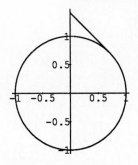

Just as predicted! Let's see a whole bunch of tangent vectors.

```
In[3]:=   Show[curveplot,
          Table[tanvector[t],{t,0,2 Pi,Pi/4}],
          AspectRatio->Automatic];
```

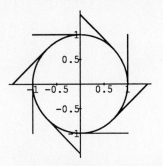

Cool. Let's see even more tangent vectors:

```
In[4]:=   Show[curveplot,
          Table[tanvector[t],{t,0,2 Pi,Pi/8}],
          AspectRatio->Automatic];
```

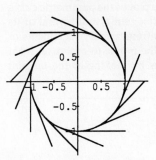

Hot. Let's try another curve:

```
In[5]:=   Clear[x,y,t]
          x[t_] = t Cos[t]
          y[t_] = t Sin[t];
```

```
In[6]:=   curveplot =
          ParametricPlot[{x[t],y[t]},
          {t,0,8 Pi},
          PlotStyle->{{Thickness[0.005],
          RGBColor[0,0,1]}},
          AspectRatio->Automatic];
```

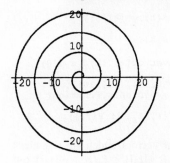

Now let's see a bunch of tangent vectors:

```
In[7]:=   Clear[tanvector]
          tanvector[t_] =
          Graphics[RGBColor[1,0,0],Line[{{x[t],y[t]},
          {x[t],y[t]}+{x'[t],y'[t]}}]]

          Show[curveplot,
          Table[tanvector[t],{t,0,8 Pi,Pi/8}],
          AspectRatio->Automatic];
```

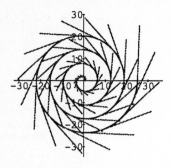

Hard to resist.

### T.2.c.i)

Make a movie showing the the plot of the parametric curve $x = x[t] = t\cos[t]$ and $y = y[t] = t\sin[t]$ and several of its tangent vectors as $t$ advances from $\pi$ to $4\pi$ increments of $\pi/2$. Use the information in the movie to report on what you can tell about the curves by looking at the tangent vectors.

**Answer:**

Here is the code for the movie:

```
In[1]:=   Clear[x,y,t,b,curveplot]
          x[t_] = t Cos[t]
          y[t_] = t Sin[t]

          curveplot[b_] :=
          ParametricPlot[{x[t],y[t]},{t,0,b},
          PlotStyle->
          {{Thickness[0.008],RGBColor[0,0,1]}},
          AspectRatio->
          Automatic,DisplayFunction->Identity]

          Clear[tanvector]
          tanvector[t_] :=
          Graphics[RGBColor[1,0,0],
          Line[{{x[t],y[t]},
          {x[t],y[t]}+{x'[t],y'[t]}}]]

          Clear[diagram,s,k]
          diagram[s_]:=
          Show[Table[tanvector[k],{k,0,s,s/8}],
          curveplot[s],
          PlotRange->{{-10,13},{-12,10}},
          DisplayFunction->$DisplayFunction]

          Do[diagram[s],{s,Pi,4 Pi,Pi/2}];
```

The direction of the tangent vectors which are the sticks from $\{x[t_0], y[t_0]\}$ to $\{x[t_0], y[t_0]\} + \{x'[t_0], y'[t_0]\}$ tell you how the curve is moving as $t$ advances from $t_0$. The length of these sticks gives an indication of how rapidly $\{x[t], y[t]\}$ changes as $t$ advances from $t_0$ at a steady rate.

### T.2.c.ii)

Illustrate your answer to the last problem by using tangent vectors to help you describe in words how the points $\{x[t], y[t]\}$ move on the parametric curve $x = x[t] = (1 - \sqrt{2}\sin[t])\cos[t]$ $y = y[t] = (1 - \sqrt{2}\sin[t])\sin[t]$ as $t$

advances from $0$ to $2\pi$ at a steady rate.

```
In[1]:=   Clear[x,y,t]
          x[t_] = (1 - Sqrt[2] Sin[t]) Cos[t]
          y[t_] = (1 - Sqrt[2] Sin[t]) Sin[t];

In[2]:=   curveplot =
          ParametricPlot[{x[t],y[t]},{t,0,2 Pi},
          AspectRatio->Automatic,
          PlotStyle->{RGBColor[0,0,1]},
          DisplayFunction->Identity]

          Clear[tanvector]
          tanvector[t_] =
          Graphics[RGBColor[1,0,0],Line[{{x[t],y[t]},
          {x[t],y[t]}+{x'[t],y'[t]}}]]

          Clear[k]
          Show[curveplot,
          Table[tanvector[k],{k,0,2 Pi,Pi/8}],
          AspectRatio->Automatic,
          DisplayFunction->$DisplayFunction];
```

The curve starts at:

```
In[3]:=   {x[0],y[0]}
Out[3]=   {1, 0}
```

Then it moves in a counterclockwise way to the left above the $x$–axis until it curves in on the inner loop. The inner loop is traversed rather slowly in a clockwise way. Once it finishes the inner loop, then it picks up speed and runs through the outer loop in a counterclockwise way.

To find the $t's$ at which the inner loop begins and ends, look at:

```
In[4]:=   Solve[{x[t],y[t]} == {0,0},Sin[t]]
Out[4]=                    Sqrt[2]
          {{Sin[t] -> -------}}
                          2
```

This tells us that the inner loop starts when $t$ is given by:

```
In[5]:=   N[ArcSin[Sqrt[2]/2]]
Out[5]=   0.785398
```

This is the same as Pi/4.

```
In[6]:=   N[Pi/4]
Out[6]=   0.785398
```

The inner loop ends when $t = \pi - \pi/4 = 3\pi/4$.

```
In[7]:=   Sin[3 Pi/4]
Out[7]=   Sqrt[2]
          -------
             2
```

### ■ T.3) Shifting the plot.

Look at the following points:

```
In[8]:=   pointA = {2,5}
```

```
pointB = {8,4}
pointC = {5,9}

allthree =
ListPlot[{pointA,pointB,pointC},
PlotStyle->{RGBColor[1,0,0],
PointSize[0.03]},
AxesLabel->{"x","y"},Axes->{0,0},
PlotRange->{{0,9},{0,10}}];
```

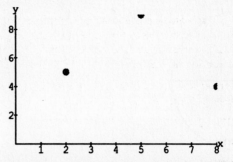

Now let's see what happens when we shift the numbers on the *x*-axis to the left and push the numbers on the the *y*-axis up so that **pointA** is at the origin.

```
In[9]:=    shiftpointA = pointA - pointA
           shiftpointB = pointB - pointA
           shiftpointC = pointC - pointA

           shiftallthree =
           ListPlot[{shiftpointA,shiftpointB,
           shiftpointC},
           PlotStyle->
           {RGBColor[1,0,0],PointSize[0.03]},
           AxesLabel->{"x","y"},
           Ticks->{{2,4,8},Automatic},
           PlotRange->{{0,9},{-2,10}}];
```

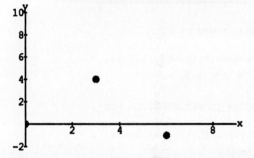

This is the way **pointB** and **pointC** look when they are charted from **pointA** . We can get back to the orginal plot easily:

```
In[10]:=   unshiftpointA = shiftpointA + pointA
           unshiftpointB = shiftpointB + pointA
           unshiftpointC = shiftpointC + pointA

           unshiftallthree =
           ListPlot[{unshiftpointA,unshiftpointB,
           unshiftpointC},
           PlotStyle->{RGBColor[1,0,0],
           PointSize[0.03]},
           AxesLabel->{"x","y"},
           PlotRange->{{0,8},{0,9}}];
```

The idea of shifting can be used to get some plots that would have driven us up the wall before this lesson.

### T.3.a)

At time $t$ minutes, **pointA** has coordinates $\{t,\sqrt{t}\}$. **PointB** is in circular orbit around **pointA** . **PointB** is always 0.7 units from **pointA** and is completing one revolution about **point** A every 5 minutes. Plot the motion of **pointB** as charted from **pointA** , and then plot the motion of **pointA** and **pointB** on the same axis as charted from $\{0,0\}$. Assume that **pointB** is directly to the right of **pointA** when $t = 0$.

### Answer:

Here is the motion of **pointB** as charted from **pointA** :

```
In[1]:=    Clear[t,pointB]
           relativepointB[t_] =
           .7 {Cos[2 Pi t/5],Sin[2 Pi t/5]}

Out[1]=            2 Pi t           2 Pi t
           {0.7 Cos[------], 0.7 Sin[------]}
                      5                5
```

Here is the motion of **pointB** as charted from **pointA** during the first 3 minutes:

```
In[2]:=    ParametricPlot[
           relativepointB[t],{t,0,3},
           AspectRatio->Automatic,
           Ticks->None];
```

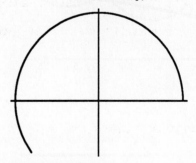

The first 5 minutes:

```
In[3]:=    ParametricPlot[relativepointB[t],
           {t,0,5},
           AspectRatio->Automatic,Ticks->None];
```

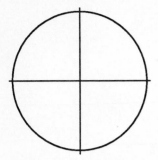

That's **pointA** at the center. Now let's see what happens when we plot from the point of view of {0,0} :

```
In[4]:=   Clear[pointA,pointB]
          pointA[t_] = {t,Sqrt[t]}

Out[4]=   {t, Sqrt[t]}
```

To get the parametric equations for **pointB** , just add **pointA** and **relativepointB**.

```
In[5]:=   pointB[t_] =
          pointA[t] + relativepointB[t]

Out[5]=             2 Pi t
          {t + 0.7 Cos[------],
                          5

                            2 Pi t
          Sqrt[t] + 0.7 Sin[------]}
                              5
```

Now we can read off the parametric equations for the path of **pointB** as charted from {0,0} : $xb[t] = t + 0.7\cos[2\pi t/5]$ and $yb[t] = \sqrt{t} + 0.7\sin[2\pi t/5]$. Here is a plot of the motion of both **pointA** and **pointB** during the first three revolutions of **pointB** around **pointA** :

```
In[6]:=   ParametricPlot[{pointA[t],
          pointB[t]},{t,0,15},
          AspectRatio->Automatic,
          PlotStyle->
          {{Thickness[0.008]},{RGBColor[0,0,1]}}];
```

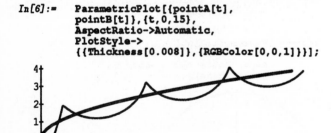

■ **T.4) Quick calculations.**

**T.4.a)**

What is the instantaneous growth rate of the parametric curve $x = x[t] = e^t + t$, $y = y[t] = t(3 - t)\log[t]$ at the point at which $x = 5$?

**Answer:**

Take a look:

```
In[1]:=   Clear[x,y,t]
          x[t_] = E^t + t
          y[t_] = t ( 3 - t ) Log[t]

          curveplot =
          ParametricPlot[{x[t],y[t]},{t,0,5},
          AxesLabel->{"x","y"}];
```

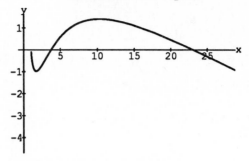

Now find the $t$ for which $x[t] = 5$ :

```
In[2]:=   Plot[{x[t],5},{t,0,4}];
```

We start the search near $t = 1.2$ :

```
In[3]:=   FindRoot[x[t] == 5, {t,1.2}]

Out[3]=   {t -> 1.30656}
```

The instantaneous growth rate at $x = 5$ is:

```
In[4]:=   (y'[t]/x'[t])/.t->1.30656

Out[4]=   0.382851
```

**T.4.b)**

What is the highest point on the parametric curve $x = x[t] = e^t + t$, $y = y[t] = t(3 - t)\log[t]$?

**Answer:**

Evidently the curve stays negative for larger $t$'s because $y[t] = t(3 - t)\log[t]$ is negative for $t \geq 3$. We search for a point $t$ with $0 \leq t \leq 3$ at which $dy/dx = 0$. At $\{x[t], y[t]\}$, we know $dy/dx$ is given by $y'[t]/x'[t]$. So we have to find the $t$'s with $0 \leq t \leq 3$ at which $y'[t] = 0$.

```
In[1]:=   Plot[y'[t],{t,0,3}];
```

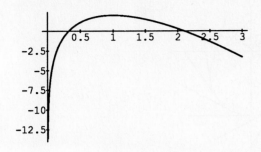

```
In[2]:=     FindRoot[y'[t] == 0, {t, .3}]
Out[2]=     {t -> 0.32105}

In[3]:=     FindRoot[y'[t] == 0, {t, 2.1}]
Out[3]=     {t -> 2.10316}
```

Look at:

```
In[4]:=     y[0.32105]
Out[4]=     -0.977184

In[5]:=     y[2.10316]
Out[5]=     1.40228
```

The point:

```
In[6]:=     high = {x[2.10316], y[2.10316]}
Out[6]=     {10.2952, 1.40228}
```

is the highest point on the curve.  Check:

```
In[7]:=     pointplot =
            Graphics[RGBColor[1, 0, 0],
            PointSize[0.03], Point[high]]
            Show[curveplot, pointplot];
```

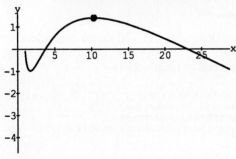

Nailed it.

### T.4.c)

If $x'[t] = \cos[t]y[t]$ with $x[0] = 5$ and $y'[t] = 4\sin[t]y[t]$ with $y[0] = 9$ then what is $dy/dx$ when $t = \pi$?

**Answer:**

When $t = \pi$, then

$$dy/dx = y'[\pi]/x'[\pi] = 4y[\pi]\sin[\pi]/(y[\pi]\cos[\pi])$$
$$= 4\sin[\pi]/\cos[\pi] = 4(0/-1) = 0.$$

### ■ T.5) Polar plots.

The usual way of specifying a point in the plane is to give its coordinates $\{x, y\}$.

```
In[1]:=     p={1.2, 1}
            point =
            Graphics[{RGBColor[1, 0, 0],
            PointSize[0.02], Point[p]}]
            label =
            Graphics[Text[{"x", "y"}, p, {0, -2}]]

            Show[point, label,
            Axes->{0, 0}, AxesLabel->{"x", "y"},
            PlotRange->{{0, 2.2}, {0, 1.2}},
            AspectRatio->Automatic,
            Ticks->{{1, 2}, {1}}];
```

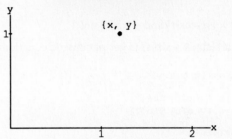

The same point can be specified by the indicated polar angle $t$ and the $d_f$ $r$ from the point to the origin. The pair $\{r, t\}$ is called the *polar coordinates* of the point $\{x, y\}$.

```
In[2]:=     dist = Graphics[Line[{{0, 0}, p}]]
            rlabel = Graphics[Text["r", p/2, {0, -2}]]
            angle =
            Graphics[Circle[{0, 0}, .17,
            {0, ArcTan[p[[2]]/p[[1]]]}]]
            anglelabel =
            Graphics[Text["t", {.22, .05}]]

            Show[point, label, dist, rlabel, angle,
            anglelabel,
            Axes->{0, 0}, AxesLabel->{"x", "y"},
            PlotRange->{{0, 2.5}, {0, 1.2}},
            AspectRatio->Automatic,
            Ticks->{{1, 2}, {1}}];
```

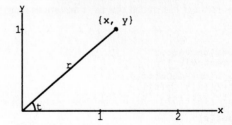

Here the polar angle $t$ is measured in the counterclockwise sense from the $x$-axis and $r$ is the distance from the origin to the point $\{x, y\}$.  Clearly if you know the polar coordinates $r$ and $t$, then you can find where the point $\{x, y\}$ is: You just leave the origin $\{0, 0\}$ in the direction specified by $t$ and you walk $r$ units out until you get to the point $\{x, y\}$.

The usual convention is: If $r > 0$, then you walk forward, but if $r < 0$, then you walk backward. If $r = 0$, then you stay put at the origin $\{0,0\}$. In fact if you know $r$ and $t$, then you know $x$ and $y$ through the easy formulas:

$$x = r \cos[t]$$

and

$$y = r \sin[t].$$

### T.5.a)

Plot the polar curve specified by $r[t] = 1 - \sin[t]$

for $t$ in the interval $[0, 2\pi]$. As $t$ advances from 0 to $2Pi$, is the movement clockwise or counterclockwise? What happens when you enlarge the plotting interval?

### Answer:

Just use $x[t] = r[t] \cos[t]$ and $y = r[t] \sin[t]$

together with $r[t] = 1 - \sin[t]$ to get parametric equations:

```
In[1]:=   Clear[r,t,x,y]
          r[t_] = 1 - Sin[t]
          x[t_] = r[t] Cos[t]
          y[t_] = r[t] Sin[t]

          cardioid =
          ParametricPlot[{x[t],y[t]},{t,0,2 Pi},
          Axes->{0,0},AxesLabel->{"x","y"},
          AspectRatio->Automatic,
          PlotStyle->RGBColor[1,0,0]];
```

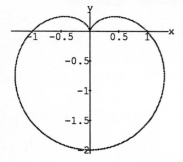

You gotta have heart. Let's see how this tomato-shaped curve (called a *cardioid*) is traced out as $t$ advances from 0 to $2\pi$.

```
In[2]:=   Clear[tanvector]
          tanvector[t_] =
          Graphics[RGBColor[0,0,1],
          Line[{{x[t],y[t]},
          {x[t],y[t]}+{x'[t],y'[t]}}]]

          Clear[s]
          Show[cardioid,
          Table[tanvector[s],{s,0,2 Pi,Pi/10}],
          AspectRatio->Automatic];
```

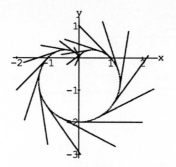

As $t$ advances from 0 to $2Pi$, the curve is traced out in a counterclockwise way. To see what happens when we make the plotting interval larger than $[0, 2\pi]$, look at:

```
In[3]:=   {x[t],y[t]}
Out[3]=   {Cos[t] (1 - Sin[t]),

          (1 - Sin[t]) Sin[t]}
```

The functions involved all repeat themselves on intervals of length $2\pi$; consequently, we expect nothing new to happen when we enlarge the plotting interval:

```
In[4]:=   cardioid =
          ParametricPlot[{x[t],y[t]},{
          t,-Pi/2,5 Pi/2},
          Axes->{0,0},AxesLabel->{"x","y"},
          AspectRatio->Automatic,
          PlotStyle->RGBColor[1,0,0]];
```

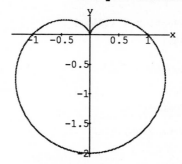

Because the functions $x[t]$ and $y[t]$ repeat themselves on intervals of length $2\pi$, we can plot the full curve by using **any** plotting interval of length $2\pi$:

```
In[5]:=   ParametricPlot[{x[t],y[t]},{t,300-Pi,300+Pi},
          Axes->{0,0},AxesLabel->{"x","y"},
          AspectRatio->Automatic,
          PlotStyle->RGBColor[1,0,0]];
```

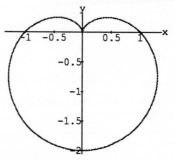

Your turn to come up with curves that look like clubs,

spades or diamonds.

### T.5.b)

Plot the polar curve specified by $r[t] = 1 - \sqrt{2}\sin[t]$ for $t$ in the interval $[0, 2\pi]$. Discuss the movement of $\{x[t], y[t]\}$ as $t$ advances from 0 to $2Pi$.

**Answer:**

This was done in T.2.c.ii) above.

### ■ T.6) Distance between curves.

### T.6.a)

Find the points on the curves

$$y = x^2 + 2x + 2$$

and

$$y = -2 + 2x - x^4/16$$

that are closest to each other. Plot the points, the curves, the tangent line to each curve at each of the closest points and the line through the two closest points on the same axes.

**Answer:**

Can we really do all this?

Let us take a look in true scale: Just for the hell of it we'll use parametric plots:

```
In[1]:=   Clear[x,y1,y2]
          y1[x_] = x^2 + 2 x + 2
          y2[x_] = -2 + 2 x - (x^4)/16
          curves =
          ParametricPlot[{{x,y1[x]},
          {x,y2[x]}}, {x,-4,6},
          PlotRange->{{-4,6},{-4,4}},
          AspectRatio->Automatic];
```

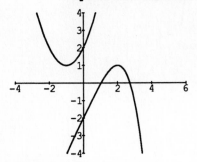

The eye has no problem detecting the approximate location of the closest points. Let's see how derivatives can be used to do what our eyes did so well. Let's call the two closest points **close1** and **close2**.

```
In[2]:=   Clear[s,t]
          close1 = {s,y1[s]}
```

```
Out[2]=
          {s, 2 + 2 s + s }
                            2
```

```
In[3]:=   close2 = {t,y2[t]}
Out[3]=
                             4
                            t
          {t, -2 + 2 t - ---}
                           16
```

The new parameters $s$ and $t$ are introduced to keep a clear-cut distinction between the two points.

```
In[4]:=   bridge[x_] =
          InterpolatingPolynomial[
          {close1, close2}, x]
Out[4]=
                      2
          2 + 2 s + s  -
                                       4
                             2        t
             s (-4 - 2 s - s  + 2 t - --)
                                      16
             ---------------------------- +
                       -s + t

                             2        t
             (-4 - 2 s - s  + 2 t - --) x
                                     16
             ---------------------------
                       -s + t
```

The slope of the line through **close1** and **close2** is:

```
In[5]:=   bridgeslope =
          Coefficient[bridge[x], x]
Out[5]=
                             2        t
          -4 - 2 s - s  + 2 t - --
                                16
          ------------------------
                   -s + t
```

The two tangent slopes are:

```
In[6]:=   tan1slope = y1'[s]
Out[6]=    2 + 2 s
```

```
In[7]:=   tan2slope = y2'[t]
Out[7]=         3
               t
          2 - --
               4
```

The points will be the closest if the line through them is perpendicular to both tangents. This says:

```
In[8]:=   eq1 = tan1slope == -1/bridgeslope
Out[8]=
                              -s + t
          2 + 2 s == -(-------------------)
                                        4
                             2        t
                    -4 - 2 s  + 2 t - --
                                      16
```

```
In[9]:=   eq2 = tan2slope == -1/bridgeslope
```

```
Out[9]=          3
               t                    -s + t
          2 - -- == -(----------------------)
               4                            2    4
                            -4 - 2 s - s  + 2 t - --
                                                  16
```

Now we can get *s* and *t* in decimals:

```
In[10]:=   sandt =
           N[Solve[{eq1,eq2},{s,t}]]
Out[10]=   {{s -> -1.90555 + 2.15575 I,

              t -> -2.05226 - 1.96976 I},

             {s -> -1.90555 - 2.15575 I,

              t -> -2.05226 + 1.96976 I},

             {s -> -1.07058 - 0.145155 I,

              t -> -0.945258 - 1.82155 I},

             {s -> -1.07058 + 0.145155 I,

              t -> -0.945258 + 1.82155 I},

             {s -> -0.277378, t -> 1.30433},

             {s -> -1.38518 + 1.03255 I,

              t -> 2.34535 - 0.508535 I},

             {s -> -1.38518 - 1.03255 I,

              t -> 2.34535 + 0.508535 I},

             {s -> 2.82843 I, t -> 2.82843 I},

             {s -> 2.82843 I, t -> 2.82843 I},

             {s -> 2.82843 I, t -> 2.82843 I},

             {s -> -2.82843 I, t -> -2.82843 I},

             {s -> -2.82843 I, t -> -2.82843 I},

             {s -> -2.82843 I, t -> -2.82843 I}}
```

That's quite a list, but so what! Only one of the listed solutions is real so that's the one we are after.

```
In[11]:=   {goods = -0.277378, goodt = 1.30433}
Out[11]=   {-0.277378, 1.30433}
```

The two closest points are:

```
In[12]:=   c1 = close1/.s->goods
Out[12]=   {-0.277378, 1.52218}
```

```
In[13]:=   c2 = close2/.t->goodt
Out[13]=   {1.30433, 0.427764}
```

Let's get ready for the plot. The equations of the two tangent lines are:

```
In[14]:=   ytan1 =
           (D[y1[x],x]/.
           x->c1[[1]]) (x - c1[[1]]) + c1[[2]]
```

```
Out[14]=   1.52218 + 1.44524 (0.277378 + x)
```

```
In[15]:=   ytan2 =
           ( D[y2[x],x]/.
           x->c2[[1]]) (x - c2[[1]]) + c2[[2]]
Out[15]=   0.427764 + 1.44524 (-1.30433 + x)
```

The line through the two closest points is:

```
In[16]:=   connectline =
           InterpolatingPolynomial[
           {c1,c2},x]
Out[16]=   1.33026 - 0.691922 x
```

Here comes a **true scale** plot:

```
In[17]:=   points =
           Graphics[{RGBColor[1,0,0],
           PointSize[.04],
           {Point[c1],Point[c2]}}]

           lines =
           Plot[{ytan1,ytan2,connectline},
           {x,-4,6},
           PlotRange->{{-4,6},{-4,4}},
           PlotStyle->RGBColor[0,0,1],
           DisplayFunction->Identity]

           bridge =
           Graphics[{Thickness[.02],
           Line[{c1,c2}]}]

           Show[points,curves,lines,
           bridge,
           PlotRange->{{-4,6},{-4,4}},
           AspectRatio->Automatic,
           DisplayFunction->$DisplayFunction];
```

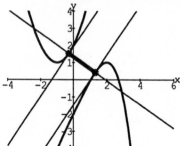

How does that grab you?

# Literacy Sheet

**1.** What are parametric equations?

**2.** What advantage does parametric plotting sometimes have over regular plotting?

**3.** If a curve is given paramterically via equations $x = x[t] = te^t$, $y = y[t] = t^2$, then what is the derivative $dy/dx$ at a point $\{x[t], y[t]\}$ on the curve. What is the derivative $dy/dx$ when $x = e$?

**4.** What is the highest point on the curve $x = te^t$, $y = 1 - t^2$?

**5** .If a curve is given paramterically via equations

$x = x[t]$, $y = y[t]$, then what does the vector running from $\{x[t_0], y[t_0]\}$ to $\{x[t_0], y[t_0]\} + \{x'[t_0], y'[t_0]\}$ represent.

**6.** What are the parametric equations of the tangent line to the parametric curve $x = x[t]$ and $y = y[t]$ t hrough the point $\{x[t_0], y[t_0]\}$?

**7.** Give parametric equations of the line through $\{2, -3\}$ that moves in the direction set by the stick (or vector) $\{-5, 6\}$. Then give the usual $xy$ -equation of this line.

**8.** What is the polar equation of the circle $x^2 + y^2 = 4$?

**9.** What are the polar coordinates of the point $\{x, y\}$ with $x = 1$ and $y = 1$?

**10.** Identify the $xy$ -curve specified in polar coordinates by $r = \sin[t]$. Hint: Remember $x = r\cos[t]$, $y = r\sin[t]$ and $r^2 = x^2 + y^2$ and multiply both sides of $r = \sin[t]$ by $r$.

**11.** Find $y$ in terms of $x$ if $x'[t] = x[t]y[t]$ with $x[0] = 1$ and $y'[t] = x[t]$ with $y[0] = e$.

**12.** Use the slope $t$ of the line from $\{0, 0\}$ to a point $\{x, y\}$ on $y^2 = x$ to generate parametric equations of $y^2 = x$ in terms of $t$.

# Circles and curvature

---

## Guide

Everybody knows what a circle is, but if you want to see how to do some neat things with circles, then you'll want to devour this lesson.

## Basics

■ **B.1) Circles: their equations and their plots.**

### B.1.a)

---

Explain why the circle of radius $r$ centered at the point $\{h, k\}$ consists of all the points $\{x, y\}$ such that

$$(x - h)^2 + (y - k)^2 = r^2.$$

---

**Answer:**

Recall that the distance between points $\{a, b\}$ and $\{c, d\}$ is given by the square root of $\sqrt{(a-c)^2 + (b-d)^2}$.

The circle of radius $r$ centered at $\{h, k\}$ consists of all the points $\{x, y\}$ that are at a distance $r$ from $\{h, k\}$.

These are precisely the points $\{x, y\}$ for which

$$\sqrt{(x-h)^2 + (y-k)^2} = r.$$

Squaring both sides, we see that these are the same points $\{x, y\}$ for which

$$(x - h)^2 + (y - k)^2 = r^2.$$

This is an example of a time when circular reasoning is correct.

### B.1.b)

---

Plot the circle of radius 5 centered at $\{3, -4\}$.

---

**Answer:**

```
In[1]:=    {h = 3, k = -4, r = 5}
Out[1]=    {3, -4, 5}
```

There are three ways of doing this. Here is the easiest:

```
In[2]:=    circle = Graphics[Circle[{h,k},r]]
           Show[circle,Axes->{0,0},
           AspectRatio->Automatic];
```

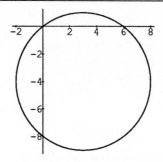

We can also include the center:

```
In[3]:=    center = Graphics[{RGBColor[1,0,0],
           PointSize[.02],Point[{h,k}]}]
           Show[circle, center,Axes->{0,0},
           AspectRatio->Automatic];
```

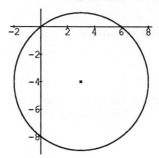

Another way of plotting this circle is:

```
In[4]:=    ParametricPlot[
           {h + r Cos[t], k + r Sin[t]},
           {t, 0, 2 Pi},
           AspectRatio->Automatic];
```

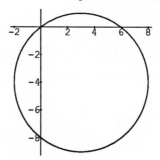

Yet another way to plot this circle is to solve the equation of the circle for $y$ :

```
In[5]:=    ysolved =
           Solve[(x - h)^2 + (y - k)^2 == r^2,y]
Out[5]=          -8 + Sqrt[64 - 4 (-6 + x) x]
```

```
                          {{y ->  ----------------------------},
                                              2

                               -8 - Sqrt[64 - 4 (-6 + x) x]
                          {y -> ----------------------------}}
                                              2
```

The first part **ysolved[[1]]** gives you the top half of the

circle:

```
In[6]:=    top = y/.ysolved[[1]]

Out[6]=    -8 + Sqrt[64 - 4 (-6 + x) x]
           ----------------------------
                        2
```

```
In[7]:=    Plot[top,{x,h - r,h + r},
               AspectRatio->Automatic];
```

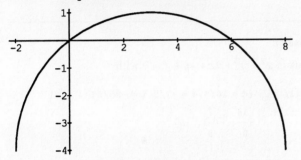

We can do the same thing with the second part **ysolved[[2]]** :

```
In[8]:=    bottom = y/.ysolved[[2]]

Out[8]=    -8 - Sqrt[64 - 4 (-6 + x) x]
           ----------------------------
                        2
```

To plot the whole circle, plot both solutions for $y$:

```
In[9]:=    Plot[{top,bottom},{x,h - r,h + r},
               AspectRatio->Automatic];
```

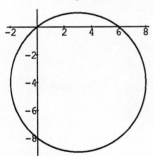

Lots of action; lots of fun

### B.1.c)

The equation of a circle of radius $r$ centered at $\{h, k\}$ is

$$(x - h)^2 + (y - k)^2 = r^2$$

. We can also look at the expanded version of this equa-

tion:

```
In[1]:=    ExpandAll[(x - h)^2 + (y - k)^2 == r^2]

Out[1]=     2    2          2              2    2
           h  + k  - 2 h x + x  - 2 k y + y  == r
```

This is the same as

$$x^2 + y^2 - 2hx - 2ky + h^2 + k^2 - r^2 = 0.$$

Use the information above to locate the center and find the radius of the circle described by

$$x^2 + y^2 + dx + ey + f = 0$$

---

**Answer:**

Equate the coefficients of like powers of $x$ and $y$ to read off

$$d = -2h,$$

$$e = -2k,$$

and

$$f = h^2 + k^2 - r^2.$$

Now we can pick off the information that we want:

$$h = -d/2$$

$$k = -e/2$$

and

$$r = \sqrt{h^2 + k^2 - f}.$$

Let's activate these formulas for future use; we can cut, paste and use them as we see fit.

The $x - y$ coordinates $\{h, k\}$ of the center of the circle

$$x^2 + y^2 + dx + ey + f = 0$$

are given by:

```
In[2]:=    h = -d/2

Out[2]=    -d
           --
            2
```

```
In[3]:=    k = -e/2

Out[3]=    -e
           --
            2
```

The radius $r$ of $x^2 + y^2 + dx + ey + f = 0$ is given by:

```
In[4]:=    r = Sqrt[ h^2 + k^2 - f]
```

*Out[4]=*
$$\text{Sqrt}\left[\frac{d^2}{4} + \frac{e^2}{4} - f\right]$$

## B.1.d)

Use the formulas derived in **B.1.c)** to locate the center, find the radius and then plot in true scale each of the following circles:

### B.1.d.i)

$$x^2 + y^2 - 4x + 8y - 5 = 0$$

**Answer:**

This is $x^2 + y^2 + dx + ey + f = 0$ with

*In[1]:=*   {d = -4, e = 8, f = -5}
*Out[1]=*   {-4, 8, -5}

Now copy, paste, edit and activate the electronic from **B.1.c)** :

The $xy$-coordinates $\{h, k\}$ of the center of the circle
$$x^2 + y^2 + dx + ey + f = 0$$
are given by:

*In[2]:=*   h = -d/2
*Out[2]=*   2

*In[3]:=*   k = -e/2
*Out[3]=*   -4

The radius $r$ of $x^2 + y^2 + dx + ey + f = 0$ is given by:

*In[4]:=*   r = Sqrt[ h^2 + k^2 - f]
*Out[4]=*   5

Here comes the true scale plot:

*In[5]:=*
```
circle = Graphics[Circle[{h,k},r]]
center =
Graphics[{RGBColor[1,0,0],PointSize[.03],
Point[{h,k}]}]
Show[circle, center,Axes->{0,0},
AspectRatio->Automatic];
```

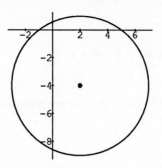

The electronic formulas are really convenient.

### B.1.d.ii)

$$5x^2 + 5y^2 + 14x - 7y - 25 = 0$$

**Answer:**

This is $x^2 + y^2 + dx + ey + f = 0$ with

*In[1]:=*   {d = 14/5, e = -7/5, f = -25/5}
*Out[1]=*
$$\left\{\frac{14}{5}, -\left(\frac{7}{5}\right), -5\right\}$$

Now copy, paste, edit and activate the electronic formulas from **B.1.c)** :

The $xy$-coordinates $\{h, k\}$ of the center of the circle
$$x^2 + y^2 + dx + ey + f = 0$$
are given by:

*In[2]:=*   h = -d/2
*Out[2]=*
$$-\left(\frac{7}{5}\right)$$

*In[3]:=*   k = -e/2
*Out[3]=*
$$\frac{7}{10}$$

The radius $r$ of $x^2 + y^2 + dx + ey + f = 0$ is given by:

*In[4]:=*   r = Sqrt[ h^2 + k^2 - f]
*Out[4]=*
$$\frac{\text{Sqrt}[149]}{2\ \text{Sqrt}[5]}$$

Here comes the true scale plot:

*In[5]:=*
```
circle = Graphics[Circle[{h,k},r]]
center =
Graphics[{RGBColor[1,0,0],PointSize[.03],
Point[{h,k}]}]
Show[circle, center,Axes->{0,0},
AspectRatio->Automatic];
```

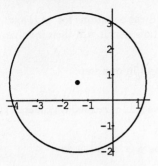

### ■ B.2) Three degrees of freedom.

The equation of the circle

$$(x - h)^2 + (y - k)^2 = r^2$$

contains **three** undetermined constants. So to determine a specific circle, we need **three** conditions. Use this fact to attack the following problems:

### B.2.a)

Find the center and the radius of the circle passing through the three points $\{1.1, 1.7\}$, $\{-1.8, 6.3\}$ and $\{4.7, 2.2\}$. Plot it and the three points on the same axes in true scale.

**Answer:**

```
In[1]:=   circleeqn = (x - h)^2 + (y - k)^2 == r^2
Out[1]=          2           2     2
          (-h + x)  + (-k + y)  == r
```

Plug in the points and solve for the center $\{h, k\}$ and the radius $r$ in the equation:

```
In[2]:=   eqn1 = circleeqn /. {x->1.1,y->1.7}
Out[2]=          2           2     2
          (1.1 - h)  + (1.7 - k)  == r
```

```
In[3]:=   eqn2 = circleeqn/.{x->-1.8,y->6.3}
Out[3]=           2           2     2
          (-1.8 - h)  + (6.3 - k)  == r
```

```
In[4]:=   eqn3 = circleeqn/.{x->4.7,y->2.2}
Out[4]=          2           2     2
          (4.7 - h)  + (2.2 - k)  == r
```

```
In[5]:=   hkrsolved = Solve[{eqn1,eqn2,eqn3},{h,k,r}]
Out[5]=   {{h -> 2.37654, k -> 5.71891,

               r -> 4.21677},

           {h -> 2.37654, k -> 5.71891,

               r -> -4.21677}}
```

The circle is centered on $\{2.37654, 5.71891\}$ and its radius is 4.21677. (Toss out the second answer because the radius $r$ cannot be negative.)

```
In[6]:=   points =
          Graphics[{RGBColor[1,0,0],PointSize[.02],
          Point[{1.1,1.7}],Point[{-1.8,6.3}],
          Point[{4.7,2.2}]}]
          circle =
          Graphics[Circle[{2.37654,5.71891},4.21677]]
          center =
          Graphics[RGBColor[0,0,1],PointSize[.03],
          Point[{2.37654,5.71891}]]
          Show[circle,points,center,
          Axes->{0,2}, AspectRatio->Automatic];
```

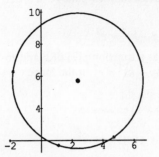

Right on target.

### B.2.b)

A circle is inscribed in the triangle with sides along the lines

$$y = 7x + 11$$

$$y = -x + 15$$

$$y = (3/2)x - 7.$$

Plot the circle and the lines in true scale on the same axes. Give the equation of the circle.

**Answer:**

Let's take a look at the situation:

```
In[1]:=   lineplot = Plot[{7 x + 11, -x + 15,
          (3/2) x - 7},{x,-30,30},
          PlotStyle->RGBColor[0,0,1],
          AspectRatio->Automatic,
          PlotRange->{-20,40}];
```

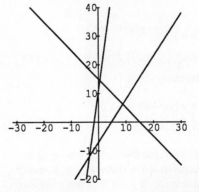

We've got to stick a circle inside this triangle so that the circle is tangent to each of the sides of the triangle. This tells us that the slopes of the circle must agree with the slope of the line at each point of contact. The equation of the circle

$$(x-h)^2 + (y-k)^2 = r^2$$

and our mission is to find $h, k$ and $r$.

This is not MISSION:IMPOSSIBLE.

Here is the key idea; some folks call this "implicit differentiation." Note that as $x$ moves, so must $y$ move in order to maintain the equality in

$$(x-h)^2 + (y-k)^2 = r^2.$$

This allows us to think of $y$ as a function $y[x]$ of $x$ and to use the equation $(x-h)^2 + (y-k)^2 = r^2$ in the form:

```
In[2]:=    circleeqn =
           (x - h)^2 + (y[x] - k)^2 == r^2
Out[2]=            2              2     2
           (-h + x)  + (-k + y[x])  == r
```

Remember that the slopes of the circle must agree with the slope of the line at each point of contact. To match up the slopes of the circle with the slopes of the lines, differentiate to get an expression for $y'[x]$:

```
In[3]:=    deriveqn =
           D[ (x - h)^2 + (y[x] - k)^2 == r^2, x]
Out[3]=    2 (-h + x) + 2 (-k + y[x]) y'[x] == 0
```

The first line $y = 7x + 11$ has a constant slope (growth rate) of 7. Therefore at the point $\{x, y[x]\}$ at which the circle touches the first line $y = 7x + 11$, we must have:

```
In[4]:=    slopeeqn1 = deriveqn/.y'[x]->7
Out[4]=    2 (-h + x) + 14 (-k + y[x]) == 0
```

Of course, this point of contact $\{x, y[x]\}$ must be on the circle

$$(x-h)^2 + (y[x] - k)^2 = r^2.$$

So it satisfies:

```
In[5]:=    circleeqn
Out[5]=            2              2     2
           (-h + x)  + (-k + y[x])  == r
```

And this point of contact $\{x, y[x]\}$ must be on the line $y = 7x + 11$. So it satisfies:

```
In[6]:=    lineeqn1 = y[x] == 7 x + 11
Out[6]=    y[x] == 11 + 7 x
```

These are **three** equations in the **five** unknowns $x, y[x], h, k$ and $r$. We cannot hope to solve these for $h, k$ and $r$. But we can hope to eliminate $5 - 3 = 2$ of the unknowns.

Having no immediate interest in the specific values of $x$ and $y[x]$, we eliminate them. This means that we take one of the equations and solve for $x$. Then we replace $x$ in the other two equations to get two equations in $y[x], h, k$,

and $r$. Next we solve one of these equations for $y[x]$ and replace $y[x]$ in the last equation which will then involve only $h, k$, and $r$.

*Mathematica* is happy to do it all in one step:

```
In[7]:=    neweqn1 =
           Eliminate[{slopeeqn1, circleeqn, lineeqn1},
           {x, y[x]}]
Out[7]=          2                        2
           50 r  == 121 + 154 h + 49 h  - 22 k -

                        2
           14 h k + k
```

Now we repeat the same procedure for the other two lines.

The second line $y = -x + 15$ has a constant slope of $-1$. Therefore at the point $\{x, y[x]\}$ where the circle hits $y = -x + 15$, we must have:

```
In[8]:=    slopeeqn2 = deriveqn/.y'[x]->-1
Out[8]=    2 (-h + x) - 2 (-k + y[x]) == 0
```

```
In[9]:=    circleeqn
Out[9]=            2              2     2
           (-h + x)  + (-k + y[x])  == r
```

```
In[10]:=   lineeqn2 = y[x] == -x + 15
Out[10]=   y[x] == 15 - x
```

```
In[11]:=   neweqn2 =Eliminate[
           {slopeeqn2, circleeqn, lineeqn2}, {x, y[x]}]
Out[11]=         2                      2
           2 r  == 225 - 30 h + h  - 30 k +

                      2
           2 h k + k
```

The third line $y = (3/2)x - 7$ has a constant slope of $3/2$. Therefore at the point $\{x, y[x]\}$ where the circle hits $y = (3/2)x - 7$, we must have:

```
In[12]:=   slopeeqn3 = deriveqn/.y'[x]->3/2
Out[12]=   2 (-h + x) + 3 (-k + y[x]) == 0
```

```
In[13]:=   circleeqn
Out[13]=            2              2     2
           (-h + x)  + (-k + y[x])  == r
```

```
In[14]:=   lineeqn3 = y[x] == (3/2)x - 7
Out[14]=                    3 x
           y[x] == -7 + ---
                         2
```

```
In[15]:=   neweqn3 =
           Eliminate[{slopeeqn3, circleeqn, lineeqn3},
           {x, y[x]}]
Out[15]=         2                       2
           13 r  == 196 - 84 h + 9 h  + 56 k -
```

$$12 \ h \ k + 4 \ k^2$$

Here is where we stand with the new equations:

```
In[16]:=  neweqn1
```

$$Out[16]= \quad 50 \ r^2 \ == \ 121 + 154 \ h + 49 \ h^2 - 22 \ k - 14 \ h \ k + k^2$$

```
In[17]:=  neweqn2
```

$$Out[17]= \quad 2 \ r^2 \ == \ 225 - 30 \ h + h^2 - 30 \ k + 2 \ h \ k + k^2$$

```
In[18]:=  neweqn3
```

$$Out[18]= \quad 13 \ r^2 \ == \ 196 - 84 \ h + 9 \ h^2 + 56 \ k - 12 \ h \ k + 4 \ k^2$$

This is great because we have **three** variables $h, k$ and $r$ to find and we've got **three** equations to use. We should be in fat city; let's go for it.

```
In[19]:=  hkrsolved =
          N[Solve[{neweqn1,neweqn2,neweqn3},{h,k,r}]]
```

```
Out[19]=  {{h -> 11.1384, k -> -17.4151,
              r -> 15.0449},
           {h -> 11.1384, k -> -17.4151,
              r -> -15.0449},
           {h -> 3.44347, k -> 5.6696,
              r -> 4.16269},
           {h -> 3.44347, k -> 5.6696,
              r -> -4.16269},
           {h -> -38.4301, k -> 1.5233,
              r -> 36.7037},
           {h -> -38.4301, k -> 1.5233,
              r -> -36.7037},
           {h -> 7.7392, k -> 16.9131,
              r -> 6.82518},
           {h -> 7.7392, k -> 16.9131,
              r -> -6.82518}}
```

So many answers? Toss out half of them because $r$ cannot be negative. This leaves just four answers-which is right? To see what's going on, let's plot all four resulting circles along with the lines. Set up the plot of the circles that

result from solutions $1, 3, 5,$ and $7$ with:

```
In[20]:=  circle1 =
          Graphics[Circle[{11.14,-17.42},15.04]]
          circle3 =
          Graphics[Circle[{3.44,5.67},4.16]]
          circle5 =
          Graphics[Circle[{-38.43,1.52},36.7]]
          circle7 =
          Graphics[Circle[{7.74,16.91},6.83]]
          Show[lineplot,circle1,circle3,
          circle5,circle7,
          AspectRatio->Automatic,
          PlotRange->{-50,50}];
```

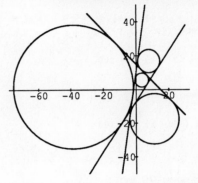

Hot plot. The circle we want seems to be the one with smaller radius; this is *circle3*:

```
In[21]:=  Show[lineplot,circle3,
          AspectRatio->Automatic,
          PlotRange->{-20,20}];
```

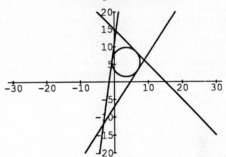

Aces! We nailed it. To two accurate decimals, the circle we are after is

$$(x - 3.44)^2 + (y - 5.67)^2 = 4.16^2$$

Incidentally, what geometric fact did the plot with all four circles reveal?

**Geometric Fact:** We found that that given three skew lines, then there are four circles all of which are tangent to all the lines.

## Tutorial

■ **T.1) Plotting circles.**

*T.1.a)*

Find the equations of the circles of radius 10 passing through the two points $\{-3,3\}$ and $\{1,-5\}$. Plot the circles and the points on the same axes in true scale.

## Answer:

The equation of the circle we are after is $(x-h)^2+(y-k)^2 = 10^2$. We have two unknowns, $h$ and $k$, to find; so we have to generate two equations. Here they are:

```
In[1]:=    circleeq = (x - h)^2 + (y - k)^2 == 10^2
Out[1]=         2          2
           (-h + x)  + (-k + y)  == 100
```

```
In[2]:=    eq1=circleeq/.{x->-3,y->3}
Out[2]=          2          2
           (-3 - h)  + (3 - k)  == 100
```

```
In[3]:=    eq2=circleeq/.{x->1,y->-5}
Out[3]=         2           2
           (1 - h)  + (-5 - k)  == 100
```

Solve them:

```
In[4]:=    Solve[{eq1,eq2},{h,k}]
Out[4]=    {{h -> -9, k -> -5}, {h -> 7, k -> 3}}
```

There are two such circles. They are $(x+9)^2+(y+5)^2 = 100$ and $(x-7)^2+(y-3)^2 = 100$. Here comes a sexy plot:

```
In[5]:=    points =
           Graphics[{RGBColor[1,0,0],PointSize[.02],
           Point[{-3,3}], Point[{1,-5}]}]
           circle1 = Graphics[Circle[{-9,-5},10]]
           circle2 = Graphics[Circle[{7,3},10]]
           Show[points,circle1,circle2,
           Axes->{0,0},AspectRatio->Automatic];
```

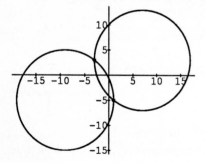

Beautiful.

## T.1.b)

Find the equations of the circles centered on $\{3,1\}$ that are tangent to the circle

$$x^2 + y^2 - 8x + 4y = 12.$$

Plot these circles and the circle $x^2 + y^2 - 8x + 4y = 12$ on the same axes in true scale.

## Answer:

First, let's get in the swim of things by plotting

$$x^2 + y^2 - 8x + 4y - 12 = 0$$

and the point $\{3,1\}$.

The circle is $x^2 + y^2 + dx + ey + f = 0$ with

```
In[1]:=    {d = -8, e = 4, f = -12}
Out[1]=    {-8, 4, -12}
```

From **B.1.c)** , we know that the $xy-$ coordinates $\{h,k\}$ of the center are given by:

```
In[2]:=    h = -d/2
Out[2]=    4
```

```
In[3]:=    k = -e/2
Out[3]=    -2
```

The radius $r$ is given by:

```
In[4]:=    r = Sqrt[(h^2 + k^2) - f]
Out[4]=    4 Sqrt[2]
```

Here comes the plot of the circle along with the point :

```
In[5]:=    basecircle = Graphics[Circle[{h,k},r]];
           basepoint =
           Graphics[{RGBColor[1,0,0],PointSize[.02],
           Point[{3,1}],Text["{3,1}",{3,1},{0,-2}]}]
           Show[basecircle,basepoint,Axes->Automatic,
           AspectRatio->Automatic];
```

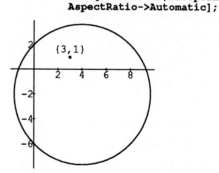

Our job is to find the circles centered on $\{3,1\}$ that are tangent to the indicated circle. This means that we have to find the $r$'s such that

$$(x - 3)^2 + (y - 1)^2 = r^2$$

touches

$$x^2 + y^2 - 8x + 4y - 12 = 0$$

at exactly one point.

```
In[6]:=    equation1 = (x - 3)^2 + ( y - 1)^2 == r^2
Out[6]=          2          2    2
           (-3 + x)  + (-1 + y)  == r
```

```
In[7]:=    equation2 = x^2 + y^2 - 8 x + 4 y - 12 == 0
```

```
Out[7]=            2         2
        -12 - 8 x + x  + 4 y + y  == 0
```

```
In[8]:=    Solve[{equation1,equation2},{x,y}]
```

```
Out[8]=                  2
        {{x -> (152 + 4 r  +

              2 2
        3 Sqrt[(-344 + 12 r )  -

                       2     4
              160 (788 - 60 r  + r )]) / 80
        , y ->

              2
        (344 - 12 r  +

                 2 2
        Sqrt[(-344 + 12 r )  -

                       2     4
              160 (788 - 60 r  + r )]) / 80}
        , {x ->

              2
        (152 + 4 r  -

              2 2
        3 Sqrt[(-344 + 12 r )  -

                       2     4
              160 (788 - 60 r  + r )]) / 80
        , y ->

              2
        (344 - 12 r  -

                 2 2
        Sqrt[(-344 + 12 r )  -

                       2     4
              160 (788 - 60 r  + r )]) / 80}}
```

The two potential solutions become one solution if and only if

$$(-344 + 12r^2)^2 - 160(788 - 60r^2 + r^4) = 0.$$

Solve this equation for $r$ to get the circles.

```
In[9]:=    Solve[
        (-344+12r^2)^2-160(788-60r^2+r^4)==0,r]
```

```
Out[9]=    {{r -> Sqrt[42 + 16 Sqrt[5]]},

           {r -> -Sqrt[42 + 16 Sqrt[5]]},

           {r -> Sqrt[42 - 16 Sqrt[5]]},

           {r -> -Sqrt[42 - 16 Sqrt[5]]}}
```

Toss out the negative $r's$. The two circles we are after are:

$$(x-3)^2 + (y-1)^2 = 42 + 16\sqrt{5}$$

and

$$(x-3)^2 + (y-1)^2 = 42 - 16\sqrt{5}.$$

Here comes the plot:

```
In[10]:=   circle1 =
        Graphics[{RGBColor[0,0,1],
        Circle[{3,1},Sqrt[42 + 16 Sqrt[5]]]}]
        circle2 =
        Graphics[{RGBColor[0,0,1],
```

```
        Circle[{3,1},Sqrt[42 - 16 Sqrt[5]]]}]
        Show[basecircle,basepoint,circle1,circle2,
        Axes->{0,0}, AspectRatio->Automatic];
```

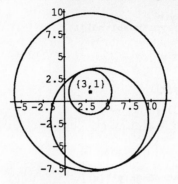

Neat plot.

### ■ T.2) Implicit differentiaition, second derivatives and the osculating circle.

#### T.2.a)

Given a circle

$$(x-h)^2 + (y-k)^2 = r^2,$$

it is certain that as $x$ varies then $y$ must also vary in order to maintain the equality. So insisting that $(x-h)^2 + (y-k)^2 = r^2$ allows us to think of $y$ as a function $y[x]$ of $x$. How do you get the formulas for $y'[x]$ and $y''[x]$ at a point $\{x, y[x]\}$ on the circle?

**Answer:**

Differentiate straight through the circle equation:

```
In[1]:=    deriveqn =
        D[(x - h)^2 + (y[x] - k)^2 == r^2,x]
```

```
Out[1]=    2 (-h + x) + 2 (-k + y[x]) y'[x] == 0
```

To get the formula for $y'[x]$ at a point $\{x, y[x]\}$ on the circle, just solve for $y'[x]$:

```
In[2]:=    yprimesolved = Solve[deriveqn,y'[x]]
```

```
Out[2]=                    -2 h + 2 x
        {{y'[x] -> -(-------------)}}
                    -2 k + 2 y[x]
```

To go after the formula for the second derivative, differentiate twice through the equation of the circle:

```
In[3]:=    secondderiveqn =
        D[(x - h)^2 + (y[x] - k)^2 == r^2,{x,2}]
```

```
Out[3]=                2
        2 + 2 y'[x]  + 2 (-k + y[x]) y''[x] == 0
```

Use the substitution rule from above:

```
In[4]:=    yprimesolved[[1]]
```

```
Out[4]=                   -2 h + 2 x
        {y'[x] -> -(-------------)}
                   -2 k + 2 y[x]
```

Put it into the second derivative equation above: equation above:

In[5]:= **bettersecondeqn =**
**secondoriveqn/.yprimesolved[[1]]**

Out[5]=
$$2 + \frac{2\,(-2\,h + 2\,x)^2}{(-2\,k + 2\,y[x])^2} +$$
$$2\,(-k + y[x])\,y''[x] \; == \; 0$$

Solve for $y''[x]$:

In[6]:= **ydoubleprime =**
**Solve[bettersecondeqn,y''[x]]**

Out[6]= {{y''[x] ->
$$-((2\,h^2 + 2\,k^2 - 4\,h\,x + 2\,x^2 -$$
$$4\,k\,y[x] + 2\,y[x]^2) /$$
$$(-2\,k^3 + 6\,k^2\,y[x] - 6\,k\,y[x]^2 +$$
$$2\,y[x]^3))}}$$

A nasty formula which nobody ever commits to memory.

### T.2.b.i)

Fix a point $\{x_0, f[x_0]\}$ on the graph of a function $f[x]$.
We seek a circle
$$(x - h)^2 + (y - k)^2 = r^2$$
such that if we think of $y$ as a function $y[x]$ of $x$, then
$$y[x_0] = f[x_0]$$
$$y'[x_0] = f'[x_0]$$
$$y''[x_0] = f''[x_0].$$

This circle is called the **osculating (=kissing) circle** to the curve $y = f[x]$ at $\{x_0, f[x_0]\}$. In principle, why should there be such an osculating circle?

---

**Answer:**

A circle is determined by three conditions and three conditions have been specified.

### T.2.b.ii)

Give formulas for the center $\{h, k\}$ and radius $r$ of the osculating circle of $f[x]$ at a point $\{x_0, f[x_0]\}$.

---

**Answer:**

Here is the circle equation:

In[1]:= **circleeqn=((x-h)^2 + (y[x]-k)^2==r^2)**

Out[1]=
$$(-h + x)^2 + (-k + y[x])^2 \; == \; r^2$$

Incorporate the condition $y[x_0] = f[x_0]$ by writing:

In[2]:= **eqn1 = circleeqn/.{x->x0,y[x]->f[x0]}**

Out[2]=
$$(-h + x0)^2 + (-k + f[x0])^2 \; == \; r^2$$

Next differentiate through the circle equation:

In[3]:= **firstderiveqn =**
**D[(x-h)^2 + (y[x]-k)^2 == r^2,x]**

Out[3]= $2\,(-h + x) + 2\,(-k + y[x])\,y'[x] \; == \; 0$

Incorporate the conditions $y[x_0] = f[x_0]$ and $y'[x_0] = f'[x_0]$ by writing:

In[4]:= **eqn2 = firstderiveqn/.**
**{x->x0,y[x]->f[x0],y'[x]->f'[x0]}**

Out[4]= $2\,(-h + x0) + 2\,(-k + f[x0])\,f'[x0] \; == \; 0$

Finally differentiate twice through the circle equation:

In[5]:= **secondoriveqn =**
**D[(x-h)^2 + (y[x]-k)^2 == r^2,{x,2}]**

Out[5]=
$$2 + 2\,y'[x]^2 + 2\,(-k + y[x])\,y''[x] \; == \; 0$$

Incorporate the conditions $y[x_0] = f[x_0]$ , $y'[x_0] = f'[x_0]$ and $y''[x_0] = f''[x_0]$ by writing:

In[6]:= **eqn3 = secondoriveqn/.**
**{x->x0,y[x]->f[x0], y'[x]->f'[x0],**
**y''[x]->f''[x0]}**

Out[6]=
$$2 + 2\,f'[x0]^2 +$$
$$2\,(-k + f[x0])\,f''[x0] \; == \; 0$$

Now solve **eqn1** , **eqn2** and **eq3** simultaneously for $h, k$ and $r$ :

In[7]:= **Solve[{eqn1,eqn2,eqn3},{h,k,r}]**

Out[7]=
$$\{\{r \to \mathrm{Sqrt}[1 + 3\,f'[x0]^2 + 3\,f'[x0]^4 +$$
$$f'[x0]^6] / f''[x0],$$
$$h \to x0 - \frac{f'[x0]}{f''[x0]} - \frac{f'[x0]^3}{f''[x0]},$$
$$k \to f[x0] + \frac{1}{f''[x0]} + \frac{f'[x0]^2}{f''[x0]}\},$$
$$\{r \to -(\mathrm{Sqrt}[1 + 3\,f'[x0]^2 +$$
$$3\,f'[x0]^4 + f'[x0]^6] / f''[x0])$$

$$, \ h \rightarrow x0 - \frac{f'[x0]}{f''[x0]} - \frac{f'[x0]^3}{f''[x0]},$$

$$k \rightarrow f[x0] + \frac{1}{f''[x0]} + \frac{f'[x0]^2}{f''[x0]}\}\}$$

There are two sets of solutions to choose from. Because $r$ cannot be negative, we pick the first set of solutions if $f''[x_0]$ is positive, but we pick the second set of solutions if $f''[x_0]$ is negative.

If $f''[x_0] = 0$, then we have to resign in disgust because there are no solutions in this unhappy case.

### T.2.b.iii)

Are these formulas for the center $\{h, k\}$ and radius $r$ of the osculating circle of $f[x]$ at a point $\{x_0, f[x_0]\}$ to be memorized?

### Answer:

No! They are not too be memorized, but the procedure that produced them is meant to be used.

### T.2.c)

Plot $f[x] = 12/x$ and its osculating circle at $\{3, 4\}$

on the same axes.

### Answer:

Copy, paste and edit the routine from part **b.ii)** immediately above and initialize it by setting $f[x] = 12/x$ and $x_0 = 3$.

```
In[1]:=    {f[x_] = 12/x, x0 = 3}
Out[1]=    12
           {--, 3}
            x
```

Here is the circle equation:

```
In[2]:=    circleeqn = ((x - h)^2 + (y[x] - k)^2 == r^2)
Out[2]=         2            2      2
           (-h + x)  + (-k + y[x])  == r
```

Incorporate the condition $y[x_0] = f[x_0]$ by writing:

```
In[3]:=    eqn1 = circleeqn/.{x->x0,y[x]->f[x0]}
Out[3]=         2        2     2
           (3 - h)  + (4 - k)  == r
```

Next differentiate through the circle equation:

```
In[4]:=    firstderiveqn =
```

```
           D[(x - h)^2 + (y[x] - k)^2 == r^2,x]
Out[4]=    2 (-h + x) + 2 (-k + y[x]) y'[x] == 0
```

Incorporate the conditions

$y[x_0] = f[x_0]$ and $y'[x_0] = f'[x_0]$

by writing:

```
In[5]:=    eqn2 = firstderiveqn/.
           {x->x0,y[x]->f[x0],y'[x]->f'[x0]}
Out[5]=                  8 (4 - k)
           2 (3 - h) -  ---------  == 0
                            3
```

Finally differentiate twice through the circle equation:

```
In[6]:=    seconderiveqn =
           D[(x-h)^2 + (y[x]-k)^2 == r^2,{x,2}]
Out[6]=                2
           2 + 2 y'[x]  + 2 (-k + y[x]) y''[x] == 0
```

Incorporate the conditions

$y[x_0] = f[x_0]$ , $y'[x_0] = f'[x_0]$ and $y''[x_0] = f''[x_0]$

by writing:

```
In[7]:=    eqn3 = seconderiveqn/.
           {x->x0,y[x]->f[x0],y'[x]->f'[x0],
           y''[x]->f''[x0]}
Out[7]=    50   16 (4 - k)
           -- + ---------- == 0
            9        9
```

Now solve $eqn1, eqn2$ and $eq3$ simultaneously for $h, k$ and $r$:

```
In[8]:=    Solve[{eqn1,eqn2,eqn3},{h,k,r}]
Out[8]=          125        43        57
           {{r -> ---, h -> --, k -> --},
                   24        6         8

                 125         43        57
           {r -> -(---), h -> --, k -> --}}
                   24         6         8
```

Here comes the true scale plot:

```
In[9]:=    osccircle =Graphics[RGBColor[0,0,1],
           Circle[{43/6,57/8},125/24]]
           curveplot = Plot[f[x],{x,.5,15},
           DisplayFunction->Identity]
           pointplot = Graphics[
           {PointSize[.02],Point[{3,f[3]}]}]
           Show[curveplot,pointplot,osccircle,
           AxesLabel->{"x","y"},
           PlotRange->{-1,15},
           AspectRatio->Automatic,
           DisplayFunction->$DisplayFunction];
```

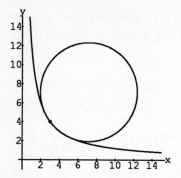

Don't you agree that there is something very satisfying
about the plot?

## Literacy Sheet

1. What does it mean to say that the equation for a circle

$$(x - h)^2 + (y - k)^2 = r^2$$

has three degrees of freedom?

.

.

2. Why does it take three distinct points to determine a circle?

.

.

3. Can you expect to impose four or more conditions on a circle $(x - h)^2 + (y - k)^2 = r^2$ and then expect to go ahead and solve for $h, k$ and $r$?

.

.

.

4. If you have seven variables and five equations, then how many variables can you hope to be able to eliminate?

.

.

.

5. Find $h, k$ and $r$ such that the circle $(x - h)^2 + (y - k)^2 = r^2$ is centered at $\{3, 4\}$ and goes through $\{0, 0\}$.

.

.

.

6. Describe the points $\{x, y\}$ that satisfy $x^2 + y^2 < 4$.

.

.

.

7. What is the slope of the tangent line to the circle $x^2 + y^2 = 25$ through the point $\{4, -3\}$?

.

.

.

8. Explain why the slope of the tangent line to the curve $x^2 - y^2 = 1$ through a point $\{x, y\}$ on the curve is $x/y$ by thinking of $y$ as function $y[x]$ of $x$ and then differentiating $x^2 - y[x]^2 = 1$ with respect to $x$ and then solving for

$y'[x]$. Some folks call this technique by the name "implicit differentiation."

.

.

.

9. A synonym for the verb " to osculate" is "to kiss" (if you don't believe this, look it up in a dictionary). Why are osculating circles aptly named?

.

.

10. Draw some osculating circles on the following plots. Start by drawing the osculating circles at the indicated points.

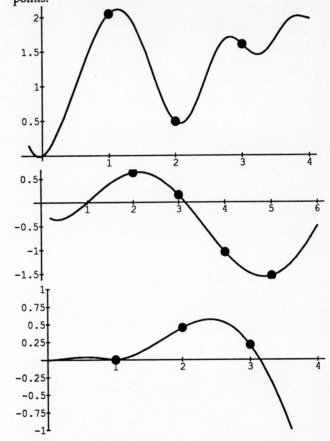

11. If all the ocsulating circles are above the graph of $y = f[x]$, then where are all the tangent lines?

.

.

.

12. Where would you look for the osculating curves to the graph of a convex function?

.

**13.** What do you think happens to the osculating circle at $\{a, f[a]\}$ if the tangent line to the curve $y = f[x]$ actually crosses the curve $y = f[x]$ at the point $\{a, f[a]\}$ of tangency? This happens, for example, if $f[x] = \cos[x]$ and $a = \pi/2$.

**14.** What is the idea of curvature supposed to convey?

# Higher derivatives, splines and approximations

## Guide

The physical meaning of the higher derivatives may not be apparent to you, but their significance and utility will not be soon forgotten after this lesson.

## Basics

■ **B.1) Some remarkable plots and an explanation.**

Sometimes functions whose formulas are strikingly different have plots that are strikingly similar.

**B.1.a.i)**

Plot $\cos[x]$ and $\sqrt{1-x^2}$ on the same axes for $-.5 \le x \le .5$. Describe what you see.

**Answer:**

```
In[1]:=    Plot[{Cos[x], Sqrt[1 - x^2]}, {x, -.5, .5}];
```

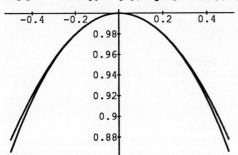

Very similar plots.

**B.1.a.ii)**

Plot $\sin[x]$ and $x(\cos[x])^{1/3}$ on the same axes for $-1 \le x \le 1$. Describe what you see.

**Answer:**

```
In[1]:=    Plot[{Sin[x], x Cos[x]^(1/3)}, {x, -1, 1}];
```

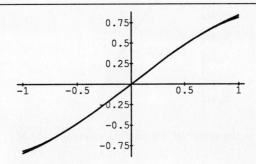

Cohabitation.

**B.1.a.iii)**

Plot $\sin[x]$ and $x(60 - 7x^2)/(60 + 3x^2)$ on the same axes for $-3 \le x \le 3$.

Describe what you see.

**Answer:**

```
In[1]:=    Plot[{Sin[x],x(60-7x^2)/(60+3x^2)},
               {x,-3,3}];
```

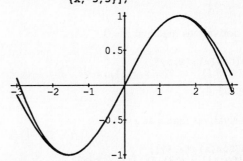

Sharing lots of ink.

**B.1.b.i)**

Are the plotting phenomena we saw in the last three plots accidents?

**Answer:**

In mathematics, there are no accidents.

**B.1.b.ii)**

What can explain what's happening?

**Answer:**

It has to do with the higher derivatives. Look at:

$In[1]:=$    Plot[{Cos[x], Sqrt[1 - x^2]},{x,-.5,.5}];

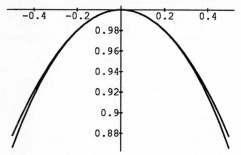

Note that 0 is the center of the plotting interval $[-.5, .5]$ and look at:

$In[2]:=$    {Cos[x], Sqrt[1 - x^2]}/.x->0

$Out[2]=$    {1, 1}

The functions match at the center of the interval, $x = 0$.

$In[3]:=$    {D[Cos[x],x], D[Sqrt[1 - x^2],x]}/.x->0

$Out[3]=$    {0, 0}

Their first derivatives match at $x = 0$.

$In[4]:=$    {D[Cos[x],{x,2}],
            D[Sqrt[1 - x^2],{x,2}]}/.x->0

$Out[4]=$    {-1, -1}

Their second derivatives match at $x = 0$.

$In[5]:=$    {D[Cos[x],{x,3}],
            D[Sqrt[1 - x^2],{x,3}]}/.x->0

$Out[5]=$    {0, 0}

Their third derivatives match at $x = 0$.

$In[6]:=$    {D[Cos[x],{x,4}],
            D[Sqrt[1 - x^2],{x,4}]}/.x->0

$Out[6]=$    {1, -3}

Their fourth derivatives do not match at $x = 0$.

We say that these two functions have **order of contact** 3 at $x = 0$.

Let's look at the next pair:

$In[7]:=$    Plot[{Sin[x],x Cos[x]^(1/3)},{x,-1,1}];

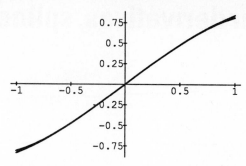

Note that $x = 0$ is the center of the plotting interval $[-1, 1]$ and look at:

$In[8]:=$    {Sin[x],x Cos[x]^(1/3)}/.x->0

$Out[8]=$    {0, 0}

$In[9]:=$    {D[Sin[x],x],D[x Cos[x]^(1/3),x]}/.x->0

$Out[9]=$    {1, 1}

$In[10]:=$    {D[Sin[x],{x,2}],
             D[x Cos[x]^(1/3),{x,2}]}/.x->0

$Out[10]=$    {0, 0}

$In[11]:=$    {D[Sin[x],{x,3}],
             D[x Cos[x]^(1/3),{x,3}]}/.x->0

$Out[11]=$    {-1, -1}

$In[12]:=$    {D[Sin[x],{x,4}],
             D[x Cos[x]^(1/3),{x,4}]}/.x->0

$Out[12]=$    {0, 0}

$In[13]:=$    {D[Sin[x],{x,5}],
             D[x Cos[x]^(1/3),{x,5}]}/.x->0

$Out[13]=$    $\{1, -(\frac{5}{3})\}$

The functions and their first **four** derivatives match at 0. We say that these two functions have **order of contact 4** at $x = 0$.

Let's look at the last pair:

$In[14]:=$    Plot[{Sin[x],x ( 60 - 7 x^2)/
             (60 + 3 x^2)},{x,-3,3}];

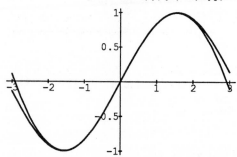

Note that 0 is the center of the plotting interval $[-3, 3]$ and

look at:

```
In[15]:=  {f[x_] = Sin[x],
          g[x_] = x (60 - 7 x^2)/(60 + 3 x^2)}
```

```
Out[15]=                   2
                  x (60 - 7 x )
          {Sin[x], -------------}
                          2
                   60 + 3 x
```

```
In[16]:=  {f[0],g[0]}
Out[16]=  {0, 0}
```

```
In[17]:=  {f'[0], g'[0]}
Out[17]=  {1, 1}
```

```
In[18]:=  {f''[0], g''[0]}
Out[18]=  {0, 0}
```

```
In[19]:=  {f'''[0], g'''[0]}
Out[19]=  {-1, -1}
```

```
In[20]:=  {f''''[0], g''''[0]}
Out[20]=  {0, 0}
```

```
In[21]:=  {f'''''[0],g'''''[0]}
Out[21]=  {1, 1}
```

```
In[22]:=  {f''''''[0],g''''''[0]}
Out[22]=  {0, 0}
```

```
In[23]:=  {f'''''''[0],g'''''''[0]}
Out[23]=            21
          {-1, -(--)}
                  10
```

The functions and their first **six** derivatives match at $x = 0$.

We say that these two functions have **order of contact** 6 at $x = 0$.

The idea of order of contact looks like it has potential.

### ■ B.2) Smooth splines.

A spline is a long flexible strip of plastic or the like that is used in drawing smooth curves. Derivatives can be used to design splines.

Suppose we have two functions $f[x]$ and $g[x]$ and we have a number $b$. Create a new function $h[x]$ by setting

$h(x) = f[x]$ for $x \le b$

and

$h[x] = g[x]$ for $x > b$.

If $f[b] = g[b]$, then we say that the new function $h[x]$ is a

*spline of $f[x]$ and $g[x]$ knotted at $\{b, f[b]\} = \{b, g[b]\}$.*

### B.2.a)

Plot a spline of $e^x$ and

$$\sqrt{(1 + x)/(1 - x)} + \sin[6x^3]$$

knotted at $\{0, 1\}$ on an interval including 0. Discuss reasons for the smoothness of the spline curve as it passes through the knot at $\{0, 1\}$.

---

**Answer:**

```
In[1]:=   f[x_] = E^x
Out[1]=   x
          E
```

```
In[2]:=   g[x_] = Sqrt[(1 + x)/(1 - x)] + Sin[6 x^3]
Out[2]=   Sqrt[1 + x]            3
          ----------- + Sin[6 x ]
          Sqrt[1 - x]
```

```
In[3]:=   h[x_] := g[x]/;x > 0
          h[x_] := f[x]/;x <= 0
```

```
In[4]:=   Plot[h[x],{x,-1,1}];
```

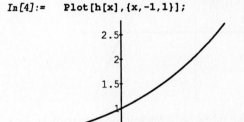

To see why the spline is so smooth at the knot at $\{0, 1\}$, look at:

```
In[5]:=   {f[0],g[0]}
Out[5]=   {1, 1}
```

```
In[6]:=   {f'[0],g'[0]}
Out[6]=   {1, 1}
```

```
In[7]:=   {f''[0],g''[0]}
Out[7]=   {1, 1}
```

```
In[8]:=   {f'''[0],g'''[0]}
Out[8]=   {1, 39}
```

The smoothness of the spline at the knot at $\{0, 1\}$ seems to be related to the fact that $f[x]$ and $g[x]$ have order of contact 2 at $\{0, 1\}$.

## B.2.b)

Plot a spline of $e^x$ and $1 - 2x^2$ knotted at $\{0, 1\}$ on an interval including 0. Discuss reasons for the lack of smoothness of the spline curve as it passes through the knot at $\{0, 1\}$.

**Answer:**

```
In[1]:=    Clear[f,g,h,x]
           f[x_] = E^x
Out[1]=     x
           E
```

```
In[2]:=    g[x_] = 1 - 2 x^2
Out[2]=            2
           1 - 2 x
```

```
In[3]:=    h[x_] := g[x]/;x > 0
           h[x_] := f[x]/;x <= 0
```

```
In[4]:=    Plot[h[x],{x,-1,1},
           AxesLabel->{"x","y"}];
```

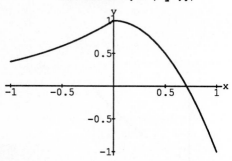

To see why the spline has a hitch at the knot at $\{0, 1\}$, look at:

```
In[5]:=    {f[0],g[0]}
Out[5]=    {1, 1}
```

```
In[6]:=    {f'[0],g'[0]}
Out[6]=    {1, 0}
```

The lack of smoothness of the spline at the knot at $\{0, 1\}$ seems to be related to the fact that $f[x]$ and $g[x]$ have order of contact 0 at $\{0, 1\}$.

A road built in the shape of this plot would be highly unsafe.

## B.2.c)

Plot a spline of $e^{x-1}$ and $3/8 + x/3 + x^2/4 + x^4/24$ knotted at $\{1, 1\}$ on an interval including 1. Discuss reasons for the smoothness of the spline curve as it passes through the knot at $\{1, 1\}$.

**Answer:**

```
In[1]:=    Clear[f,g,h,x]
           f[x_] = E^(x - 1)
Out[1]=     -1 + x
           E
```

```
In[2]:=    g[x_] = 3/8 + x/3 + x^2/4 + x^4/24
Out[2]=            2    4
           3   x   x    x
           - + - + -- + --
           8   3   4    24
```

```
In[3]:=    h[x_] := g[x]/;x > 1
           h[x_] := f[x]/;x <= 1
```

```
In[4]:=    Plot[h[x],{x,0,2},AxesLabel->{"x","y"}];
```

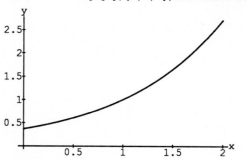

To see why the spline passes so smoothly through the knot at $\{1, 1\}$, look at:

```
In[5]:=    {f[1],g[1]}
Out[5]=    {1, 1}
```

```
In[6]:=    {f'[1],g'[1]}
Out[6]=    {1, 1}
```

```
In[7]:=    {f''[1],g''[1]}
Out[7]=    {1, 1}
```

```
In[8]:=    {f'''[1],g'''[1]}
Out[8]=    {1, 1}
```

```
In[9]:=    {f''''[1],g''''[1]}
Out[9]=    {1, 1}
```

```
In[10]:=   {f'''''[1],g'''''[1]}
Out[10]=   {1, 0}
```

The extreme smoothness of the spline at the knot at $\{1, 1\}$ seems to be related to the fact that $f[x]$ and $g[x]$ have order of contact 4 at $\{1, 1\}$.

A road built in the shape of this plot would be very safe.

**B.2.d)**

What practical good is this?

**Answer:**

Good question.

First of all, the idea of a spline smooth at its knots makes for interesting plots and nice art. But there is a practical aspect as well.

Laying out the curves on expressways and railroads amounts to joining curved plots to relatively straight plots at knots of a spline. It is unacceptable to have a hitch (corner) in the middle of the roadway.

The more degrees of contact at the knots, the safer the highway or roadbed.

Any degree of contact less than 2 is considered unsafe. And this may be pushing it.

## Tutorial

■ **T.1) Approximation via higher order contact.**

**T.1.a.i)**

Find the third degree (cubic) polynomial that has order of contact 3 with $e^x$ at $\{0,1\}$.

Plot this polynomial and $e^x$ on the same axes for $-2 \leq x \leq 2$.

**Answer:**

Set the polynomial:

```
In[1]:=    y = a + b x + c x^2 + d x^3
```
$$Out[1]=\quad a + b\,x + c\,x^2 + d\,x^3$$

The conditions that must be met are:

```
In[2]:=    eqn1 = (y /.x->0) == E^x/.x->0
Out[2]=    a == 1
```

```
In[3]:=    eqn2 = (D[y,x]/.x->0) == D[E^x,x]/.x->0
Out[3]=    b == 1
```

```
In[4]:=    eqn3 = (D[y,{x,2}]/.x->0) == D[E^x,{x,2}]/.x->0
Out[4]=    2 c == 1
```

```
In[5]:=    eqn4 = (D[y,{x,3}]/.x->0) == D[E^x,{x,3}]/.x->0
Out[5]=    6 d == 1
```

Now we nail down the coefficients:

```
In[6]:=    coeffs = Solve[{eqn1,eqn2,eqn3,eqn4}]
```
$$Out[6]=\quad \{\{a \to 1,\ b \to 1,\ c \to \frac{1}{2},\ d \to \frac{1}{6}\}\}$$

Substitute in the values of the coefficients that do the job:

```
In[7]:=    goody = y/.coeffs[[1]]
```
$$Out[7]=\quad 1 + x + \frac{x^2}{2} + \frac{x^3}{6}$$

And now the plot:

```
In[8]:=    Plot[{E^x,goody},{x,-2,2}];
```

There's a lot of snuggling going on for $-1 \leq x \leq 1$.

**T.1.a.ii)**

If you use your answer to part i) to compute values of $e^x$ for $-1 \leq x \leq 1$, then how many accurate decimals of $e^x$ do you get?

**Answer:**

Let's see how far apart they are by plotting the absolute value of their difference. We'll use the plotting option **PlotRange->All** to guarantee that we see the whole plot.

```
In[1]:=    Plot[Abs[goody - E^x],{x,-1,1},
             PlotRange->All];
```

You get at least one accurate decimal of $e^x$.

**T.1.a.iii)**

If you use your answer to part i) to calculate values of $e^x$ for $-1/4 \le x \le 1/4$, then how many accurate decimals of $e^x$ do you get?

**Answer:**

```
In[1]:=    Plot[Abs[goody - E^x], {x, -1/4, 1/4},
           PlotRange->All];
```

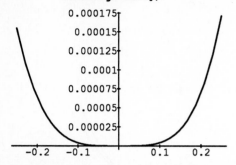

You get at least three accurate decimals of $e^x$.

**T.1.b.i)**

Find the function of the form
$$(a + bx + cx^2)/(1 + cx + dx^2)$$
that has order of contact 4 with $e^x$ at $\{0,1\}$.

Plot this function and $e^x$ on the same axes for $-2 \le x \le 2$.

**Answer:**

The function we have to determine is:

```
In[1]:=    y = (a + b x + c x^2)/(1 + d x + e x^2)
                     2
Out[1]=    a + b x + c x
           --------------
                     2
           1 + d x + e x
```

The conditions that must be met are:

```
In[2]:=    eqn1 = (y/.x->0) == E^x/.x->0
Out[2]=    a == 1

In[3]:=    eqn2 = (D[y ,x]/.x->0) == D[E^x,x]/.x->0
Out[3]=    b - a d == 1

In[4]:=    eqn3 = (D[y ,{x,2}]/.x->0) ==
                  D[E^x,{x,2}]/.x->0
                            2
Out[4]=    2 c - 2 b d + 2 a d  - 2 a e == 1

In[5]:=    eqn4 = (D[y ,{x,3}]/.x->0) ==
                  D[E^x,{x,3}]/.x->0
```

```
Out[5]=
                          2            3
           -6 c d + 6 b d  - 6 a d  - 6 b e +

           12 a d e == 1

In[6]:=    eqn5 = (D[y ,{x,4}]/.x->0) ==
                  D[E^x,{x,4}]/.x->0
                       2            3            4
Out[6]=    24 c d  - 24 b d  + 24 a d  - 24 c e +

                                2            2
           48 b d e - 72 a d  e + 24 a e  == 1
```

Now we nail down the coefficients:

```
In[7]:=    coeffs = Solve[{eqn1, eqn2,
                   eqn3, eqn4, eqn5}]
Out[7]=
                  1          1          1            1
           {{c -> --, e -> --, b -> -, d -> -(-),
                  12         12         2            2

           a -> 1}}
```

Substitute in the values of the coefficients that do the job:

```
In[8]:=    goody = y/.coeffs[[1]]
Out[8]=
                     2
                x   x
           1 + - + --
                2   12
           ----------
                     2
                x   x
           1 - - + --
                2   12
```

And now the plot:

```
In[9]:=    Plot[{E^x, goody}, {x, -2, 2}];
```

Cohabitation.

**T.1.b.ii)**

If you use your answer from part b.i) to calculate values of $e^x$ for $-1 \le x \le 1$, then how many accurate decimals of $e^x$ do you get?

**Answer:**

```
In[1]:=    Plot[Abs[goody - E^x], {x, -1, 1},
           PlotRange->All];
```

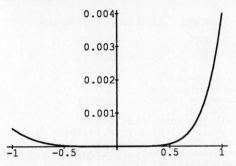

You get at least two accurate decimals of $e^x$.

### T.1.b.iii)

If you use your answer from part b.i) to calculate values of $e^x$ for $-1/4 \leq x \leq 1/4$, then how many accurate decimals of $e^x$ do you get?

**Answer:**

```
In[1]:=    Plot[Abs[goody - E^x],{x,-1/4,1/4},
           PlotRange->All];
```

```
0.00000175
0.0000015
0.00000125
0.000001
        -7
7.5 10
        -7
5. 10
        -7
2.5 10

    -0.2   -0.1        0.1    0.2
```

You get at least five accurate decimals of $e^x$.

Not bad.

### T.1.c.i)

Find the function of the form
$$a + b\cos[x] + c\sin[x] + d\cos[2x] + e\sin[2x]$$
that has order of contact 4 with $e^x$ at $\{0,1\}$.

Plot this function and $e^x$ on the same axes for $-2 \leq x \leq 2$.

**Answer:**

The function we have to determine is:

```
In[1]:=    y = a + b Cos[x] + c Sin[x] +
           d Cos[2 x] + e Sin[2 x]
Out[1]=    a + b Cos[x] + d Cos[2 x] + c Sin[x] +

           e Sin[2 x]
```

The conditions that must be met are:

```
In[2]:=    eqn1 =
```

```
           y == E^x/.x->0
Out[2]=    a + b + d == 1
```

```
In[3]:=    eqn2 =
           D[y ,x] == D[E^x,x]/.x->0
Out[3]=    c + 2 e == 1
```

```
In[4]:=    eqn3 =
           D[y,{x,2}] == D[E^x,{x,2}]/.x->0
Out[4]=    -b - 4 d == 1
```

```
In[5]:=    eqn4 = (D[y,{x,3}]/.x->0) ==
                  D[E^x,{x,3}]/.x->0
Out[5]=    -c - 8 e == 1
```

```
In[6]:=    eqn5 =
           D[y,{x,4}] == D[E^x,{x,4}]/.x->0
Out[6]=    b + 16 d == 1
```

Now we nail down the coefficients:

```
In[7]:=    coeffs = Solve[{eqn1,eqn2,
                  eqn3,eqn4,eqn5}]
Out[7]=              5         5         1         5
           {{a -> -, b -> -(-), d -> -, c -> -,
                    2         3         6         3

                     1
             e -> -(-)}}
                     3
```

Substitute in the values of the coefficients that do the job:

```
In[8]:=    goody = y/.coeffs[[1]]
Out[8]=    5   5 Cos[x]   Cos[2 x]   5 Sin[x]
           - - -------- + -------- + -------- -
           2      3          6          3

           Sin[2 x]
           --------
              3
```

And now the plot:

```
In[9]:=    Plot[{E^x,goody},{x,-2,2}];
```

```
            7
            6
            5
            4
            3
            2
            1
  -2    -1        1    2
```

There's a lot of shared ink for $-1 \leq x \leq 1$.

Isn't it amazing that sine and cosine waves can be combined in such a non-wavy fashion?

**T.1.c.ii)**

If you use your answer to part c.i) to calculate values of $e^x$ for $-1/4 \le x \le 1/4$, then how many accurate decimals of $e^x$ do you get?

**Answer:**

$In[1]:=$  `Plot[Abs[goody - E^x], {x,-1/4,1/4},`
`PlotRange->All];`

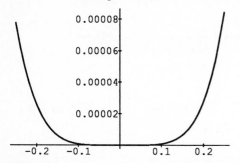

You get at least three accurate decimals of $e^x$.

Maybe sines and cosines are good for something other than boring old trigonometry.

■ **T.2) Natural cubic spline interpolation.**

**T.2.a.i)**

Try to pass a reasonable curve through the points
$$\{\{-8,-12\},\{-1,-15\},\{2,20\},\{5,-4\},$$
$$\{8,9\},\{11,3\},\{15,9\}\}.$$

**Answer:**

There are seven points in this list. One common belief is that we should pass the unique 6th degree interpolating polynomial through the points and then plot it:

$In[1]:=$  `points = {{-8,-12},{-1,-15},`
`{2,20},{5,-4},{8,9},{11,3},{15,9}}`

$Out[1]=$  `{{-8, -12}, {-1, -15}, {2, 20},`

`{5, -4}, {8, 9}, {11, 3}, {15, 9}}`

The equation of this polynomial is:

$In[2]:=$  `y = InterpolatingPolynomial[points,x]`

$Out[2]=$
```
   86999543    612132523 x
   -------- + ----------- -
   6442254     33611760

             2              3
 1637822117 x    136446911 x
 ------------- + ------------ +
   171793440      154614096

          4           5           6
 8114303 x    577061 x    974513 x
 ---------- - --------- + ----------
```

77307048     32211270     1546140960

Here comes the plot:

$In[3]:=$  `givenpoints =`
`ListPlot[points,`
`PlotStyle->{RGBColor[1,0,0],`
`PointSize[.02]},`
`DisplayFunction->Identity]`
`polythrough = Plot[y, {x,-8,15},`
`DisplayFunction->Identity]`
`Show[givenpoints,polythrough,`
`AxesLabel->{"x","y"},`
`DisplayFunction->$DisplayFunction];`

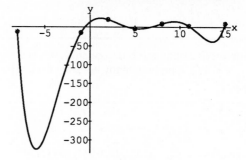

Look at that huge dip on the left!

That polynomial curve has dips and crests that are not suggested by the given points. We better give up on using this approach for this list of numbers.

A drastic measure would be to connect the consecutive points with line segments. Some folks call this procedure "linear interpolation." Let's try it:

$In[4]:=$  `sticks =`
`ListPlot[points,PlotJoined->True,`
`DisplayFunction->Identity]`
`Show[givenpoints,sticks,`
`AxesLabel->{"x","y"},`
`DisplayFunction->$DisplayFunction];`

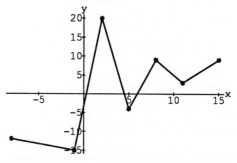

Yuk! This is rough, but it is more satisfactory than the polynomial plot. There must be a way of getting some smoothness into the curve!

**T.2.a.ii)**

What is a good way of getting smoothness into this?

**Answer:**

*"Freedom is for everyone, but using it wisely is the burden of*

*the educated person." (Argentine proverb)*

Instead of passing a line segment through consecutive points, we pass a different **cubic** curve through pair of consecutive points and make a smooth spline with knots at each of the points.

This may seem outlandish at first, because a cubic curve

$$y = ax^3 + bx^2 + cx + d$$

has **four** coefficients to determine and normally we fit a cubic through **four** points. We are going to fit the cubic through just **two** consecutive points and use the extra freedom we have to guarantee smoothness at the knots.

Start out with the points above

$$\{x[1], y[1]\}, \{x[2], y[2]\}, \{x[3], y[3]\},$$

$$\{x[4], y[4]\}, \{x[5], y[5]\}, \{x[6], y[6]\}, \{x[7], y[7]\}$$

sorted so that the $x[k]'s$ increase as $k$ increases.

Set $f[k, x]$ to be the cubic we are going to run from $\{x[k], y[k]\}$ to $\{x[k+1], y[k+1]\}$:

```
In[1]:=    Clear[f,x,y]
           f[k_,x_] =
           a[k] x^3 + b[k] x^2 + c[k] x + d[k]
Out[1]=         3        2
           x  a[k] + x  b[k] + x c[k] + d[k]
```

Here $k$ runs from $k = 1$ to $k = 6$; so we are working with 6 cubics and each cubic has 4 undetermined (so far) coefficients. This means we have 24 equations to play with.

To hit the points, we want:

$f[k, x[k]] = y[k]$ for $k = 1, 2, 3, 4, 5, 6$

and

$f[k, x[k+1]] = y[k+1]$ for $k = 1, 2, 3, 4, 5, 6$.

For smoothness at the knots, we want:

$f'[k, x[k]] = f'[k-1, x[k]]$ for $k = 2, 3, 4, 5, 6$

and

$f''[k, x[k]] = f''[k-1, x[k]]$ for $k = 2, 3, 4, 5, 6$

Here all derivatives are taken with respect to the $x$ variable.

We **cannot use** $k = 1$ in the last two equations because there is no function $f[0, x]$.

So far we have specified

$$6 + 6 + 5 + 5 = 22$$

conditions and we have 24 undetermined coefficients; as a result we can specify **two** more conditions.

Old timers at the art of curve fitting have found that good results usually come from specifying that the second derivative be 0 at the end points $\{x[1], y[1]\}$ and $\{x[7], y[7]\}$. They call the resulting spline **"the natural cubic spline."**

Here is the *Mathematica* code that does all this.

The specific rules we want *Mathematica* to follow are:

```
In[2]:=    eqn[1,1] = (f[1,x[1]] == y[1])
Out[2]=
                                         2
           d[1] + c[1] x[1] + b[1] x[1]  +

                     3
           a[1] x[1]  == y[1]
```

```
In[3]:=    eqn[2,1] = (f[1,x[2]] == y[2])
Out[3]=
                                         2
           d[1] + c[1] x[2] + b[1] x[2]  +

                     3
           a[1] x[2]  == y[2]
```

```
In[4]:=    eqn[3,1] = (D[f[1,x],{x,2}]/.x->x[1]) ==
           0
Out[4]=    2 b[1] + 6 a[1] x[1] == 0
```

```
In[5]:=    eqn[4,1] = (D[f[6,x],{x,2}]/.x->x[7]) ==
           0
Out[5]=    2 b[6] + 6 a[6] x[7] == 0
```

The general rules we want *Mathematica* to use are below. These general rules will not override the specific rules above.

```
In[6]:=    eqn[1,k_] = (f[k,x[k]] == y[k])
Out[6]=
                                         2
           d[k] + c[k] x[k] + b[k] x[k]  +

                     3
           a[k] x[k]  == y[k]
```

```
In[7]:=    eqn[2,k_] = (f[k,x[k+1]] == y[k+1])
Out[7]=
                                             2
           d[k] + c[k] x[1 + k] + b[k] x[1 + k]  +

                     3
           a[k] x[1 + k]  == y[1 + k]
```

```
In[8]:=    eqn[3,k_] = (D[f[k,x],x]/.x->x[k]) ==
                       (D[f[k-1,x],x]/.x->x[k] )
Out[8]=
                                         2
           c[k] + 2 b[k] x[k] + 3 a[k] x[k]  ==

           c[-1 + k] + 2 b[-1 + k] x[k] +

                              2
           3 a[-1 + k] x[k]
```

```
In[9]:=    eqn[4,k_] =
           (D[f[k,x],{x,2}]/.x->x[k]) ==
           (D[f[k-1,x],{x,2}]/.x->x[k])
Out[9]=    2 b[k] + 6 a[k] x[k] ==

           2 b[-1 + k] + 6 a[-1 + k] x[k]
```

Now get ready to plug the points in:

```
In[10]:=   x[k_] := points[[k,1]]
```

```
In[11]:=   y[k_] := points[[k,2]]
```

Enter the seven points sorted in order of increasing $x$-coordinates:

```
In[12]:=   points = {{-8,-12},{-1,-15},
           {2,20},{5,-4},{8,9},{11,3},{15,9}}
```
```
Out[12]=   {{-8, -12}, {-1, -15}, {2, 20},

              {5, -4}, {8, 9}, {11, 3}, {15, 9}}
```

Check for instance:

```
In[13]:=   {x[2],y[2]}
```
```
Out[13]=   {-1, -15}
```

Good.

Now solve for the correct coefficients-all 24 of them- substitute and plot:

```
In[14]:=   coeffs =
           Solve[Flatten[Table[{eqn[1,k],
           eqn[2,k],eqn[3,k],eqn[4,k]},
           {k,1,6}]]]

           Clear[ourf]
           Table[{"ourf"[k,x],ourf[k,x_] =
           (f[k,x]/.coeffs[[1]])},{k,1,6}]
           plot1 =
           Plot[ourf[1,x],{x,x[1],x[2]},
           DisplayFunction->Identity]

           plot2 =
           Plot[ourf[2,x],{x,x[2],x[3]},
           DisplayFunction->Identity]

           plot3 =
           Plot[ourf[3,x],{x,x[3],x[4]},
           DisplayFunction->Identity]

           plot4 =
           Plot[ourf[4,x],{x,x[4],x[5]},
           DisplayFunction->Identity]

           plot5 =
           Plot[ourf[5,x],{x,x[5],x[6]},
           DisplayFunction->Identity]

           plot6 =
           Plot[ourf[6,x],{x,x[6],x[7]},
           DisplayFunction->Identity]

           Show[plot1,plot2,plot3,plot4,
           plot5,plot6,givenpoints,
           AxesLabel->{"x","y"},
           DisplayFunction->$DisplayFunction];
```

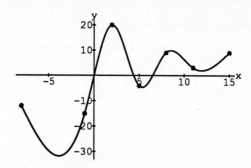

By gosh, that's just about the way that we would have connected the dots with a pencil. The old-time cubic spliners seem to know what they're talking about. And to add adventure to an already thrilling subject, show on the same plot both the interpolating polynomial and the natural spline:

```
In[15]:=   Show[plot1,plot2,plot3,plot4,
           plot5,plot6,givenpoints,polythrough,
           PlotRange->All,
           AxesLabel->{"x","y"},
           DisplayFunction->$DisplayFunction];
```

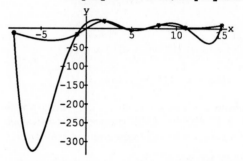

Sit back and reflect remembering that you are one of a elite group of calculus students who has ever seen a cubic spline.

### ■ T.3) Landing an airplane

The design of an elecronically-controlled airplane landing system calls for the plane to approach the runway head-on, at a constant horizontal speed $s$ and constant altitude $h$. As the plane passes over a certain point on the ground $R$ units from the designated touch-down spot on the runway, the system is to take over and bring the plane onto the runway on the trajectory of a cubic polynomial.

The constant horizontal speed $s$ is to be maintained through the whole landing procedure.

Our units will be distances $x$ and $y$ in feet and time $t$ in seconds.

Set up the cubic:

```
In[16]:=   y=a x^3 + b x^2 + c x + d
```
```
Out[16]=                   2       3
             d + c x + b x  + a x
```

Here $y$ stands for the altitude of the plane when the plane is directly above a spot on the ground $x$ feet from the designated touch-down spot on the runway. For a smooth onset of the descent we must have:

```
In[17]:=   eqn1 = (y/.x->R) == h

Out[17]=    3     2
           R  a + R  b + R c + d == h
```

```
In[18]:=   eqn2 = (D[y,x]/.x->R) == 0

Out[18]=      2
           3 R  a + 2 R b + c == 0
```

For touchdown at the designated touch-down spot on the runway, we must have:

```
In[19]:=   eqn3 = (y/.x->0) == 0

Out[19]=   d == 0
```

The runway must be tangent to the trajectory at the designated touch-down spot on the runway, so we must have:

```
In[20]:=   eqn4 = (D[y,x]/.x->0) == 0

Out[20]=   c == 0
```

To determine the trajectory, we solve:

```
In[21]:=   coeffs = Solve[{eqn1,eqn2,eqn3,eqn4},
           {a,b,c,d}]

Out[21]=          -2 h       3 h
           {{a -> ----, b -> ---, c -> 0, d -> 0}}
                    3          2
                   R          R
```

The trajectory is the graph of:

```
In[22]:=   trajectory = y/.coeffs[[1]]

Out[22]=      2        3
           3 h x    2 h x
           ------ - ------
              2        3
             R        R
```

### T.3.a)

---

Plot the trajectory for the case $R = 10000$ ft, $h = 5000$ ft.

Then plot the detail of the landing.

---

### Answer:

The plot of the landing trajectory is:

```
In[1]:=   landing =
          Plot[trajectory /.
          {R->10000,h->5000},{x,0,10000},
          PlotStyle->RGBColor[0,0,1]];
```

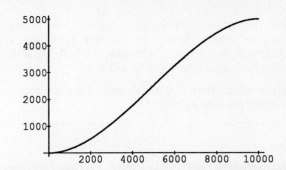

Here is a plot showing the plane's path slightly before the descent begins and slightly after the plane touches down:

```
In[2]:=   air =
          Plot[5000,{x,10000,14000},
          PlotStyle->
          RGBColor[0,0,1],
          DisplayFunction->Identity]

          land =
          Plot[0,{x,-3000,0},
          PlotStyle->
          RGBColor[0,0,1],
          DisplayFunction->Identity]

          Show[air,landing,land,
          AspectRatio->Automatic,
          DisplayFunction->$DisplayFunction];
```

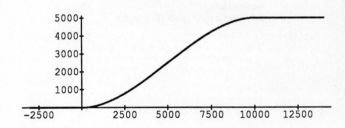

Nice.

Here is a plot of the detail of the landing in true scale:

```
In[3]:=   Show[air,landing,land,
          AspectRatio->Automatic,
          Ticks->{Range[-400,800,400],Automatic},
          PlotRange->{{-500,1000},{-10,400}},
          DisplayFunction->$DisplayFunction];
```

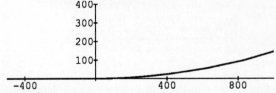

The whole path of the airplane is nothing but a smooth spline knotted at $\{0,0\}$ and $\{10000,5000\}$.

### T.3.b)

---

We show our system to a knowledgable aviator for comment. The aviator is skeptical about it because (among other issues) there is no safeguard against destructive forces. In order to hold the plane together, the vertical acceleration must be held to be less than 3.2 ft/sec per sec. Compute the vertical acceleration as a function of the constant horizontal speed $s$, $R$, $h$ and $t$.

Give the relationship $R, s$ and $h$ must have in order to stay within the safety guideline.

---

**Answer:**

We know that the plane is travelling at a constant horizontal speed of $s$ feet/sec. We also know the trajectory:

```
In[1]:=   trajectory
Out[1]=         2       3
          3 h x    2 h x
          ------ - ------
            2        3
           R        R
```

Because the plane is holding a steady horizontal speed of $s$ mph, we also know that as a function of $t$,

$$x = R - st;$$

so we can get vertical altitude $y$ (in feet) as a function of $t$:

```
In[2]:=   vertalt =
          trajectory /.x->(R - s t)
Out[2]=         2             3
          3 h (R - s t)   2 h (R - s t)
          ------------- - -------------
                2               3
               R               R
```

The vertical speed in feet per second is:

```
In[3]:=   vertspeed = D[vertalt,t]
Out[3]=                                      2
          -6 h s (R - s t)   6 h s (R - s t)
          ---------------- + ----------------
                 2                  3
                R                  R
```

The vertical acceleration feet per second per second is:

```
In[4]:=   vertaccel = Together[D[vertspeed,t]]
Out[4]=            2        3
          -6 R h s  + 12 h s  t
          ---------------------
                    3
                   R
```

From this we can see that the vertical acceleration starts out at

```
In[5]:=   vertaccel /.t->0
Out[5]=          2
           -6 h s
           -------
              2
             R
```

(pointing down) and steadily increases as $t$ increases until

the time at which the plane touches down; this time is given by $R - st = 0$ or $t = R/s$, so the vertical acceleration at touchdown is

```
In[6]:=   vertaccel /.t->R/s
Out[6]=         2
          6 h s
          ------
             2
            R
```

(pointing up). Thus to meet the safety guideline, we can get by with

$$6hs^2/R^2 \leq 3.2.$$

### T.3.c)

If the horizontal speed is 150 mph and the original altitude $h$ is 5000 feet, find the mininum safe value of $R$. Plot the trajectory for this $R$. (Remember that the plane begins its approach as it passes over a certain point on the ground $R$ units from the designated touch-down spot.)

---

**Answer:**

The speed of 150 mph translates into $s = 150(5280/3600)$ feet per second. The minimum $R$ is given by:

```
In[1]:=   Solve[((6 h s^2/R^2) /.
          {h->5000,s->150}) == 3.2,R]
Out[1]=   {{R -> 14523.7}, {R -> -14523.7}}
```

In miles this is:

```
In[2]:=   14523.7/5280
Out[2]=   2.7507
```

Here is the plot in true scale of the safe trajectory requiring the shortest $R$:

```
In[3]:=   Plot[trajectory/.{h->5000,R->14523.7},
          {x,0,14523.7},AspectRatio->Automatic,
          Ticks->{Range[0,15000,5000],Automatic}];
```

How beautiful to fly.

# Literacy Sheet

1. Complete this sentence: The higher the order of contact of $f[x]$ and $g[x]$ at 0, then the more we can expect the plots of $f[x]$ and $g[x]$ on $[-1,1]$ to . . .

.

.

.

2. What is the basic idea underlying the construction of smooth splines?

.

.

.

3. How smooth do you expect a spline of $\cos[x] - 1$ and $x$ knotted at $\{0,0\}$ to be?

.

.

.

4. How smooth do you expect a spline of $\cos[x] - 1$ and $x^2$ knotted at $\{0,0\}$ to be?

.

.

.

5. Given a constant $a$, then how would you choose another constant $k$ to try to make a spline of $f[x] = \sin[ax]$ and $g[x] = kx$ knotted at $\{0,0\}$ as smooth as possible?

.

.

.

6. Why are some scientists suspicious of interpolating polynomials but not so suspicious of the natural cubic spline?

.

.

7. Describe the advantage of a cubic spline over a stick figure.

.

.

.

8. Computers and calculators must be programmed to add, subtract, multipy and divide. All the other mathematical operations like taking logarithms, taking $e$ to a power, taking the Sine of an angle, etc. must be programmed in terms of additions, subtraction, multiplication and division. How can you use the idea of order of contact to help to program these mathematical operations?

.

.

.

9. What is the highest order of contact a straight line and a circle can have? If two circles have order of contact 2, how different can the two circles be? What is the order of conatact at 0 of the functions $x^{12}$ and $x^{19}$?

.

.

10. Fix a positive integer $n$ and give a formula for the $n$th derivative of $x^n$.

.

.

.

11. Fix a positive integer $n$ and give a formula for the $n$th derivative of $e^x$.

.

.

12. Fix a positive integer $n$ and give a formula for the $n$th derivative of $e^{-x}$.

.

.

13. Fix a positive integer $n$ and decide whether the $n$th derivative of $\sin[x]$ has the formula $\sin[x + n\pi/2]$.

This would amount to saying that each successive differentiation shifts the Sine curve by $\pi/2$. If this formula is right, then explain why you believe that it is right. If the formula is wrong, then explain why you believe that its wrong.

# Area and the integral

## Guide

Area measurements are as old as humankind's instinct for controlling territory. The origins of the integral as an attempt to measure area go back as far as written history goes. The early civilizations in China, Egypt, India and Mesopotamia all dealt with area measurements and in so doing ran into trouble handling $\pi$. The ancient Greeks, largely through the brilliance of Archimedes, made spectacular strides. Newton and Leibniz brought us the area measurements we are going to study here. More recently, Cauchy, Riemann and Lebesgue made significant contributions to this mother lode of mathematical knowledge.

## Basics

■ **B.1) Integral and area.**

**B.1.a)**

What is the integral and how is it related to area?

**Answer:**

What a beautiful question! The integral is related to area because it is nothing more and nothing less than *a tool to measure area*. Here is how it works: Take a function $y = f[x]$ and plot it on an interval $[a, b]$. It might look something like this:

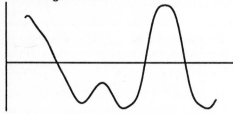

We want to measure the area between the graph of $y = f[x]$ and the $x$-axis:

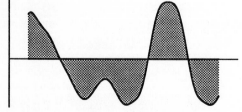

If a part of this area lies above the $x$-axis we consider it as positive. If a part lies below the $x$-axis, we consider it as negative.

This gives the signs as shown below:

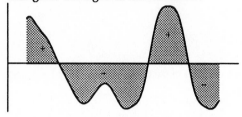

Now we define a number which we call *the integral of* $f[x]$ *from* $a$ *to* $b$. This number is signified by

$$\int_a^b f[x]\,dx$$

and its value is is the sum of these areas taken with the corresponding signs.

Accordingly, if $x_1 = a$, $x_4 = b$, and $x_2$ and $x_3$ are as indicated below:

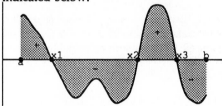

Then we see that

$$\int_a^b f[x]\,dx = \int_a^{x1} f[x]\,dx + \int_{x1}^{x2} f[x]\,dx$$

$$+ \int_{x2}^{x3} f[x]\,dx + \int_{x3}^b f[x]\,dx$$

where:

$\int_a^{x1} f[x]\,dx$ and $\int_{x3}^b f[x]\,dx$ are positive; and

$\int_{x1}^{x2} f[x]\,dx$ and $\int_{x3}^{x4} f[x]\,dx$ are negative.

**B.1.b)**

Make the indicated area measurents:

**B.1.b.i)**

$$\int_1^4 (x/2 + 1)\,dx.$$

**Answer:**

$\int_1^4 (x/2 + 1)dx$ measures the area under the curve

$$y = f[x] = (x/2 + 1)$$

and above the segment $[1,4]$ on the $x-$ axis. Here is a plot:

```
In[1]:=    Clear[f,x]
           f[x_] = (x/2 + 1)
```
```
Out[1]=        x
           1 + -
               2
```

```
In[2]:=    functionplot = Plot[f[x],{x,1,4},
           DisplayFunction->Identity];
```

```
In[3]:=    poly = Graphics[{GrayLevel[.75],
           Polygon[{{1,0},{1,f[1]},{4,f[4]},
           {4,0}}]}];
```

```
In[4]:=    Show[poly,functionplot,
           Ticks->{{1,2,3,4},{1,2,3}},Axes->{0,0},
           AspectRatio->Automatic,
           AxesLabel->{"x","y"},
           PlotRange->{{0,5},{0,4}},
           DisplayFunction->$DisplayFunction];
```

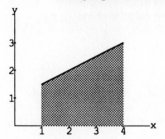

This integral is the area of a trapezoid with

```
In[5]:=    shortHeight = 1.5
           tallHeight  = 3
           width       = 3;
```

Consequently the area measurement $\int_1^4 (x/2 + 1)dx$ in square units is given by:

```
In[6]:=    ((tallHeight + shortHeight)/2) width
```
```
Out[6]=    6.75
```

*Mathematica* can also handle this measurement directly:

```
In[7]:=    Integrate[(x/2 + 1),{x,1,4}]
```
```
Out[7]=    27
           --
           4
```

This fraction has the value:

```
In[8]:=    N[27/4]
```

```
Out[8]=    6.75
```

Got it.

**B.1.b.ii)**

$$\int_{-4}^4 (x/2 + 1)dx.$$

**Answer:**

$\int_{-4}^4 (x/2 + 1)dx$ measures the area under the curve

$$y = f[x] = (x/2 + 1)$$

and above the segment $[-4,4]$ on the $x-$ axis. Take a look at the plot:

```
In[1]:=    Clear[f,x]
           f[x_] = (x/2 + 1)
```
```
Out[1]=        x
           1 + -
               2
```

```
In[2]:=    functionplot = Plot[f[x],
           {x,-4,4},PlotRange->{-2,3},
           AspectRatio->Automatic,
           AxesLabel->{"x","y"}]
```

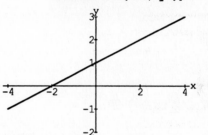

```
Out[2]=    -Graphics-
```

Put in the shading:

```
In[3]:=    poly = Graphics[{GrayLevel[.75],
           Polygon[{{-4,0},{-4,f[-4]},
           {4,f[4]},{4,0}}]}]
           Show[poly,functionplot,
           Graphics[Line[{{-4,0},{4,0}}]],
           DisplayFunction->$DisplayFunction];
```

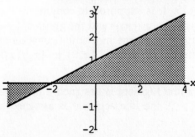

The measure $\int_{-4}^{-2} (x/2 + 1)dx$ of the triangular area on the left is negative:

```
In[4]:=    height = 1
           base = 2
           left = -(1/2) base height
Out[4]=    -1
```

The measure $\int_{-2}^{4}(x/2+1)dx$ of the triangular area on the right is positive:

```
In[5]:=    height = 3
           base = 6
           right = +(1/2) base height
Out[5]=    9
```

Consequently the area measurement

$$\int_{-4}^{4}(x/2+1)dx = \int_{-4}^{-2}(x/2+1)dx + \int_{-2}^{4}(x/2+1)dx$$

is given by:

```
In[6]:=    left + right
Out[6]=    8
```

*Mathematica* can make the area measurement $\int_{-4}^{4}(x/2+1)dx$ directly:

```
In[7]:=    Integrate[(x/2 + 1),{x,-4,4}]
Out[7]=    8
```

Got it and it wasn't very hard.

■ **B.2) The integral sign.**

**B.2.a)**

What is the strange looking notation

$$\int_{a}^{b}f[x]dx$$

supposed to convey?

**Answer:**

This whimsically suggestive notation goes all the way back to the emergence of calculus as a coherent body of study in the late 1600's. The lazy S symbol $\int$ was invented by Leibniz himself in 1675. It is supposed to conjure up the idea of an area between the $x-$ axis and the graph of $y = f[x]$ as a "moving sum" of all the individual constituent vertical line segments running from $\{x,0\}$ to $\{x,f[x]\}$ for $a \leq x \leq b$.

When $f[x] < 0$, the line segment is assigned a negative length; when $f[x] > 0$, the line segment is assigned a positive length.

So $\int_{a}^{b}f[x]dx$ means that we "sum up" the lengths of these line segments and arrive at the (signed) area that $\int_{a}^{b}f[x]dx$ measures.

This is an excellent way of visualizing what the integral is and a lot of good mathematics is based on the intuition gained by thinking this way.

■ **B.3) Three basic properties.**

**B.3.a)**

Explain the formula

$$\int_{a}^{b}f[x]dx = \int_{a}^{c}f[x]dx + \int_{c}^{b}f[x]dx$$

for any number $c$ with $a < c < b$.

**Answer:**

The integral $\int_{a}^{b}f[x]dx$ measures the (signed) area between the graph of $y = f[x]$ and the segment of the $x-$ axis between $x = a$ and $x = b$.

The sum

$$\int_{a}^{c}f[x]dx + \int_{c}^{b}f[x]dx$$

measures the same signed area but calculates it in two parts.

Let's try it for $f[x] = 2x - 4, a = 0, c = 3$ and $b = 5$. The number $\int_{0}^{5}(2x - 4)dx$ can be calculated by:

```
In[1]:=    Integrate[2 x - 4,{x,0,5}]
Out[1]=    5
```

This should be equal to $\int_{0}^{3}(2x - 4)dx + \int_{3}^{5}(2x - 4)dx$ :

```
In[2]:=    Integrate[2x - 4,{x,0,3}] + Integrate[2 x
           - 4,{x,3,5}]
Out[2]=    5
```

On the money.

**B.3.b)**

Explain the formula

$$\int_{a}^{b}Kf[x]dx = K\int_{a}^{b}f[x]dx$$

for any number $K$.

**Answer:**

The expression $K\int_{a}^{b}f[x]dx$ measures an area and then multiplies by $K$ to expand the value of the area.

The expression $\int_{a}^{b}Kf[x]dx$ expands the area vertically first (by multiplying by $K$ ) and then measures the result. Either expression leads to the same result.

Let's try it for $f[x] = 4x + 1$, $a = 1$, $b = 7$.

```
In[1]:=    Clear[K,x]
           Integrate[ K (4 x + 1), {x,1,7}]

Out[1]=    102 K
```

This should agree with:

```
In[2]:=    K Integrate[(4 x + 1), {x,1,7}]

Out[2]=    102 K
```

It does.

Let's try it for a particular value of $K$:

```
In[3]:=    Integrate[19 (x + 1), {x,1,7}]

Out[3]=    570
```

should be:

```
In[4]:=    19 Integrate[(x + 1), {x,1,7}]

Out[4]=    570
```

Got it.

**B.3.c)**

---

Explain the formula

$$\int_a^b f[x]dx = \int_a^b f[t]dt.$$

---

**Answer:**

Look at the following plots:

```
In[1]:=    Clear[x,t]
           f[x_] = 2 x^3 E^(-x)

Out[1]=        3
            2 x
            ----
              x
             E
```

```
In[2]:=    Plot[f[x],{x,0,8},
           AspectRatio->Automatic,
           AxesLabel->{"x","y"},Ticks->{{1},None}];
```

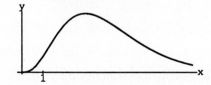

```
In[3]:=    Plot[f[t],{t,0,8},AspectRatio->Automatic,
           AxesLabel->{"t","y"},Ticks->{{1},None}];
```

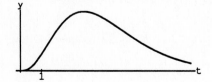

The area between horizontal axis and each curve is the same.

So for this function $f[x]$, we see that

$$\int_0^8 f[x]dx = \int_0^8 f[t]dt.$$

The same thing will happen for other functions and other intervals.

The use of the symbol $t$ instead of the symbol $x$ is a matter of bureaucracy and not a matter of mathematics or science. Only bean counters worry about it.

## Tutorial

### ■ T.1) Evaluation of integrals.

Make the indicated area measurents:

**T.1.a)**

---

$$\int_{-1}^5 (3x/2 - 6)\,dx.$$

---

**Answer:**

To see what's happening, plot:

```
In[1]:=    Clear[f,x]
           f[x_]= 3 x/2 - 6
           functionplot = Plot[f[x],{x,-1,5},
           AspectRatio->Automatic];
```

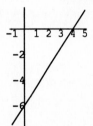

Put in some shading:

```
In[2]:=    poly = Graphics[{GrayLevel[.75],
           Polygon[{{-1,0},{-1,f[-1]},
               {5,f[5]},{5,0}}]}];
           Show[poly,functionplot,Ticks->None];
```

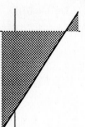

This tells us to look at

$$\int_{-1}^{5}(3x/2-6)dx = \int_{-1}^{4}(3x/2-6)dx + \int_{4}^{5}(3x/2-6)dx.$$

Here $\int_{4}^{5}(3x/2-6)dx$ is the positive area under the curve $y=3x/2-6$ and above the segment $[4,5]$ on the $x-$ axis. This is a triangle with:

```
In[3]:=    base = 5 - 4
Out[3]=    1
```

and

```
In[4]:=    height = (3 x/2 - 6)/.x->5
Out[4]=    3
           -
           2
```

So $\int_{4}^{5}(3x/2-6)dx$ is

```
In[5]:=    posarea = (1/2) (base height)
Out[5]=    3
           -
           4
```

Also $\int_{-1}^{4}(3x/2-6)dx$ is the negative of the area over the curve $y=3x/2-6$ and below the segment $[-1,4]$ on the $x-$ axis. This is a triangle with:

```
In[6]:=    base = 4 - (-1)
Out[6]=    5
```

```
In[7]:=    height = (3 x/2 - 6)/.x->-1
Out[7]=        15
           -(--)
               2
```

So $\int_{-1}^{4}(3x/2-6)dx$ is:

```
In[8]:=    negarea = (1/2) base height
Out[8]=        75
           -(--)
               4
```

The integral

$$\int_{-1}^{5}(3x/2-6)dx = \int_{-1}^{4}(3x/2-6)dx + \int_{4}^{5}(3x/2-6)dx.$$

is given by:

```
In[9]:=    posarea + negarea
Out[9]=    -18
```

*Mathematica* can handle this one too.

```
In[10]:=   Integrate[3 x/2 - 6, {x, -1, 5}]
```

```
Out[10]=   -18
```

Got it.

**T.1.b)**

$$\int_{-3}^{3}\sqrt{9-x^2}dx$$

**Answer:**

This integral measures the area of the top half of the circle $x^2+y^2=9$. If you don't believe this, then just solve $x^2+y^2=9$ for $y$ :

```
In[1]:=    Clear[x,y]
           ysolved = Solve[x^2 + y^2 == 9,y]
Out[1]=                 2
           {{y -> Sqrt[9 - x ]},

                          2
            {y -> -Sqrt[9 - x ]}}
```

The function we are integrating is $\sqrt{9-x^2}$. Here's a plot:

```
In[2]:=    functionplot=
           Plot[Sqrt[9 - x^2],{x,-3,3},
           AspectRatio->Automatic,
           PlotRange->{-1,4},
           Ticks->{{-3,3},None}];
```

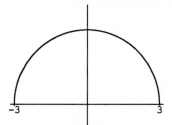

So $\int_{-3}^{3}\sqrt{9-x^2}dx$ measures the area inside this circle and above the $x-$ axis, that is half the area of the circle. The radius is $3$. As a result, the area and the integral have the value:

```
In[3]:=    (1/2) Pi 3^2
Out[3]=    9 Pi
           ----
            2
```

Let's see what *Mathematica* does with this integral:

```
In[4]:=    Integrate[Sqrt[9 - x^2],{x,-3,3}]
Out[4]=    9 Pi
           ----
            2
```

Got it.

**T.1.c)**

$$\int_0^3 \sqrt{9 - x^2}\,dx.$$

**Answer:**

This is just the area of *one quarter* of a circle of radius 3 ; so

$$\int_0^3 \sqrt{9 - x^2}\,dx = (1/4)\pi 3^2 = 9\pi/4$$

Check:

```
In[1]:=    Integrate[Sqrt[9 - x^2],{x,0,3}]
Out[1]=    9 Pi
           ----
            4
```

Got it.

■ **T.2) Integration and the instruction NIntegrate.**

**T.2.a.i)**

Plot on the same axes: $f[x] = \sin[\pi x]$ for $0 \le x \le 1$ and the broken line that joins the four points on the curve

$$\{0, f[0]\}, \{1/3, f[1/3]\}, \{2/3, f[2/3]\}, \{1, f[1]\}$$

**Answer:**

Define the function:

```
In[1]:=    Clear[f,x,jump]
           f[x_] = Sin[Pi x]
Out[1]=    Sin[Pi x]
```

Take a look at the plot:

```
In[2]:=    functionplot = Plot[f[x],{x,0,1}];
```

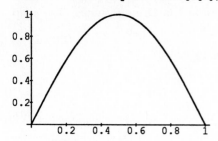

Make a table of the points

$$\{0, f[0]\}, \{1/3, f[1/3]\}, \{2/3, f[2/3]\}, \{1, f[1]\}$$

The jump from in the $x-$ coordinate from point to point is:

```
In[3]:=    jump = 1/3
```

```
Out[3]=     1
            -
            3
```

So the points

$$\{0, f[0]\}, \{1/3, f[1/3]\}, \{2/3, f[2/3]\}, \{1, f[1]\},$$

are given by:

```
In[4]:=    points = Table[{x,f[x]},{x,0,1,jump}]
Out[4]=                1  Sqrt[3]    2  Sqrt[3]
           {{0, 0}, {-, -------}, {-, -------},
                    3     2       3     2

           {1, 0}}
```

Set up the plot of the points and the broken line segments:

```
In[5]:=    pointplot =
           ListPlot[points,
           PlotStyle->{RGBColor[0,0,1],PointSize[.015]},
           DisplayFunction-> Identity]
           brokenlineplot =
           Graphics[Line[points]]
           Show[functionplot,pointplot,brokenlineplot];
```

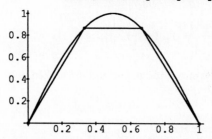

The straight-sided figure takes a healthy bite of the area under the curve measured by

$$\int_0^1 \sin[\pi x]\,dx.$$

**T.2.a.ii)**

Measure the (signed) area between the $x-$axis and broken line segments.

**Answer:**

The (signed) area between the $x-$axis and the broken line segments plotted above is the sum of the areas of consecutive trapezoids based on segments $[x, x + jump]$ on the $x-$axis. Each has base of $length = jump$. Each has two heights-namely $f[x]$ and $f[x + jump]$.

So the area of each of the trapezoids is

$$jump\big(f[x] + f[x + jump]\big)/2,$$

and we start at $x = 0$ and finish at $x = 1 - jump$. The sum of these areas is:

```
In[1]:=    Area[4] =
           N[Sum[jump (f[x] + f[x + jump])/2,
           {x,0,1-jump,jump}]]
```

*Out[1]=*    0.57735

## T.2.a.iii)

Redo (T.2.a.i) and (T.2.a.ii) for 7 equally spaced points.

### Answer:

We are going from 0 to 1 in 6 equal jumps. The jump from point to point is:

*In[1]:=*    `jump = 1/6`

*Out[1]=*    $\dfrac{1}{6}$

The points are given by:

*In[2]:=*    `points = Table[{x,f[x]},{x,0,1,jump}]`

*Out[2]=*    $\{\{0,\ 0\},\ \{\frac{1}{6},\ \frac{1}{2}\},\ \{\frac{1}{3},\ \frac{\text{Sqrt}[3]}{2}\},\ \{\frac{1}{2},\ 1\},$

$\{\frac{2}{3},\ \frac{\text{Sqrt}[3]}{2}\},\ \{\frac{5}{6},\ \frac{1}{2}\},\ \{1,\ 0\}\}$

Set up the plot of the curve, the points and the broken line segments:

*In[3]:=*
```
pointplot =
ListPlot[points,
PlotStyle->{RGBColor[0,0,1],
PointSize[.015]},
DisplayFunction-> Identity]
brokenlineplot =
Graphics[Line[points]]
Show[functionplot,pointplot,
brokenlineplot];
```

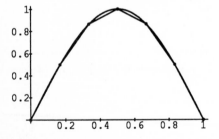

The straight-sided figure takes a healthy bite of the area under the curve. Look at one of the trapezoids:

*In[4]:=*
```
Clear[trap]
trap[x_]:=Graphics[{GrayLevel[.75],
Polygon[{{x,0},{x,f[x]},
{x+jump,f[x+jump]},{x+jump,0}}]}]
Show[trap[2 jump],functionplot,
brokenlineplot,
Ticks->{{2 jump,3 jump},Automatic}];
```

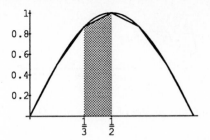

Or all of them:

*In[5]:=*
```
Show[trap[0 jump],trap[1 jump],
trap[2 jump],trap[3 jump],
trap[4 jump],trap[5 jump],
functionplot,brokenlineplot,
Ticks->{{0,jump,2 jump,3 jump,
4 jump, 5 jump, 6 jump},Automatic}];
```

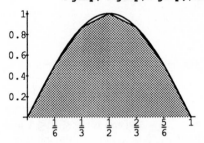

Notice that when we had four points we saw three trapezoids, but now for seven points we see six trapezoids. So for the case of $n$ points, we will see $(n-1)$ trapezoids.

The area under broken line segments is:

*In[6]:=*    `Area[7]=N[Sum[jump (f[x] + f[x+jump])/2,`
`{x,0,1-jump, jump}]]`

*Out[6]=*    0.622008

## T.2.a.iv)

Redo (T.2.a.i) and (T.2.a.ii) for 13 equally spaced points.

### Answer:

We are going from 0 to 1 in 12 equal jumps. The jump from point to point is:

*In[1]:=*    `jump = 1/12`

*Out[1]=*    $\dfrac{1}{12}$

The points are given by:

*In[2]:=*    `points = Table[{x,f[x]},{x,0,1,jump}]`

*Out[2]=*    $\{\{0,\ 0\},\ \{\frac{1}{12},\ \text{Sin}[\frac{\text{Pi}}{12}]\},\ \{\frac{1}{6},\ \frac{1}{2}\},$

$\{\frac{1}{4},\ \frac{\text{Sqrt}[2]}{2}\},\ \{\frac{1}{3},\ \frac{\text{Sqrt}[3]}{2}\},$

```
 5        5 Pi    1
{--, Sin[----]}, {-, 1},
 12       12      2

 7        7 Pi    2  Sqrt[3]
{--, Sin[----]}, {-, -------},
 12       12      3     2

 3  Sqrt[2]    5  1
{-, -------}, {-, -},
 4     2      6  2

 11       11 Pi
{--, Sin[-----]}, {1, 0}}
 12       12
```

Here is the plot:

```
In[3]:=   pointplot =
          ListPlot[points,
          PlotStyle->{RGBColor[0,0,1],
          PointSize[.015]},
          DisplayFunction-> Identity]
          brokenlineplot =
          Graphics[Line[points]]
          Show[functionplot,pointplot,
          brokenlineplot];
```

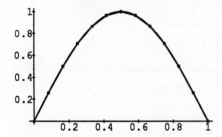

The straight-sided figure takes a very healthy bite of the area under the curve. The (shaded) area under the straight-sided figure is:

```
In[4]:=   Area[13]=N[Sum[jump (f[x]+f[x+jump])/2,
          {x,0,1-jump,jump}]]
```
```
Out[4]=   0.63298
```

### T.2.a.v)

Redo (T.2.a.i) and (T.2.a.ii) for 25 equally spaced points.

---

**Answer:**

This time the jump from point to point is:

```
In[1]:=   jump = 1/24
```
```
Out[1]=   1
          --
          24
```

and the points are given by:

```
In[2]:=   points = Table[{x,f[x]},{x,0,1,jump}]
```
```
Out[2]=                1    Pi     1    Pi
          {{0, 0}, {--, Sin[--]}, {--, Sin[--]},
                     24     24    12    12
```

```
 1      Pi    1  1
{-, Sin[--]}, {-, -},
 8      8     6  2

 5       5 Pi    1  Sqrt[2]
{--, Sin[----]}, {-, -------},
 24      24      4     2

 7       7 Pi    1  Sqrt[3]
{--, Sin[----]}, {-, -------},
 24      24      3     2

 3      3 Pi     5        5 Pi
{-, Sin[----]}, {--, Sin[----]},
 8      8        12       12

 11      11 Pi    1
{--, Sin[-----]}, {-, 1},
 24      24       2

 13      13 Pi    7        7 Pi
{--, Sin[-----]}, {--, Sin[----]},
 24      24       12       12

 5      5 Pi    2  Sqrt[3]
{-, Sin[----]}, {-, -------},
 8      8       3     2

 17      17 Pi    3  Sqrt[2]
{--, Sin[-----]}, {-, -------},
 24      24       4     2

 19      19 Pi    5  1
{--, Sin[-----]}, {-, -},
 24      24       6  2

 7      7 Pi    11       11 Pi
{-, Sin[----]}, {--, Sin[-----]},
 8      8       12       12

 23      23 Pi
{--, Sin[-----]}, {1, 0}}
 24      24
```

Set up the plot:

```
In[3]:=   pointplot =
          ListPlot[points,
          PlotStyle->{RGBColor[0,0,1],
          PointSize[.015]},
          DisplayFunction-> Identity]
          brokenlineplot =
          Graphics[Line[points]]
          Show[functionplot,pointplot,
          brokenlineplot];
```

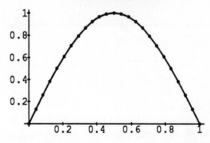

Wow! You need a microscope to tell the area under the broken line segments from the area under the curve. The area under the straight-sided figure is:

```
In[4]:=   Area[25]=N[Sum[jump (f[x]+f[x+jump])/2,
          {x,0,1-jump,jump}]]
```
```
Out[4]=   0.63571
```

### T.2.a.vi)

Measure the area under the broken line segments for 50 equally spaced points.

### Answer:

There are 49 intervals; so:

```
In[1]:=    jump = 1/49
Out[1]=    1
           --
           49
```

The area under the broken line segments is:

```
In[2]:=    Area[50]=N[Sum[jump (f[x]+f[x+jump])/2,
           {x,0,1-jump,jump}]]
Out[2]=    0.636402
```

### T.2.a.vii)

Measure the area under the broken line segments for 100 equally spaced points.

### Answer:

There are 99 intervals; so:

```
In[1]:=    jump = 1/99
Out[1]=    1
           --
           99
```

The area under the broken line segments is:

```
In[2]:=    Area[100]=N[Sum[jump (f[x]+f[x+jump])/2,
           {x,0,1-jump,jump}]]
Out[2]=    0.636566
```

### T.2.a.viii)

Review and assess your calculations.

### Answer:

The pictures show that, for large $n$ 's, the broken line segments are hard to tell from the original curve. Consequently, for large $n$ 's, the area under the straight-sided figures should be very close to the exact measurement $\int_0^1 \sin[\pi x]dx$. Let's look at the values of these areas that we calculated above:

```
In[1]:=    {Area[4],Area[7],Area[13],
           Area[25],Area[50],Area[100]}
Out[1]=    {0.57735, 0.622008, 0.63298, 0.63571,
```

```
           0.636402, 0.636566}
```

These figures strongly suggest that $\int_0^1 \sin[\pi x]dx = 0.636$ within 0.001.

### T.2.a.ix)

How are these calculations related to the *Mathematica* instruction **NIntegrate** ?

### Answer:

The *Mathematica* instruction **NIntegrate** uses an approximation procedure similar (actually it's similar but a bit better) to what we just did. Let's try this instruction to test our seat-of-the-pants estimate $\int_0^1 \sin[\pi x]dx = 0.636$ within 0.001.

```
In[1]:=    NIntegrate[Sin[Pi x],{x,0,1}]
Out[1]=    0.63662
```

Our estimate was not so bad.

### T.2.b)

Use trapezoids to get a reasonable idea of the value of the area measurement

$$\int_1^4 (x - 2x^2)e^{-x}\,dx.$$

Check your estimate with the *Mathematica* instruction **NIntegrate** .

### Answer:

The set-up involves only a few modifications to what we did in part (T.2.a) above. As we did above, we'll start with four equally spaced points to make sure everything is running smoothly and then we'll advance to more points with smaller jumps between them.

Define the function:

```
In[1]:=    Clear[f,x,jump]
           f[x_] = (x^3 - 2 x^2) E^(-x)
Out[1]=         2    3
           -2 x  + x
           ----------
               x
              E
```

Take a look at the plot:

```
In[2]:=    functionplot = Plot[f[x],{x,1,4},
           PlotRange->All,Ticks->{{1,2,3,4},
           Automatic}];
```

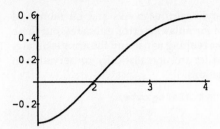

Take four equally spaced points on the $x$-axis starting with $x = 1$ and stopping with $x = 4$. The jump from point to point is:

```
In[3]:=    jump = (4 - 1)/3
Out[3]=    1
```

So the points are given by:

```
In[4]:=    points = Table[{x, f[x]}, {x, 1, 4, jump}]
```

$$Out[4]= \quad \{\{1, -(\tfrac{1}{E})\}, \{2, 0\}, \{3, \tfrac{9}{E^3}\}, \{4, \tfrac{32}{E^4}\}\}$$

Set up the plot of the curve, the points and the broken line segments:

```
In[5]:=    pointplot =
           ListPlot[points,
           PlotStyle->{RGBColor[0,0,1],
           PointSize[.015]},
           DisplayFunction-> Identity]
           brokenlineplot =
           Graphics[Line[points]]
           Show[functionplot,pointplot,
           brokenlineplot];
```

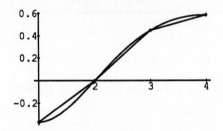

The (signed) area between the $x-$ axis and the broken line segments plotted above is the sum of the area of consecutive trapezoids based on segments $[x, x + jump]$ on the $x-$ axis. Each has base $= jump$. Each has two heights—namely $f[x]$ and $f[x + jump]$.

So the area of each of the trapezoids is

$$jump(f[x] + f[x + jump])/2.$$

The sum of these areas is:

```
In[6]:=    Area[4] =
           N[Sum[jump (f[x] + f[x+jump])/2,
           {x,1, 4 - jump, jump}]]
Out[6]=    0.557194
```

Note how the leftmost trapezoid carries a negative area:

```
In[7]:=    jump (f[1] + f[1+jump])/2
```

$$Out[7]= \quad \frac{-1}{2\,E}$$

This is the way it should be. Now let's try 13 points:

```
In[8]:=    jump = (4 - 1)/12
           points = Table[{x, f[x]}, {x,1,4,jump}]
```

$$Out[8]= \quad \{\{1, -(\tfrac{1}{E})\}, \{\tfrac{5}{4}, \tfrac{-75}{64\,E^{5/4}}\}, \{\tfrac{3}{2}, \tfrac{-9}{8\,E^{3/2}}\},$$

$$\{\tfrac{7}{4}, \tfrac{-49}{64\,E^{7/4}}\}, \{2, 0\}, \{\tfrac{9}{4}, \tfrac{81}{64\,E^{9/4}}\},$$

$$\{\tfrac{5}{2}, \tfrac{25}{8\,E^{5/2}}\}, \{\tfrac{11}{4}, \tfrac{363}{64\,E^{11/4}}\}, \{3, \tfrac{9}{E^3}\},$$

$$\{\tfrac{13}{4}, \tfrac{845}{64\,E^{13/4}}\}, \{\tfrac{7}{2}, \tfrac{147}{8\,E^{7/2}}\},$$

$$\{\tfrac{15}{4}, \tfrac{1575}{64\,E^{15/4}}\}, \{4, \tfrac{32}{E^4}\}\}$$

Here is the plot:

```
In[9]:=    pointplot =
           ListPlot[points,
           PlotStyle->{RGBColor[0,0,1],
           PointSize[.015]},
           DisplayFunction-> Identity]
           brokenlineplot =
           Graphics[Line[points]]
           Show[functionplot,pointplot,
           brokenlineplot];
```

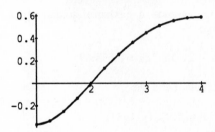

Neat; you can hardly tell the broken line segments from the curve.

The trapezoidal area is:

```
In[10]:=   Area[13]=N[Sum[jump (f[x]+f[x+jump])/2,
           {x,1,4-jump, jump}]]
Out[10]=   0.558864
```

Now try 25 points:

```
In[11]:=   jump = (4 - 1)/24
Out[11]=   1
```

$$-\over 8$$

```
In[12]:=  Area[25]=N[Sum[jump (f[x]+f[x+jump])/2,
          {x,1,4-jump,jump}]]
Out[12]=  0.558869
```

50 points:

```
In[13]:=  jump = (4 - 1)/49
Out[13]=  3
          --
          49
```

```
In[14]:=  Area[50]=N[Sum[jump (f[x]+f[x+jump])/2,
          {x,1,4-jump,jump}]]
Out[14]=  0.558869
```

100 points:

```
In[15]:=  jump = (4 - 1)/99
Out[15]=  1
          --
          33
```

```
In[16]:=  Area[100]=N[Sum[jump (f[x]+f[x+jump])/2,
          {x,1,4-jump,jump}]]
Out[16]=  0.558869
```

Review and assess:

```
In[17]:=  {Area[4],Area[13],Area[25],
          Area[50],Area[100]}
Out[17]=  {0.557194, 0.558864, 0.558869,

           0.558869, 0.558869}
```

This is pretty strong evidence. A pitcher of golden liquid refreshment bets that

$$\int_1^4 (x - 2x^2)e^{-x}\,dx = 0.558869$$

to 6 accurate decimals. Let's see:

```
In[18]:=  NIntegrate[f[x],{x,1,4}]
Out[18]=  0.558869
```

That was a safe bet.

### T.2.c)

In practice, how do we know how to set the number of points to achieve desired accuracy?

**Answer:**

That's hard to answer. Experience seems to say that the peppy functions usually require more points than lazy functions.

Some of the old style calculus books and all numerical analysis books give formulas for error estimates, but these formulas are rather revolting to use and the error estimates these formulas predict are unrealistically conservative in most practical situations.

### ■ T.3) Using the instruction NIntegrate.

### T.3.a)

Use **NIntegrate** to estimate the value of

$$\int_0^1 4/(1 + x^2)\,dx.$$

**Answer:**

```
In[1]:=  Clear[x]
         NIntegrate[4/(1 + x^2),{x,0,1}]
Out[1]=  3.14159
```

Say, that number looks familiar!

```
In[2]:=  N[Pi]
Out[2]=  3.14159
```

The smart money bets

$$\int_0^1 4/(1 + x^2)\,dx = \pi.$$

### T.3.b)

For $t \geq 0$, set

$$f[t] = \int_0^t \sin[x]\,dx - 1$$

and plot $f[t]$ for $0 \leq t \leq 2\pi$. Use your plot to guess a formula for $f[t]$.

**Answer:**

```
In[1]:=  Clear[f,x,t]
         f[t_] := NIntegrate[Sin[x],{x,0,t}] - 1
         functionplot = Plot[f[t],{t,0,2 Pi},
         AxesLabel->{"t","y"}];
```

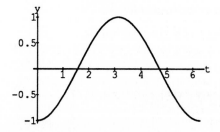

Say, that curve looks familiar!

*In[2]:=*   `negcosplot = Plot[-Cos[t],{t,0,2 Pi}];`

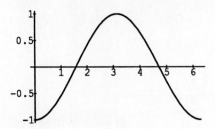

The plots appear to be identical!

Superimpose the plots:

*In[3]:=*   `Show[functionplot,negcosplot];`

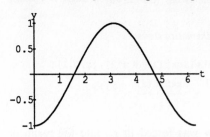

They share the same ink all the way.

The smart money bets

$$f[t] = \int_0^t \sin[x]dx - 1 = -\cos[t].$$

#### ■ T.4) Nonsense integrals.

The expression

$$\int_a^b f[x]dx$$

 is meaningless if:

$f[x]$ *becomes unbounded in* $[a,b]$;

or if

$f[x]$ *is not defined at some* $x$ *in* $[a,b]$;

or if

$f[x]$ *has a singularity (blow up or blow down) in* $[a,b]$.

Say why none of the following integrals have meaning.

#### T.4.a)

$\int_{-3}^2 (1/x^2)dx$ and $\int_0^1 (1/x^2)dx$

#### Answer:

These are meaningless because $(1/x^2)$ blows up (is unbounded) on any interval containing 0 :

*In[1]:=*   `Plot[1/x^2,{x,-.5,.2}];`

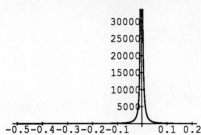

Note that *Mathematica* (version 1.2) advertises a value of:

*In[2]:=*   `Integrate[1/x^2,{x,-3,2}]`

*Out[2]=*
$$-(\frac{5}{6})$$

for $\int_{-3}^2 (1/x^2)dx$. This answer is wrong because $\int_{-3}^2 (1/x^2)dx$ is meaningless. Newer versions of *Mathematica* will not make this mistake. On the other hand *Mathematica* correctly declines to answer when you ask for $\int_0^1 (1/x^2)dx$ :

*In[3]:=*   `Integrate[1/x^2,{x,0,1}]`

*Out[3]=*   `ComplexInfinity`

#### T.4.b)

$$\int_{-1}^1 \sqrt{x}dx$$

#### Answer:

Within the realm of real numbers, this is meaningless because $\sqrt{x}$ is undefined as a real number for $-1 \le x < 0$. So there is no physical meaning to the area under the curve $y = \sqrt{x}$ for $-1 \le x \le 0$. Let's see what *Mathematica* does:

*In[1]:=*   `Integrate[Sqrt[x],{x,-1,1}]`

*Out[1]=*
$$\frac{2}{3} + \frac{2\,I}{3}$$

*Mathematica* here makes the only possible choice and interprets the integral as

$$\int_{-1}^1 \sqrt{x}dx = \int_{-1}^0 \sqrt{x}dx + \int_0^1 \sqrt{x}dx$$

$$= \int_{-1}^0 \sqrt{(-1)(-x)}dx + \int_0^1 \sqrt{x}dx$$

$$= \int_{-1}^0 \sqrt{-1}\sqrt{-x}dx + \int_0^1 \sqrt{x}dx$$

$$= \sqrt{-1}\int_{-1}^0 \sqrt{-x}dx + \int_0^1 \sqrt{x}dx.$$

Then *Mathematica* calculates (by a method that you will underestand soon)

$$\int_{-1}^{0} \sqrt{(-x)}\,dx = 2/3$$

$$\int_{0}^{1} \sqrt{x}\,dx = 2/3$$

and comes up with the overall answer:

$$(2/3) + (2/3)\sqrt{-1} = (2/3) + (2/3)I.$$

### T.4.c)

$\int_{0}^{1} \sqrt{1 - 4x^2}\,dx$.

**Answer:**

Take a look at a plot:

```
In[1]:=   Plot[Sqrt[1 - 4 x^2],{x,0,1},
          PlotRange->{{0,1},{0,1}}];
```

```
Plot::notnum:

2
Sqrt[1 - 4 x ]

does not evaluate to a real number

at x=0.541667.

Plot::notnum:

2
Sqrt[1 - 4 x ]

does not evaluate to a real number

at x=0.520833.

Plot::notnum:

2
Sqrt[1 - 4 x ]

does not evaluate to a real number

at x=0.510417.

General::stop:

Further output of Plot::notnum

will be suppressed during this

calculation.
```

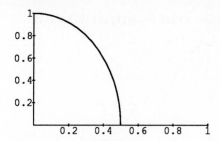

The errror messages and the plot tell you that there is no curve for $0.5 < x \leq 1$. No wonder because $1 - 4x^2 < 0$ for $x > 1/2$ and hence $\sqrt{1 - 4x^2}$ is undefined as a real number on the subinterval $(1/2, 1]$ of $[0, 1]$. Consequently $\int_{0}^{1} \sqrt{1 - 4x^2}\,dx$ is meaningless because there is no physical meaning to the area under the curve $y = \sqrt{1 - 4x^2}$ for $1/2 < x \leq 1$.

Let's see what *Mathematica* does:

```
In[2]:=   Integrate[Sqrt[1 - 4 x^2],{x,0,1}]

Out[2]=   Sqrt[-3]    ArcSin[2]
          --------  + ---------
             2            4
```

Again *Mathematica* was forced to go into the realm of complex (imaginary) numbers to handle this integral.

## Literacy Sheet

**1.** Fix numbers $a$ and $b$ with $a < b$ and fix a function $f[x]$. Explain what the symbol $\int_a^b f[x]dx$ means. Is it a number or is it a function?

**2.** Fix numbers $a$ and $b$ with $a < b$. If $f[x] \geq 0$ for $a \leq x \leq b$, then can it be that $\int_a^b f[x]dx$ ¡ 0? Why?

**3.** If $f[x] \leq g[x]$ for all $x$'s with $a \leq x \leq b$, then why it is automatic that $\int_a^b f[x]dx \leq \int_a^b g[x]dx$?

**4.** If $f[x]$ is an odd function (so $f[-x] = -f[x]$), then what is the value of $\int_{-2}^{2} f[x]dx$?

**5.** Explain the formula $\int_a^b f[x]dx = \int_a^c f[x]dx + \int_c^b f[x]dx$ for any number $c$ with $a < c < b$.

**6.** Explain the formula $\int_a^b Kf[x]dx = K\int_a^b f[x]dx$ for any number $K$.

**7.** Explain the formula $\int_a^b f[x]dx = \int_a^b f[t]dt$.

**8.** What is the idea behind integration by trapezoids?

**9.** What bothers us about writing $\int_0^2 1/(x-1)dx$?

**10.** Suppose $f[x]$ is increasing on $[a, b]$. Why is is guaranteed that $f[a](b-a) \leq \int_a^b f[x]dx \leq f[b](b-a)$?

**11.** Explain why $|\int_a^b f[x]dx| \leq \int_a^b |f[x]|dx$. Assume $a \leq b$.

# Trying to Break the Code of the Integral

## Guide

Experienced integral watchers know how to break the code of the integral. In many cases, they can calculate the exact value of an integral just by inspection. You are in an especially good position to understand the code of the integral because you have the instruction **NIntegrate** you can use to get approximate values of an integral. This lesson is a series of plotting experiments that lay the base for cracking the code of the integral.

## Basics

### ■ B.1) Exact calculation of an easy integral.

**B.1.a)**

Calculate

$$f[t] = \int_0^t x\,dx$$

as a function of $t$ for $0 \le t$. What is $f'[t]$?

**Answer:**

$f[t] = \int_0^t x\,dx$ is the area of the triangle indicated below:

```
In[1]:=   funcplot =
          Plot[x,{x,0,1.7},Ticks->None,
          PlotRange->{{-.2,2},{-.4,2}},
          DisplayFunction->Identity]
          shade=
          Graphics[{GrayLevel[.75],Polygon[{{0,0},
          {1.7,0},{1.7,1.7}}]}]
          tlabel =
          Graphics[Text["t",{1.7,-.1}]]
          ttlabel=
          Graphics[Text["{t,t}",{1.7,1.8}]]
          Show[shade,funcplot,tlabel,ttlabel,
          DisplayFunction->$DisplayFunction,
          AspectRatio->Automatic]
```

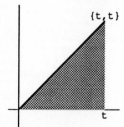

As the measurment of the area of the triangle indicated

above,

$$f[t] = \int_0^t x\,dx = t(t/2) = t^2/2.$$

As a result,

$$f'[t] = 2t/2 = t.$$

Not much to it, but this is a big clue in our effort to break the code of the integral.

### ■ B.2) A big step toward breaking the code of the integral.

**B.2.a)**

Use the instruction **NIntegrate** to give a plot of

$$f[t] = \int_0^t x^2\,dx$$

as a function of $t$ for $0 \le t \le 2$. Report anything of interest that you happen to observe.

**Answer:**

```
In[1]:=   Clear[f,t,x]
          f[t_] := NIntegrate[x^2,{x,0,t}]
          Plot[f[t],{t,0,2}]
```

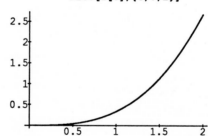

Looks like some sort of parabola or other power of $x$.

**B.2.b)**

Put

$$f[t] = \int_0^t x^2\,dx$$

and use the instruction **NIntegrate** to get an idea of what $f'[t]$ is by plotting the average growth rates

$$(f[t+h] - f[t])/h$$

for various small values of $h$ and for $0 \le t \le 2$. Report your best shot at what $f'[t]$ is.

(Recall $f'[t]$ is the limiting case of $(f[t + h] - f[t])/h$ as $h$ closes in on 0.)

---

**Answer:**

Note that for h positive

$$(f[t + h] - f[t])/h = (\int_0^{t+h} x^2 dx - \int_0^t x^2 dx)/h$$

$$= (\int_t^{t+h} x^2 dx)/h$$

because

$$f[t + h] = \int_0^{t+h} x^2 dx = \int_0^t x^2 dx + \int_t^{t+h} x^2 dx$$

and

$$f[t] = \int_0^t x^2 dx.$$

So if we fix some small values of $h$ and plot

$$f[x] = \text{NIntegrate}\ [x^2, \{x, t, t + h\}]/h$$

as a function of $t$, then we should get an idea of what the plot of the average growth rate $(f[t + h] - f[t])/h$ looks like. Here is a plot for $h = .001$:

```
In[1]:=    avgrowth001 =
           Plot[NIntegrate[x^2,{x,t,t+.001}]
           /.001,{t,0,2}]
```

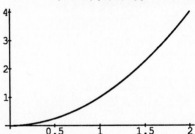

This curve seems to go right through $\{2, 4\} = \{2, 2^2\}$ and it looks a hell of a lot like the $t^2$ curve. Let's plot $t^2$ and the $(f[t + .001] - f[t])/.001$ curves on the same axes:

```
In[2]:=    tsquared =
           Plot[t^2,{t,0,2},
           DisplayFunction->Identity]
           Show[avgrowth001,tsquared,
           DisplayFunction->$DisplayFunction]
```

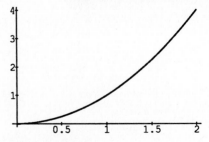

Indistinguishable!

Our best shot at $f'[t]$ is $f'[t] = t^2$.

We started with

$$f[t] = \int_0^t x^2 dx$$

and arrived at the opinion that

$$f'[t] = t^2.$$

Another big clue in breaking the code of the integral.

If anyone tells you that in mathematics you are not supposed to have opinions, then tell them your opinion on where they can go.

## Tutorial

Because this lesson consists of numerical experiments for you to do, there is no Tutorial section in this lesson.

# Literacy Sheet

**1.** What is the limiting value of $(f[t + h] - f[t])/h$ as $h$ closes in on 0?

.

.

**2.** Set $f[t] = \int_0^t g[x]dx$ and explain the formula $(f[t+h] - f[t])/h = (\int_t^{t+h} g[x]dx)/h$ for positive $h$'s.

.

.

# Breaking the code of the integral: The fundamental theorem of calculus

## Guide

History has called Isaac Newton one of the two founders of calculus because he cracked the code of the integral in 1666 by using the derivative to find the exact value of integrals. You are about to learn this fundamental contribution to science and technology.

## Basics

■ **B.1) Breaking the code of the integral: The fundamental theorem of calculus.**

### B.1.a.i)

Explain why

$$hf'[t] \leq \int_t^{t+h} f'[x]dx \leq hf'[t+h]$$

provided $f'[x]$ is increasing on $[t, t+h]$.

**Answer:**

Look at the graphic:

```
In[1]:=   Clear[fun,a,b,A,B,curveplot,frame,
          labels,dots]
          fun[x_]=2^x
          a=1
          b=2
          A={a,fun[a]}
          B={b,fun[b]}
          curveplot =
          Plot[{fun[a],fun[x],fun[b]},{x,a,b},
          DisplayFunction->Identity]
          frame=Graphics[Line[{{a,0},{a,fun[b]},
          {b,fun[b]},{b,0}}]]
          labels=
          Graphics[Text["{t,f'[t]}",A,{1.1,0}],
          Text["{t+h,f'[t+h]}",B,{-1.1,0}]]
          dots=Graphics[
          {PointSize[.03],Point[A],Point[B]}]
          Show[dots,curveplot,frame,labels,
          Ticks->{{{a,"t"},{b,"t+h"}},None},
          PlotRange->
          {{a-1,b+1},{fun[a]-4,fun[b]+2}},
          AxesLabel->{"x","y=f[x]"},
          DisplayFunction->$DisplayFunction];
```

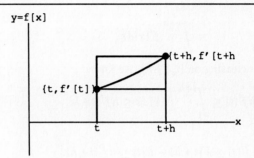

From the picture,

*area of the smaller rectangle at the bottom*

$\leq$ *area underneath the curve and over the $x$-axis*

$\leq$ *area of the overall rectangle.*

The area of the smaller rectangle at the bottom is $hf'[t]$.

The area underneath the curve and over the $x$-axis is $\int_t^{t+h} f'[x]dx$.

The area of the overall rectangle is $hf'[t+h]$.

So

$$hf'[t] \leq \int_t^{t+h} f'[x]dx \leq hf'[t+h].$$

### B.1.a.ii)

What version of the above inequality holds if $f'[x]$ is decreasing on $[t, t+h]$?

**Answer:**

$$hf'[t] \geq \int_t^{t+h} f'[x]dx \geq hf'[t+h].$$

### B.1.b.i)

For $t \geq a$, set

$$F[t] = \int_a^t f'[x]dx.$$

Why is

$$F'[t] = f'[t]?$$

**Answer:**

Take a small positive $h$ such that $f'[x]$ is either increasing or decreasing on $[t, t+h]$.

Look at

$$F[t+h] - F[t] = \int_a^{t+h} f'[x]dx - \int_a^t f'[x]dx$$

$$= \int_t^{t+h} f'[x]dx.$$

If $f'[x]$ is increasing on $[t, t+h]$, we know

$$hf'[t] \le \int_t^{t+h} f'[x]dx \le hf'[t+h].$$

Therefore

$$hf'[t] \le F[t+h] - F[t] \le hf'[t+h].$$

Consequently

$$f'[t] \le (F[t+h] - F[t])/h \le f'[t+h].$$

Hence $(F[t+h] - F[t])/h = f'[t]$ within $f'[t+h] - f'[t]$. As a result,

$$F'[t] = \lim_{h \to 0} (F[t+h] - F[t])/h = f'[t]$$

within $\lim_{h \to 0} f'[t] - f'[t+h]$. But derivatives can't jump anywhere, so $\lim_{h \to 0} f'[t] - f'[t+h] = 0$ and therefore

$$F'[t] = \lim_{h \to 0} (F[t+h] - F[t])/h = f'[t]$$

within 0. Consequently

$$F'[t] = f'[t].$$

That's IMPORTANT!

Let's run this through *Mathematica*.

```
In[1]:=    Clear[x,f]
           D[Integrate[f'[x],{x,a,t}],t]
Out[1]=    f'[t]
```

Perfect.

The same thing is true in the case that $f'[x]$ is decreasing on the short interval $[t, t+h]$ by a similar argument that uses the inequalities reversed.

### B.1.b.ii): Fundamental Theorem of Calculus.

Why is

$$f[t] - f[a] = \int_a^t f'[x]dx$$

for $t \ge a$?

**Answer:**

Hold $a$ fixed and let $t$ vary.

Put

$$G[t] = f[t] - f[a]$$

and

$$F[t] = \int_a^t f'[x]dx.$$

Both $F[t]$ and $G[t]$ have the same derivative with respect to $t$, namely $f'[t]$.

Note

$$G[a] = f[a] - f[a] = 0$$

and

$$F[a] = \int_a^a f'[x]dx = 0$$

(The reason $0 = \int_a^a f'[x]dx$ is that this integral measures the area of a (degenerate) rectangle whose base is 0 and whose height is $f[a]$. Accordingly $\int_a^a f'[x]dx = 0f[a] = 0$.)

So $F[t]$ and $G[t]$ agree for $t = a$ and both have the same derivative for $t \ge a$. The Race Track Principle tells us that $F[t] = G[t]$ for $t \ge a$. Hence

$$f[t] - f[a] = F[t] = G[t] = \int_a^t f'[x]dx$$

for $t \ge a$.

This outstanding calculation tool breaks the code of the integral. One of the main contributions of Newton and Leibniz to calculus, it is often called the *Fundamental Theorem of Calculus*.

### B.1.c)

Calculate $\int_0^{\pi/2} \cos[x]dx$ and explain what the calculation measures.

**Answer:**

Try:

```
In[1]:=    Clear[f,x]
           f[x_] = Sin[x]
Out[1]=    Sin[x]
```

Look at:

```
In[2]:=    f'[x]
Out[2]=    Cos[x]
```

Good: $f'[x] = \cos[x]$.

The **fundamental theorem** says that

$$\int_0^{\pi/2} \cos[x]dx = \int_0^{\pi/2} f'[x]dx = f[\pi/2] - f[0]$$

which is given by:

In[3]:=    **f[Pi/2] - f[0]**

Out[3]=    **1**

Check:

In[4]:=    **Integrate[Cos[x],{x,0,Pi/2}]**

Out[4]=    **1**

Nailed it.

The calculation $\int_0^{\pi/2} \cos[x]dx = 1$ means that the area under the following curve over the $x$−axis is one square unit:

In[5]:=    **Plot[Cos[x],{x,0,Pi/2},**
           **AspectRatio->Automatic,**
           **AxesLabel->{"x","y"},**
           **PlotLabel->"y = Cos[x]"];**

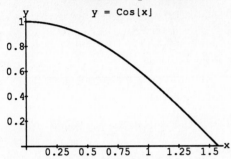

### ■ B.2) The integral of the sum.

#### B.2.a)

Why is the integral of the sum the same as the sum of the integrals?

**Answer:**

This answer boils down to understanding why

$$\int_a^t f'[x]dx + \int_a^t g'[x]dx = \int_a^t f'[x] + g'[t]dx.$$

The fundamental theorem tells us that

$$\int_a^t f'[x]dx = (f[t] - f[a])$$

$$\int_a^t g'[x]dx = (g[t]) - g[a])$$

and

$$\int_a^t (f'[x] + g'[x])dx = (f[t] + g[t]) - (f[a] + g[a])$$

(because $f'[x] + g'[x]$ is the derivative of $(f[x] + g[x])$).

Adding the first two expressions on the right hand side gives us the third expression on the right hand side.

Therefore adding the first two expressions on the left hand side gives us the third expression on the left hand side.

In other words,

$$\int_a^t f'[x]dx + \int_a^t g'[x]dx = \int_a^t (f'[x] + g'[x])dx.$$

Let's run this one through *Mathematica* :

In[1]:=    **Clear[f,g,a,x,t]**
           **first = Integrate[f'[x],{x,a,t}]**
           **second = Integrate[g'[x],{x,a,t}]**
           **third = Integrate[f'[x] + g'[x],{x,a,t}];**

Look at:

In[2]:=    **{first + second,third}**

Out[2]=    **{-f[a] + f[t] - g[a] + g[t],**

           **-f[a] + f[t] - g[a] + g[t]}**

Right!

### ■ B.3) Integrating backwards.

#### B.3.a)

In all cases we've seen so far, we've always looked at $\int_a^b g[x]dx$ where $a \leq b$. What is meant by $\int_a^b g[x]dx$ in the case where $a > b$.

**Answer:**

This is a good question because if $a > b$, then the integral

$$\int_a^b g[x]dx$$

does not have an obvious physical interpretation.

Mathematicians and scientists worldwide[1] have settled this via mutual convention by agreeing that if $b < a$, then

$$\int_a^b g[x]dx = - \int_b^a g[x]dx.$$

This agreement expedites calculation because this agreement makes it true that

$$f[t] - f[a] = \int_a^t f'[x]dx$$

regardless of whether $t \geq a$ or $t < a$.

[1] It's too bad that other worldwide issues cannot be dealt with in the same spirit of cooperation that worldwide scientific issues have received.

### ■ B.4) Differentiating integrals.

#### B.4.a)

For $t \geq a$, set

$$F[t] = \int_a^t g[x]dx.$$

Explain why

$$F'[t] = g[t]$$

provided $g[t]$ does not jump?

---

**Answer:**

The reasons are just a small modification of the discussions in **B.1)** above. For a full discussion of this, cut and paste **B.1.a.i)**, **B.1.a.ii)** and **B.1.b.i)** into this position and systematically replace $f'[t]$ by $g[t]$. The result will be a complete explanation of why $F'[t] = g[t]$.

**B.4.b)**

---

The fact that if
$$F[t] = \int_a^t g[x]\,dx,$$
then
$$F'[t] = g[t]$$

is a welcome calculational fringe benefit because it allows us to find the derivative without calculating the integral!

This is a special benefit in situations in which neither we nor *Mathematica* can calculate the integral.

Give the derivatives of the following functions $F[t]$ :

**B.4.b.i)**

---

$$F[t] = \int_0^t \sin[x^4]\,dx.$$

---

**Answer:**

$$F'[t] = \sin[t^4].$$

**B.4.b.ii)**

---

$$F[t] = \int_0^{t^2} \sin[x^4]\,dx.$$

---

**Answer:**

The chain rule gives
$$F'[t] = \sin[(t^2)^4]2t = 2t\sin[t^8].$$

# Tutorial

■ **T.1) Calculating integrals.**

The fundamental theorem of calculus
$$f[t] - f[a] = \int_a^t f'[x]\,dx$$

gives us a powerful systematic way of calculating the exact value of integrals. And it is a pleasure to use:

To find $\int_a^t g[x]\,dx$ for a given function $g[x]$, you just find another function $f[x]$ such that
$$f'[x] = g[x]$$
and then you just evaluate
$$\int_a^t g[x]\,dx = \int_a^t f'[x]\,dx = f[t] - f[a].$$

In principle, nothing could be simpler and you should not have to depend on the machine to do the simple cases.

Use this wonderful calculating tool to find exact values of the following integrals; check yourself with *Mathematica*.

**T.1.a.i)**

---

$$\int_a^t x^n\,dx$$

for any number $n \neq -1$.

---

**Answer:**

Try:

```
In[1]:=   Clear[f,x]
          f[x_] = (x^(n + 1))/(n + 1)
Out[1]=    1 + n
          x
          ------
          1 + n
```

Look at:

```
In[2]:=   f'[x]
Out[2]=    n
          x
```

Excellent; $f'[x] = x^n$.

The fundamental theorem says that
$$\int_a^t x^n\,dx = \int_a^t f'[x]\,dx = f[t] - f[a]$$
$$= t^{n+1}/(n+1) - a^{n+1}/(n+1).$$

Some folks like to use the compact notation
$$x^{n+1}/(n+1)\Big|_a^t$$
to stand for $t^{n+1}/(n+1) - a^{n+1}/(n+1)$.

So in this notation,
$$\int_a^t x^n\,dx = x^{n+1}/(n+1)\Big|_a^t$$

Check:

```
In[3]:=   Integrate[x^n, {x,a,t}]
```

*Out[3]=*

$$-\left(\frac{a^{1+n}}{1+n}\right) + \frac{t^{1+n}}{1+n}$$

Got it.

**T.1.a.ii)**

---

$$\int_1^t (1/x)\,dx.$$

---

**Answer:**

Try:

*In[1]:=*  `Clear[f,x]`
          `f[x_] = Log[x]`

*Out[1]=*  `Log[x]`

Look at:

*In[2]:=*  `f'[x]`

*Out[2]=*  $\frac{1}{x}$

Fine; $f'[x] = 1/x$.

The fundamental theorem says that

$$\int_1^t 1/x\,dx = \int_1^t f'[x]\,dx = f[t] - f[1]$$
$$= \log[t] - \log[1] = \log[t].$$

Or in the compact notation,

$$\int_1^t 1/x\,dx = \log[x]\Big|_1^t = \log[t].$$

(Remember Log[1] =0.)

Check:

*In[3]:=*  `Integrate[1/x,{x,1,t}]`

*Out[3]=*  `Log[t]`

Nailed it.

**T.1.a.iii)**

---

$$\int_1^3 x^3 - 5x^2 + 4x + 2\,dx.$$

---

**Answer:**

Try:

*In[1]:=*  `Clear[f,x]`
          `f[x_] = (x^4)/4 - 5 (x^3)/3 + 4 (x^2)/2 + 2 x`

*Out[1]=*  $$2x + 2x^2 - \frac{5x^3}{3} + \frac{x^4}{4}$$

Look at:

*In[2]:=*  `f'[x]`

*Out[2]=*  $$2 + 4x - 5x^2 + x^3$$

Just as needed: $f'[x] = x^3 - 5x^2 + 4x + 2$.

The fundamental theorem tells us that

$$\int_1^3 x^3 - 5x^2 + 4x + 2\,dx = \int_1^3 f'[x]\,dx = f[x]\Big|_1^3$$

which is given by:

*In[3]:=*  `f[3] - f[1]`

*Out[3]=*  $$-\left(\frac{10}{3}\right)$$

Check:

*In[4]:=*  `Integrate[x^3 - 5 x^2 + 4 x + 2, {x,1,3}]`

*Out[4]=*  $$-\left(\frac{10}{3}\right)$$

Got it. Not bad for beginners.

**T.1.a.iv)**

---

$$\int_0^t \cos[y[x]]y'[x]\,dx$$

---

given $y[0] = \pi/2$.

**Answer:**

$\cos[y[x]]y'[x]$ looks like the derivative of $\sin[y[x]]$.

Try it:

*In[1]:=*  `Clear[f,y,x,t]`
          `f[x_] = Sin[y[x]]`

*Out[1]=*  `Sin[y[x]]`

Look at:

*In[2]:=*  `D[f[x],x]`

*Out[2]=*  `Cos[y[x]] y'[x]`

Right on the money; $f'[x] = \cos[y[x]]y'[x]$.

The fundamental theorem tells us that

$$\int_0^t \cos[y[x]]y'[x]dx$$

$$= \int_0^t f'[x]dx = f[x]\Big|_0^t = \sin[y[t]] - \sin[y[0]]$$

$$= \sin[y[t]] - \sin[\pi/2] = \sin[y[t]] - 1.$$

Check:

```
In[3]:=    Integrate[Cos[y[x]] y'[x], {x,0,t}]/.
           y[0]->Pi/2
Out[3]=    -1 + Sin[y[t]]
```

On the bull's eye.

**T.1.a.v)**

---

$$\int_0^{\pi/2} \sin[7x]dx.$$

---

**Answer:**

Try:

```
In[1]:=    f[x_] = - Cos[7 x]
Out[1]=    -Cos[7 x]
```

Look at:

```
In[2]:=    f'[x]
Out[2]=    7 Sin[7 x]
```

But this $f[x]$ doesn't work because it carries an extra factor out front. Adjust $f[x]$ to kill the extra factor:

```
In[3]:=    f[x_] = - Cos[7 x]/7
Out[3]=    -Cos[7 x]
           ---------
               7
```

Look at:

```
In[4]:=    f'[x]
Out[4]=    Sin[7 x]
```

Good. The fundamental theorem tells us that

$$\int_0^{\pi/2} \sin[7x]dx = \int_0^{\pi/2} f'[x]dx$$

$$= f[x]\Big|_0^{\pi/2} = -\cos[7x]/7\Big|_0^{\pi/2}$$

which is given by:

```
In[5]:=    f[Pi/2] - f[0]
```

```
Out[5]=    1
           -
           7
```

Check:

```
In[6]:=    Integrate[Sin[7 x], {x,0,Pi/2}]
Out[6]=    1
           -
           7
```

Got it.

**T.1.a.vi)**

---

$$\int_{.2}^{.9} 3.1e^{-10x}dx.$$

---

**Answer:**

Try:

```
In[1]:=    Clear[f,x]
           f[x_] = (3.1) (E^(-10 x))
Out[1]=    3.1
           -----
           10 x
           E
```

Look at:

```
In[2]:=    f'[x]
Out[2]=    -31.
           -----
           10 x
           E
```

This $f[x]$ doesn't work because it carries an extra multiplicative factor of -10 out front. So we adjust $f[x]$ to kill the extra factor:

```
In[3]:=    Clear[f,x]
           f[x_] = (3.1) (E^(-10 x))/(-10)
Out[3]=    -0.31
           -----
           10 x
           E
```

Look at:

```
In[4]:=    f'[x]
Out[4]=    3.1
           -----
           10 x
           E
```

O. K.: $f'[x] = 3.1e^{-10x}$.

The fundamental theorem tells us that

$$\int_{.2}^{.9} 3.1e^{-10x}dx = \int_{.2}^{.9} f'[x]dx = f[x]\Big|_{.2}^{.9}$$

which is given by:

```
In[5]:=    f[.9] - f[.2]
Out[5]=    0.0419157
```

Check:

```
In[6]:=    Integrate[3.1 E^(-10 x),{x,.2,.9}]
Out[6]=    0.0419157
```

Got it.

**T.1.a.vii)**

---

$$\int_0^1 1/(1+9x^2)dx.$$

---

**Answer:**

Try:

```
In[1]:=    Clear[f,x]
           f[x_] = ArcTan[9 x]
Out[1]=    ArcTan[9 x]
```

Look at:

```
In[2]:=    f'[x]
Out[2]=         9
           ----------
                    2
           1 + 81 x
```

Ugh. We wanted $f'[x] = 1/(1+9x^2)$. This is way off.

Evidently the 9 was squared in the denominator. Fix this by trying:

```
In[3]:=    Clear[f,x]
           f[x_] = ArcTan[3 x]
Out[3]=    ArcTan[3 x]
```

Look at:

```
In[4]:=    f'[x]
Out[4]=        3
           --------
                  2
           1 + 9 x
```

We wanted $f'[x] = 1/(1+9x^2)$. But this $f[x]$ doesn't work because it carries an extra factor on top. Adjust $f[x]$ to kill the extra factor:

```
In[5]:=    Clear[f,x]
           f[x_] = ArcTan[3 x]/3
Out[5]=    ArcTan[3 x]
           -----------
                3
```

Look at:

```
In[6]:=    f'[x]
Out[6]=        1
           --------
                  2
           1 + 9 x
```

We wanted $f'[x] = 1/(1+9x^2)$ and we got it. The fundamental theorem tells us that

$$\int_0^1 1/(1+9x^2)dx = \int_0^1 f'[x]dx = f[x]\Big|_0^1$$

which is given by:

```
In[7]:=    f[1] - f[0]
Out[7]=    ArcTan[3]
           ---------
               3
```

If you want a decimal, then use:

```
In[8]:=    N[f[1] - f[0]]
Out[8]=    0.416349
```

Check:

```
In[9]:=    Integrate[1/(1 + 9 x^2),{x,0,1}]
Out[9]=    ArcTan[3]
           ---------
               3
```

Right on.

**T.1.a.viii)**

---

$$\int_1^t y'[x]/(1+y[x])dx$$

given $y[1] = 0$.

---

**Answer:**

$y'[x]/(1+y[x])$ looks like the derivative of $\log[1+y[x]]$.

Try it:

```
In[1]:=    Clear[f,y,x,t]
           f[x_] = Log[1 + y[x]]
Out[1]=    Log[1 + y[x]]
```

Look at:

```
In[2]:=    D[f[x],x]
Out[2]=     y'[x]
           --------
           1 + y[x]
```

Good deal; $f'[x] = y'[x]/(1 + y[x])$. The fundamental theorem tells us that

$$\int_1^t y'[x]/(1 + y[x])dx = \int_1^t f'[x]dx = f[x]\Big|_1^t.$$

$$= \log[1 + y[t]] - \log[1 + y[1]]$$

$$= \log[1 + y[t]] - \log[1 + 0] = \log[1 + y[t]].$$

Check:

```
In[3]:=    Integrate[y'[x]/(1+y[x]),{x,1,t}]/.
           y[1]->0
Out[3]=    Log[1 + y[t]]
```

Got it again.

Say, we're getting pretty good at this.

**T.1.a.ix)**

---

$$\int_0^t y[x]^4 y'[x]dx$$

given $y[0] = 2$.

---

**Answer:**

$y[x]^4 y'[x]$ looks like the derivative of $y[x]^5/5$.

Try it:

```
In[1]:=    Clear[f,y,x,t]
           f[x_] = (y[x]^5)/5
Out[1]=        5
           y[x]
           -----
             5
```

Look at:

```
In[2]:=    D[f[x],x]
Out[2]=        4
           y[x]  y'[x]
```

Good, $f'[x] = y[x]^4 y'[x]$.

The fundamental theorem tells us that

$$\int_0^t y[x]^4 y'[x]dx = \int_0^t f'[x]dx = f[x]\Big|_0^t$$

$$= y[t]^5/5 - y[0]^5/5 = y[t]^5/5 - 2^5/5$$

$$= y[t]^5/5 - 32/5.$$

Check:

```
In[3]:=    Integrate[y[x]^4 y'[x],{x,0,t}]/.y[0]->2
```

```
Out[3]=        32      y[t]
           -(--)  +  -----
            5          5
```

Got it.

**T.1.a.x)**

---

$$\int_0^1 \sin[\pi x^4]dx.$$

---

**Answer:**

No fair! We can't think of any function $f[x]$ such that $f'[x] = \sin[\pi x^4]$. Let's see whether *Mathematica* can handle it.

```
In[1]:=    Integrate[Sin[Pi x^4],{x,0,1}]
Out[1]=                  4
           Integrate[Sin[Pi x ], {x, 0, 1}]
```

*Mathematica* can't do it either. The trouble is not with us and the trouble is not with *Mathematica*. The real trouble is that $f[x]$ is not neatly expressible in terms of the everyday functions we have heard about.

It would be correct to say that

$$f[x] = \int_0^x \sin[\pi t^4]dt,$$

but this is a cop-out.

To calculate this integral, we are forced to forego an exact answer in favor of an approximate estimate:

```
In[2]:=    NIntegrate[Sin[Pi x^4],{x,0,1}]
Out[2]=    0.333639
```

This number is an approximation of the measurement of the area between the curve and the $x$−axis in the following plot:

```
In[3]:=    Plot[Sin[Pi x^4],{x,0,1},
           AspectRatio->Automatic];
```

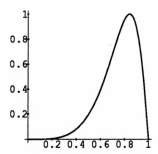

■ **T.2) Educated guessing: The method of undetermined coefficients.**

## T.2.a)

Look at *Mathematica*'s evaluation of $\int_1^t \log[x]dx$ :

```
In[1]:=    Clear[x,t]
           Integrate[Log[x],{x,0,t}]
Out[1]=    -t + t Log[t]
```

How could *Mathematica* have come up with this answer?

---

**Answer:**

Somehow *Mathematica* had to find a function $f[x]$ such that
$$f'[x] = \log[x].$$

When you think about it, then you realize that $x\log[x]$ should be involved because the product rule tells us that $\log[x]$ is involved in the derivative of $x\log[x]$ :

```
In[2]:=    D[x Log[x],x]
Out[2]=    1 + Log[x]
```

But we don't want the extra term on the left; so we knock it out by throwing an extra $x$ term onto $x\log[x]$ :

```
In[3]:=    Clear[tryf,x,a,b,c]
           tryf[x_] = a x Log[x] + b x
Out[3]=    b x + a x Log[x]
```

We want $D[tryf[x], x] = \log[x]$.

```
In[4]:=    D[tryf[x],x]
Out[4]=    a + b + a Log[x]
```

Force this to be
$$\log[x] = 0 + \log[x]$$

by equating the coefficients on the left and right and solving:

```
In[5]:=    eq0 = ( a + b == 0)
           eqlog = (a == 1)

           goodcoeffs =
               Solve[{eq0,eqlog},{a,b}]
Out[5]=    {{a -> 1, b -> -1}}
```

Here is the function $f[x]$ that we think we want:

```
In[6]:=    Clear[f]
           f[x_] = tryf[x]/.goodcoeffs[[1]]
Out[6]=    -x + x Log[x]
```

We want $f'[x]$ to be $\log[x]$ :

```
In[7]:=    f'[x]
Out[7]=    Log[x]
```

It is! This is the $f[x]$ we wanted.

Now we see how *Mathematica* might have come up with its answer:

```
In[8]:=    f[t] - f[1]
Out[8]=    1 - t + t Log[t]
```

And this agreees with:

```
In[9]:=    Integrate[Log[x],{x,1,t}]
Out[9]=    1 - t + t Log[t]
```

## T.2.b)

Look at *Mathematica*'s evaluation of $\int_0^t x^2 \sin[x]dx$ :

```
In[1]:=    Integrate[x^2 Sin[x],{x,0,t}]
Out[1]=
                              2
           -2 + 2 Cos[t] - t  Cos[t] + 2 t Sin[t]
```

How the hell could *Mathematica* have come up with this answer?

---

**Answer:**

Somehow *Mathematica* had to find a function $f[x]$ such that
$$f'[x] = x^2 \sin[x].$$

When you think about it, then you realize that $-x^2 \cos[x]$ should be involved because the product rule tells us that $x^2 \sin[x]$ is involved in the derivative of $-x^2 \cos[x]$ :

```
In[2]:=    D[-x^2 Cos[x],x]
Out[2]=
                            2
           -2 x Cos[x] + x  Sin[x]
```

Now we see that $x \sin[x]$ should also be involved because one part of its derivative can be used to knock off the $x \cos[x]$ term above.

```
In[3]:=    D[x Sin[x],x]
Out[3]=    x Cos[x] + Sin[x]
```

Now we see that $\cos[x]$ should also be involved because one part of its derivative can be used to knock off the $\sin x]$ term above.

Now, let's see how to combine the three functions:

```
In[4]:=    tryf[x_] =
               a x^2 Cos[x] + b x Sin[x] + c Cos[x]
Out[4]=
                         2
           c Cos[x] + a x  Cos[x] + b x Sin[x]
```

We want $D[tryf[x], x] = x^2 \sin[x]$.

```
In[5]:=    D[tryf[x],x]
Out[5]=    2 a x Cos[x] + b x Cos[x] + b Sin[x] -
```

```
         2
c Sin[x] - a x  Sin[x]
```

Force this to be

$$x^2 \sin[x] = 0x \cos[x] + 0 \sin[x] + x^2 \sin[x]$$

by equating the coefficients on the left and right and solving:

```
In[6]:=   eqxcos = (2 a + b == 0)
          eqsin = (b - c == 0)
          eqx2sin = (-a == 1)

          goodcoeffs =
          Solve[{eqxcos, eqsin, eqx2sin},
              {a,b,c}]
Out[6]=   {{a -> -1, b -> 2, c -> 2}}
```

Here is the function $f[x]$ that we want:

```
In[7]:=   f[x_] = tryf[x]/.goodcoeffs[[1]]
Out[7]=            2
          2 Cos[x] - x  Cos[x] + 2 x Sin[x]
```

Try it out in the expectation that $f'[x] = x^2 \sin[x]$ :

```
In[8]:=   f'[x]
Out[8]=    2
          x  Sin[x]
```

Hot Zigitty! This is the way *Mathematica* might have done this.

Now we see that $\int_0^t x^2 \sin[x]\,dx$ is given by:

```
In[9]:=   f[t] - f[0]
Out[9]=                  2
          -2 + 2 Cos[t] - t  Cos[t] + 2 t Sin[t]
```

Check:

```
In[10]:=  Integrate[x^2 Sin[x],{x,0,t}]
Out[10]=                  2
          -2 + 2 Cos[t] - t  Cos[t] + 2 t Sin[t]
```

Right on the money.

**T.2.c)**

Look at *Mathematica* 's evaluation of

$$\int_0^t (4x+1)/((x+2)^3(x+1)^2)\,dx :$$

```
In[1]:=   Integrate[(4x+1)/((x+2)^3 (x+1)^2),
          {x,0,t}]
Out[1]=    71       3          7         10
          -(--) + ----- + ----------- + ----- +
           8      1 + t         2       2 + t
                          2 (2 + t)

          13 Log[2] + 13 Log[1 + t] -
```

```
13 Log[2 + t]
```

How could *Mathematica* have come up with this answer?

**Answer:**

Hmm.

It looks like *Mathematica* took each of the factors of the denominator of $(4x+1)/((x+2)^3(x+1)^2)$, integrated each, multiplied each by certain numbers and added them up.

Let's try it for ourselves

```
In[2]:=   Clear[tryf,x,a,b,c,d,e]
          tryf[x_] =
          a/(1 + x) + b Log[1 + x] +
          c/(2 + x)^2 + d/(2 + x) +
          e Log[2 + x]
Out[2]=    a          c          d
          ----- + --------- + ----- +
          1 + x        2       2 + x
                  (2 + x)

          b Log[1 + x] + e Log[2 + x]
```

We want $D[tryf[x], x] = (4x+1)/((x+2)^3(x+1)^2)$.

```
In[3]:=   birdnest = Together[D[tryf[x],x]]
Out[3]=   (-8 a + 8 b - 2 c - 2 d + 4 e -

          12 a x + 20 b x - 4 c x - 5 d x +

                        2           2
          12 e x - 6 a x  + 18 b x  -

               2        2        2        3
          2 c x  - 4 d x  + 13 e x  - a x  +

               3      3        3      4      4
          7 b x  - d x  + 6 e x  + b x  + e x
                    2        3
          ) / ((1 + x)  (2 + x) )
```

Look at the numerator:

```
In[4]:=   num = Numerator[birdnest]
Out[4]=   -8 a + 8 b - 2 c - 2 d + 4 e - 12 a x +

          20 b x - 4 c x - 5 d x + 12 e x -
                2         2        2        2
          6 a x  + 18 b x  - 2 c x  - 4 d x  +
                2      3        3      3
          13 e x  - a x  + 7 b x  - d x  +
                3      4      4
          6 e x  + b x  + e x
```

Let us collect equal powers of $x$ together:

```
In[5]:=   Collect[num,x]
Out[5]=   -8 a + 8 b - 2 c - 2 d + 4 e +

          (-12 a + 20 b - 4 c - 5 d + 12 e) x +
```

2

$$(-6 a + 18 b - 2 c - 4 d + 13 e) x +$$

$$(-a + 7 b - d + 6 e) x^3 + (b + e) x^4$$

Force the numerator to be $4x+1$ by equating the coefficients of like powers of $x$ :

```
In[6]:=   eq1 = (-8 a + 8 b - 2 c - 2 d + 4 e == 1)
          eqx = (-12 a + 20 b - 4 c - 5 d + 12 e == 4)
          eqx2 = (-6 a + 18 b - 2 c - 4 d + 13 e == 0)
          eqx3 = (-a + 7 b - d + 6 e == 0)
          eqx4 = ( b + e == 0)

          goodcoeffs =
          Solve[{eq1,eqx,eqx2,eqx3,eqx4},{a,b,c,d,e}]
```

$$Out[6]= \quad \{\{a \to 3, \ b \to 13, \ c \to -\frac{7}{2}, \ d \to 10,$$

$$e \to -13\}\}$$

Here is the function $f[x]$ that we think we want:

```
In[7]:=   Clear[f]
          f[x_] = tryf[x]/.goodcoeffs[[1]]
```

$$Out[7]= \quad \frac{3}{1+x} + \frac{7}{2(2+x)^2} + \frac{10}{2+x} +$$

$$13 \ Log[1 + x] - 13 \ Log[2 + x]$$

Try it out in the expectation that

$$f'[x] = (4x+1)/((x+2)^3(x+1)^2) :$$

```
In[8]:=   f'[x]
```

$$Out[8]= \quad \frac{-3}{(1+x)^2} + \frac{13}{1+x} - \frac{7}{(2+x)^3} -$$

$$\frac{10}{(2+x)^2} - \frac{13}{2+x}$$

We'd better go for a common denominator:

```
In[9]:=   Together[f'[x]]
```

$$Out[9]= \quad \frac{1+4x}{(1+x)^2 (2+x)^3}$$

Oskee-wow-wow!

Now we know that $\int_0^t (4x+1)/((x+2)^3(x+1)^2)dx$ is given by:

```
In[10]:=  f[t] - f[0]
```

$$Out[10]= \quad -(\frac{71}{8}) + \frac{3}{1+t} + \frac{7}{2(2+t)^2} + \frac{10}{2+t} +$$

$$13 \ Log[2] + 13 \ Log[1 + t] -$$

$$13 \ Log[2 + t]$$

Check:

```
In[11]:=  Integrate[(4 x + 1)/((x + 2)^3 (x + 1)^2),
          {x,0,t}]
```

$$Out[11]= \quad -(\frac{71}{8}) + \frac{3}{1+t} + \frac{7}{2(2+t)^2} + \frac{10}{2+t} +$$

$$13 \ Log[2] + 13 \ Log[1 + t] -$$

$$13 \ Log[2 + t]$$

Nailed it.

Some folks call the method we used by the name *Method of Partial Fractions.*

### ■ T.3) Area between curves.

#### T.3.a)

Measure the area bounded by the curves $y = x^2$ and $y = x$.

**Answer:**

To see what's going on, look at:

```
In[1]:=   Plot[{x^2,x},{x,-1,2},
          AxesLabel->{"x","y"},
          PlotRange->{-1,2},
          AspectRatio->Automatic];
```

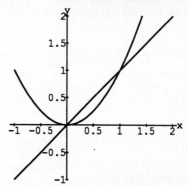

We want the area of the cresent shaped region to the right of the origin. Let's see where the curves cross.

```
In[2]:=   Solve[x == x^2,x]
Out[2]=   {{x -> 1}, {x -> 0}}
```

The measurement of the area of the cresent-shaped region is

$$\int_0^1 x \, dx - \int_0^1 x^2 \, dx = \int_0^1 (x - x^2) \, dx = x^2/2 - x^3/3 \Big|_0^1$$

which is given by:

```
In[3]:=   1/2 - 1/3
```

*Out[3]=*   $\dfrac{1}{6}$

The only trick is to be aware where the curves cross and aware of which curve is on top and which curve is on the bottom.

### T.3.b)

Measure the area bounded by the curves $y = e^{x/2}$ and $y = x^4$.

**Answer:**

Look at:

```
In[1]:=    Plot[{E^(x/2),x^4},{x,-2,2}];
```

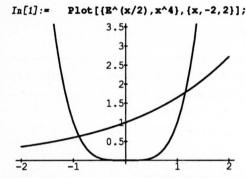

From the plot, we learn that the $e^{x/2}$ curve is on the top and that the curves cross when $x$ is near $-1$ and again when $x$ is near 1. To find an accurate value of the solution of $e^{x/2} = x^4$ near $x = -1$, use:

```
In[2]:=    FindRoot[E^(x/2) == x^4,{x,-1}]
Out[2]=    {x -> -0.894241}
```

For the root near $x = 1$, use:

```
In[3]:=    FindRoot[E^(x/2) == x^4,{x,1}]
Out[3]=    {x -> 1.15537}
```

Now we can compute the area. It is given by

```
In[4]:=    Integrate[E^(x/2) - x^4,
           {x,-0.894241,1.15537}]
Out[4]=    1.75876
```

Not too bad.

### T.3.c)

Measure the area bounded by the curves $x = y^4 - 1$ and $x = -y$.

**Answer:**

Look at:

```
In[1]:=    Plot[{y^4 - 1,-y},{y,-2,2},
           AxesLabel->{"y","x"}];
```

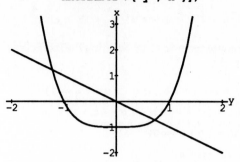

Note that the $y$-axis is in the position normally occupied by the $x$-axis.

The plot shows that the line $x = -y$ is on top and the plot indicates roots of $y^4 - 1 = -y$ near $y = -1.25$ and $y = .75$. To compute them accurately, we use:

```
In[2]:=    FindRoot[y^4 - 1 == -y,{y,-1.25}]
Out[2]=    {y -> -1.22074}
```

and

```
In[3]:=    FindRoot[y^4 - 1 == -y,{y,.75}]
Out[3]=    {y -> 0.724492}
```

Here is the area measurement:

```
In[4]:=    Integrate[
           -y-(y^4-1),{y,-1.22074,0.724492}]
Out[4]=    1.84579
```

And that's it.

### ■ T.4) Reconstructing the function from its derivative by the fundamental theorem.

In the lesson dealing with the mean value theorem, we learned a little bit about using the mean value theorem to fake the plot of $f[x]$ given $f'[x]$. In this problem, we are going to see why we no longer have to resort to fakery in many situations.

### T.4.a)

A certain function $f[x]$ has derivative

$$f'[x] = e^x - 2$$

and it is known that $f[2] = 7$. Find a formula for $f[x]$ and plot.

**Answer:**

The fundamental theorem says that

$$\int_2^x f'[t]\,dt = f[x] - f[2].$$

So

$$f[x] = f[2] + \int_2^x f'[t]dt.$$

Consequently a formula for $f[x]$ is:

```
In[1]:=    f[x_] = (f[2] + Integrate[E^t - 2,
           {t,2,x}])/.f[2]->7
Out[1]=         2    x
           11 - E  + E  - 2 x
```

Not very hard. Check:

```
In[2]:=    {f'[x],f[2]}
Out[2]=          x
           {-2 + E , 7}
```

On the money. Here is a plot:

```
In[3]:=    Plot[f[x],{x,-2,3}];
```

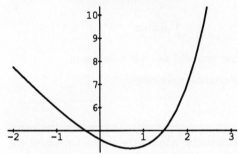

That curve has pizzazz.

**T.4.b)**

---

Another function $g[x]$ has derivative

$$g'[x] = 2x - 36/x^3.$$

and is known that $g[3] = -8$.

Find a formula for $g[x]$ and plot for $x > 0$. What can we say about a formula for $g[x]$ for $x < 0$?

---

**Answer:**

The fundamental theorem says that

$$\int_3^x g'[t]dt = g[x] - g[3].$$

So

$$g[x] = g[3] + \int_3^x g'[t]dt = g[3] + \int_3^x (2t - 36/t^3)dt$$

$$= -8 + \int_3^x (2t - 36/t^3)dt.$$

Note that $(2t - 36/t^3)$ blows up at 0 so that if $x \leq 0$, this integral does not make sense.

Thus for $x > 0$, $g[x]$ is given by:

```
In[1]:=    g[x_] =
           (g[3] + Integrate[2t - 36/t^3,
           {t,3,x}])/.g[3]->-8
Out[1]=          18    2
           -19 + -- + x
                  2
                 x
```

Not very hard. Here is a plot for $x > 0$:

```
In[2]:=    Plot[g[x],{x,0,8},PlotRange->{-12,30}];
```

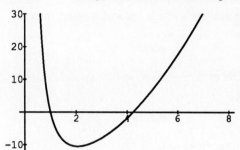

Even more pizzazz than the last plot.

Because of the singularity in $g'[x]$ at $x = 0$, it is not possible to determine a formula for $g[x]$ for $x < 0$.

■ **T.5) The "indefinite integral,"** $\int g[x]dx$.

Given a function $g[x]$, we calculate the integral

$$\int_a^b g[x]dx$$

by finding another function $f[x]$ such that

$$f'[x] = g[x]$$

and then we calculate

$$\int_a^b g[x]dx = f[b] - f[a].$$

In this set-up, the function $f[x]$ is called an **antiderivative or indefinite integral** of $g[x]$. It is common to write

$$\int g[x]dx$$

to signify a function whose derivative is $g[x]$.

**T.5.a)**

---

For given constants $a$ and $b$, one of the expression below stands for a number; the other stands for a function:

$$\int_a^b g[x]dx$$

and

$$\int g[x]dx.$$

Which is which?

**Answer:**

They are different species of animals.

If $a$ and $b$ are given numbers, then

$$\int_a^b g[x]dx$$

is a NUMBER.

If $a$ is a given number and $t$ is allowed to vary, then

$$\int_a^t g[x]dx$$

is a FUNCTION of $t$.

But

$$\int g[x]dx$$

is always a FUNCTION of $x$

because

$$D[\int g[x]dx, x] = g[x].$$

**T.5.b)**

What is the hidden danger in the notation $\int g[x]dx$?

**Answer:**

Let's look at an example: It is correct to say that

$$\int \cos[x]dx = \sin[x]$$

because $D[\sin[x], x] = \cos[x]$. It is equally correct to say that

$$\int \cos[x]dx = \sin[x] + 5$$

because $D[\sin[x] + 5, x] = \cos[x]$. In fact it is correct to say that if $C$ is any constant, then

$$\int \cos[x]dx = \sin[x] + C$$

because $D[\sin[x] + C, x] = \cos[x]$.

This is the rub: When you talk to a friend about

$$\int \cos[x]dx,$$

you and your friend may have different constants in mind and this makes communication blurry.

Matters are even worse when the function $g[x]$ has singularities:

You could say correctly that

$$\int (1/x)dx = \log[x] + C;$$

implicit in this statement is the proviso that $x > 0$.

You could also say correctly that

$$\int (1/x)dx = \log[|x|] + C$$

with the implicit proviso that $x$ is not 0.

Many traditional calculus texts insist on this. The trouble is that you could say correctly that

$$\int (1/x)dx = \log[x] + C_1$$

for $x > 0$ and

$$\int (1/x)dx = \log[-x] + C_2$$

for $x < 0$. Old style calculus books do not usually mention this.

The inherent confusion about constants introduced by the notation

$$\int g[x]dx$$

is the reason that you will see this notation in

Calculus& *Mathematica* only rarely.

**T.4.c)**

How does *Mathematica* handle the confusion about the constants in the notation $\int g[x]dx$?

**Answer:**

The *Mathematica* instruction **Integrate[g[x], x]** tells *Mathematica* to try to find a function $f[x]$ such that $f'[x] = g[x]$; i.e.

$$f[x] = \int g[x]dx.$$

Let's try it:

For $\int x^2 dx$, use:

```
In[1]:=    Integrate[x^2,x]
                  3
Out[1]=          x
                 --
                  3
```

For $\int e^x dx$, use:

```
In[2]:=    Integrate[E^x,x]
                  x
Out[2]=          E
```

For $\int \sin[x]dx$, use:

```
In[3]:=    Integrate[Sin[x],x]
```

`Out[3]=    -Cos[x]`

For $\int 1/(1 + x^2)dx$, use:

`In[4]:=    Integrate[1/(1 + x^2),x]`

`Out[4]=    ArcTan[x]`

For $\int 1/\sqrt{1 - x^2}dx$, use:

`In[5]:=    Integrate[1/Sqrt[1 - x^2],x]`

`Out[5]=    ArcSin[x]`

Evidently *Mathematica* deals with the problem of the constants by leaving the constants off altogether. Certainly not a bad compromise.

`In[6]:=`

## Literacy Sheet

1. Evaluate each of the following integrals by hand and sketch the area that is measured by each integral:

$$\int_0^2 x^3 dx$$

.

.

.

$$\int_0^{Log[3]} e^x dx$$

.

.

.

$$\int_0^{Log[2]} e^{-x} dx$$

.

.

.

$$\int_1^e 1/x\, dx$$

.

.

$$\int_{-\pi}^{\pi} \cos[x/2] dx$$

.

.

.

$$\int_0^{\pi/4} \sin[2x] dx$$

.

.

.

$$\int_0^2 1/(1+t) dt$$

.

.

$$\int_0^{-1} \sin[\pi x] dx$$

.

.

$$\int_1^8 x^{2/3} dx$$

.

.

.

$$\int_1^8 x^{-2/3} dx$$

.

.

.

$$\int_0^{\pi/4} 1/(1+t^2) dt$$

.

.

$$\int_0^{\pi/4} 1/(1+4t^2) dx$$

.

.

$$\int_{-3}^{3} x/(1+x^8) d$$

.

.

2. Evaluate the following integrals:

$$\int_0^t e^{-y[x]} y'[x] dx \text{ given that } y[0] = 6.$$

.

.

.

$$\int_0^t y'[x]/y[x] dx \text{ given that } y[0] = e.$$

.

.

.

2c. $\int_0^t \sin[y[x]] y'[x] dx$ given that $y[0] = 0$.

.

.

.

2d. $\int_1^t \sqrt{y[x]} y'[x] dx$ given that $y[1] = 4$.

.

.

.

3. Give a formula for $f[x]$ if $f'[x] = \cos[x]$ and $f[0] = \pi$.

.

.

4. How does the fundamental theorem of calculus crack the code of the integral?

How does the fundamental theorem of calculus establish a connection between the integral and the derivative?

Why is Isaac Newton often called the father of calculus?

**5. Writing**

$$\int_a^b x^3\,dx = x^4/4 + C$$

would be a sure sign of calculus illiteracy. Why?

**6.** We know that the integral of the sum equals the sum of the integrals, but does the integral of the product equal the product of the integrals?

**7.** Why does the instruction **NIntegrate[** $x^2$ , $x$ **]** fail while the instruction **Integrate[** $x^2$ , $x$ **]** works without a hitch?

**8.** Calculate $F'[t]$ by hand given that

$$F[t] = \int_0^t \cos[x^5]/(k^4 + x^6)\,dx.$$

Here $k$ is a constant.

**9.** Find $F'[t]$ by hand given that

$$F[t] = \int_1^{3t^2} Erf[x^3]\,dx.$$

**10.** Measure the area that is under the curve $y = \sqrt{x}$ and is over the curve $y = x^2$.

**11.** You are given that $f'[x] = 1/x^2$ and $f[1] = 2$. You quickly respond correctly that $f[x]$ is given by

```
In[7]:=   Integrate[1/t^2,{t,1, x}] + 2
Out[7]=         1
            3 - -
                x
```

Then you say that $f[x] = 3 - 1/x$, provided $x > 0$ and you stress that the given information does not allow you to calculate a formula for $f[x]$ for $x < 0$. Why are you right?

**12.** Complete the Formulas:

$\log[e] =$

$\log[e^t] =$

$\sin[0] =$

$\cos[0] =$

$\sin[\pi/4] =$

$\cos[\pi/4] =$

$\sin[\pi/2] =$

$\cos[\pi/2] =$

$\sin[\pi] =$

$\cos[\pi] =$

$\sin[3\pi/2] =$

$\cos[3\pi/2] =$

$\tan[\pi/4] =$

$\arctan[1] =$

$\arctan[1/\sqrt{3}] =$

# Measurements through integration

## Guide

The fundamental theorem of calculus says

$$f[b] - f[a] = \int_a^b f'[t]dt.$$

This tells us that if we have a measurement $f[b] - f[a]$ to make, then we start by finding $f'[t]$; then we can integrate to calculate $f[b] - f[a]$.

This is sort of amazing because this tells us how to use the integral to make measurements that, at first glance, have nothing to do with area!

This is why C.H. Edwards, Jr. in his book *The Historical Development of Calculus* (Springer-Verlag,1979) said:

*The contribution of Newton and Leibniz for which they are properly credited as the discoverers of calculus was not merely that they recognized the "fundamental theorem of calculus" as a mathematical fact, but that they employed it ... [as] ... a powerful algorithmic instrument for systematic calculation.*

## Basics

■ **B.1) Distance, velocity and acceleration.**

If an object moves on a straight line and is $s[t]$ units away from a reference point at time $t$, then its velocity is given by $v[t] = s'[t]$ and its acceleration is given by $a[t] = v'[t] = s''[t]$. So if we know the velocity, we can recover distance by integration and if we know the acceleration, we can recover velocity by integration.

### B.1.a)

A potted plant is thrown from a window of an apartment building with an initial velocity of 2 feet per second directly down. The sill of the window is 215 feet above the sidewalk below. How long does it take for the pot to hit the sidewalk? With what velocity does it hit the the sidewalk?

**Answer:**

Recall acceleration due to gravity is −32 ft/sec/sec.

```
In[1]:=    a[t_] = -32
Out[1]=    -32
```

Because $v'[t] = a[t]$, the fundamental theorem guarantees

that

$$v[t] - v[0] = \int_0^t v'[t]dt = \int_0^t a[t]dt.$$

This means

$$v[t] = \int_0^t a[t]dt + v[0]$$

```
In[2]:=    v[0] = -2
           v[t_] = Integrate[a[t],{t,0,t}] + v[0]
Out[2]=    -2 - 32 t
```

Because $s'[t] = v[t]$, the fundamental theorem guarantees that

$$s[t] - s[0] = \int_0^t s'[t]dt = \int_0^t v[t]dt.$$

This means

$$s[t] = \int_0^t v[t]dt + s[0].$$

```
In[3]:=    s[0] = 215
           s[t_] = Integrate[v[t],{t,0,t}] + s[0]
                            2
Out[3]=    215 - 2 t - 16 t
```

To get the time it takes for the plant to hit the sidewalk, look at:

```
In[4]:=    N[Solve[ s[t] == 0]]
Out[4]=    {{t -> 3.60375}, {t -> -3.72875}}
```

The potted plant takes 3.6 seconds to fall to the sidewalk.

To get the velocity at impact, look at:

```
In[5]:=    v[3.6]
Out[5]=    -117.2
```

It hits with a velocity of 117 ft/sec which is a pretty high speed. In miles per hour this translates to:

```
In[6]:=    v[3.6] (60^2)/5280
Out[6]=    -79.9091
```

Almost 80 mph!

Observe that the intgration and the fundamental theorem of calculus plays a crucial role in both measurements- even though neither measurement has anything to do with area.

■ **B.2) Weight and density**

If a wire or rod maintains the same thickness and same cross sectional weight all the way, then the weight of a piece of it is directly proportional to the length of the piece. This means that if $w[s]$ is the weight of the wire or rod $s$ units from an initial point at one end, then

$$w[s] = ps$$

where $p$ is a positive constant of proportionality. This constant $p$ is called the **density** of the wire or rod. Note that

$$w'[s] = p.$$

Some rods and wires look uniform on the outside, but their interior composition varies from point to point. We can still speak of density $p[s]$ of the rod at a point $s$ units from an initial point at one end by agreeing that

$$p[s] = w'[s].$$

The fundamental theorem of calculus tells us that if we know the density $p[s]$ we can measure the weight $w[s]$ of the wire or rod $L$ units from an initial point at one end via the formula

$$w[L] = w[L] - w[0] = \int_0^L w'[s]ds = \int_0^L p[s]ds.$$

**B.2.a)**

---

The density of a rod 4 feet long varies in proportion to the distance to a point $P$ that is 3 feet away from the left end of the rod in the line of the rod. If the density of the rod is 2 pounds/foot at the left endpoint, then how much does the rod weight?

How much does the left half of the rod weigh?

How much does the right half of the rod weigh?

---

**Answer:**

Let $s$ stand for distance from the left end of the rod. Here is a picture:

```
In[1]:=    rod = Graphics[{Thickness[.02],
           Line[{{.08,0},{3.95,0}}]}]
           refpoint = Graphics[
           {PointSize[.01],Point[{-3,0}]}]
           reflabel = Graphics[
           Text["P",{-3,0},{0,+2}]]
           sline = Graphics[
           Line[{{-3,.001},{1.78,.001}}]]
           slabel = Graphics[Text["s",{.89,.0025}]]
           leftvert = Graphics[
           Line[{{-3,0},{-3,.0015}}]]
           rightvert = Graphics[
           Line[{{1.78,0},{1.78,.0015}}]]
           threelabel = Graphics[
           Text["3",{-1.5,.0025}]]
           Show[rod,refpoint,reflabel,sline,slabel,
           leftvert,rightvert,threelabel,
           Axes->{0,0},
           PlotRange->{{-3.5,5},{-.01,.01}},
           Ticks->{{0,1,2,3,4},None}];
```

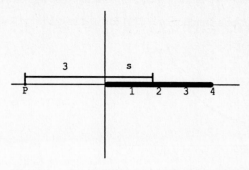

We know that the density $p[s]$ of the rod varies in proportion to the distance to a point $P$. This is the same as saying

$$p[s] = K(s + 3)$$

where K is a constant of proportionality.

The density of the rod is 2 pounds/foot at the left endpoint; so:

$$2 = p[0] = K(0 + 3) = 3K.$$

Thus $K = 2/3$ and now we have the formula for p[s]:

```
In[2]:=    p[s_] = (2/3) (s + 3)

Out[2]=    2 (3 + s)
           ---------
               3
```

Because

$$w[L] = w[L] - w[0] = \int_0^L w'[s]ds = \int_0^L p[s]ds,$$

the weight of the rod in pounds is:

```
In[3]:=    Integrate[p[s],{s,0,4}]

Out[3]=    40
           --
            3
```

The weight of the left half of the rod in pounds is:

```
In[4]:=    Integrate[p[s],{s,0,2}]

Out[4]=    16
           --
            3
```

The weight of the right half of the rod in pounds is:

```
In[5]:=    Integrate[p[s],{s,2,4}]

Out[5]=    8
```

Check:

```
In[6]:=    40/3 == 16/3 + 8

Out[6]=    True
```

The fundamental theorem at your service.

■ **B.3) Arc Length**

In this basic problem we are going to learn how to measure the length of curves.

Here is a plot of a function $y = f[x]$ for $a \leq x \leq b$.

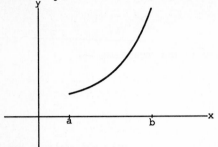

We want to measure the total length of this curve. To fix the notation, let $S[x]$ be the length of that part of the curve that is over the segment $[a, x]$ and note that total arc length is $S[b]$.

**B.3.a.i)**

---

Why is $S'[x] = \sqrt{1 + f'[x]^2}$?

---

**Answer:**

To find $S'[x]$, fix $x_0$ in $[a, b]$ and let $h > 0$. Observe that $S[x_0 + h] - S[x_0]$ is the length of that part of the curve over the interval $[x_0, x_0 + h]$.

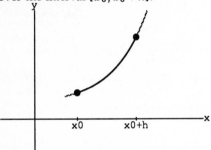

Look at the chord $AB$ joining $A = \{x_0, f[x_0]\}$ and $B = \{x_0 + h, f[x_0 + h]\}$.

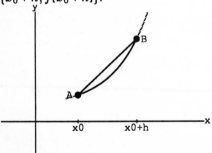

Also look at the tangent segment joining $AC$, where $C$ is the point on the line tangent to the curve at $A$ that has $x$-coordinate $x_0 + h$. The figure below illustrates the set-up:

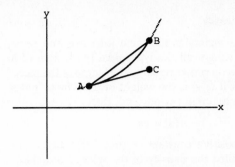

$AB$ is steeper than $AC$; so $AB$ is longer than $AC$. The straight segment $AB$ is shorter than the arc length $S[x_0 + h] - S[x_0]$; so

$$AC \leq AB \leq S[x0 + h] - S[x0].$$

Because the growth rate of $AC$ is $f'[x_0]$, the following picture tells us what we need to know about $AC$:

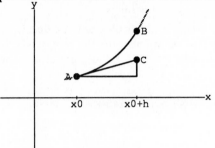

The picture tells us that

$$AC = \sqrt{h^2 + (h f'[x_0])^2} = h\sqrt{1 + f'[x_0]^2}.$$

And, because $AC \leq S[x_0 + h] - S[x_0]$, we can read off:

$$h\sqrt{1 + f'[x_0]^2} \leq S[x_0 + h] - S[x_0].$$

Divide both sides by $h$ to get

$$\sqrt{1 + f'[x_0]^2} \leq (S[x_0 + h] - S[x_0])/h.$$

Take the limit as $h \to 0$ to see that

$$\sqrt{1 + f'[x_0]^2} \leq S'[x_0].$$

We can also see that

$$\sqrt{1 + f'[x_0]^2} \geq S'[x_0]$$

as follows:

Look at the tangent segment $BD$, where $D$ is the point on the tangent line through $B$ having $x$-coordinate $x_0$:

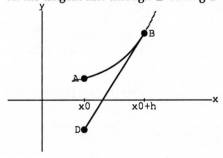

Because the line is always steeper than the arc, the picture tells us that

$$DB \geq S[x_0 + h] - S[x_0].$$

Let's calculate $DB$. Because the growth rate of $DB$ is $f'[x_0 + h]$, the following picture tells us what we need to know about $DB$:

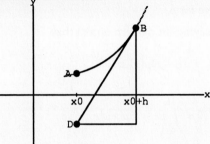

The picture tells us that

$$DB = \sqrt{h^2 + (hf'[x_0 + h])^2} = h\sqrt{1 + f'[x_0 + h]^2}.$$

And, because $DB \geq S[x_0 + h] - S[x_0]$, we can read off:

$$h\sqrt{1 + f'[x_0 + h]^2} \geq S[x_0 + h] - S[x_0].$$

Divide both sides by $h$ to get

$$\sqrt{1 + f'[x_0 + h]^2} \geq (S[x_0 + h] - S[x_0])/h.$$

Take the limit as $h \to 0$ to see that

$$\sqrt{1 + f'[x_0]^2} \geq S'[x_0].$$

The other inequality above said

$$\sqrt{1 + f'[x_0]^2} \leq S'[x_0].$$

Both inequalities together tell us that

$$S'[x_0] = \sqrt{1 + f'[x_0]^2}.$$

Similar analyses establish this formula for other curve shapes.

### B.3.a.ii)

Why is the total arc length of the curve $y = f[x]$ over the interval $[a, b]$ on the $x$-axis given by the following formula?

$$\int_a^b \sqrt{1 + f'[x]^2}\,dx$$

**Answer:**

This is just the fundamental theorem of calculus at work.

Recall that the total arc length $= S[b] = S[b] - S[a]$ (because $S[a] = 0$). So (by the fundamental theorem):

$$S[b] = \int_a^b S'[x]\,dx = \int_a^b \sqrt{1 + f'[x]^2}\,dx$$

because

$$S'[x] = \sqrt{1 + f'[x]^2}.$$

# Tutorial

■ **T.1) Archimedes's method for finding the center of gravity.**

### T.1.a)

Here is a 10 feet long see-saw set on a fulcrum. The see-saw board, when horizontal, is over the interval $[2, 12]$ on the $x$-axis.

Weights of 40 pounds and 60 pounds are hung on each end as indicated:

```
In[1]:=   seesaw = Graphics[
          {Line[{{2,4},{12,4}}]}]
          lefthanger = Graphics[
          {Line[{{2,4},{2,3.5}}]}]
          righthanger = Graphics[
          {Line[{{12,4},{12,3.5}}]}]
          leftweight = Graphics[
          {GrayLevel[0.85],Disk[{2,2.7},.8],
          GrayLevel[0],
          Thickness[.005],
          Circle[{2,2.7},.8]}]
          rightweight = Graphics[
          {GrayLevel[0.85],Disk[{12,2.5},1],
          GrayLevel[0],
          Thickness[.005],Circle[{12,2.5},1]}]
          left40 = Graphics[Text["40",{2,2.7}]]
          right60 = Graphics[Text["60",{12,2.5}]]
          fulcrum = Graphics[
          Line[{{6,0},{8,4},{10,0}}]]
          fulcrumcenter = Graphics[
          Line[{{8,-1.5},{8,4}}]]
          ground = Graphics[
          {Thickness[.01],Line[{{0,0},{14,0}}]}]
          cline = Graphics[Line[{{0,-1},{8,-1}}]]
          clabel = Graphics[
          Text["c",{4,-1},{0,1.5}]]
          Show[seesaw,lefthanger,righthanger,
          leftweight,rightweight,fulcrum,
          fulcrumcenter,left40,right60,
          ground,cline,clabel,
          PlotRange->{-2,5},
          Axes->{0,0},
          Ticks->{{2,12},None},
          AspectRatio->Automatic];
```

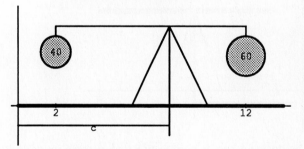

How should the indicated distance $c$ from the $y$-axis to the center of the fulcrum be set so that the see-saw is in perfect balance?

For the purposes of this problem, neglect the weight of the see-saw board.

**Answer:**

We want

$$60(12 - c) = 40(c - 2)$$

which is the same as

$$40(2 - c) + 60(12 - c) = 0.$$

```
In[2]:=    Clear[c]
           Solve[40 (2 - c) + 60 (12 - c) == 0]

Out[2]=    {{c -> 8}}
```

So we put $c = 8$. Not much to it.

### T.1.b.i)

---

Here is a rod set on a fulcrum. The rod, when horizontal, is over the interval $[a, b]$ on the $x$-axis. When the rod is held to be horizontal, the density of the rod at any point $x$ in $[a, b]$ is a given function $p[x]$ measuring pounds per unit on the x-axis.

```
In[1]:=    rod = Graphics[
           {Thickness[.02],Line[{{2,4},{15,4}}]}]
           leftlabel = Graphics[
           Text["a",{2,0},{0,+2}]]
           rightlabel = Graphics[
           Text["b",{15,0},{0,+2}]]
           atick = Graphics[
           Line[{{2,-.1},{2,.5}}]]
           btick = Graphics[
           Line[{{15,-.1},{15,.5}}]]
           fulcrum = Graphics[
           Line[{{6,0},{8,4},{10,0}}]]
           fulcrumcenter = Graphics[
           Line[{{8,-1.5},{8,4}}]]
           cline = Graphics[
           Line[{{0,-.6},{8,-.6}}]]
           clabel = Graphics[
           Text["c",{4,-1},{0,1}]]
           Show[rod,leftlabel,rightlabel,
           atick,btick,fulcrum,
           fulcrumcenter,cline,clabel,
           PlotRange->{{0,16},{-2,5}},
           PlotLabel->"Horizontal Rod",
           Axes->{0,0},Ticks->None];
```

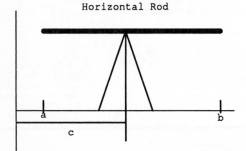

Horizontal Rod

The question here is how to use $p[x]$ to come up with a point $c$ such that if we place a fulcrum at the point $x = c$, then the rod will be in perfect balance. The point $c$ is called the **center of gravity** of the rod.

---

**Answer:**

If the density $p[x]$ is constant, then there is no problem in finding $c$ because, in this case, we just put $c$ in the middle ($c = (a + b)/2$) and the rod will be in perfect balance.

But if $p[x]$ is not constant, then this simple answer is unlikely to be correct. But we can take a clue from the last

part: there we set $c$ by solving $40(2 - c) + 60(12 - c) = 0$ when the weights were placed at the end of the see-saw.

But in the current problem, the weight is distributed throughout the rod. Archimedes said that in this case we should think of the density $p[x]$ as proportional to the weight at $x$. Then we should sum up the numbers $(x - c)p[x]$ for $x$ running from $a$ to $b$, set the result equal to 0 and then solve for $c$.

This amounts to saying (in modern terms) that

$$\int_a^b (x - c)p[x]dx = 0.$$

This means

$$0 = \int_a^b (x - c)p[x]dx$$

$$= \int_a^b xp[x]dx - \int_a^b cp[x]dx$$

$$= \int_a^b xp[x]dx - c\int_a^b p[x]dx.$$

Thus

$$\int_a^b xp[x]dx = c\int_a^b p[x]dx.$$

And this tells us that

$$c = \int_a^b xp[x]dx \Big/ \int_a^b p[x]dx.$$

Archimedes was a brilliant fellow.

### T.1.b.ii)

---

A rod is lying on the $x$-axis. Its left hand endpoint is at $x = -2$ and its right endpoint is at $x = 6$. Its density $p[x]$ is given by

$$p[x] = 2 + \sin[\pi x].$$

Where is its center of gravity?

---

**Answer:**

```
In[1]:=    Clear[x,p]
           p[x_] = 2 + Sin[Pi x]

Out[1]=    2 + Sin[Pi x]
```

The center of gravity is given by:

```
In[2]:=    c = Integrate[x p[x],{x,-2,6}] /
               Integrate[p[x],{x,-2,6}]

Out[2]=            8
            32 -  --
                  Pi
           -------
              16
```

Or in decimals:

```
In[3]:=    N[c]
Out[3]=    1.84085
```

The center of gravity is just slightly to the left of the middle
of the rod which is located at

$$x = (-2 + 6)/2 = 2.$$

# Literacy Sheet.

1.a Given that $f'[x] = \cos[x]$ and $f[0] = 4$, then give a formula for $f[x]$.

.

.

.

1.b Given that $f'[x] = \cos[x]$ and $f[\pi/4] = -1/2$, then give a formula for $f[x]$.

.

.

.

1.c Given that $f'[x] = e^x$ and $f[1] = 3$, then give a formula for $f[x]$.

.

.

.

2.a How are distance, velocity and acceleration related?

.

.

2.b How does the fundamental theorem of calculus help you to recover velocity from acceleration and distance from velocity?

.

.

3. How does the fundamental theorem of calculus help you to recover weight from density?

.

.

4. If $S[x]$ is the length of the curve $y = f[x]$ over the interval $[a, x]$, then what is $S[a]$?

How does the fundamental theorem of calculus help you to measure $S[b]$ if you have a formula for $f'[x]$ for $a \le x \le b$?

.

.

.

5. Exactly six years ago, you owed $1000. Since that time the amount you owe continuously increased at a a rate of $500e^{t/2}$ dollars per year. How much do you owe now?

.

6. Here is a 10 foot long see-saw set on a fulcrum. The see-saw board, when horizontal, is over the interval $[2, 12]$ on the x-axis.

Weights of 80 pounds and 100 pounds are hung on each end as indicated:

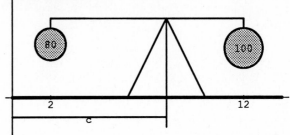

How should the indicated distance $c$ from the $y$-axis to the center of the fulcrum be set so that the see-saw is in perfect balance? For the purposes of this problem, neglect the weight of the see-saw board.

.

.

# Measurements and models through separation of variables and integrating

## Guide

At this point we have seen that the fundamental theorem is a superb tool for making exact measurements. In this lesson, we are going to learn how to widen the class of measurements we can make. In addition, we are going to see how the fundamental theorem of calculus can be used to design mathematical models of some interesting processes from the real world.

## Basics

### ▦ B.1) Separating and integrating.

The fundamental theorem of calculus is also the right tool for solving many differential equations.

### B.2.a)

Solve for $y[x]$ given that $y'[x] = x\sqrt{y[x] + 8}$ and $y[3] = 9$.

**Answer:**

Rewrite $y'[x] = x\sqrt{y[x] + 8}$ by writing

$$y'[x]/\sqrt{y[x] + 8} = x.$$

Some folks call this technique of putting all the $y[x]$ terms on one side and the x terms on the other side by the name

**"separation of variables."**

Integrate both sides from 3 to $x$ to get

$$\int_3^x y'[t]/\sqrt{y[t] + 8}\,dt = \int_3^x t\,dt,$$

using the known fact that $y[3] = 9$:

```
In[1]:=    Clear[x,y,t]
           left =
           Integrate[y'[t]/Sqrt[y[t]+8],{t,3,x}]/.
           y[3]->9
Out[1]=    -2 Sqrt[17] + 2 Sqrt[8 + y[x]]
```

```
In[2]:=    right = Integrate[t,{t,3,x}]
Out[2]=          2
              9    x
           -(-) + --
              2    2
```

To get an explicit formula for $y[x]$, just solve:

```
In[3]:=    Solve[left == right,y[x]]
Out[3]=    {{y[x] ->

                                      2
                 (225 - 72 Sqrt[17] - 18 x  +

                          2    4
                 8 Sqrt[17] x  + x ) / 16}}
```

Not bad, eh?

### B.2.b)

Solve for $y[x]$ given that $y'[x] = \sin[x]^2 y[x]$ and $y[\pi] = 2$.

**Answer:**

Separate the variables $x$ and $y[x]$ by writing $y'[x] = \sin[x]^2 y[x]$ as

$$y'[x]/y[x] = \sin[x]^2.$$

Integrate both sides from $\pi$ to $x$ to get

$$\int_\pi^x y'[t]/y[t]\,dt = \int_\pi^x \sin[t]^2\,dt,$$

remembering that $y[\pi] = 2$.

```
In[1]:=    Clear[x,y,t]
           left =
           Integrate[(1/y[t]) y'[t],{t,Pi,x}]/.
           y[Pi]->2
Out[1]=    -Log[2] + Log[y[x]]
```

```
In[2]:=    right = Integrate[Sin[t]^2,{t,Pi,x}]
Out[2]=    -Pi   x   Cos[x] Sin[x]
           --- + - - -------------
            2    2        2
```

To get an explicit formula for $y[x]$, just solve:

```
In[3]:=    Solve[left == right,y[x]]
Out[3]=    {{y[x] ->

               Power[E,

               -Pi + x + 2 Log[2] - Cos[x] Sin[x]
               ----------------------------------
                              2

               ]}}
```

Slick.

# Tutorial

■ **T.1) Solving another differential equation by separating and integrating.**

You might remember a differential equation of the following type from the section on $y' = ay$. All the problems from that section can be solved by separating and integrating.

### T.1.a)

---

For given positive constants $a, b$ and $c$, solve $y'[x] = c - ay[x]$ given that $y[0] = b$. What is the limiting value $\lim_{x \to \infty} y[x]$?

---

**Answer:**

```
In[1]:=   Clear[y,t,a,b,c,x]
          left = Integrate[y'[t]/(c - a y[t]),
          {t,0,x}]/.y[0]->b
```

```
Out[1]=   Log[-(a b) + c]     Log[c - a y[x]]
          ---------------  -  ----------------
                 a                   a
```

```
In[2]:=   right = Integrate[1,{t,0,x}]
```

```
Out[2]=   x
```

```
In[3]:=   Solve[left == right,y[x]]
```

```
Solve::tdep:

The equations appear to involve

transcendental functions of the

variables in an essentially

nonalgebraic way.
```

```
Out[3]=              Log[-(a b) + c]
          Solve[--------------- -
                       a

          Log[c - a y[x]]
          --------------- == x, y[x]]
                 a
```

We step in to help *Mathematica*.

On the left we have $\log[(c - ab)/(c - ay[x])]$ . On the right we have $x$ . So

$$\log[(c - ab)/(c - ay[x])] = x.$$

Exponentiating both sides gives

$$(c - ab)/(c - ay[x]) = e^x.$$

To solve for $y[x]$, use:

```
In[4]:=   Solve[(c - a b)/(c - a y[x]) == E^x,y[x]]
```

```
Out[4]=                           x
                   -(a b) + c - E  c
          {{y[x] -> -(------------------)}}
                             x
                            E  a
```

The limiting case of $y[x]$ as $x \to \infty$ is the quotient of the dominant terms

$$-(-e^x c)/e^x a = c/a.$$

Thus

$$\lim_{x \to \infty} y[x] = c/a.$$

There is another way to see this:

```
In[5]:=   ExpandAll[Solve[
          (c-a b)/(c-a y[x])==E^x,y[x]]]
```

```
Out[5]=                b    c     c
          {{y[x] ->   -- + - - ----}}
                       x    a    x
                      E          E  a
```

In this form you can see that as $x \to \infty$, then $y[x] \to c/a$.

■ **T.2) Iceballs.**

### T.2.a)

---

An iceball whose initial radius is 6 inches melts so that its radius decreases at a rate proportional to the area of its surface. If it takes 1/2 hour for the radius to decrease from 6 inches to 5 inches, how long will it take for the radius to decrease to 1 inch?

---

**Answer:**

All $t$ measurements are in hours. Surface area of a ball of radius $r$ is proportional to $r^2$ ; so if $r[t]$ is the radius of the iceball at time $t$, then $r'[t]$ is proportional to $r[t]^2$ . This means

$$r'[t] = Kr[t]^2$$

for some (yet to be determined) constant of proportionality $K$ .

Separate the variables $t$ and $r[t]$ by rewriting as

$$r[t]^{-2}r'[t] = K.$$

Now integrate

$$\int_0^t r[t]^{-2}r'[t]dt = \int_0^t Kdt,$$

remembering that $r[0] = 6$ :

```
In[1]:=   Clear[K,r,t]
          left =
          Integrate[r[t]^(-2) r'[t],
          {t,0,t}]/.r[0]->6
```

```
Out[1]=   1    1
          - - ----
          6   r[t]
```

```
In[2]:=   right = Integrate[ K,{t,0,t}]
```

```
Out[2]=   K t
```

To get an explicit formula for $r[t]$, just solve:

```
In[3]:=    Solve[left == right, r[t]]
```
```
Out[3]=                6
           {{r[t] -> ---------}}
                      1 - 6 K t
```

To determine $K$, recall that $r[1/2] = 5$.

```
In[4]:=    Solve[ 5 == 6/(1 - 6 K t)/.t->1/2]
```
```
Out[4]=           1
           {{K -> -(--)}}
                  15
```

This gives
$$r[t] = 6/(1 + 6t/15).$$

To find how long it takes for the radius to shrink to 1 inch, we solve:

```
In[5]:=    Solve[ 1 == 6/(1 + (6 t)/15)]
```
```
Out[5]=          25
           {{t -> --}}
                  2
```

12.5hours to wait.

■ **T.3) Making the differential equation to suit the situation: The logistic equation.**

You may recall that we made a partially successful run at the following problem earlier in the course in the section on $y' = ay$.

**T.3.a)**

---

The extension agent tells you that your pond is capable of supporting 500 perch. You stock the lake with 72 perch and after three months the extension agent comes over and estimates that there are 200 perch in the pond. Find a function that gives a reasonable estimate of the number of perch in the pond $t$ months after the original stocking. Plot.

---

**Answer:**

Let $P[t]$ be the number of perch in the pond $t$ months after stocking it. When $P[t]$ is well beneath the limiting size of 500, it is reasonable to assume that $P[t]$ grows in direct proportion to its size. Thus for $P[t]$ well below 500,
$$P'[t] = aP[t]$$
is a plausible assumption, but it can't be taken too seriously because its solution $P[t] = 72e^{at}$ allows for an unlimited population explosion. Back to the drawing board:

We want a model in which the growth rate $P'[t]$ is extremely small when $P[t]$ is close to 500 and in which the growth rate is rather generous when $P[t]$ is well below 500.

Here is one such:
$$P'[t] = a(500 - P[t]).$$

This forces the growth rate $P'[t]$ to tend to 0 as $P[t]$ sneaks up on 500. To try out this model, we'll use separation of variables.

Rewrite $P'[t] = a(500 - P[t])$ as
$$P'[t]/(500 - P[t]) = a$$
and integrate remembering that $P[0] = 72$:

```
In[1]:=    Clear[x,P,t]
           left = Integrate[
           P'[x]/(500 - P[x]),{x,0,t}]/.P[0]->72
```
```
Out[1]=    Log[428] - Log[500 - P[t]]
```

```
In[2]:=    right = Integrate[a,{x,0,t}]
```
```
Out[2]=    a t
```

This means
$$\log[428/(500 - P[t])] = at$$

Now is a good time to find out what $a$ is; recall $P[3] = 200$.

```
In[3]:=    N[Solve[ (Log[428/(500-P[t])]==a t)
           /.{t->3,P[t]->200}]]
```
```
Out[3]=    {{a -> 0.118447}}
```

Now we know $a = 0.118447$.

Exponentiating both sides of
$$\log[428/(500 - P[t])] = at$$
gives
$$428/(500 - P[t]) = e^{(\log[428/(500-P[t])])} = e^{at}.$$

```
In[4]:=    Psolved =
           Solve[(428/(500 - P[t]) == E^(a t))
           /.a->0.118447,P[t]]
```
```
Out[4]=                            0.118447 t
                         428 - 500 E
           {{P[t] -> -(----------------------)}}
                            0.118447 t
                           E
```

So our first shot at P[t] is given by:

```
In[5]:=    P1[t_] = Expand[Psolved[[1,1,2]]]
```
```
Out[5]=            428
           500. - -----------
                  0.118447 t
                 E
```

Here is a plot of the projected fish population over 3 years. The pond capacity of 500 perch is plotted as the line $y = 500$:

```
In[6]:=    fishplot1 = Plot[{P1[t],500},{t,0,36},
           AxesLabel->{"months","perch"},
           PlotRange->All];
```

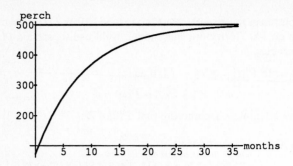

Let's include the original data.

```
In[7]:=    datapoints = ListPlot[{{0,72},{3,200}},
           PlotStyle->{RGBColor[1,0,0],PointSize[.04]},
DisplayFunction->Identity];

           Show[datapoints, fishplot1,
           AxesLabel->{"months","perch"},
           DisplayFunction->$DisplayFunction];
```

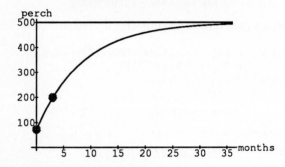

The plot looks pretty good.

But a moment of reflection is in order. We have a nice plausible answer to the original problem, but there is no guarantee that this answer is right on the target because another reasonable differential equation that might describe this set up is the **logistic equation:**

$$P'[t] = aP[t](500 - P[t]).$$

To try out the logistic model, we'll use separation of variables.

Rewrite $P'[t] = aP[t](500 - P[t])$ as

$$P'[t]/(P[t](500 - P[t])) = a$$

and integrate remembering that $P[0] = 72$ :

```
In[8]:=    Clear[x,P,t]
           left =
           Integrate[P'[x]/(P[x](500 - P[x])),
           {x,0,t}]/.P[0]->72
```
```
Out[8]=    Log[-428]   Log[72]
           --------- - ------- -
              500         500

           Log[-500 + P[t]]   Log[P[t]]
           ---------------- + ---------
                 500             500
```

```
In[9]:=    right = Integrate[a,{x,0,t}]
```

```
Out[9]=    a t
```

This means

$$\log[(428/72)\,P[t]/(500 - P[t])] = at.$$

(When you multiply 428 times $(-500 + P[t])$, then you get $428(500 - P[t])$.)

Now is a good time to find out what $a$ is; recall $P[3] = 200$.

```
In[10]:=   N[Solve[
           (Log[(428/72) (P[t]/(500 - P[t]))] == a t)
           /.{t->3,P[t]->200}]]
```
```
Out[10]=   {{a -> 0.458997}}
```

Now we know $a = 0.458997$. Exponentiating both sides of

$$\log[(428/72)\,P[t]/(500 - P[t])] = at.$$

gives

$$(428/72)(P[t]/(500 - P[t]) = e^{at}.$$

```
In[11]:=   P2solved = Solve[
           ((428/72) (P[t]/(500-P[t])) == E^(a t))
           /.a->0.458977,P[t]]
```
```
Out[11]=                       0.458977 t
                       9000 E
           {{P[t] -> --------------------}}
                                0.458977 t
                       107 + 18 E
```

So our second shot at $P[t]$ is given by:

```
In[12]:=   P2[t_] = Expand[P2solved[[1,1,2]]]
```
```
Out[12]=            0.458977 t
             9000 E
           --------------------
                      0.458977 t
           107 + 18 E
```

Here is a plot of the projected fish population over 3 years. The pond capacity of 500 perch is plotted as the line $y = 500$:

```
In[13]:=   fishplot2 = Plot[{P2[t],500},{t,0,36},
           PlotStyle->RGBColor[0,1,0],
           AxesLabel->{"months","perch"},
           PlotRange->All];
```

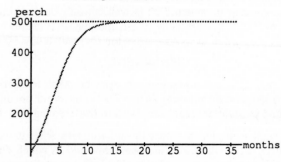

Let's include the original data.

```
In[14]:=   datapoints = ListPlot[{{0,72},{3,200}},
           PlotStyle->{RGBColor[1,0,0],PointSize[.04]},
DisplayFunction->Identity];

           Show[datapoints,fishplot2,
           AxesLabel->{"months","perch"},
           PlotRange->All,
           DisplayFunction->$DisplayFunction];
```

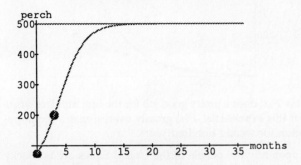

The logistic sloution predicts fast growth to 500 fish. Now we have the problem of over abundance. We have two possibilities:

We can go with $P1[t]$:

```
In[15]:=   P1[t]
```
$$Out[15]= \quad 500. - \frac{428}{E^{0.118447\,t}}$$

Or we can go with $P2[t]$:

```
In[16]:=   P2[t]
```
$$Out[16]= \quad \frac{9000\,E^{0.458977\,t}}{107 + 18\,E^{0.458977\,t}}$$

Look at both plots together:

```
In[17]:=   Show[datapoints,fishplot1,fishplot2,
           AxesLabel->{"months","perch"},
           PlotRange->All,
           DisplayFunction->$DisplayFunction];
```

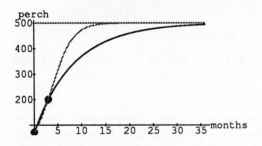

This is a helluva situation. And it's quite typical in model building.

There is nothing more we can do with the data unless we come up with another differential equation. What we might do is to make another measurement of the number

of fish and see which curve it lies on.

But what if it lies on neither? Well, then back to the drawing board.

Maybe a better answer is to call the right person in the State Fisheries Department, tell her or him what we did and ask which curve is the more typical of the two.

Give them a call.

### ■ T.4) Population growth and the logistic model.

Here are The United States population figures $\{t, P[t]\}$ where $t$ is the number of years since 1790 and $P[t]$ is the official United States Census figure in millions for year $t + 1790$. (Source: Statistical Abstracts.)

```
In[18]:=   USPopData={{0,4},{10,5},{20,7},{30,10},
           {40,13},{50,17},{60,23},{70,31},{80,40},
           {90,50},{100,63},{110,76},{120,92},
           {130,106},{140,123},{150,132},{160,151},
           {170,178},{180,203},{190,227}}
```
```
Out[18]=   {{0, 4}, {10, 5}, {20, 7}, {30, 10},

           {40, 13}, {50, 17}, {60, 23},

           {70, 31}, {80, 40}, {90, 50},

           {100, 63}, {110, 76}, {120, 92},

           {130, 106}, {140, 123}, {150, 132},

           {160, 151}, {170, 178}, {180, 203},

           {190, 227}}
```

and a plot (horizontal axis=number of years since 1790; vertical axis=population in millions):

```
In[19]:=   ListPlot[USPopData,
           PlotLabel->"US Population"];
```

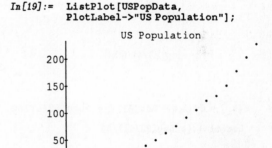

### T.4.a)

Populations lucky enough to enjoy unlimited nutrition, resources and land tend to increase at a rate proportional to the population itself. Thus in this utopian situation, we have the model

$$P'[t] = aP[t],$$

where $P[t]$ is the population of the United States (in millions) measured $t$ years after 1790.

Use the 1790 datum $\{0,4\}$ and the 1850 datum $\{60,23\}$ to find the solution of

$$P'[t] = aP[t]$$

whose plot passes through $\{0, 4\}$ and $\{60, 23\}$.

Plot $P[t]$ and the actual population data on the same axes and discuss the results.

---

**Answer:**

Separate the variables and integrate:

```
In[1]:=   Clear[P,t,x,a]
          left =
          Integrate[P'[x]/P[x],{x,0,t}]/.P[0]->4
Out[1]=   -Log[4] + Log[P[t]]
```

```
In[2]:=   right = Integrate[a,{x,0,t}]
Out[2]=   a t
```

This gives us our first shot at $P[t]$:

```
In[3]:=   firstPt = Solve[ left == right,P[t]]
Out[3]=                    a t + Log[4]
          {{P[t] -> E              }}
```

To nail down the value of $a$, plug in the 1850 datum $\{60, 23\}$ and solve for $a$: This leads to

$$23 = e^{60a + Log[4]}$$

which is the same as

$$\log[23] = 60a + \log[4];$$

$$\log[23/4] = 60a;$$

$$a = \log[23/4]/60.$$

Thus:

```
In[4]:=   P[t_] = E^(t a + Log[4])/.a->Log[23/4]/60
Out[4]=   Log[4] + (t Log[23/4])/60
          E
```

Next set up the plot of $P[t]$:

```
In[5]:=   prediction = Plot[P[t],{t,0,1990-1790},
          DisplayFunction->Identity];
```

And the plot of the data:

```
In[6]:=   datapoints = ListPlot[USPopData,
          PlotStyle->
          {PointSize[.02],RGBColor[1,0,0]},
          PlotLabel->"Actual vs Predicted",
          DisplayFunction->Identity];
```

Show them together:

```
In[7]:=   Show[datapoints,prediction,
          DisplayFunction->$DisplayFunction];
```

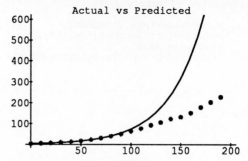

This $P[t]$ does a pretty good job for the first hundred years but this exponential $P[t]$ grossly overestimates the population for second hundred years.

Apparently the people of the United States are not lucky enough to enjoy unlimited nutrition, resources and land.

**T.4.b)**

---

How can we refine this model to try for more realistic predictions?

---

**Answer:**

We need to include something that incorporates the possible effect of overpopulation. One refinement was suggested by R. Pearl and L. J. Reed in their paper in the Proceedings of the National Academy of Sciences paper in 1920 (Proc. Nat. Acad. Sci. U.S.A. 6(1920), p.275). Their idea boils down to including an extra term in the model

$$P'[t] = aP[t]$$

to quantify the effect of overpopulation. Their idea is that the effect of overpopulation should be proportional to the number of random encounters between individuals within the population.

The total number of random encounters should be roughly proportional to the square of the population. Here is their model:

$$P'[t] = aP[t] - bP[t]^2 = P[t](a - bP[t])$$

where $a$ and $b$ are positive constants.

Nowadays many folks call this the logistic model. We already had a crack at the logistic model in T.3)

**T.4.c.i)**

---

Use the 1790 datum $\{0, 4\}$, the 1850 datum $\{60, 23\}$ and the 1910 datum $\{120, 92\}$ to find the solution of

$$P'[t] = aP[t] - bP[t]^2$$

whose plot passes through $\{0, 4\}$, $\{60, 23\}$ and $\{120, 92\}$.

Plot $P[t]$ and the actual population data on the same axes and discuss the results.

## Answer:

Separate the variables by writing

$$P'[t] = aP[t] - bP[t]^2$$

as

$$P'[t]/(aP[t] - bP[t]^2) = 1$$

and integrate both sides from 0 to $t$ remembering that $P[0] = 4$:

```
In[1]:=   Clear[a,b,c,x,t,P]
          left =
          Integrate[P'[x]/( a P[x] - b P[x]^2),
          {x,0,t}]/.P[0]->4
```

$$Out[1]= -\left(\frac{Log[4]}{a}\right) + \frac{Log[-a + 4 b]}{a} + \frac{Log[P[t]]}{a} - \frac{Log[-a + b P[t]]}{a}$$

Now make a hand simplification based on the identity

$$-\log[u] + \log[v] + \log[w] - \log[y] = \log[vw/(uy)]$$

to see that on the left, we have

$$(1/a)\log[P[t](4b - a)/(4bP[t] - 4a)].$$

```
In[2]:=   right = Integrate[1,{x,0,t}]
Out[2]=   t
```

As a result,

$$(1/a)\log[P[t](4b - a)/(4bP[t] - 4a)] = t.$$

Multiply both sides by $a$ to get

$$\log[P[t](4b - a)/(4bP[t] - 4a)] = at.$$

Exponentiating both sides gives

$$P[t](4b - a)/(4bP[t] - 4a) = e^{at}.$$

Divide the numerator and denominator of the left side by $a$ to get

$$P[t]4(b/a) - 1/4(b/a)P[t] - 4 = e^{at}.$$

This is the same as

$$P[t]4c - 1/4cP[t] - 4 = e^{at}$$

**where** $c = b/a$. To get a shot at a formula for $P[t]$, look at:

```
In[3]:=   Solve[P[t](4c-1)/(4 c P[t]-4)
          ==E^(a t),P[t]]
```

$$Out[3]= \left\{\left\{P[t] \rightarrow \frac{-4 E^{a t}}{-1 + 4 c - 4 E^{a t} c}\right\}\right\}$$

Divide top and bottom by $-4e^{at}$ to get for our first shot:

```
In[4]:=   firstP[t_] =
          1/((-1 + 4 c)/(-4 E^(a t)) + c)
```

$$Out[4]= \frac{1}{c - \dfrac{-1 + 4 c}{4 E^{a t}}}$$

Now use the data points $\{60, 23\}$ and $\{120, 92\}$ to determine $a$ and $c$:

```
In[5]:=   eqn1 = (firstP[60] == 23)
```

$$Out[5]= \frac{1}{c - \dfrac{-1 + 4 c}{4 E^{60 a}}} == 23$$

```
In[6]:=   eqn2 = (firstP[120]== 92)
```

$$Out[6]= \frac{1}{c - \dfrac{-1 + 4 c}{4 E^{120 a}}} == 92$$

These equations have many extraneous complex solutions introduced by the high powers 60 and 120 found in these equations. To lighten the load, temporarily replace $e^{60a}$ by $z$ and $e^{120a}$ by $z^2$.

```
In[7]:=   eqn11 = 1/(c - (-1 + 4 c)/(4 z)) == 23
```

$$Out[7]= \frac{1}{c - \dfrac{-1 + 4 c}{4 z}} == 23$$

```
In[8]:=   eqn22 = 1/(c - (-1 + 4 c)/(4 z^2)) == 92
```

$$Out[8]= \frac{1}{c - \dfrac{-1 + 4 c}{4 z^2}} == 92$$

Now solve for $z$ and $c$:

```
In[9]:=   Solve[{eqn11,eqn22},{z,c}]
```

$$Out[9]= \left\{\left\{z \rightarrow \frac{19}{3}, c \rightarrow \frac{7}{1472}\right\}, \left\{z \rightarrow 0, c \rightarrow \frac{1}{4}\right\}\right\}$$

The second solution is a red herring. Go with $z = 19/3$ and $c = 7/1472$.

Recall $e^{60a} = z$; so $e^{60a} = 19/3$. This means

$$a = \log[19/3]/60.$$

Now we've got $P[t]$ nailed down.

```
In[10]:=  P[t_] = firstP[t]/.
          {a->N[Log[19/3]/60],c->N[7/1472]}
```

$Out[10]=$

$$\cfrac{1}{0.00475543 + \cfrac{0.245245}{E^{0.0307638\,t}}}$$

Set up the plot of $P[t]$.

```
In[11]:=   prediction = Plot[P[t],{t,0,1990-1790},
           DisplayFunction->Identity];
```

The data:

```
In[12]:=   datapoints = ListPlot[USPopData,
           PlotStyle->
           {PointSize[.02],RGBColor[1,0,0]},
           PlotLabel->"Actual vs Predicted",
           DisplayFunction->Identity];
```

Show them all together:

```
In[13]:=   Show[datapoints,prediction,
           DisplayFunction->$DisplayFunction];
```

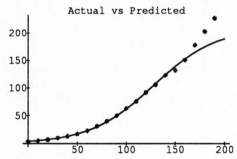

This $P[t]$ models the actual data very well for the years 1790 – 1950. For the years since 1950, this $P[t]$ is decidedly on the light side. But still the plot indicates that the logistic model

$$P'[t] = aP[t] - bP[t]^2$$

may be more than just a theoretical exercise.

### T.4.c.ii)

Plot the $P[t]$ you found above for $t$ corresponding to the years 1790 through 2290.

What is the limiting global scale population prediction

$$\lim_{t\to\infty} P[t]?$$

**Answer:**

```
In[1]:=   Plot[P[t],{t,0,500},
          PlotLabel->"Predicted"];
```

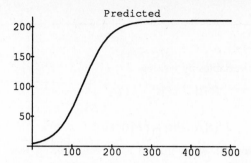

This model predicts a dramatic leveling off of population starting **right now!**

We know this model needs further refinement, but still there is an eerie realism about this plot that makes us wonder what the future will be.

We can spot the limiting value of $P[t]$ by looking at:

```
In[2]:=   P[t]
```

$Out[2]=$

$$\cfrac{1}{0.00475543 + \cfrac{0.245245}{E^{0.0307638\,t}}}$$

As $t$ gets very large the second term in the denominator goes to 0; so the limiting value of $P[t]$ (in millions) is:

```
In[3]:=   1/.00475543
```

$Out[3]=$    210.286

That's something to think about.

### T.4.d)

Where can we read more about this?

**Answer:**

The book *Calculus with Applications in Computing* by P. Lax, S. Bernstein and A. Lax (Springer, New York 1974 ) is an excellent starting point.

### ■ T.5) The predator-prey model.

*This mathematical model was orginally studied by Lotka and Volterra.*

Two species coexist in a closed environment. One species,. *the predator* , feeds on the other, *the prey* . There is always plenty of food for the prey, but the predators eat nothing but the prey.

Let $x[t]$ = population of prey at time $t$.

Let $y[t]$ = population of predators at time $t$.

It is reasonable to assume that there are positive constants $a$ and $b$ such that:

$$x'[t] = ax[t] - bx[t]y[t]$$

because the abundance of food for the prey allows the birth rate of the prey to be proportional to their current number and the death rate of prey is proportional to both the current number of prey and the current number of predators.

It is also makes some sense to assume that there are positive constants

$c$ and $d$ such that:

$$y'[t] = cx[t]y[t] - dy[t]$$

because it is reasonable to assume that the birth rate of the predators is proportional to both the current number of the predators and the size of the food supply (the prey) and the death rate of the predators is likely to be proportional to the current population of predators.

Dividing the second equation by the first equation gives

$$y'[t]/x'[t] =$$

$$(cx[t]y[t] - dy[t])/(ax[t] - bx[t]y[t]).$$

Supressing the time variable $t$ gives

$$dy/dx = (cxy - dy)/(ax - bxy)$$

$$= y(cx - d)/(x(a - by)).$$

Separate the variables $x$ and $y$ to get

$$(a - by)/y \, dy/dx = (cx - d)/x.$$

Integrate both sides from $x_0$ to $x$ where $x_0$ is held constant:

```
In[1]:=    left =
           Integrate[((a - b y[s])/y[s]) y'[s],
           {s,x0,x}]

Out[1]=    a Log[y[x]] - a Log[y[x0]] - b y[x] +

           b y[x0]


In[2]:=    right = Integrate[(c s - d )/s , {s,x0,x}]

Out[2]=    c x - c x0 - d Log[x] + d Log[x0]
```

This results in the relationship

$$a\log[y] - by+ = cx - d\log[x] + K,$$

where

$$K = -cx_0 + d\log[x_0] + a\log[y[x_0] - by[x_0]$$

is a constant.

A little algebra gives:

$$a\log[y] + d\log[x] - (by + cx) = K$$

$$\log[y^a] + \log[x^d] - (by + cx) = K$$

$$\log[y^a x^d] - (by + cx) = K$$

$$\log[y^a x^d] - by - cx = K.$$

Exponentiating both sides gives

$$y^a x^d e^{-by-cx} = e^K = constant.$$

Not even mighty *Mathematica* can solve this for $y$ in terms of $x$ but for given choices of $a, b, c$ and $d$, *Mathematica* can show how the graph of $y$ in terms of $x$ looks for various constants $K$:

```
In[3]:=    a = 4
           b = 3
           c = 1
           d = 3
           constant = y^a x^d E^(- b y - c x)
           graphs =
           ContourPlot[constant,{x,0,9},{y,0,5},
           PlotPoints->30,
           AxesLabel->{"prey","predators"},
           AspectRatio->Automatic];
```

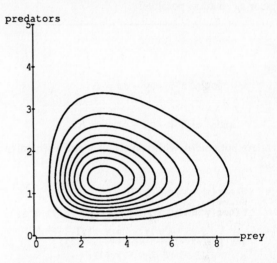

The graphs of $y$ ( predators) in terms of $x$ (prey) seem to be distorted ellipses.

**T.5.a.i)**

What values of $x$ (prey) are likely to make $y$ (predators) as large or as small as possible?

**Answer:**

It is useful to go back to the equation

$$\log[y^a x^d] - by - cx = K.$$

To calculate $y'[x]$, change this to

$$\log[y[x]^a x^d] - by[x] - cx = K;$$

then differentiate both sides with respect to x and finally solve for $y'[x]$:

```
In[1]:=    Clear[a,b,c,d,x,y,K]
           Solve[
           D[Log[y[x]^a x^d]-b y[x]-c x == K,x],
           y'[x]]
```

$$
Out[1]= \quad \{\{y'[x] \to -(\frac{d\ y[x]\ -\ c\ x\ y[x]}{a\ x\ -\ b\ x\ y[x]})\}\}
$$

From this we can spot that $y'[x] = 0$ provided $y[x] = 0$ (extinction of predator) or

$$d - cx = 0.$$

This means $y$ (predators) is likely to be as large or as small as possible when

$$x = d/c.$$

### T.5.a.ii)

What values of $y$ (predators) are likely to make $x$ (prey) as large or as small as possible?

**Answer:**

Change

$$\log[y^a x^d] - by - cx = K.$$

to

$$\log[y^a x[y]^d] - by - cx[y] = K,$$

differentiate both sides with respect to $y$ and then solve for $x'[y]$:

```
In[1]:=    Clear[a,b,c,d,x,y,K]
           Solve[
           D[Log[y^a x[y]^d]-b y - c x[y]==K,y],x'[y]]
```

$$
Out[1]= \quad \{\{x'[y] \to -(\frac{a\ x[y]\ -\ b\ y\ x[y]}{d\ y\ -\ c\ y\ x[y]})\}\}
$$

From this we can spot that $x'[y] = 0$ provided $x[y] = 0$ (extinction of the prey) or when

$$y = a/b.$$

This means $x$ (prey) is likely to be as large or as small as possible when

$$y = a/b.$$

### T.5.b)

Try this out for several choices of $a, b, c$ and $d$.

**Answer:**

Let's look at the example we did above except this time we'll put the axes right through the point $\{d/c, a/b\}$ because $x = d/c$ and $y = a/b$ are where we hope to find the highs and the lows:

```
In[1]:=    a = 4
           b = 3
           c = 1
           d = 3
           constant =  y^a x^d E^(- b y - c x)
           ContourPlot[constant,{x,0,9},{y,0,5},
           PlotPoints->30,
           AxesLabel->{"prey","predators"},
           Axes->{d/c,a/b},
           AspectRatio->Automatic];
```

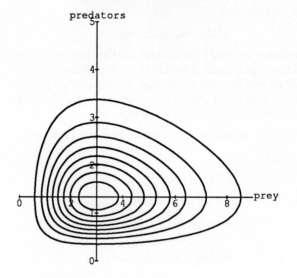

It worked!

The predators ($= y$) are as large and as small as possible when

$$x = d/c = 3.$$

The prey ($= x$) are as large and as small as possible when

$$y = a/b = 4/3.$$

Let's try again with new parameters $a, b, c$ and $d$:

```
In[2]:=    a = 5
           b = 4
           c = 2
           d = 4
           constant = E^(- b y - c x) y^a x^d
           ContourPlot[constant,{x,0,6},{y,0,3},
           PlotPoints->30,
           AxesLabel->{"prey","predators"},
           Axes->{d/c,a/b},
           AspectRatio->Automatic];
```

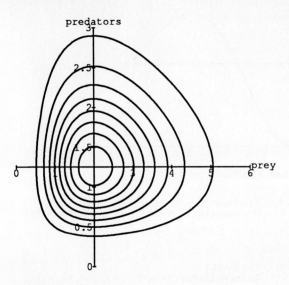

Great!

Again the axes $x = d/c$ and $y = a/b$ go right through the highs and the lows.

Lets look at another case:

```
In[3]:=    a = 6
           b = 2
           c = 1
           d = 3
           constant = E^(- b y - c x) y^a x^d
           graphs = ContourPlot[constant, {x, 0, 8}, {y, 0, 8},
           PlotPoints->30,
           AxesLabel->{"prey", "predators"},
           Axes->{d/c, a/b},
           AspectRatio->Automatic];
```

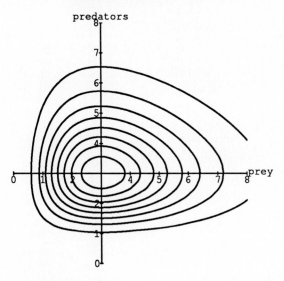

Yet again the axes $x = d/c$ and $y = a/b$ go right through the highs and the lows.

**T.5.c)**

___

How can we interpret the closed curves? What happens if for some time to we observe that $x[t_0]$ is close to $d/c$ and $y[t_0]$ is close to $a/b$?

**Answer:**

Take a look at a typical example:

```
In[1]:=    a = 5
           b = 4
           c = 2
           d = 4
           constant = E^(- b y - c x) y^a x^d
           plot = ContourPlot[constant, {x, 0, 6}, {y, 0, 3},
           PlotPoints->30,
           AxesLabel->{"prey", "predators"},
           Axes->{d/c, a/b},
           AspectRatio->Automatic];
```

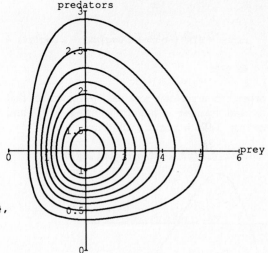

The closed curves suggest that both predator and prey populations are periodic; they go through repeated cycles. And the time cycle for each is the same. The four sectors above indicate four phases depicting trends that reverse themselves as the curves cross the axes at $x = d/c$ and $y = a/b$.

If for some time $t_0$ we observe that $x[t_0]$ is close to $d/c$ and $y[t_0]$ is close to $a/b$, then it appears that $x[t]$ stays near $d/c$ and $y[t]$ stays near $a/b$ at all other times $t$; so, in this case, the populations of both predators and prey remain fairly stable.

It also appears that if $x[t_0] = d/c$ and $y[t_0] = a/b$ for some time $t_0$, then both populations remain constant at all other times.

**T.5.d)**

___

Given $a = 3.9, b = 1.1, c = 0.7, d = 4.1$, and given that when $x = 8.2$, then $y = 2.5$, what are the largest and smallest that $x$ (prey) and $y$ (predators) can be?

___

**Answer:**

Recall from above:

$$e^{-by-cx}y^a x^d = e^K$$

where $K = -cx_0 + d\log[x_0] + a\log[y[x_0] - by[x_o]$.

Plug in the given data:

```
In[1]:=    a = 3.9
           b = 1.1
           c = 0.7
           d = 4.1
           x0 = 8.2
           y0 = 2.5;
```

```
In[2]:=    left = E^(-b y -c x) y^a x^d
```

```
Out[2]=     -0.7 x - 1.1 y  4.1  3.9
           E                x    y
```

```
In[3]:=    right = N[E^(-c x0 + d Log[x0] + a Log[y0] -
           b y0)]
```

```
Out[3]=    40.8736
```

To find when the largest and smallest $y$, remember that the largest and smallest $y$ happen when $x = d/c$ and remember that **left = right** at all times:

```
In[4]:=    Plot[{left/.x->d/c,right},{y,0,10}];
```

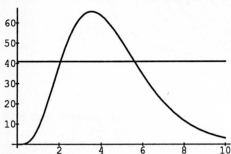

Now use **FindRoot** :

```
In[5]:=    FindRoot[
           Release[left/.x->d/c] == right,{y,2}]
```

```
Out[5]=    {y -> 2.07336}
```

```
In[6]:=    FindRoot[
           Release[left/.x->d/c] == right,{y,5.8}]
```

```
Out[6]=    {y -> 5.58938}
```

The number of predators oscillates between 2.07 and 5.59.

To find when the largest and smallest $x$, remember that the largest and smallest $y$ happen when $y = a/b$.

```
In[7]:=    Plot[{left/.y->a/b,right},{x,0,11}];
```

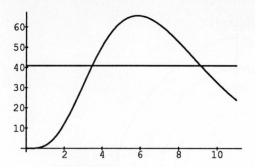

Now use **FindRoot** :

```
In[8]:=    FindRoot[
           Release[left/.y->a/b] == right,{x,3.5}]
```

```
Out[8]=    {x -> 3.47489}
```

```
In[9]:=    FindRoot[
           Release[left/.y->a/b] == right,{x,9}]
```

```
Out[9]=    {x -> 9.1381}
```

The number of prey oscillates between 3.47 and 9.14.

We've come a long way; it's time to break off.

# Literacy Sheet

**1.** Solve $y' = 3y$ for $y$ as a function of $x$;

given that when $x = 0, y = 8$.

**2.** Solve $y' = x^3/y$ for $y$ as a function of $x$ given that when $x = 0, y = 5$.

**3.** Why won't the method of separation of variables and intergating work to obtain a formula for $y[x]$ given

$$y'[x] = x^2 y[x] + 4\sin[x]$$

and given that when $x = 0, y = 3$.

**4.** Use separation of variables to derive the formula $y = ke^{ax}$ for a function $y = y[x]$ satisfying $y' = ay$ with $y[0] = k$.

**5.** Homer Lake has a bluegill population of 1000 right now. Let $P[t]$ be the number of bluegills in Homer Lake in $t$ months. Given that the birth and death rates of the bluegills are both proportional to the square root of the bluegill population, derive the formula

$$P[t] = (kt/2 + \sqrt{1000})^2$$

If in five months, there are $1600 = 40^2$ bluegills in Homer Lake, then how many bluegills will there be in 12 months?

**6.** Comment on the quotation: *The derivative does not display its full strength until it is allied with the integral.* Michael Spivak.

**7.** Write a short essay on how population (of humans, of combat troops, of infected persons, etc) can be modeled with the help of separating and integrating.

# Volume Measurements

## Guide

Even though the integral was orginally conjured up to measure area, you now understand that it can be used to measure a lot of other quantities.

It should be no surprise that the integral can also be used to measure volume and this is what we concentrate on in this lesson.

Along the way you will meet a new idea- that of the double integral

$$\int_a^d \int_c^d f[x,y]dxdy.$$

Here is a double integral in action:

```
In[1]:=    Clear[x,y]
           Integrate[Sin[x y] + 1,
           {x,-2 Pi,2 Pi},{y,-2 Pi,2 Pi}]
Out[1]=            2
           16 Pi
```

This is the volume in cubic units of the solid under the surface and over the floor in:

```
In[2]:=    Plot3D[Sin[x y]+1,
           {x,-Pi,Pi},{y,-Pi,Pi},
           PlotPoints->20,
           Lighting->True,Mesh->False];
```

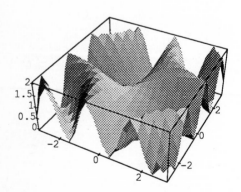

Intrigued?

## Basics

■ **B.1) Volumes by integration.**

Take a solid sitting on a floor. Paint an $x$-axis on the floor through it or next to it and fix numbers $a$ and $b$:

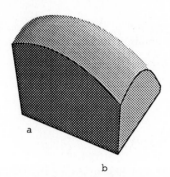

For each $x$ with $a \leq x \leq b$, let $A[x]$ be the area measurement of the cross section of the solid sliced off by a cutting plane perpendicular to the floor and perpendicular to the $x$-axis and through the point $\{x,0\}$ on the $x$-axis.

**B.1.a.i)**

What is the intuitive basis for saying that:

Volume of the solid = $\int_a^b A[x]dx$?

**Answer:**

When we are calculating the area under a curve $y = f[x]$ for $a \leq x \leq b$, we envision that $\int_a^b f[x]dx$ takes the lengths of the constiuent line segments from $\{x,0\}$ to $\{x,f[x]\}$ for $a \leq x \leq b$ and "sums "them up to arrive at the area under the curve.

By analogy, $\int_a^b A[x]dx$ should be the "summation" of the areas of the cross sections constituent to the solid and the "sum" of these areas should be the solid's volume in much the same way that a telephone book is the "sum" of its constituent pages.

This intuition does lead to the right formula. We can put this formula on a firm basis by setting $V[x]$ (for $a \leq x \leq b$) equal to the volume of the part of the solid indicated below:

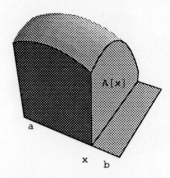

Note that $V[a] = 0$ and $V[b]$ measures the entire volume that we are computing. Thus:

Volume of solid $= V[b] = V[b] - V[a] = \int_a^b V'[x]dx$.

Consequently if we can see why $V'[x] = A[x]$, then we'll have the formula:

Volume of the solid $= \int_a^b A[x]dx$

on a firm basis.

### B.1.a.ii)

Explain why $V'[x] = A[x]$.

### Answer:

We'll give the full discussion under the simplifying assumption that $A[x]$ gets smaller as $x$ gets larger. For $h > 0$, $V[x + h] - V[x]$ is the volume of the segment of the solid between the level $x$ and $x + h$.

Because $A[x + h]$ is the area of the smallest cross section of the indicated segment, we see that

$$V[x + h] - V[x] \geq A[x + h]h$$

because $A[x + h]h$ is the volume of a cylindrical solid no larger than the indicated solid. Similarly because $A[x]$ is the area of the largest cross section of the indicated segment, we see that

$$A[x]h \geq V[x + h] - V[x].$$

Thus

$$A[x]h \geq V[x + h] - V[x] \geq A[x + h]h.$$

Dividing by $h$ gives

$$A[x] \geq (V[x + h] - V[x])/h \geq A[x + h].$$

Accordingly

$$(V[x + h] - V[x])/h = A[x]$$

within $A[x] - A[x + h]$. Taking limits gives

$$V'[x] = \lim_{h \to 0} (V[x + h] - V[x])/h = A[x]$$

within $\lim_{h \to 0}(A[x] - A[x + h]) = 0$ provided $A[x]$ does not jump. This explains why $V'[x] = A[x]$.

### B.1.b)

The base of a certain solid is a circle of radius 4. All cross sections perpendicular to a fixed diameter are squares. Find the volume of the solid.

### Answer:

Here is a rough sketch:

Let's take a look at the base remembering that the equation of the circle of radius 4 centered at $\{0,0\}$ is $x^2 + y^2 = 16$. The top half circle has the equation $y = \sqrt{16 - x^2}$. Put the origin in the center and put the $x$-axis through the fixed diameter described above.

```
In[1]:=    Clear[x,y]
           base =
           Graphics[Circle[{0,0},4]]
           Show[base,
           PlotLabel->"Top view",
           Axes->{0,0},
           AxesLabel->{"x","y"},
           AspectRatio->Automatic];
```

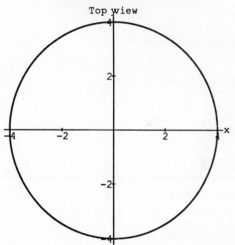

Put in a sample cross section perpendicular to the $x$-axis and to the screen.

```
In[2]:=    crosssection =
           Graphics[RGBColor[0,0,1],
           Line[{{0.8,-Sqrt[16 - (0.8)^2]},
           {0.8,Sqrt[16 - (0.8)^2]}}]]
           labels =

           Show[base,crosssection,
           Axes->{0,0},
           AxesLabel->{"x","y"},
           AspectRatio->Automatic];
```

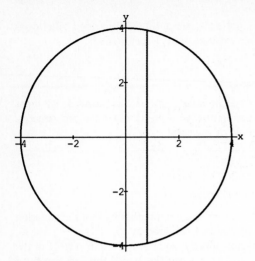

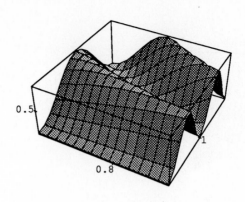

The segment running parallel to the $y$-axis is the base of the square cross section of the solid. So the area $A[x]$ of the cross section is:

```
In[3]:=    Clear[A]
           A[x_] = (y + y)^2/.y->Sqrt[4 - x^2]

Out[3]=           2
           4 (4 - x )
```

The volume is:

```
In[4]:=    Integrate[A[x],{x,-2,2}]

Out[4]=    128
           ---
            3
```

Not too hard.

### ■ B.2) Double integrals.

#### B.2.a.i)

---

Measure the volume under the surface $f[x,y] = \sin[\pi x y]^2$ and over the rectangle $1/2 \le x \le 1, 0 \le y \le 2$ in the $xy$-plane.

**Answer:**

---

Take a look:

```
In[1]:=    Clear[x,y,f]
           f[x_,y_] = Sin[Pi x y]^2

           Plot3D[f[x,y],{x,1/2,1},{y,0,2},
           Lighting->True,
           Ticks->{{.8},{1},{.5}}];
```

The solid whose volume we are trying to measure is under the curved surface on top and over the flat floor at the bottom of the box. Here the $x$-axis runs from 0.5 to 1 and the $y$-axis runs from 0 to 2.

The area $A[x]$ of the cross section of the solid sliced off by a cutting plane perpendicular to the $x$-axis and through the point $\{x,0\}$ on the $x$-axis is given by

$$A[x] = \int_0^2 f[x,y]dy.$$

You can see the tops of some of these cross sections in the plot.

```
In[2]:=    Clear[A,x,y]
           A[x_] = Integrate[f[x,y],{y,0,2}]

Out[2]=          Cos[2 Pi x] Sin[2 Pi x]
           1 - -----------------------
                      2 Pi x
```

And the volume under the surface and over the rectangle $1/2 \le x \le 1$ and $0 \le y \le 2$ in the $xy$-plane is (in cubic units) $\int_{1/2}^1 A[x]dx$ :

```
In[3]:=    Volume =
           Integrate[A[x],{x,1/2,1}]

Out[3]=        -I
               -- ExpIntegralEi[-4 I Pi]
           1   8
           - + ------------------------- +
           2            Pi


            I
            - ExpIntegralEi[-2 I Pi]
            8
           ------------------------ +
                     Pi


           -I
           -- ExpIntegralEi[2 I Pi]
            8
           ------------------------ +
                     Pi


            I
            - ExpIntegralEi[4 I Pi]
            8
           ------------------------
                     Pi
```

Or in decimal terms in cubic units:

```
In[4]:=    N[Volume]
Out[4]=    0.49411
```

We can do this in one step by calculating the **double integral**

$$\int_{1/2}^{1} \int_{0}^{2} f[x,y] dy dx = \int_{1/2}^{1} \int_{0}^{2} f[x,y] dy dx :$$

```
In[5]:=    check = Integrate[f[x,y],
           {x,1/2,1},{y,0,2}]

Out[5]=          -I
                 -- ExpIntegralEi[-4 I Pi]
            1    8
            - + ------------------------- +
            2             Pi

                 I
                 - ExpIntegralEi[-2 I Pi]
                 8
                 ------------------------ +
                          Pi

                 -I
                 -- ExpIntegralEi[2 I Pi]
                 8
                 ------------------------ +
                          Pi

                 I
                 - ExpIntegralEi[4 I Pi]
                 8
                 -----------------------
                          Pi
```

```
In[6]:=    N[check]
Out[6]=    0.49411
```

Not too hard, but we did see something new: **the double integral** .

**B.2.a.ii)**

What does the double integral

$$\int_{-2}^{2} \int_{0}^{3} x^2 + y^2 dy dx = \int_{-2}^{2} (\int_{0}^{3} (x^2 + y^2) dy) dx$$

measure?

**Answer:**

This is a measurement of the volume of a solid whose top runs along with the surface $f[x,y] = x^2 + y^2$. Take a look at this surface:

```
In[1]:=    Clear[x,y,f]
           f[x_,y_] = x^2 + y^2
           plot =
           Plot3D[f[x,y],
           {y,0,3},{x,-2,2},
           Lighting->True];
```

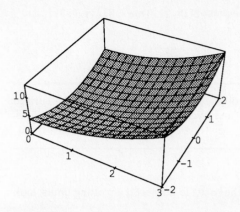

This is the surface $f[x,y] = x^2 + y^2$ plotted above the rectangle $-2 \le x \le 2$, $0 \le y \le 3$ in the $xy$ -plane. Think of the solid region below this surface and above the floor where the axes are.

For a fixed $x$ between -2 and 2, cut this solid with a plane perpendicular to the $x$ -axis (the $x$ -axis is the axis running from -2 to 2). Call the area of the resulting cross section $A[x]$. You can see the tops of some of these cross sections in the plot.

Now because integration measures area, we know that

$$A[x] = \int_{0}^{3} f[x,y] dy = \int_{0}^{3} (x^2 + y^2) dy$$

$$= x^2 y + y^3/3 \Big[_{y=0}^{y=3} = 3x^2 + 9.$$

As a result, the **volume** (in cubic units) of the solid region below this surface and above the floor where the axes are is given by

$$\int_{-2}^{2} A[x] dx = \int_{-2}^{2} (\int_{0}^{3} f[x,y] dy) dx$$

$$= \int_{-2}^{2} (\int_{0}^{3} (x^2 + y^2) dy) dx = \int_{-2}^{2} 3x^2 + 9 dx.$$

And this is given by:

```
In[2]:=    Integrate[3 x^2 + 9, {x,-2,2}]
Out[2]=    52
```

This cumulative answer the same as an the instant answer:

```
In[3]:=    Integrate[
           Integrate[f[x,y],{y,0,3}],
           {x,-2,2}]
Out[3]=    52
```

You can get the same result with one integration instruction corresponding to the original notation

$$\int_{-2}^{2} \int_{0}^{3} (x^2 + y^2) dy dx :$$

```
In[4]:=    Integrate[f[x,y],{x,-2,2},{y,0,3}]
Out[4]=    52
```

### B.2.b.i)

Measure the volume of the solid underneath the surface $f[x,y] = x^2 y^2$ and over the region bounded by the lines $y = 3x + 4$, $y = -x - 8$ and $x = 2$ in the $xy$-plane

**Answer:**

Take a look at the base: To help set the plotting limits, look at:

```
In[1]:=    Clear[x]
           Solve[3 x + 4 == -x - 8]
Out[1]=    {{x -> -3}}
```

```
In[2]:=    { (3 x+4)/.x->-3, (-x-8)/.x->-3}
Out[2]=    {-5, -5}
```

```
In[3]:=    { (3 x+4)/.x->2, (-x-8)/.x->2}
Out[3]=    {10, -10}
```

Here is a plot of the base in the $xy$-plane:

```
In[4]:=    lines =
           Plot[{-x-8 ,3 x + 4},{x,-3,2},
           DisplayFunction->Identity]

           right =
           Graphics[Line[{{2,-10},{2,10}}]]

           Show[lines,right,
           AxesLabel->{"x","y"},
           DisplayFunction->$DisplayFunction];
```

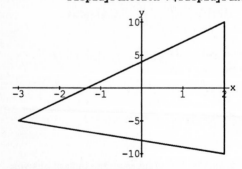

Here is a plot of the surface:

```
In[5]:=    Clear[x,y,f]
           f[x_,y_] = x^2 y^2

           Plot3D[f[x,y],{y,-10,10},{x,-3,2},
           PlotRange->All,Lighting->True];
```

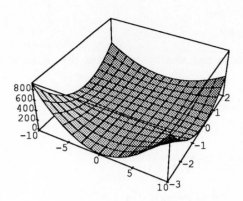

Slice the solid with a plane perpendicular to the $x$-axis through a point $\{x,0\}$ and let $A[x]$ be the area of the cross section. Here is the top view when we slice through $\{0.7,0\}$ :

```
In[6]:=    x0 = 0.7
           slice =
           Graphics[Thickness[.01],
           Line[{{x0,-x0 -8},{x0,3 x0 + 4}}]]

           Show[lines,right,slice,
           AxesLabel->{"x","y"},
           DisplayFunction->$DisplayFunction];
```

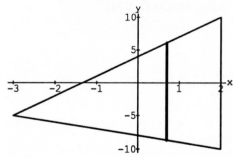

The area $A[x]$ of a cross section obtained by slicing through $\{x,0\}$ is given by

$$A[x] = \int_{-x-8}^{3x+4} f[x,y]\,dy$$

```
In[7]:=    Clear[A]
           A[x_] =
           Integrate[f[x,y],{y,-x-8,3 x + 4}]
Out[7]=          (-8 - x)^3 x^2     x^2 (4 + 3 x)^3
           -(---------------) + ---------------
                  3                   3
```

The slices start at $x = -3$ and stop at $x = 2$; so the volume in cubic units is given by:

```
In[8]:=    Integrate[A[x],{x,-3,2}]
Out[8]=    16250
           -----
             9
```

You can get the same result by evaluating the double integral

$$\int_{-3}^{2}\int_{-x-8}^{3x+4} f[x,y]\,dy\,dx$$

in one step:

```
In[9]:=    Integrate[f[x,y],
           {x,-3,2},{y,-x-8,3 x + 4}]

Out[9]=    16250
           -----
             9
```

Spiffy.

A common notation is to let $R$ stand for the region bounded by the lines $y = 3x$, $y = -x$ and $x = 2$ in the $xy$ -plane and to write

$$\int\int_{R} f[x,y]\,dy\,dx = \int_{0}^{2}\int_{-x}^{3x} f[x,y]\,dy\,dx.$$

### B.2.b.ii)

What does the double integral

$$\int_{0}^{2}\int_{x^2}^{2x} 8x^2 e^{-x^2-y}\,dy\,dx$$

measure?

**Answer:**

Evidently this is a measurement of the volume of a solid whose top runs along with the surface $f[x,y] = 8x^2 e^{-x^2-y}$. Peek at this surface:

```
In[1]:=    Clear[x,y,f]
           f[x_,y_] = 8 x^2 E^(-x^2 - y/3)
           Plot3D[f[x,y],{x,0,2},{y,0,4},
           Lighting->True,
           PlotPoints->25,PlotRange->All];
```

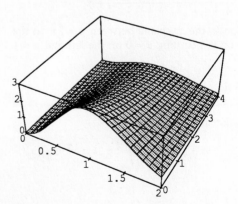

The $x$ -axis is the axis running from $0$ to $2$ and the $y$ -axis is the axis $0$ to $4$. Now look at the double integral $\int_{0}^{2}\int_{x^2}^{2x} f[x,y]\,dy\,dx$. The inner integral $\int_{x^2}^{2x} f[x,y]\,dy$ measures the area $A[x]$ of a cross section of this solid slices by a plane perpendicular the the the $x$ -axis through the point $\{x,0\}$. This cross section runs from $y = x^2$ to $y = 2x$.

Here is a top view of the cross section for $x = 1.3$ :

```
In[2]:=    x0 = 1.3
           curves =
           Plot[{x^2 ,2 x},{x,0,2},
           DisplayFunction->Identity]
           toplabel =
           Graphics[Text["y = 2 x",{1,2.5}]]
           bottomlabel =
           Graphics[Text["y = x^2",{1,1/2}]]
           slice =
           Graphics[Thickness[.01],
           Line[{{x0,x0^2},{x0,2 x0}}]]

           Show[curves,toplabel,
           bottomlabel,slice,
           AxesLabel->{"x","y"},
           DisplayFunction->$DisplayFunction];
```

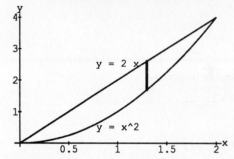

The limits of integration $0$ and $2$ in the outer integral $\int_{0}^{2}\int_{x^2}^{2x} f[x,y]\,dy\,dx$ tell us that the slices start at $x = 0$ and stop at $x = 2$.

The upshot is that the double integral

$$\int_{0}^{2}\int_{x^2}^{2x} 8x^2 e^{-x^2-y}\,dy\,dx$$

measures the volume under the surface

$$f[x,y] = 8x^2 e^{-x^2-y}$$

and over the region between the curves $y = x^2$ and $y = 2x$ in the $xy$ -plane.

Here is the measurement in cubic units:

```
In[3]:=    volume =Integrate[f[x,y],
           {x,0,2},{y,x^2,2 x}]

Out[3]=                      1/9              1
                         4 E     Sqrt[Pi] Erf[-]
              8                               3
          8 - ----- + -------------------- +
             16/3              3
            E


                        1/9              1
                       E     Sqrt[Pi] Erf[-]
             1                              3
          24 (-(-) + -------------------- ) -
             6                4


             1/9              7
          4 E     Sqrt[Pi] Erf[-]
                               3
          -------------------- -
```

```
                              3
                            1/9            7
                          E      Sqrt[Pi] Erf[-]
              -7                              3
        24 (-------  +  ------------------------) +
             16/3                   4
           6 E


              -3
        24 (-------  +
             16/3
           4 E


                                              4
              3 Sqrt[3] Sqrt[Pi] Erf[-------]
                                     Sqrt[3]
            -------------------------------------)
                          32
```

Or in decimals:

*In[4]:=*    N[volume]

*Out[4]=*    1.66577

**B.2.b.iii)**

What do the double integrals

$\int_1^2 \int_{x^2}^{2x} 8x^2 e^{-x^2-y} dy dx$

and

$\int_0^1 \int_{x^2}^{2x} 8x^2 e^{-x^2-y} dy dx$

measure?

**Answer:**

$$\int_1^2 \int_{x^2}^{2x} 8x^2 e^{-x^2-y} dy dx :$$

This double integral measures the volume of a solid under the surface $f[x,y] = 8x^2 e^{-x^2-y}$ and above the following region in the $xy$-plane:

*In[1]:=*    ```
            Clear[x]
            curveplot1 =
            Plot[{x^2,2 x},{x,1,2},
            DisplayFunction->Identity]

            leftvertical =
            Graphics[Line[{{1,x^2/.x->1},
            {1,2 x/.x->1}}]]

            Show[curveplot1,leftvertical,
            PlotRange->{{0,2},{0,4}},
            AxesLabel->{"x","y"},
            DisplayFunction->$DisplayFunction];
            ```

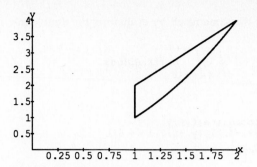

This is the region in the $xy$-plane between $y = 2x$ on the top and $y = x^2$ on the bottom, $x = 1$ on the left and $x = 2$ on the right.

$$\int_0^1 \int_{x^2}^{2x} 8x^2 e^{-x^2-y} dy dx :$$

This double integral measures the volume of a solid under the surface

$f[x,y] = 8x^2 e^{-x^2-y}$ and above the following region in the $xy$-plane:

*In[2]:=*    ```
            Clear[x]
            curveplot2 =
            Plot[{x^2,2 x},{x,0,1},
            DisplayFunction->Identity]

            rightvertical =
            Graphics[Line[{{1,x^2/.x->1},
            {1,2 x/.x->1}}]]

            Show[curveplot2,rightvertical,
            PlotRange->{{0,2},{0,4}},
            AxesLabel->{"x","y"},
            DisplayFunction->$DisplayFunction];
            ```

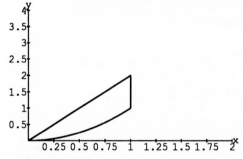

This is the region in the $xy$-plane between $y = 2x$ on the top and $y = x^2$ on the bottom, $x = 0$ on the left and $x = 1$ on the right.

**B.2.b.iv)**

What is the relationship between

$$\int_0^1 \int_{x^2}^{2x} 8x^2 e^{-x^2-y} dy dx + \int_1^2 \int_{x^2}^{2x} 8x^2 e^{-x^2-y} dy dx$$

and

$$\int_0^2 \int_{x^2}^{2x} 8x^2 e^{-x^2-y} dy dx?$$

**Answer:**

They are equal. If you don't believe this, then look at:

```
In[1]:=  N[Integrate[f[x,y],{x,0,1},
         {y,x^2,2 x}] +
         Integrate[f[x,y],{x,1,2},
         {y,x^2,2 x}]]
Out[1]=  1.66577
```

```
In[2]:=  N[Integrate[f[x,y],{x,0,2},
         {y,x^2,2 x}]]
Out[2]=  1.66577
```

# Tutorial

■ **T.1) Volume measurements.**

### T.1.a)

The curve $y = 3\sqrt{x}$ with $0 \le x \le 7$ (centimeters) is rotated about the $y$-axis and the resulting container is filled to the brim with our favorite cool beverage. How many fluid ounces of refreshing beverage does the container hold?

**Answer:**

Let's take a true scale look from the side.

```
In[1]:=  Clear[x,f]
         f[x_] = 3 Sqrt[x]
         sideview = ParametricPlot[
         {{x,f[x]},{-x,f[x]}},{x,0,7},
         PlotRange->All,
         AspectRatio->Automatic,
         AxesLabel->{"x","y"}];
```

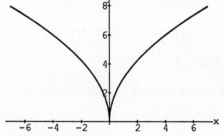

This is the outline of the container. Next embellish the plot by putting in a cross section perpendicular to the $y$-axis and to the screen.

```
In[2]:=  crosssection =
         Graphics[Line[{{-5,f[5]},
         {5,f[5]}}]],
         Text[x,{2.5,f[5]+.2}],
         Text[{"x","y"},{6,f[5]}]];
         Show[sideview,crosssection];
```

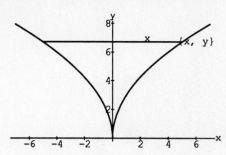

The segment running parallel to the $x$-axis is the diameter of a circular cross section of the container. The area of this cross section is $\pi x^2$.

```
In[3]:=  Clear[x,y]
         Solve[y == 3 Sqrt[x],x]
Out[3]=            2
                  y
         {{x -> --}}
                  9
```

The area $A[y]$ of the cross section is:

```
In[4]:=  Clear[A]
         A[y_] = Pi x^2/.x->(y^2)/9
Out[4]=        4
         Pi y
         -----
          81
```

The volume in cubic centimeters is:

```
In[5]:=  capacitycc =
         Integrate[A[y],{y,0,f[7]}]
Out[5]=    5/2
         3 7    Pi
         ---------
             5
```

Or in decimals:

```
In[6]:=  N[capacitycc]
Out[6]=  244.369
```

There are 1000 cubic centimeters in a liter and there are 33.814 fluid ounces in a liter. The capacity of the container in fluid ounces is:

```
In[7]:=  N[(capacitycc/1000) 33.814]
Out[7]=  8.2631
```

A little more than half a pint.

If you want to get an idea of how the container looks, look at the following plot and smooth off the rough edges:

```
In[8]:=  plot =
         Plot3D[f[Sqrt[x^2 + y^2]],
         {x,-7,7},{y,-7,7},
         PlotRange->{0,f[7]},
         Lighting->True,
         PlotPoints->25,
         ClipFill->None];
```

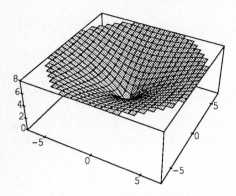

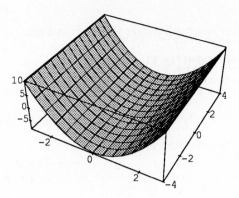

Here is a plot of the surface $f[x, y] = 4 - y^2$ :

```
In[1]:=   Clear[x,y,f]
          f[x_,y_] = 4 - y^2

          fplot =
          Plot3D[f[x,y],{x,-3,3},{y,-4,4},
          Lighting->True];
```

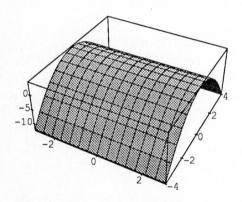

Here is a plot of the surface $g[x, y] = 2x^2 - 8$ :

```
In[2]:=   Clear[g]
          g[x_,y_] = 2 x^2 - 8

          gplot =
          Plot3D[g[x,y],{x,-3,3},{y,-4,4},
          Lighting->True];
```

What is the volume of the solid whose underneath is made the from the $g[x, y]$ surface and whose top is made the $f[x, y]$ surface?

**Answer:**

Let's go after a top view of the solid. The first step is to find the relationships between $x$ and $y$ for the $x$'s and $y$'s at which the two surfaces meet:

```
In[3]:=   ysolved =
          Solve[f[x,y] == g[x,y],y]

Out[3]=                          2
          {{y -> Sqrt[12 - 2 x ]},

                             2
           {y -> -Sqrt[12 - 2 x ]}}
```

```
In[4]:=   yupper[x_] = ysolved[[1,1,2]]

Out[4]=              2
          Sqrt[12 - 2 x ]
```

```
In[5]:=   ylower[x_] = ysolved[[2,1,2]]

Out[5]=               2
          -Sqrt[12 - 2 x ]
```

The next step is to plot **yupper** [$x$] and **ylower** [$x$].

To set the plot range, look at:

```
In[6]:=   Solve[yupper[x] == ylower[x],x]

Out[6]=   {{x -> Sqrt[6]}, {x -> -Sqrt[6]}}
```

Here comes a plot of the top view of the solid:

```
In[7]:=   topview =
          Plot[{yupper[x],ylower[x]},
          {x,-Sqrt[6],Sqrt[6]},
          AxesLabel->{"x","y"},
          DisplayFunction->Identity]

          Show[topview,
          AxesLabel->{"x","y"},
          AspectRatio->Automatic,
          DisplayFunction->$DisplayFunction];
```

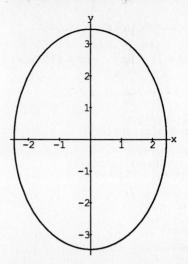

Slice the solid with a plane perpendicular to the $x$-axis through a point $\{x,0\}$ and let $A[x]$ be the area of the cross section.

Here is the top view when we slice through $\{0.7, 0\}$ :

```
In[8]:=   x0 = 0.7
          slice =
          Graphics[Thickness[.01],
          Line[{{x0,ylower[x0]},
          {x0,yupper[x0]}}]]

          Show[topview, slice,
          AxesLabel->{"x","y"},
          AspectRatio->Automatic,
          DisplayFunction->$DisplayFunction];
```

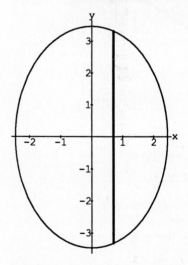

Recalling that the $f[x,y]$ surface provides the top of the solid and that the $g[x,y]$ surface provides the bottom of the solid , we know that area $A[x]$ of a cross section obtained by slicing through $\{x,0\}$ is given by

$$A[x] = \int_{ylower[x]}^{yupper[x]} (f[x,y] - g[x,y])dy.$$

```
In[9]:=   Clear[A]
          A[x_] =
          Integrate[f[x,y] - g[x,y],
```

```
          {y,ylower[x],yupper[x]}]
Out[9]=
          24 Sqrt[12 - 2 x ] -

             2           2
          4 x  Sqrt[12 - 2 x ] -

                    2 3/2
          2 (12 - 2 x )
          ----------------
                 3
```

The slices start at $x = -\sqrt{6}$ and stop at $x = \sqrt{6}$; so the volume in cubic units is given by:

```
In[10]:=  volume =
          Integrate[A[x],{x,-Sqrt[6],Sqrt[6]}]
Out[10]=           -(Sqrt[2] Sqrt[6])
          -72 ArcSin[------------------]
                        2 Sqrt[3]
          ----------------------------- +
                    Sqrt[2]

                     Sqrt[2] Sqrt[6]
           72 ArcSin[---------------]
                       2 Sqrt[3]
          --------------------------
                   Sqrt[2]
```

Or in decimals:

```
In[11]:=  N[volume]
Out[11]=  159.944
```

This problem was not as hard as it seemed at the start.

■ **T.2) What went wrong?**

The lab adventures of "Calculus Cal."

**T.2.a)**

A Calculus& *Mathematica* student named "Calculus Cal" is in a hurry and attempts to measure the volume under the surface $f[x,y] = (x - \pi)^2$ over the region between the curves

$$y = \sin[x]$$

and

$$y = -\sin[x]$$

for $0 \le x \le 2\pi$ in the $xy$-plane. He plots:

```
In[1]:=   Clear[x,y]
          Plot3D[(x - Pi)^2,
          {x,0,2 Pi},{y,-1,1},
          Lighting->True];
```

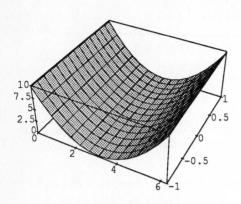

In his haste, he goes after this volume measurement in one step:

```
In[2]:=    Clear[x,y]
           Integrate[(x - Pi)^2,{x,0,2 Pi},
           {y,-Sin[x],Sin[x]}]

Out[2]=    0
```

Cal says, "(expletive deleted) " because he knows that the **volume cannot be 0** and he asks Accuracy Ann, who is at the next machine, "What went wrong." What did Ann tell him?

---

### Answer:

Ann tells Cal to do what he should have done in the first place; she tells him to plot the curves $y = \sin[x]$ and $y = -\sin[x]$ for $0 \le x \le 2\pi$ in the $xy$-plane: (The curve $y = \sin[x]$ is the thicker of the two curves.)

```
In[3]:=    Plot[{Sin[x],-Sin[x]},{x,0,2 Pi},
           AxesLabel->{"x","y"},
           PlotStyle->{{Thickness[0.01]},
           {RGBColor[1,0,0]}}];
```

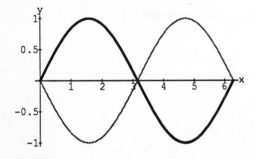

Then Ann says, "The trouble is that the curves cross each other."

The volume cannot be $\int_0^{2\pi} \int_{-Sin[x]}^{Sin[x]} (x - \pi)^2 dy dx$ because the $y = -\sin[x]$ curve **is not always below** the $y = \sin[x]$ curve.

To fix the calculation, calculate in two parts and add

$$\int_0^\pi \int_{-Sin[x]}^{Sin[x]} (x - \pi)^2 dy dx + \int_\pi^{2\pi} \int_{Sin[x]}^{-Sin[x]} (x - \pi)^2 dy dx :$$

```
In[4]:=    leftvol =
           Integrate[(x - Pi)^2,
           {x,0,Pi},{y,-Sin[x],Sin[x]}]
           rightvol =
           Integrate[(x - Pi)^2,
           {x,Pi,2 Pi},{y,Sin[x],-Sin[x]}]
           totalvol = leftvol + rightvol
           N[totalvol]

Out[4]=    23.4784
```

Ann says, "Cal, you owe me."

### T.2.b)

---

Cal thanks Ann and goes on to the next problem which is to measure the volume under the surface $f[x,y] = x^2 + 4y^2$ over the region between the curves $y = x/3$ and $y = \sqrt{x}$. Cal looks at the curves:

```
In[1]:=    Clear[x,y]
           Plot[{x/3, Sqrt[x]},{x,0,12},
           AxesLabel->{"x","y"},
           PlotStyle->{{Thickness[0.01]},
           {RGBColor[1,0,0]}}];
```

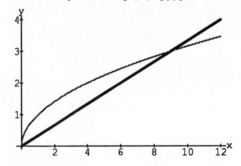

Then he solves:

```
In[2]:=    Solve[x/3 == Sqrt[x]]

Out[2]=    {{x -> 9}, {x -> 0}}
```

He says that the volume measurement is

$$\int_0^9 \int_{x/3}^{Sqrt[x]} (x^2 + 4y^2) dy dx$$

and turns on *Mathematica* :

```
In[3]:=    Integrate[x^2 + 4 y^2,
           {y,x/3,Sqrt[x]},{x,0,9}]

Out[3]=
                                         3
                                 3/2   4 x
           243 Sqrt[x] - 81 x + 12 x   - ----
                                          9
```

Cal issues another expletive because he knows the **volume must be a number and not a function.** Afraid to ask Ann about this, he asks Mathematical Maria, "What went wrong?" Maria is no slouch of a student. What does Maria tell Cal?

---

**Answer:**

Maria says that everything Cal did was correct until the last step. She goes on to say that his *Mathematica* code was wrong because to calculate

$$\int_0^9 \int_{x/3}^{Sqrt[x]} (x^2 + 4y^2)\,dy\,dx,$$

Cal should have typed:

```
In[4]:=    Integrate[x^2 + 4 y^2,
           {x,0,9},{y,x/3,Sqrt[x]}]
```

```
Out[4]=    17739
           -----
            140
```

Instead of:

```
In[5]:=    Integrate[x^2 + 4 y^2,
           {y,x/3,Sqrt[x]},{x,0,9}]
```

```
Out[5]=                               3
                          3/2   4 x
           243 Sqrt[x] - 81 x + 12 x   - ----
                                         9
```

Maria says, "Cal, next time you have a problem, ask Ann."

**T.2.c)**

---

Cal's spirits are getting better because there is only one more problem to do. The problem is to measure the volume between the surface

$$f[x,y] = 5.6(x - y^2)e^{-.013x}$$

and the rectangle whose sides are the lines $x = 0$, $x = 4$, $y = 0$ and $y = 3$ in the $xy$-plane.

Cal says to himself that calculating the volume of a solid whose base is a rectangle is easy and types:

```
In[1]:=    Clear[x,y,f]
           f[x_,y_] = 5.6 (x - y^2) E^(-.013 x)
           Integrate[f[x,y],{x,0,4},{y,0,3}]
```

```
Out[1]=    -66.6177
```

Cal says to himself, "Dammit, another wrong answer because **volume can't be negative!**"

It's time to join his friends for a little meeting at a popular campus hangout and Cal wants to get out of the lab soon. He turns to Ann and asks "What went wrong?" What does Ann tell Cal?

---

**Answer:**

Ann says, "Cal, at least this time you have an interesting question. Take a look at the surface:"

```
In[2]:=    Plot3D[f[x,y],{x,0,4},{y,0,3},
           Lighting->True,
           PlotRange->All];
```

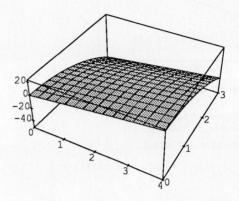

She says, "The trouble is that sometimes the surface is below the $xy$-plane and sometimes the surface is above the $xy$-plane. We'll have to see where it crosses the $xy$-plane:"

```
In[3]:=    Solve[f[x,y] == 0,y]
```

```
Out[3]=    {{y -> 1. Sqrt[x]}, {y -> -1. Sqrt[x]}}
```

Ann says, "Look at this:"

```
In[4]:=    f[x,y]
```

```
Out[4]=               2
           5.6 (x - y )
           ------------
              0.013 x
           E
```

```
In[5]:=    N[f[2,Sqrt[2] - .9]]
```

```
Out[5]=    9.46983
```

```
In[6]:=    N[f[2,Sqrt[2] + .9]]
```

```
Out[6]=    -18.309
```

She goes on to say that the surface is below or on the $xy$-plane when $y \geq \sqrt{x}$ and is above or on the $xy$-plane when $y \leq \sqrt{x}$.

```
In[7]:=    curveplot =
           Plot[{Sqrt[x],0,3},{x,0,4},
           AxesLabel->{"x","y"},
           DisplayFunction->Identity]

           verticals =
           Graphics[Line[{{4,0},{4,3}}]]

           labels =
           Graphics[Text["f[x,y] on bottom",{1.5,2}],
           Text["f[x,y] on top",{2.5,0.5}],
           Text["y = Sqrt[x]",{1,1}]]

           Show[curveplot, verticals, labels,
           DisplayFunction->$DisplayFunction];
```

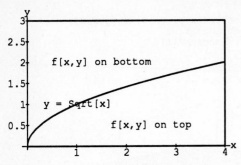

Ann says, "Now we can calculate the volume."

```
In[8]:=    part1 = -Integrate[f[x,y],
           {x,0,4},{y,Sqrt[x],3}]
```
```
Out[8]=    -(-66.6177 - 7.46667

               (-8718.1 + 4922.14 Sqrt[Pi]))
```

```
In[9]:=    N[part1]
```
```
Out[9]=    112.665
```

```
In[10]:=   part2 = Integrate[f[x,y],
           {x,0,4},{y,0,Sqrt[x]}]
```
```
Out[10]=   7.46667 (-8718.1 + 4922.14 Sqrt[Pi])
```

```
In[11]:=   N[part2]
```
```
Out[11]=   46.0471
```

Ann says, "The total volume is:"

```
In[12]:=   N[part1 + part2]
```
```
Out[12]=   158.712
```

Cal invites Ann and Maria to join him at the campus hangout. Ann decides to go with him and Maria says that she'll meet them in an hour.

Cal is happy.

■ **T.3) Setting them up.**

**T.3.a.i)**

---

Measure the volume of the solid under the surface $f[x,y] = x^2 y^2$ and over the region between the curves $y = x/2$ and $y^2 = -x + 8$ in the $xy$-plane.

---

**Answer:**

Look the base using curves with $x$ as a function of $y$:

```
In[1]:=    Clear[x,y]
           f[x_,y_] = x^2 y^2
```
```
Out[1]=     2  2
           x  y
```

```
In[2]:=    Solve[{y == x/2,y^2 == -x + 8},{x,y}]
```

```
Out[2]=    {{x -> -8, y -> -4}, {x -> 4, y -> 2}}
```

Note the axes labels.

```
In[3]:=    topview =
           Plot[{2 y, 8 - y^2},{y,-4,2},
           AxesLabel->{"y","x"}];
```

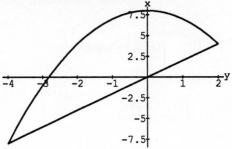

Put in a slice:

```
In[4]:=    y0 = -1.5
           slice =
           Graphics[Thickness[.01],
           Line[{{y0,2 y0},{y0,8 - y0^2}}]]
           Show[topview,slice];
```

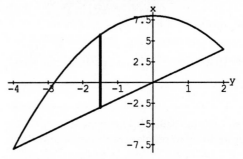

The area of a cross section through $\{y,0\}$ perpendicular to the $y$-axis is $\int_{2y}^{8-y^2} f[x,y]dx$.

The volume is

$$\int_{-4}^{2} \int_{2y}^{8-y^2} f[x,y]dxdy:$$

```
In[5]:=    vol =
           Integrate[f[x,y],{y,-4,2},{x,2 y,8-y^2}]
```
```
Out[5]=    52992
           -----
            35
```

Or in decimals:

```
In[6]:=    N[vol]
```
```
Out[6]=    1514.06
```

**T.3.a.ii)**

---

What was the advantage in plotting $x$ as a function of $y$ to set up the plot of the base?

---

**Answer:**

The main advantage was simplicity. Watch what happens when we do the same problem but plot $y$ as a function of $x$: We had to measure the volume of the solid under the surface

$$f[x, y] = x^2 y^2$$

and over the region between the curves $y = x/2$ and $y^2 = -x+8$ in the $xy$-plane. Look at the base using curves with $y$ as a function of $x$:

```
In[1]:=    Clear[x,y]
           f[x_,y_] = x^2 y^2
Out[1]=    2  2
           x  y
```

```
In[2]:=    Solve[{y == x/2,y^2 == -x + 8},{x,y}]
Out[2]=    {{x -> -8, y -> -4}, {x -> 4, y -> 2}}
```

Note the axes labels.

```
In[3]:=    topview =
           Plot[{x/2,Sqrt[-x + 8],-Sqrt[-x + 8]},
           {x,-8,4},AxesLabel->{"x","y"}];
```

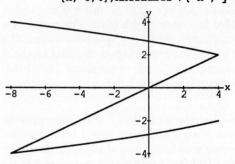

This picture is confusing. Extend the plotting interval to the right:

```
In[4]:=    topview =
           Plot[{x/2,Sqrt[-x + 8],-Sqrt[-x + 8]},
           {x,-8,8},AxesLabel->{"x","y"}];
```

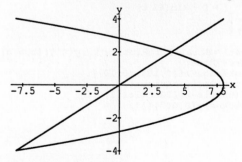

Put in two generic slices:

```
In[5]:=    x0 = -1.5
           slice1 =
           Graphics[Thickness[.01],
           Line[{{x0,x0/2},
           {x0,-Sqrt[-x0 + 8]}}]]

           x1 = 6
```

```
slice2 =

Graphics[Thickness[.01],
Line[{{x1,Sqrt[-x1 + 8]},
{x1,-Sqrt[-x1 + 8]}}]]

Show[topview,slice1,slice2];
```

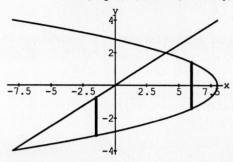

Cross sections to the left of $x = 4$ have different descriptions than cross sections to the right of $x = 4$. As a rusult we will have to calculate the volume by adding two different double integrals:

$$\int_{-8}^{4}\int_{-Sqrt[8-x]}^{x/2} f[x,y]dydx + \int_{4}^{8}\int_{-Sqrt[8-x]}^{Sqrt[8-x]} f[x,y]dydx$$

```
In[6]:=    leftvol =
           Integrate[f[x,y],
           {x,-8,4},{y,-Sqrt[8 - x],x/2}]
Out[6]=    1211648
           -------
             945
```

```
In[7]:=    rightvol =
           Integrate[f[x,y],
           {x,4,8},{y,-Sqrt[8 - x],Sqrt[8 - x]}]
Out[7]=    219136
           ------
            945
```

The total volume is:

```
In[8]:=    N[leftvol + rightvol]
Out[8]=    1514.06
```

This is the same answer we got in part i). And we did much less work in part i). This explains the advantage in plotting $x$ as a function of $y$ to set up the plot of the base.

■ **T.4) Interchanging the order of integration.**

**T.4.a)**

Calculate

$$\int_0^1 \int_{x^2}^x e^{-x} dy dx.$$

Then calculate the same volume measurement with the order of integration interchanged.

**Answer:**

Calculating $\int_0^1 \int_{x^2}^x e^{-x} dy\,dx$ is easy:

```
In[1]:=    Clear[x,y]
           Integrate[E^(-x),{x,0,1},{y,x^2,x}]

Out[1]=         3
           -1 + -
                E
```

To interchange the order of integration, look at the base of the solid whose volume is calculated by $\int_0^1 \int_{x^2}^x e^{-x} dy\,dx$ :

```
In[2]:=    topview =
           Plot[{x^2,x},{x,0,1},
           AxesLabel->{"x","y"}];
```

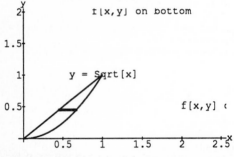

Slice with a plane perpendicular to the $y$−axis:

```
In[3]:=    slice =
           Graphics[Thickness[.01],
           Line[{{0.45,0.45},
           {Sqrt[0.45],0.45}}]]
           Show[topview,labels,slice];
```

The area $A[y]$ of a slice of the solid cut by a plane perpendicular to the $y-$ axis through $\{0,y\}$ is

$$A[y] = \int_y^{Sqrt[x]} e^{-x} dx.$$

The volume of the whole solid is

$$\int_0^1 \int_y^{Sqrt[y]} e^{-x} dx\,dy :$$

Check:

```
In[4]:=    Integrate[E^(-x),{y,0,1},{x,y,Sqrt[y]}]

Out[4]=         3
           -1 + -
                E
```

This is the same answer we got at the beginning. So

$$\int_0^1 \int_{x^2}^x e^{-x} dy\,dx = \int_0^1 \int_y^{Sqrt[y]} e^{-x} dx\,dy$$

because they both measure the same volume.

### ■ T.5) Linear dimension and volumes, areas and cones.

The volume $V[r]$ and the surface area $S[r]$ of a sphere of radius $r$ are given by

$$V[r] = 4\pi r^3/3$$
$$S[r] = 4\pi r^2.$$

This is the same as saying

$$V[r] = r^3 V[1]$$
$$S[r] = r^2 S[1].$$

This tells us that $V[r]$ is proportional to $r^3$ and $S[r]$ is proportional to $r^2$.

For other three dimensional objects, the formulas for volume and surface area are not so easy to come by, but the idea of proportionality survives. Here is the idea: A **linear dimension** of a given solid or shape is any length between specified locations on the solid. The radius of a sphere or the radius of a cirlce is a linear dimension. The total length of a solid, the total width or the total height of a solid are all examples of linear dimensions. The diameter of the finger loop on a coffee cup is a linear dimension of the cup.

Next take a given shape for a solid. *If the shape stays the same but the linear dimensions change, then it is still true that the volume is proportional to the cube of any linear dimension and it is still true that the surface area is proportional to the square of any linear dimension.*

Here is a shape in the $xy$-plane:

```
In[5]:=    Clear[x,y,t]
           x[t_] = 4 t ( 2 - t) E^(t/4)
           y[t_] = 2 - Sin[Pi t]
           a = 0
           b = 2

           plot1=ParametricPlot[{x[t],y[t]},{t,a,b},
           Axes->None,
           PlotRange->{{0,17},{0,10}},
           AspectRatio->Automatic,
           PlotStyle->{{RGBColor[1,0,0],
           Thickness[0.007]}}];
```

Here is the same shape with all its linear dimensions increased by a factor of 3 :

```
In[6]:=    plot2 =
           ParametricPlot[{3 x[t],3 y[t]},{t,a,b},
           Axes->None,PlotRange->{{0,17},{0,10}},
           AspectRatio->Automatic,
           PlotStyle->{{RGBColor[1,0,0],
           Thickness[0.007]}}];
```

Here they are together:

```
In[7]:=    Show[plot1,plot2,
           AspectRatio->Automatic];
```

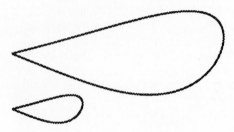

The area of the larger region is $3^2 = 9$ times the area of the smaller region.

The idea of linear dimension leads to some intriguing biological implications. E Batschelet ( *Introduction to Mathematics for Life Scientists*, Springer-Verlag, Berlin, 1971) points out that, for example, a giant mouse with linear dimension 10 times larger than the usual mouse would not be viable because the volume of its body would be larger than the volume of the usual mouse by a factor of $10^3$, but the surface area of some of its critical supporting organs like lungs, intestines and skin would be larger only by a factor of $10^2$. That big mouse would be out of breath at all times! Similarly, there will never be a twelve-feet-tall basketball player in the Big Ten or even the Atantic Coast Conference. The approximate size of an adult mammal is dictated by its shape!

The same common sense applies to buildings and other structures. An architect or engineer does not design a 200 foot tall building by taking a proven design for a 20 foot tall building and multiplying all the linear dimensions by 10.

### T.5.a)

A crystal grows in such a way that all the linear dimen-

sions increase by 25 percent. What are the percentages of the increases in the surface area and the volume?

**Answer:**

The percentage increase in surface area is:

```
In[1]:=    100 (1.25)^2 - 100
Out[1]=    56.25
```

The surface area increases by about 56 per cent. The percentage increase in volume is:

```
In[2]:=    100 (1.25)^3 - 100
Out[2]=    95.3125
```

So an increase of the linear dimensions by 1/4 nearly doubles the volume.

### T.5.b)

A cone of height $h$ has a (possibly irregular base) irregular planar base of area $A$.

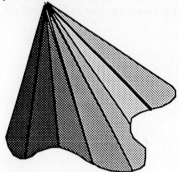

Find the volume of the cone.

**Answer:**

Let $h$ be the distance from the apex (= top point) of the cone to the plane of the base. Look at the cross section of the cone sliced by a plane parallel to the base at a distance $x$ from the apex.

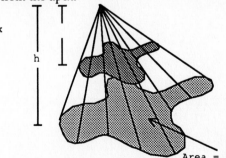

The linear dimensions of this slice are proportional to $x$. So The area of the cross section is proportional to $x^2$. Thus the area $A[x]$ of a cross section parallel to the base $x$ down from the apex is given by $A[x] = Kx^2$ where $K$ is a constant of proportionality.

To find $K$, remember that the area of the base is $A$ and hence $Kh^2 = A[h] = A$ because the base is the cross section corresponding to $x = h$. Solve this for $K$ to see that $K = A/h^2$. Consequently:

```
In[1]:=    Clear[A,x,h]
           A[x_] = K x^2/.K->A/h^2
```
```
Out[1]=      2
           A x
           ----
             2
           h
```

And the volume of the cone is given by:

```
In[2]:=    Integrate[A[x],{x,0,h}]
```
```
Out[2]=    A h
           ---
            3
```

This formula is beautiful for its pure simplicity: *No matter how irregular the planar base, the cone has volume equal to one third times the product of the area of the base and the height.*

## Literacy Sheet:

**1.** The base of a certain solid is the circle $x^2 + y^2 = 4$ in the $x - y$ plane. Each cross section cut by a plane perpendicular to the $x-$ axis is a square. Measure the volume of the solid.

$\cdot$

$\cdot$

$\cdot$

**2.** Set up and evaluate an integral that measures the volume below the surface $f[x, y] = e^{-x-y}$ and above the rectangle in the $x - y$ plane whose vertices are $\{0,0\}, \{1,0\}, \{1,1\}$ and $\{0,1\}$.

$\cdot$

$\cdot$

$\cdot$

**3.** Set up and evaluate an integral that measures the volume below the surface $f[x, y] = 1 - x - y$ and above the triangle in the $x-y$ plane whose vertices are $\{0,0\}, \{1,0\}$ and $\{0,1\}$.

$\cdot$

$\cdot$

$\cdot$

**4.** Set up an equivalent double integral with the order of integration reversed for each of the following double integrals. Check yourself by evaluating all four of the resulting double integrals.

$\int_0^2 \int_1^x xy\,dy\,dx$

$\cdot$

$\cdot$

$\int_0^2 \int_{Sqrt[y]}^1 xy\,dx\,dy$

$\cdot$

$\cdot$

$\cdot$

**5.** Change the order of intergration to explain why the double integral

$$\int_0^x \int_0^u e^{m(x-t)} f[t]\,dt\,du$$

reduces to the single integral

$$\int_0^x (x - t)e^{m(x-t)} f[t]\,dt.$$

$\cdot$

$\cdot$

$\cdot$

**6.** A region is sitting in the $x - y$ plane. All of its linear dimensions are doubled. What is the ratio of the new area to the original area?

$\cdot$

$\cdot$

$\cdot$

**7.** All the linear dimensions of a certain solid are doubled.

What is the ratio of the new volume to the original volume?

What is the ratio of the new surface area to the original surface area?

$\cdot$

$\cdot$

**8a.** The volume of a sphere of radius 1 is $4\pi/3$. Use the idea of linear dimension to give a formula for the surface area of a sphere of radius 2. How about the volume of a sphere of radius $r$?

$\cdot$

$\cdot$

$\cdot$

**8b.** The surface area of a sphere of radius 1 is $4\pi$. Use the idea of linear dimension to give a formula for the surface area of a sphere of radius 3. How about the surface area of a sphere of radius $r$?

$\cdot$

$\cdot$

$\cdot$

# Integration by parts and integration over an infinite interval

## Guide

A time-honored proverb in calculus is: "When you can't do anything else, then integrate by parts." The technique of integration by parts is a devilish idea that pops up often when it is least expected. What it does is to convert the product rule of differentiation into an instrument for calculation of integrals via the fundamental theorem of calculus. In addition to being a strong calculational technique, it is also a valuable theoretical technique. Watch for it from time to time.

Another topic of this lesson is the idea of integrating over infinite intervals. Over the years, this idea has acquired the unfortunate label of "improper integartion." But there is nothing improper or obscene about this idea. You will like it.

## Basics

■ **B.1) Integration by parts.**

### B.1.a)

Look at the derivative of $xe^x$ to find how to calculate

$$\int_0^5 xe^x \, dx.$$

**Answer:**

Calculate:

```
In[1]:=    D[x E^x,x]
Out[1]=     x    x
           E  + E  x
```

This tells us that

$$xe^x + e^x = D[xe^x, x]$$

So

$$xe^x = D[xe^x, x] - e^x.$$

As a result,

$$\int_0^5 xe^x \, dx = \int_0^5 D[xe^x, x] \, dx - \int_0^5 e^x \, dx.$$

Consequently $\int_0^5 xe^x \, dx$ is given by:

```
In[2]:=    ((x E^x/.x->5) - (x E^x /.x->0)) -
           (E^5 - E^0)
Out[2]=             5
           1 + 4 E
```

Check:

```
In[3]:=    Integrate[x E^x, {x,0,5}]
Out[3]=             5
           1 + 4 E
```

Got it.

Now let's reflect on what we did. Our success in calculating $\int_0^5 xe^x \, dx$ by the method we used above was based on our ability to replace $\int_0^5 xe^x \, dx$ by two easily calculated integrals.

You are entitled to protest that what we did was just an isolated trick. The response is that unless this idea is systematically developed, then it will remain just an isolated trick. The next problem develops the idea in full.

### B.1.b)

Use the product rule for differentiation to explain why

$$\int_a^b u[x]v'[x] \, dx = u[x]v[x] \Big|_a^b - \int_a^b v[x]u'[x] \, dx.$$

This is called the integration by parts formula.

**Answer:**

The product rule says

$$d(u[x]v[x])/dx = u[x]v'[x] + u'[x]v[x].$$

Integrating this from $x = a$ to $x = b$ gives:

$$\int_a^b (d(u[x]v[x])/dx) \, dx = \int_a^b u[x]v'[x] \, dx + \int_a^b u'[x]v[x] \, dx.$$

This is the same as:

$$u[x]v[x] \Big|_a^b = \int_a^b u[x]v'[x] \, dx + \int_a^b u'[x]v[x] \, dx.$$

Rearranging gives

$$\int_a^b u[x]v'[x] \, dx = u[x]v[x] \Big|_a^b - \int_a^b v[x]u'[x] \, dx$$

and this tells us why the advertised formula holds up.

**B.1.c.i)**

Use integration by parts to calculate

$$\int_0^\pi x\cos[x]dx.$$

**Answer:**

The integration by parts formula is

$$\int_0^\pi u[x]v'[x]dx = u[x]v[x]\Big|_0^\pi - \int_0^\pi v[x]u'[x]dx.$$

Make the assignments:

```
In[1]:=   u = x
Out[1]=   x
```

```
In[2]:=   vprime = Cos[x]
Out[2]=   Cos[x]
```

To get $v[x]$, we just need a function $v[x]$ whose derivative is *vprime*; here is one such:

```
In[3]:=   v = Integrate[vprime,x]
Out[3]=   Sin[x]
```

To get $u'$, we just differentiate:

```
In[4]:=   u' = D[u,x]
Out[4]=   1
```

The integration by parts formula tells us that

$$\int_0^\pi x\cos[x]dx = \int_0^\pi u[x]v'[x]dx$$

is given by

$$= u[x]v[x]\Big|_0^\pi - \int_0^\pi v[x]u'[x]dx$$

$$= x\sin[x]\Big|_0^\pi - \int_0^\pi \sin[x]1\,dx$$

$$= x\sin[x]\Big|_0^\pi + \cos[x]\Big|_0^\pi.$$

The final answer is:

```
In[5]:=   (x Sin[x]/.x->Pi) - (x Sin[x]/.x->0) +
          (Cos[x]/.x->Pi) - (Cos[x]/.x->0)
Out[5]=   -2
```

Check:

```
In[6]:=   Integrate[x Cos[x],{x,0,Pi}]
```

```
Out[6]=   -2
```

You betcha- we got it.

**B.2.c.ii)**

In what specific way did integration by parts expedite the calculation of

$$\int_0^\pi x\cos[x]dx?$$

**Answer:**

Review that last answer to see that the integration by parts formula let us arrive at

$$\int_0^\pi x\cos[x]dx = x\sin[x]\Big|_0^\pi - \int_0^\pi \sin[x]dx.$$

At first, we did not see how to calculate the integral $\int_0^\pi x\cos[x]dx$ on the left, but this did not hold us back because integration by parts allowed us to calculate the easy integral $\int_0^\pi \sin[x]dx$ on the right instead.

This is the potential value of integration by parts: Integration by parts gives us the hope of being able to replace a hard integral by an easy integral.

■ **B.2) Integration over an infinite interval.**

**B.2.a)**

How can we make sense of an integral like

$$\int_1^\infty x^2 e^{-x}dx?$$

**Answer:**

You can't say that $\int_1^\infty x^2 e^{-x}dx$ measures a concrete area because the horizontal mesurement would be infinite.

So we do the next best thing by agreeing that $\int_1^\infty x^2 e^{-x}dx$ is the limiting case of $\int_1^t x^2 e^{-x}dx$ as $t \to \infty$. In short,

$$\int_1^\infty x^2 e^{-x}dx = \lim_{t\to\infty}\int_1^t x^2 e^{-x}dx.$$

Some folks regard this as a cop-out and call an integral like $\int_1^\infty x^2 e^{-x}dx$ an *improper* integral. But this terminology, although in common use is unnecessarily pejorative.

Now the calculation: We want to look at:

```
In[1]:=   Integrate[x^2 E^(-x),{x,1,t}]
                             2
          5    -2 - 2 t - t
Out[1]=   -  + ------------
          E          t
                     E
```

This tells us

$$\int_1^t x^2 e^{-x} dx = 5/e - (2 + 2t + t^2)/e^t.$$

So

$$\int_1^\infty x^2 e^{-x} dx = \lim_{t\to\infty} \int_1^t x^2 e^{-x} dx$$

$$= \lim_{t\to\infty} (5/e - (2 + 2t + t^2)/e^t) = 5/e + 0 = 5/e$$

because exponential growth dominates polynomial growth.

Check:

```
In[2]:=    Integrate[x^2 E^(-x),{x,1,Infinity}]
Out[2]=    5
           -
           E
```

Got it.

You might ask how $\int_1^\infty x^2 e^{-x} dx$ could be finite. If this question bothers you, then look at the plot:

```
In[3]:=    Plot[x^2 E^(-x),{x,1,40},PlotRange->All]
```

The reason $\int_1^\infty x^2 e^{-x} dx$ is finite is that most of the area under the $x^2 e^{-x}$ curve for $1 \le x < \infty$ is concentrated over the interval $1 \le x \le 20$. For $x \ge 20$ very little extra area comes in. To confirm this, look at the folowing decimal approximations of $\int_1^\infty x^2 e^{-x} dx$ and $\int_1^{20} x^2 e^{-x} dx$:

```
In[4]:=    N[Integrate[x^2 E^(-x),
           {x,1,Infinity}],10]
Out[4]=    1.839397206

In[5]:=    N[Integrate[x^2 E^(-x),{x,1,20}],10]
Out[5]=    1.839396295
```

Almost the same. For most practical purposes and especially for government work, there is no difference between $\int_1^\infty x^2 e^{-x} dx$ and $\int_1^{20} x^2 e^{-x} dx$.

**B.2.b)**

Find a number $b$ with $0 < b < \infty$ such that

$$\int_1^b x^2 e^{-x} dx = \int_1^\infty x^2 e^{-x} dx$$

to nine accurate decimals.

**Answer:**

Well,

$$\int_1^b x^2 e^{-x} dx + \int_b^\infty x^2 e^{-x} dx = \int_1^\infty x^2 e^{-x} dx.$$

So

$$\int_1^\infty x^2 e^{-x} dx = \int_1^b x^2 e^{-x} dx$$

within $\int_b^\infty x^2 e^{-x} dx$.

Consequently if $b$ is picked so that $\int_b^\infty x^2 e^{-x} dx < 10^{-9}$, then $\int_1^b x^2 e^{-x} dx = \int_1^\infty x^2 e^{-x} dx$ to nine accurate decimals. To find a good $b$, look at:

```
In[1]:=    Table[{b,N[Integrate[x^2 E^(-x),
           {x,b,Infinity}],12]},{b,10,50,10}]
Out[1]=    {{10, 0.00553879143102},
           {20, 9.11029901118 10^-7},
           {30, 9.00203329602 10^-11},
           {40, 7.1457318574 10^-15},
           {50, 5.0186071044 10^-19}}
```

This tells us that $b = 30$ will do very well. Check:

```
In[2]:=    N[Integrate[x^2 E^(-x),
           {x,1,Infinity}],12]
Out[2]=    1.83939720586

In[3]:=    N[Integrate[x^2 E^(-x),{x,1,30}],12]
Out[3]=    1.83939720577
```

The first nine decimals agree as predicted.

## Tutorial

■ **T.1) Integration by parts.**

**T.1.a)**

Use integration by parts to calculate

$$\int_0^{\pi/4} \arctan[x] dx.$$

**Answer:**

The integration by parts formula is

$$\int_0^{\pi/4} u[x]v'[x] dx = u[x]v[x]\Big[_0^{\pi/4} - \int_0^{\pi/4} v[x]u'[x] dx.$$

Make the assignments:

$In[1]:=$    `u = ArcTan[x]`

$Out[1]=$    `ArcTan[x]`

$In[2]:=$    `vprime = 1`

$Out[2]=$    `1`

To get $v[x]$, we just need a function $v[x]$ whose derivative is *vprime*; here is one such:

$In[3]:=$    `v = Integrate[vprime,x]`

$Out[3]=$    `x`

To get $u'$, we just differentiate:

$In[4]:=$    `uprime = D[u,x]`

$Out[4]=$

$$\frac{1}{1+x^2}$$

The integration by parts formula tells us that

$$\int_0^{\pi/4} \arctan[x]dx = \int_0^{\pi/4} u[x]v'[x]dx$$

$$= u[x]v[x]\Big[_0^{\pi/4} - \int_0^{\pi/4} v[x]u'[x]dx$$

$$= x\arctan[x]\Big[_0^{\pi/4} - \int_0^{\pi/4} (1/(1+x^2))x\,dx$$

$$= x\arctan[x]\Big[_0^{\pi/4} - \int_0^{\pi/4} (x/(1+x^2))dx$$

$$= x\arctan[x]\Big[_0^{\pi/4} + \log[1+x^2]/2\Big[_0^{\pi/4}$$

The final answer is:

$In[5]:=$    `(x ArcTan[x]/.x->Pi/4) -`
           `(x ArcTan[x]/.x->0) +`
           `(Log[1 + x^2]/2/.x->Pi/4) -`
           `(Log[1 + x^2]/2/.x->0)`

$Out[5]=$

$$\frac{Pi\ ArcTan[\frac{Pi}{4}]}{4} + \frac{Log[1 + \frac{Pi^2}{16}]}{2}$$

Check:

$In[6]:=$    `Integrate[ArcTan[x],{x,0,Pi/4}]`

$Out[6]=$

$$\frac{Pi\ ArcTan[\frac{Pi}{4}]}{4} - \frac{Log[1 + \frac{Pi^2}{16}]}{2}$$

Got it.

### T.1.b)

The natural way to calculate $\int_0^\pi x\sin[x]dx$ is to use the method of undetermined coefffficients. Just for the heck of it, slam out

$$\int_0^\pi x\sin[x]dx$$

by the method of integration by parts.

**Answer:**

The integration by parts formula is

$$\int_0^\pi u[x]v'[x]dx = u[x]v[x]\Big[_0^\pi - \int_0^\pi v[x]u'[x]dx.$$

Make the assignments:

$In[1]:=$    `u = x`

$Out[1]=$    `x`

$In[2]:=$    `vprime = Sin[x]`

$Out[2]=$    `Sin[x]`

To get $v[x]$, we just need a function $v[x]$ whose derivative is *vprime*; here is one such:

$In[3]:=$    `v = Integrate[vprime,x]`

$Out[3]=$    `-Cos[x]`

To get $u'$, we just differentiate:

$In[4]:=$    `uprime = D[u,x]`

$Out[4]=$    `1`

The integration by parts formula tells us that

$$\int_0^\pi x\sin[x]dx = \int_0^\pi u[x]v'[x]dx$$

$$= u[x]v[x]\Big[_0^\pi - \int_0^\pi v[x]u'[x]dx$$

$$= x(-\cos[x])\Big[_0^\pi - \int_0^\pi (-\cos[x])1\,dx$$

$$= x(-\cos[x])\Big[_0^\pi + \int_0^\pi \cos[x]dx$$

$$= -x\cos[x]\Big[_0^\pi + \sin[x]\Big[_{0x}^\pi.$$

The final answer is:

$In[5]:=$    `(-x Cos[x]/.x->Pi)-(-x Cos[x]/.x->0) +`
           `(Sin[x]/.x->Pi)-(Sin[x]/.x->0)`

$Out[5]=$    `Pi`

Check:

*In[6]:=*     `Integrate[x Sin[x],{x,0,Pi}]`

*Out[6]=*    `Pi`

Nailed it.

### T.1.c): Laplace Transform.

Suppose $p[x]$ is a polynomial. Use integration by parts to explain why

$$\int_0^\infty e^{-sx}p'[x]dx = s\int_0^\infty e^{-sx}p[x]dx - p[0]$$

provided $s > 0$.

**Answer:**

Recall

$$\int_0^\infty e^{-sx}p'[x]dx = \lim_{t\to\infty}\int_0^t e^{-sx}p'[x]dx$$

Lets calculate $\int_0^t e^{-sx}p'[x]dx$ using integration by parts.

The integration by parts formula is

$$\int_0^t u[x]v'[x]dx = u[x]v[x]\Big|_0^t - \int_0^t v[x]u'[x]dx.$$

Make the assignments: $v'[x] = p'[x]$ and $u[x] = e^{-sx}$. This gives $v[x] = p[x]$ and $u'[x] = -se^{-sx}$.

Thus

$$\int_0^t e^{-sx}p'[x]dx = e^{-sx}p[x]\Big|_0^t - \int_0^t p[x](-se^{-sx})dx.$$

Accordingly,

$$\int_0^\infty e^{-sx}p'[x]dx = \lim_{t\to\infty}\int_0^t e^{-sx}p'[x]dx$$

$$= \lim_{t\to\infty}\left(e^{-sx}p[x]\Big|_0^t - \int_0^t p[x](-se^{-sx})dx\right)$$

$$= \lim_{t\to\infty}\left(p[x]/e^{sx}\Big|_0^t + s\int_0^t p[x]e^{-sx}dx\right)$$

$$= \lim_{t\to\infty}\left(p[x]/e^{sx}\Big|_0^t\right) + s\int_0^\infty p[x]e^{-sx}dx$$

$$= \lim_{t\to\infty}p[t]/e^{st} - p[0]/e^0 + s\int_0^\infty p[x]e^{-sx}dx$$

$$= \lim_{t\to\infty}p[t]/e^{st} - p[0] + s\int_0^\infty p[x]e^{-sx}dx$$

$$= 0 - p[0] + s\int_0^\infty p[x]e^{-sx}dx$$

because exponential growth dominates polynomial growth.

This tells why

$$\int_0^\infty e^{-sx}p'[x]dx = s\int_0^\infty p[x]e^{-sx}dx - p[0],$$

as advertised.

Try it:

*In[1]:=*    `p[x_] = x^3; s = 7`
           `Integrate[E^(-s x) p'[x],`
           `{x,0, Infinity}]`

*Out[1]=*    $\dfrac{6}{343}$

*In[2]:=*    `s Integrate[p[x] E^(-s x),`
           `{x,0,Infinity}] - p[0]`

*Out[2]=*    $\dfrac{6}{343}$

Throw some other choices of positive $s's$ and other $p[x]'s$ in and run the machine.

If someone ever asks you whether you have seen the Laplace transform, now you can say that you've seen a little bit about it. The Laplace transform of a function $f[x]$ is another function $L[f,s]$ whose formula is

$$L[f,s] = \int_0^\infty f[x]e^{-sx}dx.$$

The little calculation we did above can be interpreted to say

$$L[f',s] = sL[f,s] - f[0].$$

Most of you will encounter the Laplace transform again as you progress in your studies. And you can tell them that you saw it first in Calculus& *Mathematica*.

### ■ T.2) Integration by iteration.

### T.2.a)

Use integration by parts and iteration to prepare a little table of the values of

$$\int_0^1 x^n e^x dx$$

for $n = 0,1,2,3,...,12$ by evaluating only one actual integral.

**Answer:**

Start by looking at

$$\int_0^1 x^n e^x dx.$$

The integration by parts formula is

$$\int_0^1 u[x]v'[x]dx = u[x]v[x]\Big|_0^1 - \int_0^1 v[x]u'[x]dx.$$

Make the assignments:

*In[1]:=*    `u = x^n`
*Out[1]=*    $x^n$

*In[2]:=*    `vprime = E^x`

*Out[2]=*    $\dfrac{x}{E}$

To get $v[x]$, we just need a function $v[x]$ whose derivative is *vprime*; here is one such:

*In[3]:=*    `v = Integrate[vprime, x]`

*Out[3]=*    $\dfrac{x}{E}$

To get $u'$, we just differentiate:

*In[4]:=*    `uprime = D[u, x]`

*Out[4]=*    $\dfrac{-1+n}{n\ x}$

The integrate by parts formula tells us that

$$\int_0^1 x^n e^x dx = \int_0^1 u[x]v'[x]dx$$

$$= u[x]v[x]\Big[\Big]_0^1 - \int_0^1 v[x]u'[x]dx$$

$$= x^n e^x \Big[\Big]_0^1 - \int_0^1 nx^{n-1}e^x dx.$$

$$= e - n\int_0^1 x^{n-1}e^x dx.$$

Accordingly,

$$\int_0^1 x^n e^x dx = e - n\int_0^1 x^{n-1}e^x dx.$$

Something great just happened. Letting

$$Int[n] = \int_0^1 x^n e^x dx,$$

we see that the formula immediately above takes the form

$$Int[n] = e - nInt[n-1].$$

This iteration formula will let us build the whole table on the basis of evaluating just one lonesome integral:

Integrate to find $Int[0] = \int_0^1 x^0 e^x dx$ :

*In[5]:=*    `Int[0] = Integrate[ E^x, {x, 0, 1}]`

*Out[5]=*    `-1 + E`

Enter the iteration formula:

*In[6]:=*    `Int[n_] := Int[n] = E - n Int[n - 1]`

Now we can get the whole table

$$\{Int[0], Int[1], Int[2], ..., Int[12]\}$$

$$= \{\int_0^1 e^x dx, \int_0^1 xe^x dx, \int_0^1 x^2 e^x dx, ..., \int_0^1 x^{12} e^x dx\}$$

with no additional work on our part:

*In[7]:=*    `ColumnForm[Table[Expand[`
　　　　　　`{"Int"[n], Int[n]}], {n, 0, 12}]]`

*Out[7]=*    `{Int[0], -1 + E}`
　　　　　　`{Int[1], 1}`
　　　　　　`{Int[2], -2 + E}`
　　　　　　`{Int[3], 6 - 2 E}`
　　　　　　`{Int[4], -24 + 9 E}`
　　　　　　`{Int[5], 120 - 44 E}`
　　　　　　`{Int[6], -720 + 265 E}`
　　　　　　`{Int[7], 5040 - 1854 E}`
　　　　　　`{Int[8], -40320 + 14833 E}`
　　　　　　`{Int[9], 362880 - 133496 E}`
　　　　　　`{Int[10], -3628800 + 1334961 E}`
　　　　　　`{Int[11], 39916800 - 14684570 E}`
　　　　　　`{Int[12], -479001600 + 176214841 E}`

Now we can read off, for instance,

$$\int_0^1 x^9 e^x dx = Int[9] = 3628800 - 14684570e.$$

We can try for machine accuracy:

*In[8]:=*    `N[Int[9]]`

*Out[8]=*    `0.249028`

A lot more accuracy:

*In[9]:=*    `N[Int[9], 30]`

*Out[9]=*    `0.249028031297260340306372`

## T.2.b)

Use integration by parts to prepare a table of the values of

$$\int_0^{\pi/2} \sin[x]^n dx$$

for $n = 1, 2, 3, ..., 18$ by evaluating only two actual integrals. Comment on the calculational advantage of iteration over direct calculation in this situation.

**Answer:**

Start by looking at

$$\int_0^{\pi/2} \sin[x]^n dx.$$

The integration by parts formula is

$$\int_0^{\pi/2} u[x]v'[x]dx = u[x]v[x]\Big[\Big]_0^{\pi/2} - \int_0^{\pi/2} v[x]u'[x]dx.$$

Here is a little trick:

$$\int_0^{\pi/2} \sin[x]^n dx = \int_0^{\pi/2} \sin[x]^{n-1}\sin[x]dx.$$

Run through the mill:

*In[1]:=*    `u = Sin[x]^(n-1)`

*Out[1]=*    $\text{Sin}[x]^{-1+n}$

```
In[2]:=    vprime = Sin[x]
Out[2]=    Sin[x]

In[3]:=    v = Integrate[vprime,x]
Out[3]=    -Cos[x]

In[4]:=    uprime = D[u,x]
                                        -2 + n
Out[4]=    (-1 + n) Cos[x] Sin[x]
```

Now

$$\int_0^{\pi/2} \sin[x]^n dx = \int_0^{\pi/2} \sin[x]^{n-1} \sin[x] dx$$

$$= \int_0^{\pi/2} u[x] v'[x] dx$$

$$= u[x]v[x]\Big|_0^{\pi/2} - \int_0^{\pi/2} v[x]u'[x] dx.$$

This is given by:

```
In[5]:=    ((u v)/.x->Pi/2) - ((u v)/.x->0) -
           Integrate[v uprime, {x,0,Pi/2}]
                                          2
Out[5]=    -Integrate[-((-1 + n) Cos[x]

                  -2 + n       Pi
           Sin[x]      ), {x, 0, --}]
                                  2
```

As a result,

$$\int_0^{\pi/2} \sin[x]^n dx = (n-1)\int_0^{\pi/2} \sin[x]^{n-2}\cos[x]^2 dx$$

$$= (n-1)\int_0^{\pi/2} \sin[x]^{n-2}(1-\sin[x]^2) dx$$

$$= (n-1)\int_0^{\pi/2} (\sin[x]^{n-2} - \sin[x]^n) dx$$

$$= (n-1)\int_0^{\pi/2} \sin[x]^{n-2} dx - (n-1)\int_0^{\pi/2} \sin[x]^n dx.$$

This looks bad, but it feels good because we now see that

$$\int_0^{\pi/2} \sin[x]^n dx$$

$$= (n-1)\int_0^{\pi/2} \sin[x]^{n-2} dx - (n-1)\int_0^{\pi/2} \sin[x]^n dx.$$

And this is great because we can solve for $\int_0^{\pi/2}\sin[x]^n dx$. Just watch:

$$\int_0^{\pi/2} \sin[x]^n dx + (n-1)\int_0^{\pi/2} \sin[x]^n dx$$

$$= (n-1)\int_0^{\pi/2} \sin[x]^{n-2} dx.$$

This is the same as

$$n\int_0^{\pi/2} \sin[x]^n dx = (n-1)\int_0^{\pi/2} \sin[x]^{n-2} dx.$$

Divide both sides by $n$ to see the iteration formula:

$$\int_0^{\pi/2} \sin[x]^n dx = ((n-1)/n)\int_0^{\pi/2} \sin[x]^{n-2} dx.$$

Now we are ready to crank up *Mathematica*. Let $Int[n]$ stand for $\int_0^{\pi/2}\sin[x]^n dx$. The formula immediately above tells us that

$$Int[n] = ((n-1)/n)Int[n-2].$$

Integrate to find $Int[1]$ and $Int[2]$:

```
In[6]:=    Int[1] =
           Integrate[Sin[x]^1,{x,0,Pi/2}]
Out[6]=    1

In[7]:=    Int[2] =
           Integrate[Sin[x]^2,{x,0,Pi/2}]
           Pi
Out[7]=    --
           4
```

Enter the iteration formula:

```
In[8]:=    Int[n_] :=
           Int[n] = ((n - 1)/n) Int[n - 2]
```

Now we can get the whole table

$$\{Int[1], Int[2], ..., Int[18]\} =$$

$$\{\int_0^{\pi/2}\sin[x]dx, \int_0^{\pi/2}\sin[x]^2 dx..., \int_0^{\pi/2}\sin[x]^{18}dx\}:$$

```
In[9]:=    Table[Expand[{"Int"[n],Int[n]}],{n,1,18}]
                                    Pi
Out[9]=    {{Int[1], 1}, {Int[2], --},
                                    4

                  2              3 Pi
           {Int[3], -}, {Int[4], ----},
                  3               16

                  8              5 Pi
           {Int[5], --}, {Int[6], ----},
                  15              32

                  16             35 Pi
           {Int[7], --}, {Int[8], -----},
                  35             256

                  128            63 Pi
           {Int[9], ---}, {Int[10], -----},
                  315            512

                   256             231 Pi
           {Int[11], ---}, {Int[12], ------},
                   693             2048

                   1024            429 Pi
           {Int[13], ----}, {Int[14], ------},
                   3003            4096

                   2048            6435 Pi
```

{Int[15], $\dfrac{}{6435}$----}, {Int[16], $\dfrac{}{65536}$-------},

{Int[17], $\dfrac{32768}{109395}$------},

{Int[18], $\dfrac{12155\ Pi}{131072}$--------}}

That was quick. Spot check:

*In[10]:=* `Integrate[Sin[x]^16,{x,0,Pi/2}]`

*Out[10]=* $\dfrac{6435\ Pi}{65536}$

Good. The whole table was calculated in slightly over 2 seconds on a Macintosh IIcx, but the evaluation of the individual spot check

$$Int[16] = \int_0^{\pi/2} \sin[x]^{16} dx$$

took more than 6 seconds on the same machine. This shows off the calculational advantage of iteration.

If we were to attempt $\int_0^{\pi/2} \sin[x]^{18} dx$ by hand using integration by parts, then we would have needed 18 additional integrations by parts to succeed in evaluating this integral.

If we had wanted to use trigonometric identities to solve this integral (this is how *Mathematica* did this one), then we would have had to come up with the staggering identity for $\sin[x]^{18}$:

*In[11]:=* `TrigExpand[Sin[x]^18]`

*Out[11]=* $\dfrac{12155}{65536} - \dfrac{21879\ Cos[2\ x]}{65536} +$

$\dfrac{1989\ Cos[4\ x]}{8192} - \dfrac{4641\ Cos[6\ x]}{32768} +$

$\dfrac{1071\ Cos[8\ x]}{16384} - \dfrac{765\ Cos[10\ x]}{32768} +$

$\dfrac{51\ Cos[12\ x]}{8192} - \dfrac{153\ Cos[14\ x]}{131072} +$

$\dfrac{9\ Cos[16\ x]}{65536} - \dfrac{Cos[18\ x]}{131072}$

When all is said and done, the iterative attack we used on this problem is the best attack.

■ **T.3) Approximate calculation of improper integrals.**

For $s > 0$, calculating $\int_0^\infty e^{-sx} \sin[x] dx$ is easy. We just look at the limiting case of $\int_0^t e^{-sx} \sin[x] dx$ as $t \to \infty$:

*In[12]:=* `Integrate[E^(- s x) Sin[x] ,{x,0,t}]`

*Out[12]=* $\dfrac{1}{}$------ - $\dfrac{Cos[t]}{}$-------------- - $\dfrac{s\ Sin[t]}{}$-------------

$1 + s^2$        $E^{st} (1 + s^2)$        $E^{st} (1 + s^2)$

So

$$\int_0^\infty e^{-sx} \sin[x] dx = \lim_{t \to \infty} \int_0^t e^{-sx} \sin[x] dx = 1/(1 + s^2)$$

because the exponential $e^{st}$ in the denominators of the second and third terms above dominates (owing to the fact that $s > 0$) and sends these terms to $0$ as $t \to \infty$.

**T.3.a)**

Trying the same approach to calculate

$$\int_0^\infty e^{-x^2}/\sqrt{1 + \sin[x]^2} dx$$

does not work because we cannot obtain the exact value of

$$\int_0^t e^{-x^2}/\sqrt{1 + \sin[x]^2} dx :$$

*In[1]:=* `Integrate[`
`E^(-x^2)/Sqrt[1 + Sin[x]^2],{x,0,t}]`

*Out[1]=* $Integrate[\dfrac{1}{E^{x^2}\ Sqrt[1 + Sin[x]^2]},$

`{x, 0, t}]`

In spite of this setback, come up with a reasonably accurate calculation of

$$\int_0^\infty e^{-x^2}/\sqrt{1 + \sin[x]^2} dx$$

**Answer:**

At first we might like to try to examine what happens to

$$\int_0^t e^{-x^2}/\sqrt{1 + \sin[x]^2} dx$$

for some large values of t:

*In[2]:=* `NIntegrate[`
`E^(-x^2)/Sqrt[1 + Sin[x]^2],{x,0,5}]`

*Out[2]=* `0.786831`

*In[3]:=* `NIntegrate[`
`E^(-x^2)/Sqrt[1 + Sin[x]^2],{x,0,10}]`

*Out[3]=* `0.786831`

*In[4]:=* `NIntegrate[`
`E^(-x^2)/Sqrt[1 + Sin[x]^2],{x,0,100}]`

*Out[4]=* `0.786831`

*In[5]:=*    `NIntegrate[`
            `E^(-x^2)/Sqrt[1 + Sin[x]^2],{x,0,1000}]`

*Out[5]=*    `0.`

This last answer is clearly wrong because the last integral measures more area than the others. Thus the last calculation above should have been the largest of the calculations. To see what the trouble is, plot:

*In[6]:=*    `Plot[E^(-x^2)/Sqrt[1 + Sin[x]^2],{x,0,10},`
            `PlotRange->All]`

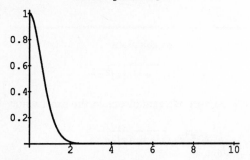

The trouble is that when we use **NIntegrate** over a really long interval, then the automatic approximation routine built into **NIntegrate** misses the spike on the left. Now look again ata the values we got for $t = 5, 20, 100, 1000$. An educated guess is that to about six decimals,

$$\int_0^\infty e^{-x^2}/\sqrt{1 + \sin[x]^2}\,dx$$

$$= \int_0^5 e^{-x^2}/\sqrt{1 + \sin[x]^2}\,dx = 0.786831.$$

Here is a way to get even more confidence in this educated guess: Note that

$$\int_0^\infty e^{-x^2}/\sqrt{1 + \sin[x]^2}\,dx$$

$$= \int_0^5 e^{-x^2}/\sqrt{1 + \sin[x]^2}\,dx + \int_5^\infty e^{-x^2}/\sqrt{1 + \sin[x]^2}\,dx.$$

So

$$\int_0^5 e^{-x^2}/\sqrt{1 + \sin[x]^2}\,dx$$

estimates

$$\int_0^\infty e^{-x^2}/\sqrt{1 + \sin[x]^2}\,dx$$

within an error of no more than

$$\int_5^\infty e^{-x^2}/\sqrt{1 + \sin[x]^2}\,dx.$$

But the error

$$\int_5^\infty e^{-x^2}/\sqrt{1 + \sin[x]^2}\,dx < \int_5^\infty e^{-x^2}/\sqrt{2}\,dx$$

and this last integral is easy to calculate:

*In[7]:=*    `esterror = Integrate[`
            `E^(-x^2)/Sqrt[2],{x,5,Infinity}]`

*Out[7]=*    $\dfrac{\text{Sqrt[Pi]}}{2\ \text{Sqrt[2]}} - \dfrac{\text{Sqrt[Pi] Erf[5]}}{2\ \text{Sqrt[2]}}$

Go to decimals:

*In[8]:=*    `N[esterror,10]`

*Out[8]=*    $9.634600518\ 10^{-13}$

This means

$$0 < \int_5^\infty e^{-x^2}/\sqrt{1 + \sin[x]^2}\,dx$$

$$< \int_5^\infty e^{-x^2}/\sqrt{2}\,dx < 10^{-12}.$$

So

$$\int_0^5 e^{-x^2}/\sqrt{1 + \sin[x]^2}\,dx$$

estimates

$$\int_0^\infty e^{-x^2}/\sqrt{1 + \sin[x]^2}\,dx$$

to at least twelve accurate decimals. This makes us really comfortable with the following estimate of

$$\int_0^\infty e^{-x^2}/\sqrt{1 + \sin[x]^2}\,dx :$$

*In[9]:=*    `NIntegrate[`
            `E^(-x^2)/Sqrt[1 + Sin[x]^2],{x,0,5}]`

*Out[9]=*    `0.786831`

**T.3.b)**

Come up with a reasonably accurate estimate of

$$\int_\pi^\infty \sin[x]/x^2\,dx.$$

**Answer:**

Look at:

*In[1]:=*    `Integrate[Sin[x]/x^2,{x,Pi,t}]`

*Out[1]=*    $-\dfrac{\text{ExpIntegralEi}[-I\ \text{Pi}]}{2} -$

            $\dfrac{\text{ExpIntegralEi}[I\ \text{Pi}]}{2} +$

            $\dfrac{\text{ExpIntegralEi}[-I\ t]}{2} +$

            $\dfrac{\text{ExpIntegralEi}[I\ t]}{2} - \dfrac{\text{Sin}[t]}{}$

2        t

Good, *Mathematica* seems to be able to handle $\int_\pi^t \sin[x]/x^2 dx$. Let's see whether *Mathematica* can handle the whole calculation:

```
In[2]:=    Integrate[Sin[x]/x^2,{x,Pi,Infinity}]
```

*Limit::nlm:*

*Could not find definite limit.*

*Limit::nlm:*

*Could not find definite limit.*

*Limit::nlm:*

*Could not find definite limit.*

*General::stop:*

*Further output of Limit::nlm*

*will be suppressed during this*

*calculation.*

```
Out[2]=    -ExpIntegralEi[-I Pi]
           --------------------- -
                    2

           ExpIntegralEi[I Pi]
           ------------------- +
                    2

           Limit[ExpIntegralEi[-I x],

             x -> Infinity] / 2 +

           Limit[ExpIntegralEi[I x],

             x -> Infinity] / 2
```

Too bad. To get an idea of the value of $\int_\pi^\infty \sin[x]/x^2 dx$, look at:

```
In[3]:=    N[Integrate[Sin[x]/x^2,{x,Pi,100}]]
Out[3]=    -0.0737531
```

```
In[4]:=    N[Integrate[Sin[x]/x^2,{x,Pi,1000}]]
Out[4]=    -0.0736685
```

```
In[5]:=    N[Integrate[Sin[x]/x^2,{x,Pi,5000}]]
Out[5]=    -0.0736679
```

A good seat-of-the-pants estimate of $\int_\pi^\infty \sin[x]/x^2 dx$ is $-0.074$. Now let's try to check this: Note that

$$\int_\pi^\infty \sin[x]/x^2 dx = \int_\pi^a \sin[x]/x^2 dx + \int_a^\infty \sin[x]/x^2 dx.$$

So $\int_\pi^a \sin[x]/x^2 dx$ estimates $\int_\pi^\infty \sin[x]/x^2 dx$ within an error of no more than $\int_a^\infty \sin[x]/x^2 dx$.

Now to get an idea of the size of the possible error

$$\int_\pi^\infty \sin[x]/x^2 dx,$$

note that $|\sin[x]/x|^2 \le 1/x^2$ so the possible error $\int_a^\infty \sin[x]/x^2 dx$ is no larger than:

```
In[6]:=    Integrate[1/x^2,{x,a,Infinity}]
Out[6]=    1
           -
           a
```

This tells us that

$$\int_\pi^{10000} \sin[x]/x^2 dx$$

estimates

$$\int_\pi^\infty \sin[x]/x^2 dx$$

within an error of no more than:

```
In[7]:=    N[(1/a)/.a->10000]
Out[7]=    0.0001
```

As a result, the number:

```
In[8]:=    estimate =
           Integrate[Sin[x]/x^2,{x,Pi,10000}]
Out[8]=    ExpIntegralEi[-10000 I]
           ----------------------- +
                     2

           ExpIntegralEi[10000 I]
           ---------------------- -
                     2

           ExpIntegralEi[-I Pi]
           -------------------- -
                    2

           ExpIntegralEi[I Pi]     Sin[10000]
           -------------------  -  ----------
                    2                10000
```

whose decimal evaluation is:

```
In[9]:=    N[estimate]
Out[9]=    -0.0736679
```

is accurate estimate of to three decimal places. This makes us very confident in writing

$$\int_1^\infty \sin[x]/x^2 dx = -0.074$$

## T.3.c)

Come up with a reasonably accurate estimate of

$$\int_1^\infty \sin[x]/x\, dx.$$

**Answer:**

Look at:

```
In[1]:=    Integrate[Sin[x]/x,{x,1,t}]
Out[1]=    -Pi   I
           --- + - (-ExpIntegralE[1, -I] +
            2    2

                ExpIntegralE[1, I]) +

           SinIntegral[t]
```

Good, *Mathematica* seems to be able to handle $\int_1^t \sin[x]/x\,dx$. Let's see whether *Mathematica* can handle the whole calculation:

```
In[2]:=    Integrate[Sin[x]/x,{x,1,Infinity}]
```

```
Limit::nlm:
```

```
Could not find definite limit.
Out[2]=    -Pi   I
           --- + - (-ExpIntegralE[1, -I] +
            2    2

                ExpIntegralE[1, I]) +

           Limit[SinIntegral[x], x -> Infinity]
```

Too bad. To get an idea of the value of $\int_1^\infty \sin[x]/x\,dx$, look at:

```
In[3]:=    N[Integrate[Sin[x]/x,{x,1,100}]]
Out[3]=    0.616142
```

```
In[4]:=    N[Integrate[Sin[x]/x,{x,1,1000}]]
Out[4]=    0.62415
```

```
In[5]:=    N[Integrate[Sin[x]/x,{x,1,5000}]]
Out[5]=    0.624682
```

```
In[6]:=    N[Integrate[Sin[x]/x,{x,1,10000}]]
Out[6]=    0.624808
```

A good seat-of-the-pants estimate of $\int_1^\infty \sin[x]/x\,dx$ is 0.625. Now let's try to check this: Note that

$$\int_1^\infty \sin[x]/x\,dx = \int_1^a \sin[x]/x\,dx + \int_a^\infty \sin[x]/x\,dx.$$

So $\int_1^a \sin[x]/x\,dx$ estimates $\int_1^\infty \sin[x]/x\,dx$ within an error of no more than $\int_a^\infty \sin[x]/x\,dx$. Now to get an idea of the size of the possible error $\int_a^\infty \sin[x]/x\,dx$, integrate by parts:

$$\int_a^t u[x]v'[x]\,dx = u[x]v[x]\Big[_a^t - \int_a^t v[x]u'[x]\,dx.$$

Make the assignments:

```
In[7]:=    u = 1/x
Out[7]=    1
           -
           x
```

```
In[8]:=    vprime = Sin[x]
Out[8]=    Sin[x]
```

Get $v$ :

```
In[9]:=    v = Integrate[vprime,x]
Out[9]=    -Cos[x]
```

Get $u'$ :

```
In[10]:=   uprime = D[u,x]
Out[10]=      -2
           -x
```

The integration by parts formula tells us that

$$\int_a^t \sin[x]/x\,dx = \int_a^t u[x]v'[x]\,dx$$

$$= u[x]v[x]\Big[_{a\,a}^t - \int_a^t v[x]u'[x]\,dx$$

$$= -\cos[x]/x\Big[_a^t - \int_a^t \cos[x]/x^2\,dx$$

$$= -\cos[t]/t + \cos[a]/a - \int_a^t \cos[x]/x^2\,dx.$$

Letting $t \to \infty$ gives

$$\int_a^\infty \sin[x]/x\,dx$$

$$= 0 + \cos[a]/a - \int_a^\infty \cos[x]/x^2\,dx.$$

But $|\cos[x]/x|^2 \le 1/x^2$; so the absolute error

$$|\int_a^\infty \sin[x]/x\,dx|$$

$$\le 0 + |\cos[a]/a| + \int_a^\infty 1/x^2\,dx$$

$$\le \cos[a]/a + 1/a$$

because $\int_a^\infty 1/x^2\,dx$ is given by:

```
In[11]:=   Integrate[1/x^2,{x,a,Infinity}]
Out[11]=   1
           -
           a
```

Consequently, $\int_1^a (\sin[x]/x)\,dx$ estimates $\int_1^\infty (\sin[x]/x)\,dx$ within $|\cos[a]|/a + 1/a$. Thus $\int_1^{10000}(\sin[x]/x)\,dx$ estimates $\int_1^\infty (\sin[x]/x)\,dx$ within an error of no more than:

```
In[12]:=   N[(Abs[Cos[a]]/a + 1/a)/.a->10000]
Out[12]=   0.000195216
```

As a result, the number:

```
In[13]:=    estimate =
            Integrate[Sin[x]/x,{x,1,10000}]
```

```
Out[13]=    I
            - (-ExpIntegralE[1, -I] +
            2

                ExpIntegralE[1, I]) +

             -I
             -- (-ExpIntegralE[1, -10000 I] +
             2

                ExpIntegralE[1, 10000 I])
```

whose decimal evaluation is:

```
In[14]:=    N[estimate]
```
```
Out[14]=    0.624808
```

is accurate to three decimal places. This makes us very confident is writing

$$\int_1^\infty (\sin[x]/x)dx = 0.625$$

# Literacy Sheet

**1.** Calculate the following by integration by parts:

$\int_1^e \log[x]dx$

$\int_1^e x \log[x]dx$

$\int_0^\pi x \sin[x]dx$

$\int_0^1 \arctan[x]dx$.

**2.** Calculate the exact values of:

$\int_1^\infty 1/x^2 dx$

$\int_2^\infty 1/x^p dx$ (assume $p > 1$).

$\int_1^\infty xe^{-2x}dx$.

**3.** What is the integration by parts formula and how is it related to the product rule of differentiation?

**4.** $\int_a^\infty f[x]dx$ is the limiting case of what?

**5.** If $f[x]$ is always positive and $\int_a^m f[x]dx = 6\pi - 10^{-m}$ for every positive integer $m$, then what is the value of $\int_a^\infty f[x]dx$?

**6.** How do you know that $\int_1^{100} \sin[x]^4/x^3 dx$ estimates $\int_1^\infty \sin[x]^4/x^3 dx$ within an error of no more than $1/20000$? If you use the value of $\int_1^{100} \sin[x]^4/x^3 dx$ in place of the true value of $\int_1^\infty \sin[x]^4/x^3 dx$, then how many accurate decimals can you feel confident of?

**7.** Can you calculate $\int_1^\infty (1/x)dx$ with a finite number?

**8.** Let $Int[n] = \int_0^\infty x^n e^{-x} dx$ and find a formula that expresses $Int[n]$ in terms of $Int[n-1]$. Use you formula and a pencil to write out the values of

$$Int[0], Int[1], Int[2], ..., Int[10].$$

**9.** An infinitely long straight rod of variable density is laid out on the non-negative $x$–axis. If its density is $10xe^{-x}$ pounds per foot $x$ units to the right of 0, then what is the total weight of the rod?

# Substitutions

## Guide

This lesson capitalizes on the fundamental theorem and the chain rule for differentiation to allow us to see how to calculate some complicated integrals by replacing them with simpler integrals. The techniques involved are basic to hand calculation of integrals –and they are sometimes fun to do.

In addition we can use these techniques to learn how to handle area and length measurements for curves given parametrically or in polar form.

## Basics

■ **B.1) Breaking more of the code of the integral: Integration by substitution.**

#### B.1.a)

How do you know that the integrals

$$\int_{aa}^{b} f'[u[x]]u'[x]dx$$

and

$$\int_{u[a]}^{u[b]} f'[u]du$$

are equal?

**Answer:**

The best way to see why they are equal is to calculate both of them.

The second is the easier because the fundamental theorem of calculus tells us that

$$\int_{u[a]}^{u[b]} f'[u]du = f[u]\Big|_{u[a]}^{u[b]} = f[u[b]] - f[u[a]].$$

For the first integral, the chain rule tells us that the derivative of $f[u[x]]$ is $f'[u[x]]u'[x]$. The fundamental theorem of calculus tells us that

$$\int_{aa}^{b} f'[u[x]]u'[x]dx = f[u[x]]\Big|_{a}^{b} = f[u[b]] - f[u[a]].$$

Thus the two integrals are equal because **they both have the same value,** namely

$$f[u[b]] - f[u[a]].$$

A point of anxiety might come up: In the integral

$$\int_{u[a]}^{u[b]} f'[u]du,$$

the lone symbol $u$ is treated as a variable and $u[a]$ and $u[b]$ are treated as numbers. In the integral,

$$\int_{aa}^{b} f'[u[x]]u'[x]dx,$$

$u[x]$ is a function, $x$ is a variable and $a$ and $b$ are numbers. This should not cause you any trouble.

The fact that

$$\int_{aa}^{b} f'[u[x]]u'[x]dx = \int_{u[a]}^{u[b]} f'[u]du$$

is a fortunate mathematical coincidence which will be exploited again and again throughout this lesson.

#### B.1.b)

Of what practical use is the fact that the two integrals

$$\int_{aa}^{b} f'[u[x]]u'[x]dx$$

and

$$\int_{u[a]}^{u[b]} f'[u]du$$

are equal?

**Answer:**

Notational magic allows us to take the more complicated integral and replace it with the less complicated but equal integral.

We pair them up as follows:

$f'[u[x]]$ <-------> $f'[u]$

$u'[x]dx$ <-------> $du$

$\int_{aa}^{b}$ <-------> $\int_{u[a]}^{u[b]}$ .

#### B.1.c)

Use the pairings

$f'[u[x]]$ <-------> $f'[u]$

$$u'[x]dx \longleftrightarrow du$$

$$\int_{aa}^{b} \longleftrightarrow \int_{u[a]}^{u[b]}$$

to calculate

$$\int_{0}^{\pi} \cos[x^2]2x\,dx.$$

Check yourself with *Mathematica* .

---

**Answer:**

The key is that $2x$ is the derivative of $x^2$; so we use

$$u[x] = x^2.$$

This gives the pairings

$$\cos[x^2] \longleftrightarrow \cos[u]$$

$$2x\,dx = u'[x]dx \longleftrightarrow du$$

$$\int_{0}^{\pi} = \int_{aa}^{b} \longleftrightarrow \int_{u[a]}^{u[b]} = \int_{0^2}^{\pi^2}.$$

So

$$\int_{0}^{\pi} \cos[x^2]2x\,dx = \int_{0^2}^{\pi^2} \cos[u]du$$

$$= \sin[u]\Big[_{0^2}^{\pi^2} = \sin[\pi^2] - \sin[0] = \sin[\pi^2].$$

Check:

```
In[1]:=    Integrate[Cos[x^2] 2 x, {x,0,Pi}]
                   2
Out[1]=    Sin[Pi ]
```

Got it.

**B.1.d)**

---

Calculate

$$\int_{0}^{\pi/2} e^{Sin[x]} \cos[x]dx.$$

Check yourself with *Mathematica.*

---

**Answer:**

$\cos[x]$ is the derivative of $\sin[x]$; so we use $u[x] = \sin[x]$.

This gives

$$e^{sin[x]} \longleftrightarrow e^{u}$$

$$\cos[x]dx = u'[x]dx \longleftrightarrow du$$

$$\int_{0}^{\pi/2} \longleftrightarrow \int_{Sin[0]}^{Sin[\pi/2]} = \int_{0}^{1}.$$

So

$$\int_{0}^{\pi/2} e^{Sin[x]} \cos[x]dx$$

$$= \int_{0}^{1} e^{u}du = e^{u}\Big[_{0}^{1} = e^1 - e^0 = e - 1.$$

Check:

```
In[1]:=    Integrate[E^(Sin[x]) Cos[x],{x,0,Pi/2}]
Out[1]=    -1 + E
```

Got it.

**B.1.e)**

---

Calculate

$$\int_{0}^{\pi/2} e^{sin[3x]} \cos[3x]dx.$$

Check yourself with *Mathematica.*

---

**Answer:**

$3\cos[3x]$ is the derivative of $\sin[3x]$; so we use $u[x] = \sin[3x]$. The $3\cos[3x]$ term we want is not immediately avaiable; so rewrite the integral as

$$(1/3)\int_{0}^{\pi/2} e^{Sin[3x]}3\cos[3x]dx.$$

Now there is a $3\cos[3x]$ term right where we want it.

This gives the pairings

$$e^{sin[3x]} \longleftrightarrow e^{u}$$

$$3\cos[3x]dx = u'[x]dx \longleftrightarrow du$$

$$(1/3)\int_{0}^{\pi/2} \longleftrightarrow (1/3)\int_{0}^{Sin[3\pi/2]} = (1/3)\int_{0}^{-1}.$$

So

$$\int_{0}^{\pi/2} e^{Sin[3x]} \cos[3x]dx$$

$$= (1/3)\int_{0}^{\pi/2} e^{Sin[3x]}3\cos[3x]dx$$

$$= (1/3)\int_{0}^{-1} e^{u}du = (1/3)e^{u}\Big[_{0}^{-1} = (e^{-1}-e^0)/3 = (e^{-1}-1)/3.$$

Check:

```
In[1]:=    Integrate[E^(Sin[3 x]) Cos[3 x],{x,0,Pi/2}]
                1      1
Out[1]=    -(-)  +  ---
                3     3 E
```

Got it.

**■ B.2) Integrals for curves given parametrically.**

**B.2.a)**

---

Suppose we find a curve in parametric equations $x = x[t]$ and $y = y[t]$ and our job is to measure the area under the curve over the interval $[a, b]$ on the $x$ -axis.

If we know that there is a function $f[x]$ such that $y[t] = f'[x[t]]$ and if we can find a formula for $f[x]$, then we can just make the measurement by calculating

$$\int_a^b f'[x]dx = f[b] - f[a].$$

If we cannot find a formula for $f[x]$, it might seem that we are out of luck in calculating

$$\int_a^b f'[x]dx = \int_a^b ydx.$$

But even if we are *not* able to come up with a formula for $f[x]$, there is an easy way to calculate this integral.

How?

---

**Answer:**

Find the numbers $A$ and $B$ such that $x[A] = a$ and $x[B] = b$. Make the substitution $x = x[t]$.

This gives the pairings (these pairings are actually the reverse of the pairings we did in B.1):

$$f'[x] \longleftarrow\!\text{-------} \longrightarrow f'[x[t]]$$

$$dx \longleftarrow\!\text{-------} \longrightarrow x'[t]dt$$

$$\int_{aa}^b \longleftarrow\!\text{-------} \longrightarrow \int_{x[a]}^{x[b]} = \int_A^B \ .$$

So

$$\int_a^b f'[x]dx = \int_{x[a]}^{x[b]} f'[x[t]]x'[t]dt = \int_A^B f'[x[t]]x'[t]dt.$$

This does not look promising because we don't have a formula for $f[x]$.

But wait a minute! We know $f'[x[t]] = y[t]$, and this gives us the formula we want:

$$\int_a^b f'[x]dx = \int_A^B f'[x[t]]x'[t]dt = \int_{Ay}^B [t]x'[t]dt.$$

This tells us how to calculate the area measurement $\int_a^b f'[x]dx$ directly from the parametric equations by calculating

$$\int_A^B y[t]x'[t]dt$$

*without bothering about a formula for* $f[x]$. All we need are the formulas for $x[t]$ and $y[t]$.

Neat trick.

**B.2.b)**

---

Plot and measure the area under one arch of the cycloid

$$x = x[t] = 10t - 10\sin[t]$$

and

$$y = y[t] = 10 - 10\cos[t].$$

---

**Answer:**

As usual, we take a look at the situation by opening one of *Mathematica* 's magic boxes:

```
In[1]:=  x[t_] = 10 t - 10 Sin[t]
         y[t_] = 10  - 10 Cos[t]
         ParametricPlot[{x[t],y[t]},{t,0,2 Pi},
         AspectRatio->Automatic,PlotStyle->
         {{RGBColor[0,0,1],Thickness[0.005]}}];
```

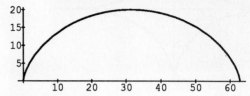

Look at:

```
In[2]:=  x[0]
Out[2]=  0
```

```
In[3]:=  x[2 Pi]
Out[3]=  20 Pi
```

The area measurement we are after is

$$\int_0^{20\pi} ydx = \int_0^{2\pi} y[t]x'[t]dt :$$

```
In[4]:=  Integrate[y[t] x'[t],{t,0,2 Pi}]
Out[4]=  300 Pi
```

Easy.

But think of how hard this would have been if we had to find the formula for the function $f[x]$ with $y[t] = f[x[t]]$. Until we got to this lesson, finding this formula would have been necessary.

**B.2.c)**

---

Plot and measure the area enclosed by the parametric curve

$$x = x[t] = 7\cos[t]^3$$

and

$$y = y[t] = 4\sin[t]^5.$$

---

**Answer:**

Take a look:

```
In[1]:=  x[t_] = 7 Cos[t]^3
         y[t_] = 4 Sin[t]^5

         ParametricPlot[{x[t],y[t]},{t,0,2 Pi},
         AspectRatio->Automatic,PlotStyle->
         {{RGBColor[0,0,1],Thickness[0.005]}}];
```

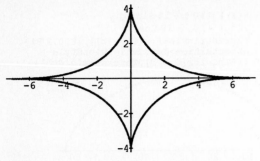

Evidently the area the curve encloses is 4 times the area under the curve and over $[0,7]$ on the $x$-axis.

Let's see which $t's$ give us the plot in the first quadrant: It looks like $0 \leq t \leq \pi/2$ will do the trick, but we check with a plot:

$In[2]:=$ **ParametricPlot[{x[t],y[t]},{t,0,Pi/2},**
**AspectRatio->Automatic];**

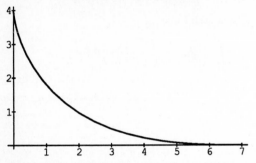

Good. Now check to see which $t's$ correspond to $x = 0$ and $x = 7$ :

$In[3]:=$ **x[0]**

$Out[3]=$ **7**

$In[4]:=$ **x[Pi/2]**

$Out[4]=$ **0**

The measurement we want is

$$4 \int_0^7 y\,dx = 4 \int_{\pi\pi/2}^0 y[t]x'[t]dt :$$

$In[5]:=$ **4 Integrate[y[t] x'[t],{t,Pi/2,0}]**

$Out[5]=$ $\dfrac{105\ Pi}{16}$

Looks good!

# Tutorial

### ■ T.1) Integration by substitution.

Calculate the following integrals by hand. Check with *Mathematica*.

**T.1.a)**

$$\int_0^1 \cos^4[\pi x]\sin[\pi x]dx$$

**Answer:**

$-\pi\sin[\pi x]$ is the derivative of $\cos[\pi x]$; so we use $u[x] = \cos[\pi x]$. Rewrite the integral as

$$(-1/\pi)\int_0^1 \cos^4[x\pi](-\pi\sin[\pi x])dx.$$

This gives the pairings

$\cos^4[\pi x] <-------> u^4$

$-\pi\sin[\pi x]dx = u'[x]dx <------> du-(1/\pi)\int_0^1 <---$
$---> -(1/\pi)\int_{Cos[0]}^{Cos[\pi]} = -(1/\pi)\int_{+1}^{-1}$ .

So

$$\int_0^1 \cos^4[\pi x]\sin[\pi x]dx$$

$$= (-1/\pi)\int_0^1 \cos^4[\pi x](-\pi\sin[\pi x])dx$$

$$= (-1/\pi)\int_{+1}^{-1} u^4 du = (-1/\pi)(u^5)/5\Big[_1^{-1}.$$

$In[1]:=$ **upper = (-1/Pi) (u^5)/5/.u->(-1)**

$Out[1]=$ $\dfrac{1}{5\ Pi}$

$In[2]:=$ **lower = (-1/Pi) (u^5)/5/.u->(1)**

$Out[2]=$ $\dfrac{-1}{5\ Pi}$

The integral is:

$In[3]:=$ **upper - lower**

$Out[3]=$ $\dfrac{2}{5\ Pi}$

Check:

$In[4]:=$ **Integrate[Cos[Pi x]^4 Sin[Pi x],{x,0,1}]**

$Out[4]=$ $\dfrac{2}{5\ Pi}$

Got it.

**T.1.b)**

$$\int_{2.7}^{6.5} x/^3 \sqrt{1 + x^4}\,dx$$

---

**Answer:**

$4x^3$ is the derivative of $1 + x^4$; so we use $u[x] = 1 + x^4$.

Rewrite the integral as

$$(1/4)\int_{2.7}^{6.5}(1/\sqrt{1 + x^4})4x^3\,dx.$$

This gives the pairings

$(1/\sqrt{1 + x^4})$ <-------> $1/\sqrt{u}$

$4x^3 dx = u'[x]dx$ <-------> $du$

$(1/4)\int_{2.7}^{6.5}$ <-------> $(1/4)\int_{1+2.7^4}^{1+6.5^4}$ .So

$$\int_{2.7}^{6.5} x/^3 \sqrt{1 + x^4}\,dx = (1/4)\int_{2.7}^{6.5} 1/\sqrt{1 + x^4}\,4x^3\,dx$$

$$= (1/4)\int_{1+2.7^4}^{1+6.5^4} 1/\sqrt{u}\,du = (1/4)\int_{1+2.7^4}^{1+6.5^4} u^{-1/2}\,du$$

$$= (1/2)u^{1/2}\Big[_{11+2.7^4}^{1+6.5^4}.$$

```
In[1]:=    upper = (1/2) Sqrt[u]/.u->(1 + 6.5^4)
Out[1]=    21.1309
```

```
In[2]:=    lower = (1/2) Sqrt[u]/.u->(1 + 2.7^4)
Out[2]=    3.67913
```

The integral is:

```
In[3]:=    upper - lower
Out[3]=    17.4518
```

Check:

```
In[4]:=    Integrate[x^3/(Sqrt[1 + x^4]),
           {x,2.7,6.5}]
Out[4]=    17.4518
```

Got it. If you want extra accuracy, then you can get it:

```
In[5]:=    N[upper - lower, 9]
Out[5]=    17.4517826
```

```
In[6]:=    integral = Integrate[
           x^3/(Sqrt[1 + x^4]),{x,27/10,65/10}]
Out[6]=    41 Sqrt[17]     Sqrt[541441]
           -----------  -  ------------
               8              200
```

```
In[7]:=    N[integral,9]
Out[7]=    17.4517826
```

**T.1.c)**

---

$$\int_{2.7}^{6.5} x/^2 \sqrt{1 + x^4}\,dx$$

---

**Answer:**

We cannot go with $u[x] = 1 + x^4$ in this one because in the pairing

$u'[x]dx$ <-------> $du$

we must have an $x^3$ term available in the numerator. The only term we have to go with is $x^2$; so we are pissed off.

Let's try the machine:

```
In[1]:=    Integrate[
           x^2/(Sqrt[1 + x^4]),{x,2.7,6.5}]
Out[1]=
                         2
                        x
           Integrate[------------, {x, 2.7, 6.5}]
                               4
                     Sqrt[1 + x ]
```

No luck; the machine is also stumped. In this case (unless someone comes up with a brilliant suggestion), we have no choice but to compromise with a numerical approximation:

```
In[2]:=    NIntegrate[
           x^2/(Sqrt[1 + x^4]),{x,2.7,6.5}]
Out[2]=    3.79219
```

At least we got an answer.

**T.1.d)**

---

$$\int (e^{2x})/(1 + e^{2x})\,dx$$

---

**Answer:**

$2e^{2x}$ is the derivative of $(1+e^{2x})$; so we use $u[x] = (1+e^{2x})$.

Rewrite the integral as

$$(1/2)\int (1/(1 + e^{2x}))2e^{2x}\,dx.$$

This gives the pairings

$1/(1 + e^{2x})$ <-------> $1/u$

$2e^{2x}dx = u'[x]dx$ <-------> $du$

$(1/2)\int$ <-------> $(1/2)\int$ .

So

$$\int e^{2x}/(1+e^{2x})dx = (1/2)\int (1/(1+e^{2x}))2e^{2x}dx$$

$$= (1/2)\int (1/u)du = (1/2)\log[u] = \log[1+e^{2x}]/2.$$

You can add a constant if you like. Check:

```
In[1]:=    D[Log[1 + E^(2 x)]/2,x]
```

```
Out[1]=        2 x
           E
          --------
                2 x
          1 + E
```

or

```
In[2]:=    Integrate[
           (E^(2 x))/(1 + E^(2 x)),x]
```

```
Out[2]=          2 x
          Log[1 + E   ]
          -------------
                2
```

Got it coming and going.

■ **T.2) Squashing the geeks.**

**T.2.a)**

Use a substitution to calculate

$$\int_1^{16} \sqrt{x}/(1+x^{1/4})dx.$$

Check with *Mathematica*.

**Answer:**

The geek in this integral is $x^{1/4}$.

Squash the geek by setting $x = t^4$. This gives the pairings

$$\sqrt{x}/(1+x^{1/4}) \longleftrightarrow \sqrt{t^4}/(1+(t^4)^{1/4}) = t^2/(1+t)$$

$$dx \longleftrightarrow 4t^3 dt$$

$$\int_1^{16} \longleftrightarrow \int_1^2.$$

So

$$\int_1^{16} \sqrt{x}/(1+x^{1/4})dx = \int_1^2 (t^2/(1+t))4t^3 dt$$

$$= \int_1^2 4t^5/(1+t))dt.$$

The process of going from $x's$ to $t's$ can be automated a bit:

```
In[1]:=    subs = x->t^4
```

```
Out[1]=          4
          x -> t
```

```
In[2]:=    xintegrand = Sqrt[x]/(1 + x^(1/4))
```

```
Out[2]=    Sqrt[x]
          --------
                1/4
          1 + x
```

```
In[3]:=    tintegrand =
           (xintegrand/.subs) D[x/.subs,t]
```

```
Out[3]=        5
           4 t
          -----
          1 + t
```

Next divide $4t^5$ by $(1+t)$ :

```
In[4]:=    Apart[tintegrand]
```

```
Out[4]=               2       3       4      4
          4 - 4 t + 4 t  - 4 t  + 4 t  - -----
                                         1 + t
```

Now we can read off

$$\int_1^{16} \sqrt{x}/(1+x^{1/4})dx = \int_1^2 4t^5/(1+t))dt$$

$$= (4t - 2t^2 + 4t^3/3 - t^4 + 4t^5/5 - 4\log[1+t])\Big|_1^2.$$

```
In[5]:=    upper = (4t - 2 t^2 + 4 (t^3)/3 -
           t^4 + 4 (t^5)/5 - 4 Log[1+t])/.t->2
```

```
Out[5]=    304
          --- - 4 Log[3]
          15
```

```
In[6]:=    lower = (4t -2 t^2 + 4 (t^3)/3 -
           t^4 + 4 (t^5)/5 - 4 Log[1+t])/.t->1
```

```
Out[6]=    47
          -- - 4 Log[2]
          15
```

So $\int_1^{16} \sqrt{x}/(1+x^{1/4})dx = \int_1^2 4t^5/(1+t))dt$ is:

```
In[7]:=    upper - lower
```

```
Out[7]=    257
          --- + 4 Log[2] - 4 Log[3]
          15
```

Check:

```
In[8]:=    Integrate[
           Sqrt[x]/(1 + x^(1/4)),{x,1,16}]
```

```
Out[8]=    257
          --- + 4 Log[2] - 4 Log[3]
          15
```

Got it.

**T.2.b.i)**

Use a substitution to calculate

$$\int_0^{Sqrt[5]} x^3/\sqrt{16-x^2}dx.$$

Check with *Mathematica*.

---

**Answer:**

The geek in this integral is the term $\sqrt{16 - x^2}$. Squash the geek by setting $16 - x^2 = t^2$. This means $x^2 = 16 - t^2$. It also means

$2x\,dx \longleftarrow\!\!\!-\!\!\!-\!\!\!-\!\!\!\longrightarrow -2t\,dt$

which is the same as $x\,dx \longleftarrow\!\!\!-\!\!\!-\!\!\!-\!\!\!\longrightarrow -t\,dt$. This substitution also gives the pairings

$x^2/\sqrt{16 - x^2} \longleftarrow\!\!\!-\!\!\!-\!\!\!-\!\!\!\longrightarrow (16 - t^2)/\sqrt{t^2} = (16 - t^2)/t$

$\int_0^{Sqrt[5]} \longleftarrow\!\!\!-\!\!\!-\!\!\!-\!\!\!\longrightarrow \int_4^{Sqrt[11]}$

because $t = \sqrt{11}$ when $x = \sqrt{5}$ and $t = 4$ when $x = 0$. Let the computer handle the algebra:

```
In[1]:=   subs = x->Sqrt[16 - t^2]
                       2
Out[1]=   x -> Sqrt[16 - t ]
```

```
In[2]:=   xintegrand = (x^3)/Sqrt[16 - x^2]
                   3
                  x
Out[2]=   -------------
                     2
          Sqrt[16 - x ]
```

```
In[3]:=   tintegrand = ExpandAll[
            (xintegrand/.subs) D[x/.subs,t]]
                   2
Out[3]=   -16 + t
```

At this point the integral is a dead duck because, now we can read off

$$\int_0^{Sqrt[5]} x^3/\sqrt{16 - x^2}\,dx = \int_4^{Sqrt[11]} (t^2 - 16)\,dt$$

$$= (t^3/3 - 16t)\Big|_4^{Sqrt[11]}$$

```
In[4]:=   upper = ((t^3)/3 - 16 t)/.t->Sqrt[11]
                          3/2
                       11
Out[4]=   -16 Sqrt[11] + -----
                          3
```

```
In[5]:=   lower = ((t^3)/3 - 16 t)/.t->4
              128
Out[5]=   -(---)
               3
```

So

$$\int_0^{Sqrt[5]} x^3/\sqrt{16 - x^2}\,dx = \int_4^{Sqrt[11]} (t^2 - 16)\,dt$$

is:

```
In[6]:=   upper - lower
                               3/2
          128                11
Out[6]=   --- - 16 Sqrt[11] + -----
           3                   3
```

Check:

```
In[7]:=   Apart[Simplify[Integrate[
            (x^3)/Sqrt[16 - x^2],{x,0,Sqrt[5]}]]]
                               3/2
          128                11
Out[7]=   --- - 16 Sqrt[11] + -----
           3                   3
```

Got it.

**T.2.b.ii)**

---

Use a trig substitution to calculate

$$\int_0^{Sqrt[5]} x^3/\sqrt{16 - x^2}\,dx.$$

Check with *Mathematica*.

---

**Answer:**

The geek in this integral still is the term $\sqrt{16 - x^2}$. This time we squash the geek by setting $x = 4\sin[t]$. This is a good idea because it gives the pairing

$x^3/\sqrt{16 - x^2} \longleftarrow\!\!\!-\!\!\!-\!\!\!-\!\!\!\longrightarrow 4^3 \sin[t]^3/\sqrt{16 - 16\sin[t]^2}$

and this is also $= 4^3 \sin[t]^3/4\cos[t[$ This substitution also gives the pairings

$dx \longleftarrow\!\!\!-\!\!\!-\!\!\!-\!\!\!\longrightarrow 4\cos[t]\,dt \quad \int_0^{Sqrt[5]} \longleftarrow\!\!\!-\!\!\!-\!\!\!-\!\!\!\longrightarrow \int_0^{ArcSin[Sqrt[5]/4]}$

because $\sqrt{5} = 4\sin[\arcsin[5]/4]$. Let the computer handle the algebra:

```
In[1]:=   subs = x->4 Sin[t]
Out[1]=   x -> 4 Sin[t]
```

```
In[2]:=   xintegrand = (x^3)/Sqrt[16 - x^2]
                   3
                  x
Out[2]=   -------------
                     2
          Sqrt[16 - x ]
```

```
In[3]:=   tintegrand =
            ExpandAll[(xintegrand/.subs)
            D[x/.subs,t]]
                             3
          256 Cos[t] Sin[t]
Out[3]=   --------------------
                            2
          Sqrt[16 - 16 Sin[t] ]
```

We recognize this as:

```
In[4]:=    tintegrand =
           256 Cos[t] Sin[t]^3/(4 Cos[t])
```
```
Out[4]=           3
           64 Sin[t]
```

Now apply some trig identities:

```
In[5]:=    tintegrand = TrigExpand[tintegrand]
```
```
Out[5]=    48 Sin[t] - 16 Sin[3 t]
```

The integral is a dead duck again because

$$\int_0^{Sqrt[5]} x^3/\sqrt{16-x^2}dx$$

$$= \int_0^{ArcSin[Sqrt[5]/4]} (48\sin[t] - 16\sin[3t])dt$$

$$= (-48\cos[t] + (16/3)\cos[3t]) \Big|_0^{ArcSin[Sqrt[5]/4]}$$

Here is the value in decimals:

```
In[6]:=    f[t_] = -48 Cos[t] + (16/3) Cos[3 t]
           N[f[ArcSin[Sqrt[5]/4]] - f[0]]
```
```
Out[6]=    1.76163
```

Check:

```
In[7]:=    NIntegrate[
           x^3/Sqrt[16 - x^2],{x,0,Sqrt[5]}]
```
```
Out[7]=    1.76163
```

That was a big-time integral.

Your friends in the old-fashioned calculus classes expend a lot of blood, sweat and tears doing this kind of integral with pencil and paper.

But in the new style courses like this one, you can do what they are doing without working up even a bead of perspiration.

You might wonder why they are even doing thus sort of thing.

Maybe it is just stay busy.

■ **T.3) Area inside a closed parametric curve.**

Here is a parametric curve:

```
In[8]:=    x[t_] = 4 t ( 2 - t ) E^(t/4)
           y[t_] = 2 - Sin[Pi t]
           a = 0
           b = 2

           ParametricPlot[{x[t],y[t]},{t,a,b},
           AxesLabel->{"x","y"},Axes->{0,0},
           PlotRange->{{0,6},{0,4}},
           PlotStyle->RGBColor[1,0,0]];
```

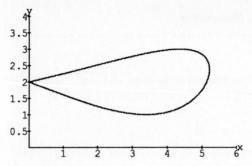

Next look at:

```
In[9]:=    ParametricPlot[{x[t],y[t]},{t,0,1},
           AxesLabel->{"x","y"},Axes->{0,0},
           PlotRange->{{0,6},{0,4}},
           PlotStyle->RGBColor[1,0,0]];
```

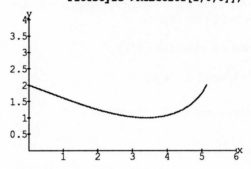

As $t$ leaves $a = 0$ and advances to 1, the points $\{x[t], y[t]\}$ initially ride on the lower part of the curve.

```
In[10]:=   ParametricPlot[{x[t],y[t]},{t,1,2},
           AxesLabel->{"x","y"},Axes->{0,0},
           PlotRange->{{0,6},{0,4}},
           PlotStyle->RGBColor[1,0,0]]
```

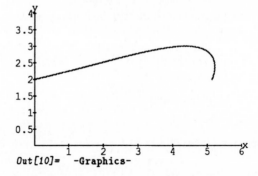

```
Out[10]=   -Graphics-
```

As $t$ advances from 1 to $b = 2$, the point $\{x[t], y[t]\}$ rides on the lower part until $t$ reaches an intermediate point $c$ at which $\{x[t], y[t]\}$ stops moving to the right and the begins moving to the left on its return trip to $\{0,0\}$ on the higher part of the curve.

Thus the point $\{x[t], y[t]\}$ moves in **counterclockwise** fashion as $t$ advances from $a$ to $b$.

If we knew the precise value of $c$, then we could measure the area enclosed by this curve by writing (see B.2)

$$Area = \int_b^c y[t]x'[t]dt - \int_a^c y[t]x'[t]dt.$$

**T.3.a)**

What good is this if we don't know the value of $c$?

---

**Answer:**

Good question. And like many good questions this question has a good answer.

Look again at the formula

$$Area = \int_b^c y[t]x'[t]dt - \int_a^c y[t]x'[t]dt$$

$$= -\int_c^b y[t]x'[t]dt - \int_a^c y[t]x'[t]dt$$

$$= -\left(\int_c^b y[t]x'[t]dt + \int_a^c y[t]x'[t]dt\right)$$

$$= -\left(\int_a^c y[t]x'[t]dt + \int_c^b y[t]x'[t]dt\right)$$

$$= -\int_a^b y[t]x'[t]dt.$$

A miracle just happened.

To calculate the area, we don't give two hoots in hell what the value of $c$ is because we can calculate the area inside this curve via the formula

$$Area = -\int_a^b y[t]x'[t]dt.$$

where no $c$ appears. Hot damn!

**T.3.b)**

---

Measure the area inside the curve described above.

---

**Answer:**

Use the formula

$$Area = -\int_a^b y[t]x'[t]dt.$$

*Mathematica* will probably take a long time to get an exact answer, so we'll settle for a numerical approximation:

```
In[1]:=   - NIntegrate[y[t]x'[t],{t,a,b}]
Out[1]=   6.4952
```

When all is said and done, this was not too hard.

**T.3.c)**

---

For what other curves parameterized by $x = x[t]$, $y = y[t]$ with $a \le t \le b$ and

$$\{x[a], y[a]\} = \{x[b], y[b]\}$$

can we use the formula

$$Area = -\int_a^b y[t]x'[t]dt$$

to accurately measure the area inside the curve?

---

**Answer:**

Scroll back to the answer to part i) above. The argument leading to the formula

$$Area = -\int_a^b y[t]x'[t]dt$$

did not involve the specific functions or the specific values of $a$ and $b$.

Only two considerations were important:

i) The curve must have no interior loops.

ii) As $t$ advances from $a$ to $b$, the point $\{x[t], y[t]\}$ must go around the curve in a counterclockwise way and must trace out the curve only once.

If you have a situation that stays within these ground rules, then you may use the formula

$$Area = -\int_a^b y[t]x'[t]dt$$

with confidence.

**T.3.d.i)**

---

What happens if everything else is O.K., but the curve is traced out in a clockwise way?

---

**Answer:**

The formula becomes

$$Area = +\int_a^b y[t]x'[t]dt.$$

**T.3.d.ii)**

---

Are there other easy formulas for the area?

---

**Answer:**

Yes. For *Counterclockwise* movement we have:

$$Area = -\int_a^b y[t]x'[t]dt$$

$$Area = +\int_a^b x[t]y'[t]dt$$

(You get this one by calculating area along the $y$-axis instead of the $x$-axis.)

Or we can use half of each:

$$Area = \int_a^b (x[t]y'[t] - y[t]x'[t])/2dt$$

For *Clockwise* movement we have:

$$Area = + \int_a^b y[t]x'[t]dt$$

$$Area = - \int_a^b x[t]y'[t]dt$$

Or we can use half of each:

$$Area = \int_a^b (y[t]x'[t]dt - x[t]y'[t])/2dt$$

**T.3.e)**

---

Measure the area inside the parametric curve

$$x = x[t] = \sin[t]^3$$

and

$$y = y[t] = \cos[t]^3$$

for $0 \le t \le 2\pi$.

---

**Answer:**

```
In[1]:=    x[t_] = Sin[t]^3
           y[t_] = Cos[t]^3
           a = 0
           b = 2 Pi

           ParametricPlot[{x[t],y[t]},{t,a,b},
           AxesLabel->{"x","y"},Axes->{0,0},
           AspectRatio->Automatic];
```

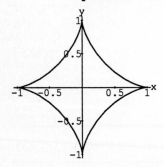

Let's see whether $\{x[t], y[t]\}$ traces ou the curve in a clockwise or counterclockwise way as $t$ advances from $a = 0$ :

```
In[2]:=    ParametricPlot[{x[t],y[t]},{t,0,1},
           AxesLabel->{"x","y"},
           PlotRange->
           {{-1.5,1.5},{-1.5,1.5}}];
```

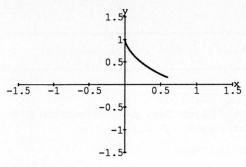

Clockwise.

Let's see whether it is traced out more than once as $t$ advances from 0 to $2\pi$ :

```
In[3]:=    ParametricPlot[{x[t],y[t]},
           {t,0,2 Pi - 1},
           AxesLabel->{"x","y"}];
```

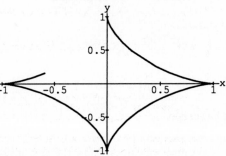

Good. The curve is traced out only once. A formula for the area is

$$Area = + \int_a^b y[t]x'[t]dt :$$

```
In[4]:=    Integrate[y[t]x'[t],{t,a,b}]
```
```
Out[4]=    3 Pi
           ----
            8
```

Another correct formula is

$$Area = - \int_a^b x[t]y'[t]dt :$$

```
In[5]:=    -Integrate[x[t]y'[t],{t,a,b}]
```
```
Out[5]=    3 Pi
           ----
            8
```

Convinced?

■ **T.4) Polar plots and area measurements.**

The usual way of specifying a point in the plane is to give its coordinates

$$\{x, y\}.$$

```
In[6]:=    p={1.2,1}
           point = Graphics[{RGBColor[1,0,0],
           PointSize[0.02],Point[p]}]
           label =
           Graphics[Text[{"x","y"},p,{0,-2}]]

           Show[point,label,
           Axes->{0,0},AxesLabel->{"x","y"},
           PlotRange->{{0,2.2},{0,1.2}},
           AspectRatio->Automatic,
           Ticks->{{1,2},{1}}];
```

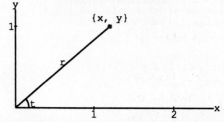

The same point can be specified by the indicated polar angle $t$ and the distance $r$ from the point to the origin. The pair $\{r, t\}$ is called the polar coordinates of the point $\{x, y\}$.

```
In[7]:=    dist = Graphics[Line[{{0,0},p}]]
           rlabel =
           Graphics[Text["r",p/2,{0,-2}]]
           angle = Graphics[Circle[{0,0},.17,
           {0,ArcTan[p[[2]]/p[[1]]]}]]
           anglelabel =
           Graphics[Text["t",{.22,.05}]]

           Show[point,label, dist,rlabel,
           angle,anglelabel,Axes->{0,0},
           AxesLabel->{"x","y"},
           PlotRange->{{0,2.5},{0,1.2}},
           AspectRatio->Automatic,
           Ticks->{{1,2},{1}}];
```

Here the polar angle $t$ is measured in the counterclockwise sense from the $x$-axis and $r$ is the distance from the origin to the point $\{x, y\}$.

Clearly if you know the polar coordinates $r$ and $t$, then you can find where the point $\{x, y\}$ is:

You just leave the origin $\{0, 0\}$ in the direction specified by $t$ and you walk $r$ units out until you get to the point $\{x, y\}$.

The usual convention is: If $r > 0$, then you walk forward, but if $r < 0$, then you walk backward. If $r = 0$, then you stay put at the origin $\{0, 0\}$.

In fact if you know $r$ and $t$, then you know $x$ and $y$ through the easy formulas:

$$x = r \cos[t]$$

and

$$y = r \sin[t].$$

## T.4.a)

Measure the area enclosed by the polar curve

$$r[t] = 1 - \sin[t].$$

## Answer:

First let's look at a plot. To get it, just use:

```
In[1]:=    r[t_] = 1 - Sin[t]
           x[t_] = r[t] Cos[t]
           y[t_] = r[t] Sin[t]

           cardioid =
           ParametricPlot[{x[t],y[t]},{t,0,2 Pi},
           Axes->{0,0},AxesLabel->{"x","y"},
           AspectRatio->Automatic];
```

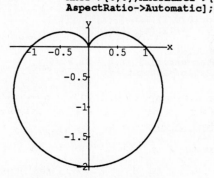

Faint heart never won fair lady!

Nothing venture, nothing win-

Blood is thick but water's thin-

In for a penny, in for a pound-

It's Love that makes the world go round!

-W. S. Gilbert in *The Mikado*

Let's see how this heart-shaped curve called a "cardioid" is traced out as $t$ advances from 0 to $2\pi$.

```
In[2]:=    ParametricPlot[{x[t],y[t]},{t,0,1},
           Axes->{0,0},AxesLabel->{"x","y"},
           PlotRange->{{-1.5,1.5},{-2,0.5}},
           AspectRatio->Automatic,
           PlotStyle->RGBColor[1,0,0]]
```

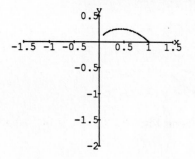

```
In[3]:=   ParametricPlot[{x[t],y[t]},
          {t,2 Pi-1,2 Pi},
          Axes->{0,0},AxesLabel->{"x","y"},
          PlotRange->{{-1.5,1.5},{-2,0.5}},
          AspectRatio->Automatic,
          PlotStyle->RGBColor[1,0,0]]
```

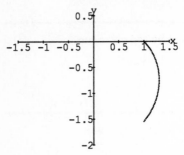

Counterclockwise.

This can also be seen by looking at some tangent vectors that point in the direction of increasing $t$ :

```
In[4]:=   tanvector[t_] = Graphics[
          RGBColor[0,0,1],Line[{{x[t],y[t]},
          {x[t],y[t]}+{x'[t],y'[t]}}]]

          Show[cardioid,
          Table[tanvector[s],{s,0,2 Pi,Pi/6}],
          AspectRatio->Automatic];
```

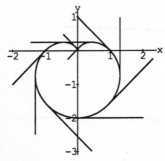

Again we see that as $t$ advances from 0 to $2Pi$, the curve is traced out in a counterclockwise way. Let's be sure that the curve is not traced out more than once as $t$ advances from 0 to $2\pi$ :

```
In[5]:=   ParametricPlot[{x[t],y[t]},{t,0,2 Pi-1},
          Axes->{0,0},AxesLabel->{"x","y"},
          PlotRange->{{-1.5,1.5},{-2,0.5}},
          AspectRatio->Automatic];
```

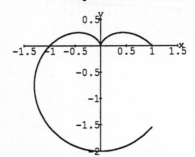

Good. The area measurement is:

```
In[6]:=   -Integrate[y[t] x'[t],{t,0,2 Pi}]
Out[6]=   3 Pi
          ----
          2
```

This is the same as:

```
In[7]:=   Integrate[x[t] y'[t],{t,0,2 Pi}]
Out[7]=   3 Pi
          ----
          2
```

And this is the same as:

```
In[8]:=   Integrate[(x[t] y'[t] - y[t] x'[t])/2,
          {t,0,2 Pi}]
Out[8]=   3 Pi
          ----
          2
```

**T.4.b.i)**

Measure the area enclosed by the polar curve

$$r[t] = 1 - 2\sin[t].$$

**Answer:**

First let's look at a plot. To get it, just use:

```
In[1]:=   r[t_] = 1 - 2 Sin[t]
          x[t_] = r[t] Cos[t]
          y[t_] = r[t] Sin[t]

          ParametricPlot[{x[t],y[t]},{t,0,2 Pi},
          Axes->{0,0},AxesLabel->{"x","y"},
          PlotRange->{{-2.5,2.5},{-3.5,1.5}},
          AspectRatio->Automatic];
```

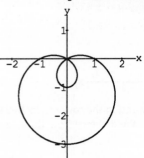

A loop.

We want the area enclosed by the outer curve. Let's find what $t$'s correspond to the start and finish of the outer curve. The loop begins and ends when

$$r[t] = 1 - 2\sin[t] = 0.$$

This means $t = \pi/6$ and $t = \pi - \pi/6 = 5\pi/6$.

Take a look:

```
In[2]:=   ParametricPlot[{x[t],y[t]},
          {t,Pi/6,5 Pi/6},
```

```
        Axes->{0,0},AxesLabel->{"x","y"},
        PlotRange->{{-2.5,2.5},{-3.5,1.5}},
        AspectRatio->Automatic];
```

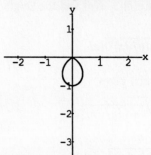

Dammit; that's the inner loop.

Let's try:

```
In[3]:=  outercurve = ParametricPlot[
         {x[t],y[t]},{t,-Pi - Pi/6, Pi/6},
         Axes->{0,0},AxesLabel->{"x","y"},
         PlotRange->{{-2.5,2.5},{-3.5,1.5}},
         AspectRatio->Automatic];
```

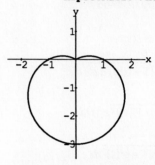

Good; that's the outer curve. Now let's check clockwise versus counterclockwise by looking at some tangent vectors that point in the direction of increasing $t$ :

```
In[4]:=  tanvector[t_] =Graphics[RGBColor[0,0,1],
         Line[{{x[t],y[t]},
         {x[t],y[t]}+{x'[t],y'[t]}}]]

         Show[outercurve,Table[tanvector[s],
         {s,-Pi - Pi/6, Pi/6, Pi/6}],
         AspectRatio->Automatic];
```

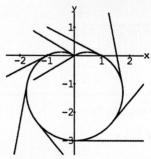

Counterclockwise.

The area measurement is:

```
In[5]:=  meas1 =
         -Integrate[y[t] x'[t],{t,-Pi-Pi/6, Pi/6}]
```

```
Out[5]=                          3/2
                                3
               -(2 Sqrt[3] + ---- - 2 Pi)
                               6
```

This is the same as:

```
In[6]:=  meas2 =
         Integrate[x[t] y'[t],{t,-Pi - Pi/6, Pi/6}]
Out[6]=     3/2
           3
           ---- + 2 Pi
            2
```

And this is the same as:

```
In[7]:=  meas3 = Integrate[ (x[t] y'[t] -
         y[t] x'[t])/2,{t,-Pi - Pi/6 ,Pi/6}]
Out[7]=               3/2
                     3
           Sqrt[3] + ---- + 2 Pi
                      6
```

If you don't see why all three are the same, look at:

```
In[8]:=  N[{meas1,meas2,meas3}]
Out[8]=  {8.88126, 8.88126, 8.88126}
```

**T.4.b.ii)**

---

Measure the area enclosed the inner loop of the polar curve

$$r[t] = 1 - 2\sin[t].$$

**Answer:**

---

First let's look at a plot. To get it, just use:

```
In[1]:=  r[t_] = 1 - 2 Sin[t]
         x[t_] = r[t] Cos[t]
         y[t_] = r[t] Sin[t]

         innerloop = ParametricPlot[
         {x[t],y[t]},{t,Pi/6, 5 Pi/6},
         Axes->{0,0},AxesLabel->{"x","y"},
         PlotRange->{{-2.5,2.5},{-3.5,1.5}},
         AspectRatio->Automatic];
```

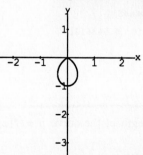

Good; that's the inner loop. Now let's check clockwise versus counterclockwise by looking at some tangent vectors that point in the direction of increasing $t$ :

```
In[2]:=    tanvector[t_] =Graphics[RGBColor[0,0,1],
           Line[{{x[t],y[t]},
           {x[t],y[t]}+{x'[t],y'[t]}}]]

           Show[innerloop,Table[tanvector[s],
           {s,Pi/6,5 Pi/6, Pi/12}],
           AspectRatio->Automatic];
```

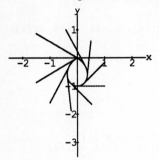

Counterclockwise. The area measurement is:

```
In[3]:=    meas1 =
           -Integrate[y[t] x'[t],{t,Pi/6,5 Pi/6}]

Out[3]=                    3/2
                          3
           -(2 Sqrt[3] - ---- - Pi)
                          6
```

This is the same as:

```
In[4]:=    meas2 =
           Integrate[x[t] y'[t],{t,Pi/6,5 Pi/6}]

Out[4]=      3/2
            -3
           ----- + Pi
             2
```

And this is the same as:

```
In[5]:=    meas3 =
           Integrate[(x[t] y'[t] - y[t] x'[t])/2,
           {t,Pi/6,5 Pi/6}]

Out[5]=                    3/2
                          3
           -Sqrt[3] - ---- + Pi
                       6
```

If you don't see why all three are the same, look at:

```
In[6]:=    N[{meas1,meas2,meas3}]

Out[6]=    {0.543516, 0.543516, 0.543516}
```

■ **T.5) Arc Length.**

**T.5.a.i)**

What is a formula for the length of the curve $y = f[x]$ specified parametrically by

$$x = x[t]$$

and

$$y = y[t]$$

with $a \le t \le b$?

**Answer:**

If $x'[t] > 0$ for $a \le t \le b$, then $x[b] > x[a]$ and the length is given by

$$\int_{x[a]}^{x[b]} \sqrt{1 + (dy/dx)^2} \, dx.$$

Substituting $y = y[t]$ and $x = x[t]$ gives

$$length = \int_{x[a]}^{x[b]} \sqrt{1 + (dy/dx)^2} \, dx$$

$$= \int_{a}^{b} \sqrt{1 + (y'[t]/x'[t])^2} \, x'[t] \, dt$$

$$= \int_{a}^{b} \sqrt{x'[t]^2 + y'[t]^2} \, dt.$$

If $x'[t] < 0$ for $a \le t \le b$, then $x[b] < x[a]$ and the length is given by

$$\int_{x[b]}^{x[a]} \sqrt{1 + (dy/dx)^2} \, dx.$$

Substituting $y = y[t]$ and $x = x[t]$ gives

$$length = \int_{x[b]}^{x[a]} \sqrt{1 + (dy/dx)^2} \, dx$$

$$= \int_{b}^{a} \sqrt{1 + (y'[t]/x'[t])^2} \, x'[t] \, dt$$

$$= \int_{b}^{a} -\sqrt{x'[t]^2 + y'[t]^2} \, dt$$

$$= -\int_{b}^{a} \sqrt{x'[t]^2 + y'[t]^2} \, dt$$

$$= \int_{a}^{b} \sqrt{x'[t]^2 + y'[t]^2} \, dt.$$

This is the same formula we got in the case that $x'[t] > 0$.

As a result, regardless of the sign of $x'[t]$, the formula for the arc length is

$$length = \int_{a}^{b} \sqrt{x'[t]^2 + y'[t]^2} \, dt.$$

**T.5.a.ii)**

Plot the curve specified parametrically by

$$x = x[t] = e^t + t$$

and

$$y = y[t] = t(3 - t) \log[t]$$

Measure the area bounded on the top by the curve and on the bottom by the $x$-axis.

Then measure the length of the part of this curve that runs above the $x$-axis.

**Answer:**

Here is the plot of the curve showing $y$ as a function of $x$:

```
In[1]:=   x[t_] = E^t + t
          y[t_] = t (3 - t) (Log [t])
          ParametricPlot[{x[t],y[t]},{t,0,5},
          PlotRange->{-2,2}];
```

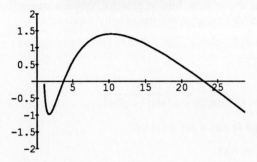

To find the $t$'s at which the curve crosses the $x$−axis, plot $y$ as a function of $t$:

```
In[2]:=   Plot[y[t],{t,0,5}];
```

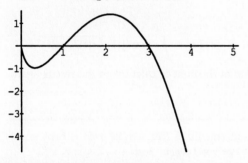

We should have been able to guess! Because $y[t] = t(3 - t) \log[t]$, we see $y[t]$ goes through 0 at $t = 0, t = 3$ and $t = 1$. The area measurement we want is:

```
In[3]:=   Integrate[y[t]x'[t],{t,1,3}]

Out[3]=      28                   t
          -(--) + Integrate[3 E  t Log[t] -
             9

              t  2                 9 Log[3]
            E  t  Log[t], {t, 1, 3}] + --------
                                          2
```

*Mathematica* bailed out; so we try:

```
In[4]:=   NIntegrate[y[t]x'[t],{t,1,3}]

Out[4]=   17.4272
```

Not too bad. To measure the length,use:

```
In[5]:=   arclength = NIntegrate[
          Sqrt[x'[t]^2 + y'[t]^2],{t,1,3}]

Out[5]=   19.6783
```

**T.5.a.iii)**

Measure the length of the inner loop and the outer curve in the plot of the polar curve

$$r[t] = 1 - 2\cos[t].$$

**Answer:**

First let's look at a plot.

```
In[1]:=   r[t_] = 1 - 2 Cos[t]
          x[t_] = r[t] Cos[t]
          y[t_] = r[t] Sin[t]

          ParametricPlot[{x[t],y[t]},{t,0,2 Pi},
          Axes->{0,0},AxesLabel->{"x","y"},
          PlotRange->{{-3,1},{-2.5,2.5}},
          Ticks->{Range[-3,1],Range[-2,2]},
          AspectRatio->Automatic];
```

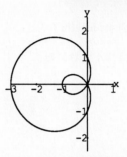

We want the length of the inner loop and the outer curve. Let's find what $t$'s correspond to the start and finish of the inner loop. The loop begins and ends when

$$r[t] = 1 - 2\cos[t] = 0.$$

This means $t = \pi/3$ and $t = -\pi/3$.

Take a look:

```
In[2]:=   ParametricPlot[
          {x[t],y[t]},{t,-Pi/3,Pi/3},
          Axes->{0,0},AxesLabel->{"x","y"},
          PlotRange->{{-3,1},{-2.5,2.5}},
          Ticks->{Range[-3,1],Range[-2,2]},
          AspectRatio->Automatic];
```

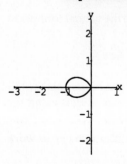

Nice; that's the inner loop. Its length is:

```
In[3]:=   NIntegrate[
          Sqrt[x'[t]^2 + y'[t]^2],{t,-Pi/3,Pi/3}]
```

*Out[3]=* 2.68245

Now go for the outer curve:

*In[4]:=* ```
outercurve = ParametricPlot[
{x[t],y[t]},{t,Pi/3,2 Pi - Pi/3},
Axes->{0,0},AxesLabel->{"x","y"},
PlotRange->{{-3,1},{-2.5,2.5}},
Ticks->{Range[-3,1],Range[-2,2]},
AspectRatio->Automatic];
```

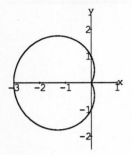

Good; that's the outer curve. Here's its length:

*In[5]:=* ```
NIntegrate[Sqrt[x'[t]^2 + y'[t]^2],
{t,Pi/3,2 Pi - Pi/3}]
```

*Out[5]=* 10.6824

### ■ T.6) Density and weight.

#### T.6.a)

A wire of length $L$ inches is bent to duplicate the graph of a parametric curve

$$x = x[t]$$
$$y = y[t]$$

for $a \le t \le b$ with all measurements in inches.

At a point $\{x[t], y[t]\}$ on the wire, the density is given by a function $dens[t]$ pounds per inch. Find a formula for the weight of the wire.

**Answer:**

This appears vexing at first because the natural formula to use is

$$weight = \int_0^L p[s]ds$$

where s represents the length along the curve from the left end and p[s] is the density of the wire $s$ units along the wire from the left .

To say the least, the variable $s$ is inconvenient to work with given the information we have.

There is no easy way for expressing $p[s]$ in terms of $dens[x]$.

(Here $p[s]$ and $dens[t]$ are not the same function; one is a function of $s$ and the other is a function of the parameter $t$. )

If you want to see more about how they are connected, then click the box.

To go from $dens[t]$ to $p[s]$, fix $s0$. Then find $t0$ so that

$$\int_a^{t0} \sqrt{x'[t]^2 + y'[t]^2}\,dt = s0.$$

Then $p[s0]$ is given by

$$p[s0] = dens[t0].$$

In principle, this is easy; but, in practice, this is a bastard. We rebound by changing the variable by a substitution.

We know (see T.5 above) that $s$ is calculated by integrating

$$\sqrt{x[t]^2 + y[t]^2}.$$

So

$$ds/dt = \sqrt{x[t]^2 + y[t]^2}.$$

And this fact is just what we need.

Make the pairings

$$p[s] \longleftrightarrow dens[t]$$

$ds \longleftrightarrow s'[t]dt = \sqrt{x[t]^2 + y[t]^2}\,dt$ $\int_0^L \longleftrightarrow \int_a^b$ .

So

$$weight = \int_0^L p[s]ds = \int_a^b dens[t]\sqrt{x[t]^2 + y[t]^2}\,dt.$$

The fundamental theorem of calculus to the rescue again!

#### T.1.b)

With all measurements in feet, a steel arch is bent to duplicate the curve $y = 3\cos[x]$ with $-\pi/2 \le x \le \pi/2$.

*In[1]:=* ```
Clear[x]
Plot[3 Cos[x], {x,-Pi/2, Pi/2},
PlotLabel->"Meet me in St. Louis.",
AspectRatio->Automatic,
PlotStyle->{{Thickness[0.017],
GrayLevel[0.7]}}];
```

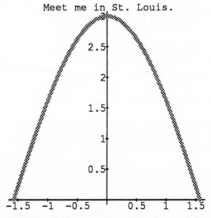

Its density at the point $\{x, \cos[x]\}$ is proportional to the product $(x - \pi)^2(x + \pi)^2$ and its density at $x = 0$ is 200 pounds per foot. How much does it weigh?
```

Out[3]=    1123.58

**Answer:**

Better than half a ton.

First let's set up the parameterization:

```
In[2]:=    x[t_] = t
           y[t_] = 3 Cos[t]
           a = -Pi/2
           b = Pi/2
           ParametricPlot[{x[t],y[t]},{t,a,b},
           PlotLabel->
           "Stan Musial played for the Cards.",
           AspectRatio->Automatic,
           PlotStyle->
           {{Thickness[0.017],GrayLevel[0.7]}}];
```

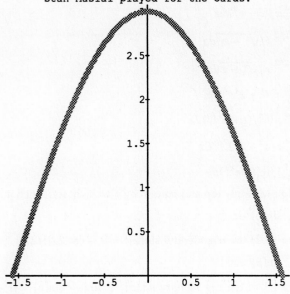

Stan Musial played for the Cards.

Good.

Now let's work on $dens[x] = K(x - \pi)^2(x + \pi)^2$.

Because $dens[0] = 200$, we have

$$200 = K(0 - \pi)^2(0 + \pi)^2 = K\pi^4.$$

So

$$K = 200/\pi^4.$$

and

$$dens[x] = (200/\pi^4)(x - \pi)^2(x + \pi)^2.$$

Because $x = x[t] = t$, we see that

$$dens[t] = (200/\pi^4)(t - \pi)^2(t + \pi)^2.$$

Now we plug into the formula

$$weight = \int_a^b dens[t]\sqrt{x[t]^2 + y[t]^2}\,dt.$$

The weight of the arch is given by:

```
In[3]:=    dens[t_] =
           (200/Pi^4) (t - Pi)^2 (t + Pi)^2
           NIntegrate[
           dens[t] Sqrt[x'[t]^2 + y'[t]^2],
           {t,-Pi/2,Pi/2}]
```

## Literacy Sheet

**1.** Calculate the folloing integrals by hand:

$\int_0^x e^{Cos[t]} \sin[t]dt$

.

$\int_0^x \cos[2t]^5 \sin[2t]dt$

.

$\int_0^a e^x/(1 + e^x)dx$

$\int_0^a x\sqrt{3 + x^2}dx$

$\int_0^a x/\sqrt{3 + x^2}dx$

$\int_0^a x/(3 + x^2)dx$

$\int_0^{1/2} 1/\sqrt{1 - x^2}dx$

$\int_0^{1/3} x/\sqrt{1 - x}dx$

$\int_0^a x/(1 + x^{1/4})dx$

$\int_0^a x^{1/3}/(1 + x^{1/2})dx$

$\int_0^a x(a^2 + x^2)^{7/3}dx$

$\int_0^a 1/(c + de^{kx})dx$

**Tip** : Multiply top and bottom by a well chosen function.

$\int_0^\pi \sin[x]^2 dx$

**Tip** : Use the trig identity $\sin[x]^2 = (1 - \cos[2x[)/2$.

$\int_0^\pi \sin[x]^3 dx$

**Tip** : Use the trig identity $\sin[x]^3 = \sin[x]^2 \sin[x] = (1 - \cos[x]^2)\sin[x]$.

**2.a** Suppose we find a curve in parametric equations $x = x[t]$ and $y = y[t]$ and our job is to measure the area under the curve over the interval $[a, b]$ on the $x$-axis. If we know that there is a function $f[x]$ such that $y[t] = f'[x[t]]$ and if we can find a formula for $f[x]$, then we can just make the measurement by calculating

$$\int_a^b f'[x]dx = f[b] - f[a].$$

If we cannot find a formula for $f[x]$, it might seem that we are out of luck in calculating

$$\int_a^b f'[x]dx = \int_a^b ydx.$$

But even if we are not able to come up with a formula for $f[x]$, there is an easy way to calculate this integral. How?

.

.

.

.

Measure the area under the parametric curve

$$x = x[t] = t^2 + 1$$

and

$$y = y[t] = t^{5/3}$$

over the interval $1 \le x \le 5$ on the $x$-axis.

.

.

.

**3.**  Here is a parametric curve:

*In[4]:=*     `x[t_] = 4 t^3 ( 2 - t )^5 E^(t + 2)`
            `y[t_] = 5 - 3 E^t Sin[Pi t]`
            `a = 0`
            `b = 2`

            `ParametricPlot[{x[t],y[t]},{t,a,b},`
            `AxesLabel->{"x","y"},Axes->{0,0},`
            `PlotRange->{{0,90},{-5,20}}]`

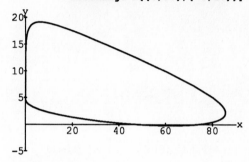

Next look at:

*In[5]:=*     `ParametricPlot[{x[t],y[t]},{t,0,.8},`
            `AxesLabel->{"x","y"},Axes->{0,0},`
            `PlotRange->{{0,90},{-5,20}}]`

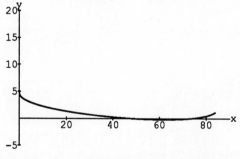

As $t$ leaves $a = 0$ and advances to 0.8, the points $\{x[t], y[t]\}$ initially ride on the lower part of the curve.

*In[6]:=*     `ParametricPlot[{x[t],y[t]},{t,0.8,2},`
            `AxesLabel->{"x","y"},Axes->{0,0},`
            `PlotRange->{{0,90},{-5,20}}]`

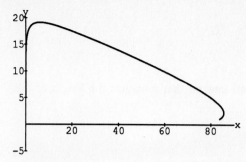

As $t$ advances from 0. to $b = 2$, the point $\{x[t], y[t]\}$ rides on the lower part until $t$ reaches an intermediate point $c$ at which $\{x[t], y[t]\}$ stops moving to the right and the begins moving to the left on its return trip to $\{0, 0\}$ on the higher part of the curve.

Thus the point $\{x[t], y[t]\}$ moves in  **counterclockwise** fashion as $t$ advances from $a$ to $b$. If we knew the precise value of $c$, then we could measure the area enclosed by this closed curve by writing (see B.2)

$$Area = \int_b^c y[t]x'[t]dt - \int_a^c y[t]x'[t]dt.$$

Explain how we know that can we measure the area without ever bothering to find the value of $c$.

.

.

.

.

**4.a** How do you know that the parametric curve

$$x[t] = t(1 - t)$$

and

$$y[t] = 1 + t^2(1 - t)^4(t - 1/2)$$

for $0 \le t \le 1$ is a plots out in a single loop ( no inner loops)?

**Tip:** $y[t] > 1$ for for $0 < t < 1/2$ and $y[t] < 1$ for $1/2 < t < 1$.

.

.

As $t$ advances from 0 to 1, does the curve trace out in a clockwise or counterclockwise manner?

**Tip:** $y[t] > 1$ for for $0 < t < 1/2$ and $y[t] < 1$ for $1/2 < t < 1$.

.

.

Write down an integral that measures the area enclosed by this curve.

.

.

.

Write down an integral that measures the length of this curve.

.

.

.

**5.** What are the parametric equations for the polar curve $r[t] = 2 - 2\cos[t]$?

.

.

.

**6.** Measure the area inside the polar curve $r[t] = 2 - 2\cos[t]$.

.

.

.

**7.** Which of the following polar curves has an inner loop?

$$r[t] = 3 - 2\sin[t]$$

$$r[t] = 3 - 3\sin[t]$$

$$r[t] = 2 - 3\sin[t]$$

$$r[t] = 2 - 2\sin[t]$$

.

.

.

**6.** If $a > 0$, then explain why

$$\int_0^a f[x]dx = \int_0^a f[a-x]dx.$$

.

.

.

# Empirical approximations

## Guide

In this lesson, you are going to learn how to run curves and lines through a given bunch of points. Here is a bunch of points:

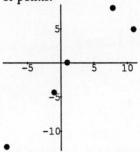

Here is a curve through them all:

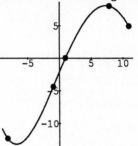

And here is a broken line through them all:

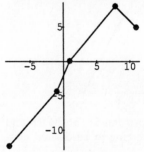

In this case there is no single straight line that does the job. The endeavor of trying to fit straight lines to given data points comes up in every quantative field of inquiry. Here is the idea: We take the data points and plot them.

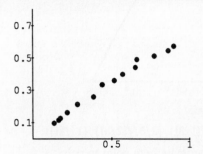

Then we fit a straight line by trying to miss the points as little as possible

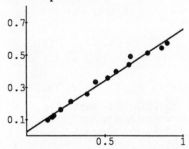

The art of fitting curves through or near data points is a time-honored mathematical activity. The whole subject is rife with successes and failures. You are invited to participate in the joy of victory and the agony of defeat. Some sweat will be required.

## Basics

■ **B.1) Approximation by linear interpolation: running broken line segments through data lists.**

**B.1.a.i)**

Plot on the same axes: $f[x] = \sin[\pi x]$ for $0 \le x \le 1$ and the broken line that joins the four points on the curve

$$\{0, f[0]\}, \{1/3, f[1/3]\}, \{2/3, f[2/3]\}, \{1, f[1]\}.$$

Assess the quality of the approximation by the broken lines.

**Answer:**

Define the function:

```
In[1]:=    Clear[f,x,jump]
           f[x_] = Sin[Pi x]
```

```
Out[1]=    Sin[Pi x]
```

Take a look at the plot:

```
In[2]:=    curveplot = Plot[f[x],{x,0,1}];
```

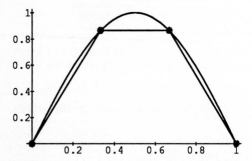

Look at the points again:

$$\{0, f[0]\}, \{1/3, f[1/3]\}, \{2/3, f[2/3]\}, \{1, f[1]\}.$$

The jump in the $x$-coordinate from point to point is:

```
In[3]:=    jump = 1/3
Out[3]=    1
           -
           3
```

So the list of points is:

```
In[4]:=    points = Table[{x,f[x]},{x,0,1,jump}]
Out[4]=              1  Sqrt[3]   2  Sqrt[3]
           {{0, 0}, {-, -------}, {-, -------},
                     3     2       3     2

               {1, 0}}
```

Set up the plot of the points and the broken line segments:

```
In[5]:=    pointplot =
           ListPlot[points,
           PlotStyle->
           {RGBColor[1,0,0],PointSize[.03]},
           DisplayFunction-> Identity]
           brokenlineplot =
           Graphics[Line[points]]

           Show[curveplot,pointplot,
           brokenlineplot];
```

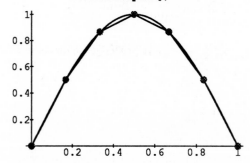

The the broken lines give a nice approximation at the left and at the right but leave something to be desired else-where.

**B.1.a.ii)**

Redo (B.1.a.i) for 7 equally spaced points.

**Answer:**

We are going from 0 to 1 in 6 equal jumps. The jump from point to point is:

```
In[1]:=    jump = 1/6
Out[1]=    1
           -
           6
```

The points are given by:

```
In[2]:=    points =
           Table[{x,f[x]},{x,0,1,jump}]
Out[2]=              1  1    1  Sqrt[3]    1
           {{0, 0}, {-, -}, {-, -------}, {-, 1},
                     6  2    3     2       2

               2  Sqrt[3]   5  1
           {-, -------}, {-, -}, {1, 0}}
            3     2       6  2
```

Set up the plot of the curve, the points and the broken line segments:

```
In[3]:=    pointplot =
           ListPlot[points,
           PlotStyle->
           {RGBColor[1,0,0],PointSize[.03]},
           DisplayFunction-> Identity]
           brokenlineplot =
           Graphics[Line[points]]

           Show[curveplot,pointplot,
           brokenlineplot];
```

The broken lines give a nice approximation at the left and at the right but still leave something to be desired else-where.

**B.1.a.iii)**

Redo (B.1.a.i) for 13 equally spaced points.

**Answer:**

We are going from 0 to 1 in 12 equal jumps. The jump from point to point is:

```
In[1]:=    jump = 1/12
Out[1]=     1
           --
           12
```

The list of points is:

```
In[2]:=    points = Table[{x,f[x]},{x,0,1,jump}]
```

$$Out[2]= \{\{0, 0\}, \{\frac{1}{12}, Sin[\frac{Pi}{12}]\}, \{\frac{1}{6}, \frac{1}{2}\},$$

$$\{\frac{1}{4}, \frac{Sqrt[2]}{2}\}, \{\frac{1}{3}, \frac{Sqrt[3]}{2}\},$$

$$\{\frac{5}{12}, Sin[\frac{5 Pi}{12}]\}, \{\frac{1}{2}, 1\},$$

$$\{\frac{7}{12}, Sin[\frac{7 Pi}{12}]\}, \{\frac{2}{3}, \frac{Sqrt[3]}{2}\},$$

$$\{\frac{3}{4}, \frac{Sqrt[2]}{2}\}, \{\frac{5}{6}, \frac{1}{2}\},$$

$$\{\frac{11}{12}, Sin[\frac{11 Pi}{12}]\}, \{1, 0\}\}$$

Set up the plot of curve, the points and the broken line segments:

```
In[3]:=    pointplot =
           ListPlot[points,
           PlotStyle->{RGBColor[1,0,0],
           PointSize[.03]},
           DisplayFunction-> Identity];
           brokenlineplot =
           Graphics[Line[points]];
           Show[curveplot,pointplot,
           brokenlineplot];
```

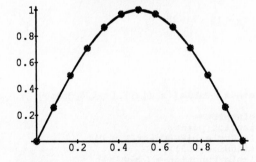

Sharing a lot of ink. This looks fairly good.

### B.1.a.iv)

---

Redo (B.1.a.i) for 25 equally spaced points.

---

### Answer:

We are going from 0 to 1 in 24 equal jumps.  Define the jump and the points are:

```
In[1]:=    jump = 1/24
           points = Table[{x,f[x]},{x,0,1,jump}];
```

And here is the plot:

```
In[2]:=    pointplot =
           ListPlot[points,
           PlotStyle->{RGBColor[1,0,0],
           PointSize[.03]},
           DisplayFunction-> Identity];
           brokenlineplot =
           Graphics[Line[points]];
           Show[curveplot,pointplot,brokenlineplot];
```

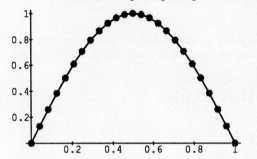

Sharing ink all the way across.  This is a beautiful approximation.

### B.1.b)

---

Give a reasonably good broken line appoximation of the curve

$$y = f[x] = (4x^3 - 9x^2)e^{-x}$$

with broken line segments for $1 \le x \le 6$.

Use your broken line approximation to estimate $f[2.7]$.

---

### Answer:

The set-up involves only a few modifications to what we did in part ($B.1.a$) above.  As we did above, we'll start with four equally spaced points to make sure everything is running smoothly and then we'll advance to more points with smaller jumps between them.

Define the function:

```
In[1]:=    Clear[f,x,jump]
           f[x_] = (4 x^3 - 9 x^2) E^(-x)
```

$$Out[1]= \frac{-9 x^2 + 4 x^3}{E^x}$$

Take a look at the plot:

```
In[2]:=    curveplot = Plot[f[x],{x,1,6},
           PlotRange->{{1,6},{-2,2.5}}];
```

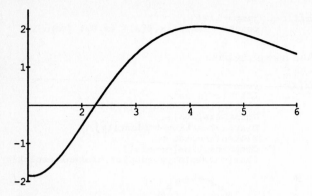

Take four equally spaced points on the $x$−axis starting with $x = 1$ and stopping with $x = 4$. The jump from point to point is:

```
In[3]:=    jump = (6 - 1)/3

Out[3]=    5
           -
           3
```

So the points are given by:

```
In[4]:=    points = Table[{x,f[x]},{x,1,6,jump}]

Out[4]=         -5       8    320        13    4225
          {{1, --}, {-, -------}, {--, --------},
                E    3    8/3      3      13/3
                        27 E              27 E

                540
          {6, ---}}
                6
                E
```

Set up the plot of the curve, the points and the broken line segments:

```
In[5]:=    pointplot =
           ListPlot[points,
           PlotStyle->{RGBColor[0,0,1],
           PointSize[.03]},
           DisplayFunction-> Identity]
           brokenlineplot =
           Graphics[Line[points]]

           Show[curveplot,pointplot,brokenlineplot,
           PlotRange->{{1,6},{-2,2.5}}];
```

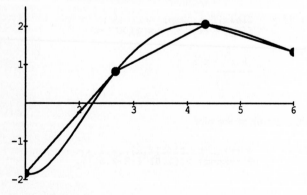

This approximation sucks.

Now let's try 13 equally spaced points:

```
In[6]:=    jump = (6 - 1)/12

Out[6]=    5
           --
           12
```

We want to set the points but we have no interest in seeing a long list of numbers, so we tack a semicolon to the next line:

```
In[7]:=    points = Table[{x,f[x]},{x,1,6,jump}];
```

Here is the plot:

```
In[8]:=    pointplot =
           ListPlot[points,
           PlotStyle->{RGBColor[1,0,0],
           PointSize[.03]},
           DisplayFunction-> Identity]
           brokenlineplot =
           Graphics[Line[points]]

           Show[curveplot,pointplot,brokenlineplot,
           PlotRange->{{1,6},{-2,2.5}}];
```

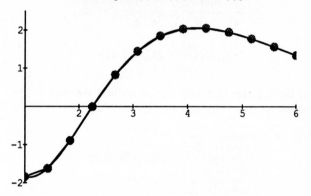

Not bad except at the far left. Let's try 26 equally spaced points:

```
In[9]:=    jump = (6 - 1)/25

Out[9]=    1
           -
           5
```

```
In[10]:=   points = Table[{x,f[x]},{x,1,6,jump}]

           pointplot =
           ListPlot[points,
           PlotStyle->{RGBColor[1,0,0],
           PointSize[.03]},
           DisplayFunction-> Identity]
           brokenlineplot =
           Graphics[Line[points]]

           Show[curveplot,pointplot,brokenlineplot,
           PlotRange->{{1,6},{-2,2.5}}];
```

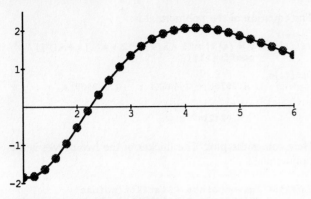

Sharing ink almost all the way.

Let's use these broken lines to estimate $f[2.7]$. First take a look at the points:

```
In[11]:=  N[points]
Out[11]=  {{1., -1.8394}, {1.2, -1.82162},
          {1.4, -1.64332}, {1.6, -1.34382},
          {1.8, -0.964023}, {2., -0.541341},
          {2.2, -0.107257}, {2.4, 0.313521},
          {2.6, 0.702925}, {2.8, 1.04885},
          {3., 1.34425}, {3.2, 1.58614},
          {3.4, 1.77466}, {3.6, 1.91222},
          {3.8, 2.00281}, {4., 2.05135},
          {4.2, 2.06327}, {4.4, 2.04413},
          {4.6, 1.99935}, {4.8, 1.93406},
          {5., 1.85294}, {5.2, 1.76018},
          {5.4, 1.65946}, {5.6, 1.55393},
          {5.8, 1.44623}, {6., 1.33853}}
```

Run a line through:

```
In[12]:=  specialpoints =
          {{2.6,f[2.6]},{2.8,f[2.8]}}
Out[12]=  {{2.6, 0.702925}, {2.8, 1.04885}}
```

```
In[13]:=  linearestimator =
          InterpolatingPolynomial[specialpoints,x]
Out[13]=  -3.79412 + 1.72963 x
```

The linear interpolation estimate forestimate for f[2.7] is:

```
In[14]:=  linearestimator/.x->2.7
Out[14]=  0.875889
```

The precise value of $f[2.7]$ is:

```
In[15]:=  f[2.7]
Out[15]=  0.881871
```

```
In[16]:=
```
Our estimate is not great, but it's not way off.

### ■ B.2) Approximation by polynomial interpolation: running polynomials through data lists.

If you are given three data points $\{x_1, y_1\}, \{x_2, y_2\}$ and $\{x_3, y_3\}$ with all the $x$-coordinates different, then you can pass a second degree polynomial $y = ax^2 + bx + c$ through these points by solving the three equations

$$y_1 = ax_1^2 + bx_1 + c$$
$$y_2 = ax_2^2 + bx_2 + c$$
$$y_3 = ax_3^2 + bx_3 + c$$

for the three constants $a, b$ and $c$. Similarly if you are given $k$ data points, $\{x_1, y_1\}, \{x_2, y_2\}, ..., \{x_k, y_k\}$ with all the $x$-coordinates different, then you can set up and solve the $k$ equations that determine a polynomial of degree $k-1$ that passes through the given data points.

No polynomial of smaller degree than this one can do this job. That's why this polynomial gets a name. Its name is the *interpolationg polynomial* for the given data points.

### B.2.a.i)

Plot on the same axes: $f[x] = (4x^3 - 9x^4)e^{-x}$ for $1 \le x \le 6$ and the third degree interpolating polynomial that runs through the points $\{1, f[1]\}, \{3, f[3]\}, \{5, f[5]\}$ and $\{7, f[7]\}$. Assess the quality of the approximation by the interpolating polynomial.

### Answer:

```
In[1]:=   Clear[f,x,jump]
          f[x_] = (4 x^3 - 9 x^2) E^(-x)
Out[1]=       2      3
          -9 x  + 4 x
          ------------
               x
              E
```

Take a look at the plot:

```
In[2]:=   curveplot = Plot[f[x],{x,1,7},
          PlotRange->All];
```

Take four equally spaced points on the $x$−axis starting with $x = 1$ and stopping with $x = 7$. The jump from point to point is:

```
In[3]:=    jump = (7 - 1)/3
Out[3]=    2
```

So the points are given by:

```
In[4]:=    points = N[Table[{x, f[x]}, {x, 1, 7, jump}]]
Out[4]=    {{1., -1.8394}, {3., 1.34425},

              {5., 1.85294}, {7., 0.848962}}
```

There are four points; so we can determine four coefficients. This explains why it is possible to fit a third degree polynomial through the four points.

```
In[5]:=    polyeqn =
           (y == a[3] x^3 + a[2] x^2 + a[1] x + a[0])
Out[5]=                    2         3
           y == a[0] + x a[1] + x  a[2] + x  a[3]
```

Substitute the given points in to the polynomial equation to get three equations involving the three yet-to be determined coefficients.

```
In[6]:=    eqn[1] = (polyeqn)/.
           {x->points[[1,1]],y->points[[1,2]]}
Out[6]=    -1.8394 == a[0] + 1. a[1] + 1. a[2] +

              1. a[3]
```

```
In[7]:=    eqn[2] = polyeqn/.
           {x->points[[2,1]],y->points[[2,2]]}
Out[7]=    1.34425 == a[0] + 3. a[1] + 9. a[2] +

              27. a[3]
```

```
In[8]:=    eqn[3] = polyeqn/.
           {x->points[[3,1]],y->points[[3,2]]}
Out[8]=    1.85294 == a[0] + 5. a[1] + 25. a[2] +

              125. a[3]
```

```
In[9]:=    eqn[4] = polyeqn/.
           {x->points[[4,1]],y->points[[4,2]]}
Out[9]=    0.848962 == a[0] + 7. a[1] + 49. a[2] +

              343. a[3]
```

Determine the coefficients by solving the equations.

```
In[10]:=   coeffs =
           Solve[{eqn[1],eqn[2],eqn[3],eqn[4]},
           {a[0],a[1],a[2],a[3]}]
Out[10]=   {{a[0] -> -4.79755, a[1] -> 3.48624,

              a[2] -> -0.552303, a[3] -> 0.0242147}

              }
```

The equation of the polynomial is:

```
In[11]:=   y = (a[3] x^3 + a[2] x^2 + a[1]x + a[0])/.
           coeffs[[1]]
Out[11]=                                         2
           -4.79755 + 3.48624 x - 0.552303 x  +

                           3
           0.0242147 x
```

Here comes the plot: The thicker of the two curves is the polynomial.

```
In[12]:=   givenpoints = ListPlot[points,
           PlotStyle->{PointSize[.03],
           RGBColor[1,0,0]},
           AxesLabel->{x,"y"},
           DisplayFunction->Identity]
           polythrough = Plot[y,{x,1,7},
           PlotStyle->{{RGBColor[0,0,1],
           Thickness[.01]}},
           DisplayFunction->Identity]

           Show[curveplot,givenpoints,polythrough,
           DisplayFunction->$DisplayFunction];
```

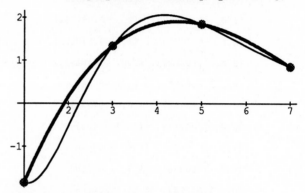

Say, that's not so bad.

**B.2.a.ii)**

How could we have part a.i) immediately above with less typing?

**Answer:**

We'll start over.

```
In[1]:=    f[x_] = (4 x^3 - 9 x^2) E^(-x)
Out[1]=         2      3
           -9 x  + 4 x
           ------------
                x
               E
```

Take a look at the plot:

```
In[2]:=    curveplot = Plot[f[x],{x,1,7},
           PlotRange->All];
```

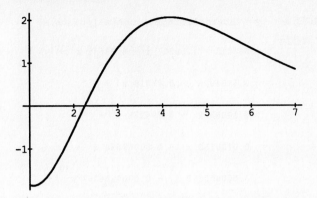

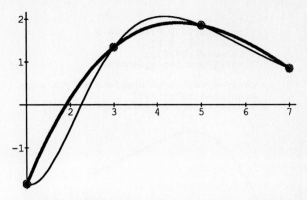

Take four equally spaced points on the $x$−axis starting with $x = 1$ and stopping with $x = 7$. The jump from point to point is:

```
In[3]:=    jump = (7 - 1)/3
Out[3]=    2
```

So the points are given by:

```
In[4]:=    points = N[Table[{x,f[x]},{x,1,7,jump}]]
Out[4]=    {{1., -1.8394}, {3., 1.34425},

            {5., 1.85294}, {7., 0.848962}}
```

Instead of laying out and solving all the equations as we did above, we can obtain the same polynomial with one *Mathematica* instruction:

```
In[5]:=    y = InterpolatingPolynomial[points,x]
Out[5]=
                                            2
           -4.79755 + 3.48624 x - 0.552303 x  +

                  3
           0.0242147 x
```

Here comes the plot:

The thicker of the two curves is the polynomial.

```
In[6]:=    givenpoints =
           ListPlot[points,
           PlotStyle->{PointSize[.03],
           RGBColor[1,0,0]},
           AxesLabel->{x,"y"},
           DisplayFunction->Identity]
           polythrough =
           Plot[y,{x,1,7},PlotStyle->
           {{RGBColor[0,0,1],Thickness[.01]}},
           DisplayFunction->Identity]

           Show[curveplot,givenpoints,polythrough,
           DisplayFunction->$DisplayFunction];
```

It's the same plot we got above. And we got it with very little work on our part.

### B.2.a.iii)

Repeat part a.i) but this time use six equally spaced points on the $x$−axis instead of using only four equally spaced points on the $x$−axis.

**Answer:**

Take six equally spaced points on the $x$−axis starting with $x = 1$ and stopping with $x = 7$. The jump from point to point is:

```
In[1]:=    jump = (7 - 1)/5
Out[1]=    6
           -
           5
```

So the points are given by:

```
In[2]:=    points = N[Table[{x,f[x]},{x,1,7,jump}]]
Out[2]=    {{1., -1.8394}, {2.2, -0.107257},

            {3.4, 1.77466}, {4.6, 1.99935},

            {5.8, 1.44623}, {7., 0.848962}}
```

The fifth degree polynomial through the six points is:

```
In[3]:=    Clear[x]
           y = InterpolatingPolynomial[points,x]
Out[3]=
                                          2
           1.88541 - 8.70205 x + 6.57851 x  -

                  3              4
           1.80701 x  + 0.215221 x  -

                  5
           0.00948474 x
```

Here comes the plot: The thicker of the two curves is the polynomial.

```
In[4]:=    givenpoints = ListPlot[points,
           PlotStyle->{PointSize[.03],
           RGBColor[1,0,0]},
           AxesLabel->{x,"y"},
           DisplayFunction->Identity]
```

```
polythrough =
Plot[y,{x,1,7},PlotStyle->
{{RGBColor[0,0,1],
Thickness[.005]}},
DisplayFunction->Identity]

Show[curveplot,givenpoints,polythrough,
AxesLabel->{"x","y"},
DisplayFunction->$DisplayFunction];
```

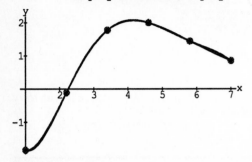

Nice! The interpolating polynomial shares ink with $f[x]$ almost all the way.

### B.2.a.iv)

Repeat part a.i) but this time use twelve equally spaced points on the $x$–axis .

**Answer:**

Take twelve equally spaced points on the $x$–axis starting with $x = 1$ and stopping with $x = 7$. The jump from point to point is:

```
In[1]:=     jump = (7 - 1)/11
Out[1]=     6
            --
            11
```

So the points are given by:

```
In[2]:=     points = N[Table[{x,f[x]},{x,1,7,jump}]]
Out[2]=     {{1., -1.8394}, {1.54545, -1.43516},

            {2.09091, -0.3438},

            {2.63636, 0.769323},

            {3.18182, 1.56638},

            {3.72727, 1.97501},

            {4.27273, 2.05961},

            {4.81818, 1.92727},

            {5.36364, 1.67822},

            {5.90909, 1.38736},

            {6.45455, 1.10239}, {7., 0.848962}}
```

The eleventh degree polynomial through the twelve points is:

```
In[3]:=     y = InterpolatingPolynomial[points,x]
Out[3]=                                        2
            0.366413 - 1.59207 x - 5.93532 x  +

                    3              4
            9.51649 x  - 5.87318 x  +

                    5              6
            2.10089 x  - 0.494279 x  +

                     7               8
            0.0799155 x  - 0.00887586 x  +

                       9                10
            0.000650668 x  - 0.0000284357 x   +

                    -7 11
            5.61826 10   x
```

Here comes the plot: The thicker of the two curves is the polynomial.

```
In[4]:=     givenpoints =
            ListPlot[points,
            PlotStyle->{PointSize[.03],
            RGBColor[1,0,0]},
            AxesLabel->{x,"y"},
            DisplayFunction->Identity]
            polythrough =
            Plot[y,{x,1,7},PlotStyle->
            {{RGBColor[0,0,1],
            Thickness[0.005]}},
            DisplayFunction->Identity]

            Show[curveplot,givenpoints,polythrough,
            AxesLabel->{"x","y"},
            DisplayFunction->$DisplayFunction];
```

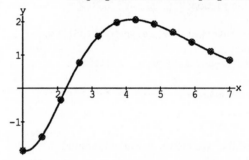

Spectacular results! The interpolating polynomial shares ink with $f[x]$ all the way across.

### ■ B.3) Least square approximation of functions: running functions near data lists.

This time instead of running polynomials or broken lines through given points we are going to be content to run them **near** given points.

Here is the idea for approximating by quadratic functions: Given a function $f[x]$, we take points $\{x[k], f[x[k]]\}$ for $k = 1, ..., n$. Then we determine numbers $a, b$ and $c$ so that $a + bx[k] + cx[k]^2$ is as close as it can be to $f[x[k]]$ for the aggregate of $k's$ running from 1 to $n$.

Here is what "as close as it can be" means: Each $a + bx[k] + cx[k]^2$ misses $f[x[k]]$ by an error of

$$(f[x[k]] - a - bx[k] - cx[k]^2).$$

We square the error to get

$$(f[x[k]] - a - bx[k] - cx[k]^2)^2$$

and add these square errors together.

The sum is measure of the overall error for a given choice of $a, b$ and $c$ because the squaring process and the addition prevents negative error from canceling positive errors.

To find the best $a, b$ and $c$, we differentiate with respect to $a$ then differentiate with respect to $b$ and finally differentiate with repect to $c$.

Then we take each of the three derivatives, set them equal to 0 and then solve for $a, b$ and $c$. The resulting solutions are the constants we go with for the approximation.

This approximation is called the quadratic **least square approximation** of $f[x]$ "through" the given points

$$\{x[k], f[x[k]]\}$$

for $k = 1, ..., n$.

You can do these steps using *Mathematica* can do these steps very nicely. But *Mathematica* can also do all this in one step via the instruction **Fit**.

### B.3.a.i)

---

Plot $f[x] = (4x^3 - 9x^4)e^{-x}$ for $1 \le x \le 6$ and least square quadratic approximation of f[x] "through" the points

$$\{1, f[1]\}, \{3, f[3]\}, \{5, f[5]\}$$

and $\{7, f[7]\}$. Assess the quality of the approximation by the least square approximation.

---

**Answer:**

```
In[1]:=    Clear[f,x,jump]
           f[x_] = (4 x^3 - 9 x^2) E^(-x)
Out[1]=        2       3
           -9 x  + 4 x
           ------------
                x
                E
```

Take a look at the plot:

```
In[2]:=    curveplot = Plot[f[x],{x,1,7},
           PlotRange->All];
```

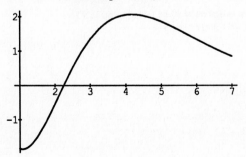

Take four equally spaced points on the $x$-axis starting with $x = 1$ and stopping with $x = 7$. The jump from point to point is:

```
In[3]:=    jump = (7 - 1)/3
Out[3]=    2
```

So the points are given by:

```
In[4]:=    points = N[Table[{x, f[x]},{x,1,7,jump}]]
Out[4]=    {{1., -1.8394}, {3., 1.34425},

           {5., 1.85294}, {7., 0.848962}}
```

Now we want to find the constants $a, b$ and $c$ such that $ax^2 + bx + c$ is the least square approximation of $f[x]$ "through" these points.

```
In[5]:=    leastsqquadratic=Fit[points,{1,x,x^2},x]
Out[5]=                                         2
           -4.04205 + 2.5225 x - 0.261726 x
```

(The way we tell the **fit** instruction that we want a quadratic is to include the list $\{1, x, x^2\}$ .) This means a = - 0.261726, b = 2.5225 and c = -4.04205 are the coefficints we want.

Here comes the plot: The thicker of the two curves is the least square quadratic.

```
In[6]:=    givenpoints = ListPlot[points,
           PlotStyle->{PointSize[.03],
           RGBColor[1,0,0]},
           AxesLabel->{x,"y"},
           DisplayFunction->Identity]
           leastsquare =
           Plot[leastsqquadratic,{x,1,7},
           PlotStyle->{{RGBColor[0,0,1],
           Thickness[.01]}},
           DisplayFunction->Identity]

           Show[curveplot,givenpoints,leastsquare,
           DisplayFunction->$DisplayFunction];
```

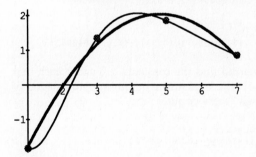

Not half bad.

### B.3.a.ii)

---

Repeat part a.i) but this time use twelve equally spaced points on the $x$-axis.

---

**Answer:**

We'll start over.

Take a look at the plot:

```
In[1]:=    curveplot = Plot[f[x], {x,1,7},
            PlotRange->All];
```

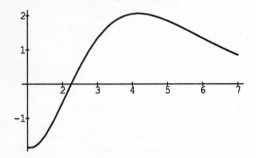

Take twelve equally spaced points on the $x$–axis starting with $x = 1$ and stopping with $x = 7$. The jump from point to point is:

```
In[2]:=    jump = (7 - 1)/11
```

```
Out[2]=    6
           --
           11
```

So the points are given by:

```
In[3]:=    points = N[Table[{x,f[x]}, {x,1,7,jump}]]
```

```
Out[3]=    {{1., -1.8394}, {1.54545, -1.43516},

           {2.09091, -0.3438},

           {2.63636, 0.769323},

           {3.18182, 1.56638},

           {3.72727, 1.97501},

           {4.27273, 2.05961},

           {4.81818, 1.92727},

           {5.36364, 1.67822},

           {5.90909, 1.38736},

           {6.45455, 1.10239}, {7., 0.848962}}
```

Now we want to find the constants $a, b$ and $c$ such that $ax^2 + bx + c$ is the least square qudratic approximation of $f[x]$ "through" these points.

```
In[4]:=    Clear[x,y,a]
            leastsqquadratic = Fit[points,{1,x,x^2},x]
```

```
Out[4]=                                         2
           -4.60917 + 2.77305 x - 0.290349 x
```

This means $a = -0.290349$, $b = 2.77305$ and $c = -4.60917$ are the coefficents we want.

Here comes the plot: The thicker of the two curves is the least square quadratic.

```
In[5]:=    givenpoints = ListPlot[points,
            PlotStyle->{PointSize[.03],
```

```
            RGBColor[1,0,0]},
            AxesLabel->{x,"y"},
            DisplayFunction->Identity]
            leastsquare =
            Plot[leastsqquadratic, {x,1,7},
            PlotStyle->{{RGBColor[0,0,1],
            Thickness[.01]}},
            DisplayFunction->Identity];

            Show[curveplot,givenpoints,leastsquare,
            DisplayFunction->$DisplayFunction];
```

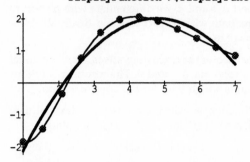

Not a whole lot better than we got using four points. But then there's only so much you can do with a quadratic.

### B.3.a.iii)

---

Least square approximations are not confined to quadratics.

Repeat part a.i) but this time use twelve equally spaced points on the $x$–axis and instead of finding the least squares quadratic approximation "through" these points, see what happens when you find the fourth degree polynomial (quartic) least squares approximation "through" these points.

---

**Answer:**

We'll start over.

```
In[1]:=    Clear[f,x,jump]
            f[x_] = (4 x^3 - 9 x^2) E^(-x)
```

```
Out[1]=        2      3
           -9 x  + 4 x
           ------------
                x
               E
```

Take a look at the plot:

```
In[2]:=    curveplot = Plot[f[x], {x,1,7},
            PlotRange->All];
```

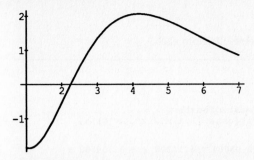

Take twelve equally spaced points on the $x$-axis starting with $x = 1$ and stopping with $x = 7$. The jump from point to point is:

```
In[3]:=    jump = (7 - 1)/11
Out[3]=    6
           --
           11
```

So the points are given by:

```
In[4]:=    points = N[Table[{x,f[x]},{x,1,7,jump}]]
Out[4]=    {{1., -1.8394}, {1.54545, -1.43516},

           {2.09091, -0.3438},

           {2.63636, 0.769323},

           {3.18182, 1.56638},

           {3.72727, 1.97501},

           {4.27273, 2.05961},

           {4.81818, 1.92727},

           {5.36364, 1.67822},

           {5.90909, 1.38736},

           {6.45455, 1.10239}, {7., 0.848962}}
```

Now we want to find the constants $a, b, c, d$ and $e$ such that

$$ax^4 + bx^3 + cx^2 + dx + e$$

is the least square quartic approximation of $f[x]$ "through" these points.

```
In[5]:=    leastsqquartic =
           Fit[points, {1,x,x^2,x^3,x^4},x]
Out[5]=
                                             2
           -1.78141 - 1.67895 x + 1.8972 x  -

                      3              4
           0.415878 x  + 0.0267194 x
```

(The way we tell the **Fit** instruction that we want a *quartic* is to include the list $\{1, x, x^2, x^3, x^4\}$ .)

Here comes the plot: The thicker of the two curves is the least square quartic.

```
In[6]:=    givenpoints =
           ListPlot[points,
           PlotStyle->{PointSize[.03],
```

```
           RGBColor[1,0,0]},
           AxesLabel->{x, "y"},
           DisplayFunction->Identity]
           leastsquare =
           Plot[leastsqquartic,{x,1,7},
           PlotStyle->{{RGBColor[0,0,1],
           Thickness[.01]}},
           DisplayFunction->Identity]

           Show[curveplot,givenpoints,leastsquare,
           DisplayFunction->$DisplayFunction];
```

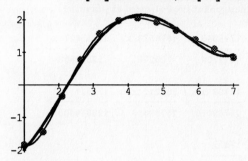

Not so bad.

# Tutorial

■ **T.1) Empirical functions: Approximating the function when you don't know the formula for the function.**

Sometimes all the information you have is a list of in the form $\{x, f[x]\}$ for various $x$'s. Here is one example of such a table::

```
In[7]:=    data =
           {{-8,-12}, {-1,-15}, {2,2}, {5,5},
           {8,4}, {11,3}, {15,9}}
Out[7]=    {{-8, -12}, {-1, -15}, {2, 2}, {5, 5},

           {8, 4}, {11, 3}, {15, 9}}
```

The trouble is that you don't have a formula for the function $f[x]$, and it is your job to come up with a formula for $f[x]$ that does a reasonably good job of cranking out the data and whose plot is in harmony with the flow of the data that you have.

**T.1.a)**

Find the interpolating polynomial through:

$$\{\{-8, -12\}, \{-1, -15\}, \{2, 2\}, \{5, 5\},$$
$$\{8, 4\}, \{11, 3\}, \{15, 9\}\}$$

Plot this interpolating polynomial and the points on the same axes. Say whether you think the interpolating polynomial describes the flow of the data.

**Answer:**

Set the data points:

```
In[1]:=    data =
           {{-8,-12}, {-1,-15}, {2,2},
           {5,5}, {8,4}, {11,3}, {15,9}}
```

*Out[1]=*    {{-8, -12}, {-1, -15}, {2, 2}, {5, 5},

{8, 4}, {11, 3}, {15, 9}}

There are seven points in the list; so we expect the interpolating polynomial through them to be a sixth degree ploynomial.

*In[2]:=*    y =
InterpolatingPolynomial[data,x]

*Out[2]=*

$$-\left(\frac{7798948}{1073709}\right) + \frac{4676573\ x}{700245} - \frac{13837027\ x^2}{12884508} +$$

$$\frac{12709\ x^3}{25769016} + \frac{387211\ x^4}{25769016} - \frac{179617\ x^5}{128845080} +$$

$$\frac{79\ x^6}{1982232}$$

Here comes the plot:

*In[3]:=*    givenpoints =
ListPlot[data,
PlotStyle->{PointSize[.03],
RGBColor[1,0,0]},
AxesLabel->{x,"y"},
DisplayFunction->Identity]
polythrough = Plot[y,{x,-11,16},
DisplayFunction->Identity]

Show[givenpoints,polythrough,
DisplayFunction->$DisplayFunction,
PlotRange->{-60,20}];

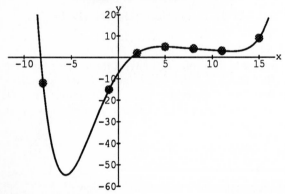

Right through all of them.

Note the **deep dip** on the left. The data points don't suggest this dip, but the interpolating polynomial is compelled to dip in this way. For $2 \leq x \leq 15$, the interpolating polynomial looks pretty good; for the $x$'s on the far left, the interpolating polynomial does not seem to capture the flow of the data.

**T.1.b)**

See what happens when you use a fourth degree polynomial least square approximation "through" the same points instead of the sixth degree interpolating polyno-

mial.

Compare the results from each.

**Answer:**

*In[1]:=*    leastsqfourth =
Fit[data,{1,x,x^2,x^3,x^4},x]

*Out[1]=*

$$-9.03054 + 4.50398\ x - 0.104565\ x^2 -$$

$$0.0531656\ x^3 + 0.00303647\ x^4$$

*In[2]:=*    leastsqplot =
Plot[leastsqfourth,{x,-11,16},
PlotStyle->{{RGBColor[0,0,1],
Thickness[.006]}},
DisplayFunction->Identity]

Show[givenpoints,leastsqplot,
AxesLabel->{"x","y"},
PlotRange->{-30,20},
DisplayFunction->$DisplayFunction];

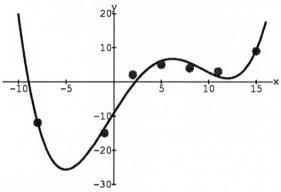

Not half bad.

The fourth degree least square polynomial misses some of the points, but this is not a fatal flaw because the points may have had some built in errors resulting from inaccurate measurements.

The fourth degree least square polynomial does seem to give a reasonably good picture of the flow of the data.

Let's look at the points, the sixth degree interplolating polynomial between them and the fourth degree least square polynomial all together: The thicker of the two curves is the fourth degree least square polynomial "through" the points:

*In[3]:=*    Show[givenpoints,leastsqplot,
polythrough,AxesLabel->{"x","y"},
PlotRange->{-60,20},
DisplayFunction->$DisplayFunction];

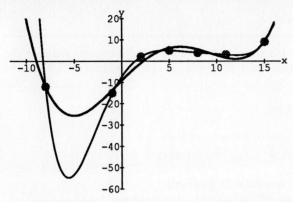

You've got to hand it to the least square polynomial. It really does seem to do a better job at describing the flow of the data than the interpolating polynomial. Unless we get more information, then

```
In[4]:=   leastsqfourth
Out[4]=
          -9.03054 + 4.50398 x - 0.104565 x  -
                                            3              4
                  0.0531656 x  + 0.00303647 x
```

seems to be a reasonable formula for $f[x]$.

This outcome is quite common: The inexact least squares approximation with a low degree polynomial can give more pleasing results than the exact (high degree) polynomial fit.

Of course, for a beautiful fit the natural cubic spline will be better (but harder to work with) than either of these two polynomials.

### T.1.c)

What happens when we fit the seven points with a least square sixth degree polynomial?

**Answer:**

We should get the interpolating polynomial because a sixth degree polynomial can be passed through the seven points with no error at all. Let's see:

```
In[1]:=   Fit[data,{1,x,x^2,x^3,x^4,x^5,x^6},x]
Out[1]=
                                          2
          -7.26356 + 6.67848 x - 1.07393 x  +
                    3              4
          0.000493189 x  + 0.0150262 x  -
                    5              6
          0.00139405 x  + 0.0000398541 x
```

```
In[2]:=   N[InterpolatingPolynomial[data,x]]
Out[2]=
                                          2
          -7.26356 + 6.67848 x - 1.07393 x  +
                    3              4
```

```
          0.000493189 x  + 0.0150262 x  -
                    5              6
          0.00139405 x  + 0.0000398541 x
```

Right.

### ■ T.2) Data Analysis: Empirical line functions.

### T.2.a.i)

A weight $x$ pounds extends a spring balance $y[x]$ inches. Measurements for a particular spring balance for $x = 2, 5, 7, 10$ and $12$ pounds resulted in $y = 0.58, 1.58, 1.88, 2.87$ and $3.15$ inches respectively.

Analyze the given data and try to estimate a formula for $y[x]$.

**Answer:**

Here are the the data points in the form $\{x, y[x]\}$:

```
In[1]:=   data = {{2,0.58},{5,1.58},
          {7,1.88},{10,2.87},{12,3.15}}
Out[1]=   {{2, 0.58}, {5, 1.58}, {7, 1.88},
          {10, 2.87}, {12, 3.15}}
```

Look at a broken line plot through the data points:

```
In[2]:=   pointplot =
          ListPlot[data,
          PlotStyle->{RGBColor[1,0,0],
          PointSize[.03]},
          DisplayFunction-> Identity]

          Show[pointplot,
          AxesLabel->{"x","y"},
          DisplayFunction->$DisplayFunction];
```

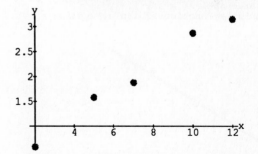

Looks like almost like a straight line. The data indicate that the length $y[x]$ seems to vary linearly with the weights $x$. Probably the deviations from a straight line are caused by errors in the measurements. Let's see what the interpolating polynomial does for us:

```
In[3]:=   y = InterpolatingPolynomial[data,x]
Out[3]=
                                          2
          -2.465 + 2.42313 x - 0.554963 x  +
                    3              4
```

$$0.0562548 \; x \; - \; 0.00196548 \; x$$

Here comes the plot:

```
In[4]:=    polythrough = Plot[y,{x,0,14},
           DisplayFunction->Identity]

           Show[pointplot,polythrough,
           DisplayFunction->$DisplayFunction,
           PlotRange->{0,4}];
```

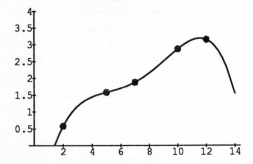

This looks nice, but it is hardly in harmony with our idea that the data are lining up in a near straight line. As matters sit now, no straight line will hit all of the points. As a response, we do the next best thing by finding the line function least squares approximation "through" the given data:

```
In[5]:=    leastsqline = Fit[data,{1,x},x]

Out[5]=    0.144586 + 0.259363 x
```

Let's see how this line does:

```
In[6]:=    leastsqplot =
           Plot[leastsqline,{x,0,14},
           PlotStyle->{{RGBColor[0,0,1],
           Thickness[.01]}},
           DisplayFunction->Identity]

           Show[pointplot,leastsqplot,
           AxesLabel->{"x","y"},
           DisplayFunction->$DisplayFunction];
```

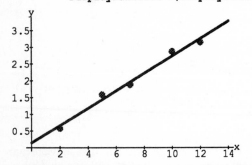

Pretty good fit. It would have been impossible to hit all of these points with a line, but the best least square line is not all that far off.

To describe these data, we'll go with the formula:

```
In[7]:=    y = leastsqline

Out[7]=    0.144586 + 0.259363 x
```

**T.2.a.ii)**

On the basis of the above measurements, estimate the weight that will extend the spring balance by 3.00 inches.

---

**Answer:**

```
In[1]:=    Solve[y == 3,x]

Out[1]=    {{x -> 11.0093}}
```

The answer is 11.0 pounds.

■ **T.3) Data Analysis: Empirical power and exponential fits- "log paper" and "semi-log paper".**

As in T.1) and T.2) above, information between the variable $x$ and the function $y[x]$ may consist of nothing more than some measurements indicating the observed value of $y[x]$ for various values of $x$.

A quick plot of the points usually reveals to the alert eye whether the functional relationship of $y$ in terms of $x$ is of the form $y = ax + b$.

If the plot of the points looks like:

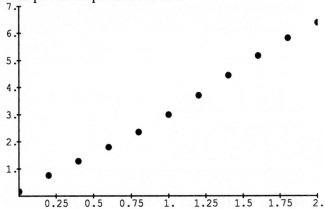

then we are happy with a least squares line "through" the points. The slight deviation from a true straight line can be charged off to errors in measurement. (There is no such thing as an exact measurement except in pure mathematics.)

Sometimes the points do not line up in a straight line:

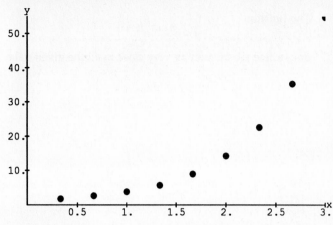

In this situation, a least squares linear fit is certainly **not** indicated.

### T.3.a)

How can we detect a relationship of the form

$$y = ae^{bx}$$

for the given points?

**Answer:**

$$y = ae^{bx}$$

is the same as

$$\log[y] = \log[a] + bx.$$

This means that if $y = ae^{bx}$, then $\log[y]$ is a line function of $x$. If we plot the points $\{x, \log[y]\}$ instead of the usual $\{x, y\}$ and we observe that they line up in nearly straight line, then the relationship $y = ae^{ax}$ is indicated.

### T.3.b)

How can we detect a relationship of the form

$$y = ax^b$$

for the given points?

■ **Answer:**

$$y = ax^b$$

is the same as

$$\log[y] = \log[a] + b\log[x].$$

This means that if $y = ax^b$, then $\log[y]$ is a line function of $\log[x]$. If we plot the points $\{\log[x], \log[y]\}$ instead of the usual $\{x, y\}$ and we observe that they line up in nearly straight line, then the relationship

$$y = ax^b$$

is indicated.

### T.3.c)

Find a good fit of the data points

$$\{\{0.33, 1.77\}, \{0.667, 2.66\}, \{1.00, 3.83\},$$
$$\{1.33, 5.73\}, \{1.67, 9.05\}, \{2.00, 14.3221\},$$
$$\{2.33, 22.61\}, \{2.67, 35.25\}, \{3.00, 54.17\}\}$$

by functions of the form $y = ax + b$, $y = ae^{bx}$ or $y = ax^b$. Plot your function and the points on the same axes.

**Answer:**

```
In[1]:=   data =
          {{0.33,1.77},{0.667,2.66},{1.00,3.83},
          {1.33,5.73},{1.67,9.05},{2.00,14.3221},
          {2.33,22.61},{2.67,35.25},{3.00,54.17}}

Out[1]=   {{0.33, 1.77}, {0.667, 2.66},
            {1., 3.83}, {1.33, 5.73},
            {1.67, 9.05}, {2., 14.3221},
            {2.33, 22.61}, {2.67, 35.25},
            {3., 54.17}}
```

Look at the plot.

```
In[2]:=   dataplot =
          ListPlot[data,
          PlotStyle->{PointSize[.02],
          RGBColor[1,0,0]},
          PlotRange->{{0,3},{0,55}},
          AxesLabel->{"x","y"}];
```

The linear relationship $y = ax + b$ is not indicated. What about $y = ae^{bx}$? To check it, we have to do the plot in the form $\{x, \log[y]\}$. This requires a little data processing.

```
In[3]:=   logdata =
          Table[{data[[j,1]],Log[data[[j,2]]]},
          {j,1,Length[data]}]

Out[3]=   {{0.33, 0.57098}, {0.667, 0.978326},
            {1., 1.34286}, {1.33, 1.74572},
            {1.67, 2.20276}, {2., 2.6618},
            {2.33, 3.11839}, {2.67, 3.56247},
            {3., 3.99213}}
```

Note how the $x$−values were not changed, but the $y$−values were changed to their natural logarithms.

Look at the plot: This is the same plot you would get if you plotted the the original data points on "semi-log paper".

```
In[4]:=    ListPlot[logdata,
           PlotStyle->{PointSize[.013],
           RGBColor[1,0,0]},
           AxesLabel->{"x","Log[y]"}];
```

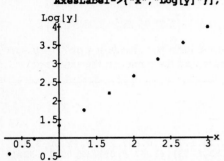

Aces! They are nearly lining up in a straight line. The empirical relationship $y = ae^{bx}$ is indicated. But how do we nail down $a$ and $b$?

Here is a way; we pass the least square line "through" the *logdata* points:

```
In[5]:=    Fit[logdata,{1,x},x]
Out[5]=    0.0855448 + 1.29396 x
```

This means a good fit comes from

$$\log[y] = 1.294x + 0.086.$$

This is the same as

$$y = e^{Log[y]} = e^{1.294x + 0.086} = e^{0.086}e^{1.294x}.$$

```
In[6]:=    y = E^(0.086) E^(1.294 x)
                        1.294 x
Out[6]=    1.08981 E
```

Here comes the plot:

```
In[7]:=    goodfit =
           Plot[y,{x,0,3},
           PlotRange->{0,55},
           DisplayFunction->Identity]

           Show[dataplot,goodfit,
           DisplayFunction->$DisplayFunction];
```

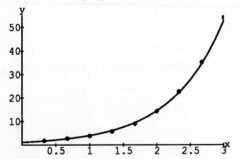

Hot dog!

The function

$$y = e^{0.086}e^{1.294x}$$

does a fine job of running very close to all the given data.

## Literacy Sheet

**1.** Explain the idea behind approximation by broken line segments.

.

.

**2.** Broken line approximations lack the style and pizzazz of polynomial approximations, but they rarely cause terrible problems because you can always fix them by doing what?.

.

.

.

**3.** If you have six data points and want to run the polynomial $p[x]$ of lowest possible degree through them, then what is the largest that the degree of $p[x]$ can be? Is the resulting lowest degree polynomial the same as the interpolating polynomial for this list of six data points?

.

.

.

**4.** What outcome sometimes makes us less than happy with an interpolating polynomial? What polynomials can we try in place of interpolating polynomials?

.

.

**5.** What is the idea behind least squares approximation?

.

.

**6.** Describe a situation in which you might be happier with least squares approximation than with approximation by interpolating polynomials.

.

.

.

**6.** Take out pencil and paper and turn the machine off.

Find by hand calculation the formula for the interpolating polynomial through the points $\{0,1\}$, $\{1,4\}$ and $\{2,9\}$.

.

.

Given the three data points $\{0,0.0\}$, $\{1,2.2\}$ and $\{2,4.0\}$ find by a hand calculation the least squares line function "through" these three points.

.

.

.

Sketch the points and the line function on your paper with your pencil.

.

.

.

7. Essentially no physical measurement is ever made without some error. Why does this fact make least squares approximations very important?

.

.

8. Given a list of points $\{\{x_1, y\}_1, \{x_2, y\}_2, ..., \{x_k, y\}_k\}$, then:

What would you do it it to see whether an empirical fit by a least squares line is indicated?

.

.

What would you do it it to see whether an empirical fit by an exponential $y = be^{ax}$ is indicated?

.

.

What would you do it it to see whether an empirical fit by an power function $y = bx^a$ is indicated?

.

.

What would you do it it to see whether an empirical fit by an shifted logarithm $y = log[ax + b]$ indicated?

.

.

.

9.    Why do science labs like to keep a stock of log-log paper and semi-log paper? What is the idea behind these special forms of graphing paper? How can you use *Mathematica* to get the effect of these forms of graphing paper on your machine?

# Approximations by expansions: quotients of polynomials

## Guide

The decimal form of a number is useful because it gives us an easy way of understanding the size of some numbers.

```
In[1]:=    N[355/113,20]
Out[1]=    3.1415929203539823009
```

```
In[2]:=    N[112358132134/6287,10]
Out[2]=    17871501.85
```

```
In[3]:=    N[387/1323591,15]
Out[3]=    0.000292386394286453
```

Some numbers have a decimal form that stops after a few digits:

```
In[4]:=    N[1445/8,20]
Out[4]=    180.625
```

Others are not so obliging:

```
In[5]:=    N[ Pi,20]
Out[5]=    3.1415926535897932385
```

No matter how many accurate decimals of $\pi$ that we look at, we find new digits appearing later on.

```
In[6]:=    N[Pi,50]
Out[6]=    3.14159265358979323846264338327950288419
            71693993751
```

```
In[7]:=    N[Pi,100]
Out[7]=    3.14159265358979323846264338327950288419
            7169399375105820974944592307816406286
            20899862803482534211706B
```

An analogous situation occurs in the realm of functions. Some functions are in the form of a polynomial:

```
In[8]:=    LegendreP[6,x]
                        2        4        6
Out[8]=    -5 + 105 x  - 315 x  + 231 x
           --------------------------------
                        16
```

```
In[9]:=    ChebyshevT[12,x]
                    2        4        6
Out[9]=
```

```
           1 - 72 x  + 840 x  - 3584 x  +

                 8          10          12
           6912 x  - 6144 x   + 2048 x
```

Functions are similar to numbers in the sense that not all functions are in the form of a polynomial

$$a[0] + a[1]x + a[2]x^2 + a[3]x^3 + ... + a[n]x^n$$

where the number of terms is finite. For example:

```
In[10]:=   1/(1 - x)
Out[10]=      1
           -----
           1 - x
```

is a quotient of polynomials, but not a polynomial.

But many functions can be put in the form

$$a[0] + a[1]x + a[2]x^2 + a[3]x^3 + ... + a[k]x^k + ...$$

where the number of terms is not finite. Look at these:

```
In[11]:=   Series[1/(1 - x),{x,0,10}]
                      2   3   4   5   6   7
Out[11]=   1 + x + x + x + x + x + x + x +

               8   9   10      11
               x + x + x   + 0[x]
```

```
In[12]:=   Series[1/(1 - x),{x,0,20}]
                      2   3   4   5   6   7
Out[12]=   1 + x + x + x + x + x + x + x +

               8   9   10    11    12    13
               x + x + x   + x   + x   + x   +

               14    15    16    17    18    19
               x   + x   + x   + x   + x   + x   +

               20      21
               x   + 0[x]
```

Just as decimals are useful in allowing us to understand the size of various numbers, this form has proven to be a good form to allow the human mind to grasp the true nature of non-polynomial functions.

In this section, we are going to deal with non-polynomial functions that are quotients of two polynomials. The procedure is straightforward; to write the quotient $P[x]/Q[x]$ of two polynomials $P[x]$ and $Q[x]$ in the form

$$a[0] + a[1]x + a[2]x^2 + a[3]x^3 + ... + a[k]x^k + ...$$

we simply divide the denominator into the numerator.

# Basics

■ **B.1) Dividing polynomials.**

### B.1.a)

Write a *Mathematica* routine that accepts as inputs

a polynomial $P[x]$

another polynomial $Q[x]$

a positive integer $n$

and uses the method of undetermined cofficients to produce the first $n$ terms of the divided out quotient $P[x]/Q[x]$.

**Answer:**

Take the input; say for example:

```
In[1]:=    P = 3 x^2 - 7
Out[1]=            2
           -7 + 3 x
```

```
In[2]:=    Q = x^2 + x - 1
Out[2]=              2
           -1 + x + x
```

```
In[3]:=    n = 7
Out[3]=    7
```

We want to find what constants $a[0], a[1], a[2], ..., a[n]$ result from dividing $Q$ into $P$ where

$$P/Q = a[0] + a[1]x + a[2]x^2 + ..... + a[n]x^n + O[x]^{n+1}.$$

Here $O[x]^{n+1}$ stands for the remainder after the $x^n$ term. (This remainder term is usually called "big oh of $x$ to the $n+1$.")

```
In[4]:=    PoverQ =
           Sum[a[j] x^j, {j,0,n}] + O[x]^(n + 1)
Out[4]=                    2         3
           a[0] + a[1] x + a[2] x + a[3] x +

                4         5         6
           a[4] x  + a[5] x  + a[6] x  +

                7        8
           a[7] x  + O[x]
```

We have to determine what the coefficients

$$a[0], a[1], a[2], ..., a[n]$$

must be. Remember that $P = Q(P/Q)$. So, write:

```
In[5]:=    left = P
Out[5]=            2
           -7 + 3 x
```

```
In[6]:=    right = Q PoverQ
```

```
Out[6]=    -a[0] + (a[0] - a[1]) x +

                                2
           (a[0] + a[1] - a[2]) x +

                                3
           (a[1] + a[2] - a[3]) x +

                                4
           (a[2] + a[3] - a[4]) x +

                                5
           (a[3] + a[4] - a[5]) x +

                                6
           (a[4] + a[5] - a[6]) x +

                                7        8
           (a[5] + a[6] - a[7]) x  + O[x]
```

Now find the equations that result from equating the coefficients of like powers of $x$ on the left and right sides of the equation $P == Q(PoverQ)$.

You can spot a few of them:

$$a[0] == 7, a[0] - a[1] == 0, a[0] + a[1] - a[2] == 3$$
$$a[1] + a[2] - a[3] == 0, a[2] + a[3] - a[4] == 0,$$

etc.

Now let *Mathematica* set up these equations for you:

```
In[7]:=    coeffequations =
           LogicalExpand[P == Q PoverQ]
Out[7]=    -7 + a[0] == 0 &&

           -(a[0] - a[1]) == 0 &&

           3 - (a[0] + a[1] - a[2]) == 0 &&

           -(a[1] + a[2] - a[3]) == 0 &&

           -(a[2] + a[3] - a[4]) == 0 &&

           -(a[3] + a[4] - a[5]) == 0 &&

           -(a[4] + a[5] - a[6]) == 0 &&

           -(a[5] + a[6] - a[7]) == 0
```

Now solve these equations for $a[0], a[1], ...a[n]$.

```
In[8]:=    coeffs = Solve[coeffequations]
Out[8]=    {{a[0] -> 7, a[1] -> 7, a[2] -> 11,

           a[3] -> 18, a[4] -> 29, a[5] -> 47,

           a[6] -> 76, a[7] -> 123}}
```

Insert these now known values of $a[0], a[1], ..., a[n]$ into the expression for $PoverQ$:

```
In[9]:=    quotient = PoverQ/.coeffs[[1]]
Out[9]=                    2        3        4
           7 + 7 x + 11 x + 18 x + 29 x +

                5        6         7        8
           47 x + 76 x + 123 x + O[x]
```

This is the result of dividing $Q$ into $P$ and stopping with the remainder at the stage when $x^n$ turns up.

Let's try a larger $n$:

```
In[10]:=  n = 12
          PowerQ =
          Sum[a[j] x^j, {j,0,n}] + O[x]^(n + 1)
          left = P
          right = Q PowerQ
          coeffequations =
          LogicalExpand[P == Q PowerQ]
          coeffs = Solve[coeffequations]
          quotient = PowerQ/.coeffs[[1]]
```

```
Out[10]=               2       3        4
          7 + 7 x + 11 x + 18 x + 29 x +

                5       6        7        8
            47 x + 76 x + 123 x + 199 x +

                 9        10         11
            322 x + 521 x  + 843 x   +

                 12       13
            1364 x   + O[x]
```

This is the result of dividing $Q$ into $P$ and stopping with the remainder at the stage when $x^n$ turns up.

Let's try it for different polynomials $P$ and $Q$:

```
In[11]:=  P = x
          Q = 4 + x^2
          n = 20
          PowerQ =
          Sum[a[j] x^j, {j,0,n}] + O[x]^(n + 1)
          left = P
          right = Q PowerQ
          coeffequations =
          LogicalExpand[P == Q PowerQ]
          coeffs = Solve[coeffequations]
          quotient = PowerQ/.coeffs[[1]]
```

```
Out[11]=     3    5    7     9      11
           x    x    x    x      x
           - - -- + -- - --- + ---- - ---- +
           4   16   64   256   1024   4096

             13      15      17       19
           x       x       x        x
           ----- - ----- + ------ - ------- +
           16384   65536   262144   1048576

              21
           O[x]
```

This is the result of dividing $Q$ into $P$ and stopping with the remainder at the stage $x^n$ turns up.

Can you spot any pattern?

Play with this routine by changing $P, Q$ and $n$.

### ■ B.2) Finding some expansions.

If $m$ and $n$ are positive integers, then the decimal expansion of the quotient $m/n$ is what we get when we continue to divide $n$ into $m$ without stopping.

For instance, if we look at what happens when we divide 23 by 11, we can easily spot the decimal expansion of 23/11:

```
In[12]:=  N[23/11,19]
Out[12]=  2.090909090909090909
```

```
In[13]:=  N[23/11,39]
Out[13]=  2.09090909090909090909090909090909090909090
          9
```

```
In[14]:=  N[23/11,59]
Out[14]=  2.09090909090909090909090909090909090909090
          90909090909090909090909
```

Thus

$$23/11 = 2.0909090909090909090909090909090909... $$

where the pattern 09 repeats forever.

If $P[x]$ and $Q[x]$ are polynomials, then the **expansion of the quotient** $P[x]/Q[x]$ **in powers of** $x$ is what we get when we continue to divide $Q[x]$ into $P[x]$ without stopping.

### B.2.a)

Find the expansion of $1/(1-x)$ in powers of $x$.

This expansion is the most imortant of them all.

### Answer:

Copy, paste and edit the routine from B.1) above:

```
In[1]:=  P = 1
Out[1]=  1
```

```
In[2]:=  Q = 1 - x
Out[2]=  1 - x
```

```
In[3]:=  n = 10
Out[3]=  10
```

We want to find what constants $a[0], a[1], a[2], ..., a[n]$ result from dividing $Q$ into $P$ where

$$P/Q = a[0] + a[1]x + a[2]x^2 + ..... + a[n]x^n + O[x]^{n+1}.$$

```
In[4]:=  PowerQ = Sum[a[j] x^j, {j,0,n}] + O[x]^(n + 1)
Out[4]=
                               2       3
          a[0] + a[1] x + a[2] x + a[3] x +

                 4        5        6
          a[4] x  + a[5] x + a[6] x +

                 7        8        9
          a[7] x  + a[8] x + a[9] x +

                  10         11
          a[10] x   + O[x]
```

We have to determine what the coefficients

$$a[0], a[1], a[2], ..., a[n]$$

must be. Remember that $P = Q(P/Q)$.

So, write:

```
In[5]:=    left = P
Out[5]=    1
```

```
In[6]:=    right = Q PowerQ
Out[6]=    a[0] + (-a[0] + a[1]) x +

                               2
            (-a[1] + a[2]) x  +

                               3
            (-a[2] + a[3]) x  +

                               4
            (-a[3] + a[4]) x  +

                               5
            (-a[4] + a[5]) x  +

                               6
            (-a[5] + a[6]) x  +

                               7
            (-a[6] + a[7]) x  +

                               8
            (-a[7] + a[8]) x  +

                               9
            (-a[8] + a[9]) x  +

                                10        11
            (-a[9] + a[10]) x   + O[x]
```

Now find the equations that result from equating the co-efficients of like powers of $x$ on the left and right sides of the equation $P == Q(PowerQ)$.

You can spot a few of them:

$$a[0] = 1, -a[0] + a[1] = 0, -a[1] + a[2] = 0$$
$$-a[2] + a[3] = 0, -a[3] + a[4] == 0,$$

etc.

Now let *Mathematica* set up these equations for you:

```
In[7]:=    coeffequations =
           LogicalExpand[P == Q PowerQ]
Out[7]=    1 - a[0] == 0 &&

            -(-a[0] + a[1]) == 0 &&

            -(-a[1] + a[2]) == 0 &&

            -(-a[2] + a[3]) == 0 &&

            -(-a[3] + a[4]) == 0 &&

            -(-a[4] + a[5]) == 0 &&

            -(-a[5] + a[6]) == 0 &&

            -(-a[6] + a[7]) == 0 &&

            -(-a[7] + a[8]) == 0 &&

            -(-a[8] + a[9]) == 0 &&

            -(-a[9] + a[10]) == 0
```

Now solve these equations for $a[0], a[1], ..., a[n]$.

```
In[8]:=    coeffs = Solve[coeffequations]
Out[8]=    {{a[0] -> 1, a[1] -> 1, a[2] -> 1,

              a[3] -> 1, a[4] -> 1, a[5] -> 1,

              a[6] -> 1, a[7] -> 1, a[8] -> 1,

              a[9] -> 1, a[10] -> 1}}
```

If you look at the individual equations, you can see why all the coefficients $a[k] = 1$.

Insert these now known values of $a[0], a[1], ..., a[n]$ into the expression for **PowerQ**:

```
In[9]:=    quotient = PowerQ/.coeffs[[1]]
Out[9]=              2    3    4    5    6    7
            1 + x + x  + x  + x  + x  + x  + x  +

             8    9    10        11
            x  + x  + x   + O[x]
```

This is the result of dividing $Q$ into $P$ and stopping with the remainder at the stage $x^n$ turns up.

Let's go for a larger $n$:

```
In[10]:=   n = 20
           PowerQ =
           Sum[a[j] x^j, {j,0,n}] + O[x]^(n + 1)
           left = P
           right = Q PowerQ
           coeffequations =
           LogicalExpand[P == Q PowerQ]
           coeffs = Solve[coeffequations]
           quotient = PowerQ/.coeffs[[1]]
Out[10]=             2    3    4    5    6    7
            1 + x + x  + x  + x  + x  + x  + x  +

             8    9    10    11    12    13
            x  + x  + x   + x   + x   + x   +

             14    15    16    17    18    19
            x   + x   + x   + x   + x   + x   +

             20        21
            x   + O[x]
```

The general pattern is now clear: The expansion of $1/(1 - x)$ in powers of $x$ is

$$1 + x + x^2 + x^3 + x^4 + ... + x^k + ...$$

Check it out:

```
In[11]:=   Sum[x^k, {k,0,20}] + O[x]^21
Out[11]=             2    3    4    5    6    7
            1 + x + x  + x  + x  + x  + x  + x  +

             8    9    10    11    12    13
            x  + x  + x   + x   + x   + x   +

             14    15    16    17    18    19
            x   + x   + x   + x   + x   + x   +

             20        21
            x   + O[x]
```

Good.

**B.2.b)**

Find the expansion of $x^3/(1 + x^2)$ in powers of $x$.

**Answer:**

If it worked once it will work again:

```
In[1]:=    P = x^3
           Q = (1 + x^2)
           n = 9
           Clear[a,j,x]
           PoverQ =
           Sum[a[j] x^j,{j,0,n}] + O[x]^(n + 1)
           left = P
           right = Q PoverQ
           coeffequations =
           LogicalExpand[P == Q PoverQ]
           coeffs = Solve[coeffequations]
           quotient = PoverQ/.coeffs[[1]]
Out[1]=    3   5   7   9      10
           x - x + x - x + O[x]
```

Go for a larger $n$:

```
In[2]:=    n = 25
           Clear[a,j,x]
           PoverQ =
           Sum[a[j] x^j,{j,0,n}] + O[x]^(n + 1)
           left = P
           right = Q PoverQ
           coeffequations =
           LogicalExpand[P == Q PoverQ]
           coeffs = Solve[coeffequations]
           quotient = PoverQ/.coeffs[[1]]
Out[2]=    3   5   7   9   11   13   15
           x - x + x - x + x - x + x -

           17   19   21   23   25    26
           x + x - x + x - x + O[x]
```

The general pattern is now fairly clear: The expansion of $x^3/(1 + x^2)$ in powers of $x$ is

$$x^3 - x^5 + x^7 - x^9 + ...(-1)^k x^{3+2k} + ....$$

Check it out:

```
In[3]:=    Sum[ (-1)^k x^(3 + 2 k),{k,0,11}] +
              O[x]^(3 + 2*11 + 1)
Out[3]=    3   5   7   9   11   13   15
           x - x + x - x + x - x + x -

           17   19   21   23   25    26
           x + x - x + x - x + O[x]
```

Looking good.

■ **B.3) A** *Mathematica* **shortcut.**

As is the case with most mathematical chores, *Mathematica* has an instruction that helps us get expansions. The instruction `Series[P[x]/Q[x],x,0,n]` gives the sum of the first terms of the expansion through the $x^n$ term together with the remainder term.

The instuction `Normal[Series[P[x]/Q[x],x,0,n]]` gives the sum of the first terms of the expansion through the $x^n$ term without the remainder term.

Use these instructions to obtain expansions of the following quotients of polynomials:

**B.3.a)**

$$1/(1 + x^2)$$

**Answer:**

The sum of the expansion through the $x^{16}$ term with the remainder term is:

```
In[1]:=    Series[1/(1 + x^2),{x,0,16}]
Out[1]=        2   4   6   8   10   12
           1 - x + x - x + x - x  + x  -

           14   16     17
           x  + x  + O[x]
```

The sum of the expansion through the $x^{24}$ term without the remainder term is:

```
In[2]:=    Normal[Series[1/(1 + x^2),{x,0,24}]]
Out[2]=        2   4   6   8   10   12
           1 - x + x - x + x - x  + x  -

           14   16   18   20   22   24
           x  + x  - x  + x  - x  + x
```

The expansion of $1/(1 + x^2)$ in powers of $x$ is

$$1 - x^2 + x^4 - x^6 + ... + (-1)^k x^{2k} + ...$$

Check it out:

```
In[3]:=    Sum[(-1)^k x^( 2 k),{k,0,12}]
Out[3]=        2   4   6   8   10   12
           1 - x + x - x + x - x  + x  -

           14   16   18   20   22   24
           x  + x  - x  + x  - x  + x
```

Looks good and feels good.

**B.3.b)**

$$(x + 2)/(2x + 1)$$

**Answer:**

The sum of the expansion through the $x^7$ term with the remainder term is:

```
In[1]:=    Series[x/(2x+1),{x,0,7}]
Out[1]=            2     3      4      5
           x - 2 x + 4 x - 8 x + 16 x -

                6      7      8
           32 x + 64 x + O[x]
```

The sum of the expansion through the $x^7$ term without the remainder term is:

```
In[2]:=   Normal[Series[x/(2x+1),{x,0,7}]]
Out[2]=          2      3       4        5
          x - 2 x  + 4 x  - 8 x  + 16 x   -

                  6       7
             32 x  + 64 x
```

The expansion of $x/(2x+1)$ in powers of $x$ is

$$1 - x^2 + x^4 - x^6 + ... + (-1)^k x^{2k} + ...$$

Check it out:

```
In[3]:=   Sum[(-1)^k 2^k x^(k+1),{k,0,6}]
Out[3]=          2      3       4        5
          x - 2 x  + 4 x  - 8 x  + 16 x   -

                  6       7
             32 x  + 64 x
```

Yes!

### ■ B.4) Approximations via expansions.

#### B.4.a)

Give the expansion of $1/(1+x^2)$ in powers of $x$.

**Answer:**

The infinite series expansion of $1/(1+x^2)$ is

$$1 - x^2 + x^4 - x^6 + x^8 - x^{10} + ... + (-1)^k x^{2k} + ...$$

Check it out:

```
In[1]:=   Series[1/(1 + x^2),{x,0,18}]
Out[1]=        2    4    6    8     10    12
          1 - x  + x  - x  + x  - x   + x   -

               14    16    18        19
              x   + x   - x   + O[x]
```

or:

```
In[2]:=   Sum[(-1)^k x^(2 k), {k,0,9}] + O[x]^19
Out[2]=        2    4    6    8     10    12
          1 - x  + x  - x  + x  - x   + x   -

               14    16    18        19
              x   + x   - x   + O[x]
```

#### B.4.b.i)

Now you know that the expansion of $1/(1+x^2)$ in powers of $x$ is

$$1 - x^2 + x^4 - x^6 + x^8 - x^{10} + ... + (-1)^n x^{2n} + ...$$

Plot $1/(1+x^2)$ and and the part $1 - x^2$ of the expansion on the same axes for $-1 \le x \le 1$. Then use more of the expansion and plot $1/(1+x^2)$ and $1 - x^2 + x^4$ on the same axes for $-1 \le x \le 1$.

Describe what you see.

**Answer:**

Here is the plot of $1/(1+x^2)$ and $1 - x^2$ for $-1 \le x \le 1$:

```
In[1]:=   Plot[{1/(1 + x^2), 1 - x^2},{x,-1,1}]
```

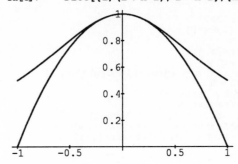

Nice match for $x's$ near 0.

Here is the plot of $1/(1+x^2)$ and $1 - x^2 + x^4$ for $-1 \le x \le 1$:

```
In[2]:=   Plot[{1/(1 + x^2), 1 - x^2 + x^4},{x,-1,1}]
```

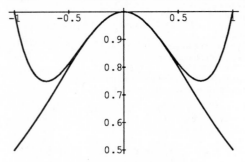

A better match for $x's$ near 0.

#### B.4.b.ii)

See what happens when you use even more of the expansion by plotting $1/(1+x^2)$ and $1 - x^2 + x^4 - x^6$ on the same axes for $-1 \le x \le 1$. Then use even more of the expansion and plot $1/(1+x^2)$ and $1 - x^2 + x^4 - x^6 + x^8$ on the same axes for $-1 \le x \le 1$.

Describe what you see.

**Answer:**

Here is the plot of $1/(1+x^2)$ and $1 - x^2 + x^4 - x^6$ for $-1 \le x \le 1$.

```
In[1]:=   Plot[{1/(1 + x^2), 1 - x^2 + x^4 - x^6},
          {x,-1,1}]
```

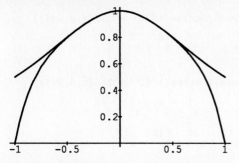

Sharing some ink.

Here comes the plot of $1/(1 + x^2)$ and $1 - x^2 + x^4 - x^6 + x^8$ for $-1 \le x \le 1$.

```
In[2]:=    Plot[{1/(1 + x^2),
           1 - x^2 + x^4 - x^6 + x^8}, {x,-1,1}]
```

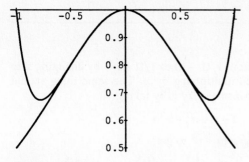

Sharing more ink.

### B.4.b.iii)

Plot $1/(1 + x^2)$ and the sum of the terms of its expansion through the $x^{12}$ term on the same axes for $-1 \le x \le 1$.

Describe what you see.

### Answer:

```
In[1]:=    expansion12 = Normal[
           Series[1/(1 + x^2), {x, 0, 12}]]
```

$$Out[1]= \quad 1 - x^2 + x^4 - x^6 + x^8 - x^{10} + x^{12}$$

```
In[2]:=    Plot[{1/(1 + x^2), expansion12 }, {x,-1,1}]
```

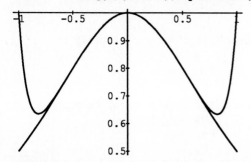

Cohabitation between $x = -0.6$ and $x = 0.6$.

### B.4.b.iv)

Plot $1/(1 + x^2)$ and the sum of the terms of its expansion through the $x^{18}$ term on the same axes for $-1 \le x \le 1$.

Describe what you see.

### Answer:

```
In[1]:=    Clear[x]
           expansion18 = Normal[
           Series[1/(1 + x^2), {x, 0, 18}]]
```

$$Out[1]= \quad 1 - x^2 + x^4 - x^6 + x^8 - x^{10} + x^{12} - x^{14} + x^{16} - x^{18}$$

```
In[2]:=    Plot[{1/(1 + x^2), expansion18 }, {x,-1,1}]
```

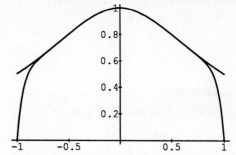

Cohabitation most of the way.

### B.4.b.v)

Interpret your plots.

### Answer:

The cohabitation interval grows as we use more of the expansion.

### B.4.b.vi)

What could be the reason for this interesting behavior?

### Answer:

It is hard to give an irrefutable reason for what's happening. But an explanation by analogy does suggest itself:

If $m$ and $n$ are positive integers, then the decimal expansion of the quotient $m/n$ is what we get when we continue to divide $n$ into $m$ without stopping. The reason we continue the dividing process is because the more decimals increase the accuracy.

```
In[1]:=    N[409/99,37]
```

$$Out[1]= \quad 4.131313131313131313131313131313131313$$

is a more accurate decimal description of 409/99 than:

$In[2]:=$    `N[409/99,14]`

$Out[2]=$    `4.1313131313131`

Maybe the same thing happens when we divide polynomials: The longer we carry out the dividing process, the more accuracy we pick up. This may indeed be the reason that:

$In[3]:=$    `Normal[Series[1/(1 + x^2),{x,0,18}]]`

$Out[3]=$
$$1 - x^2 + x^4 - x^6 + x^8 - x^{10} + x^{12} - x^{14} + x^{16} - x^{18}$$

plots out closer to $1/(1 + x^2)$ than does:

$In[4]:=$    `Normal[Series[1/(1 + x^2),{x,0,8}]]`

$Out[4]=$
$$1 - x^2 + x^4 - x^6 + x^8$$

# Tutorial

## ■ T.1) Expansions by substitution.

The expansion of $1/(1 - x)$ in powers of $x$ is

$$1 + x + x^2 + x^3 + ... + x^k + ...$$

Use this basic expansion to give the expansions in powers $x$ of the following quotients of polynomials.

Use *Mathematica* to check yourself.

### T.1.a)

$$1/(1 - x^2)$$

### Answer:

You can go from $1/(1 - x)$ to $1/(1 - x^2)$ by changing $x$ to $x^2$. The same thing should work for expansions. Try it by taking

$$1 + x + x^2 + x^3 + ... + x^k + ...$$

and changing $x$ to $x^2$ to get

$$1 + x^2 + x^4 + x^6 + ... + x^{2k} + ...$$

This should be the expansion of $1/(1 - x^2)$ in powers of $x$
.

Check:

$In[1]:=$    `Normal[Series[1/(1 - x^2),{x,0,8}]]`

$Out[1]=$
$$1 + x^2 + x^4 + x^6 + x^8$$

$In[2]:=$    `Normal[Series[1/(1 - x^2),{x,0,14}]]`

$Out[2]=$
$$1 + x^2 + x^4 + x^6 + x^8 + x^{10} + x^{12} + x^{14}$$

$In[3]:=$    `Normal[Series[1/(1 - x^2),{x,0,20}]]`

$Out[3]=$
$$1 + x^2 + x^4 + x^6 + x^8 + x^{10} + x^{12} + x^{14} + x^{16} + x^{18} + x^{20}$$

It checks out.

### T.1.b)

$$x/(1 + x^4)$$

### Answer:

You can go from $1/(1 - x)$ to $1/(1 + x^4)$ by changing $x$ to $-x^4$ and then multiplying by $x$. The same thing should work for expansions. Try it by taking

$$1 + x + x^2 + x^3 + ... + x^k + ...$$

and changing $x$ to $-x^4$ to get

$$1 + (-x^4) + (-x^4)^2 + (-x^4)^3 + ... + (-x^4)^k + ...$$
$$= 1 - x^4 - x^8 - x^{12} + ... + (-1)^k x^{4k} + ...$$

Now multiply by $x$ to get

$$x - x^5 - x^9 - x^{13} + ... + (-1)^k x^{4k+1} + ...$$

This should be the expansion of $x/(1 + x^4)$ in powers of $x$.

Check:

$In[1]:=$    `Clear[k,x]`
               `Sum[ (-1)^k x^(4 k + 1), {k,0,2}]`

$Out[1]=$
$$x - x^5 + x^9$$

$In[2]:=$    `Normal[Series[x/(1 + x^4),{x,0,9}]]`

$Out[2]=$
$$x - x^5 + x^9$$

$In[3]:=$    `Sum[ (-1)^k x^(4 k + 1), {k,0,3}]`

$Out[3]=$
$$x - x^5 + x^9 - x^{13}$$

$In[4]:=$    `Normal[Series[x/(1 + x^4),{x,0,13}]]`

$Out[4]=$
$$x - x^5 + x^9 - x^{13}$$

$In[5]:=$    `Sum[ (-1)^k x^(4 k + 1), {k,0,8}]`

$Out[5]=$
$$x^5 - x^9 + x^{13} - x^{17} + x^{21} - x^{25} + x^{29} + x^{33}$$

$In[6]:=$ `Normal[Series[x/(1 + x^4), {x, 0, 33}]]`
$Out[6]=$
$$x^5 - x^9 + x^{13} - x^{17} + x^{21} - x^{25} + x^{29} + x^{33}$$

It checks out.

■ **T.2) Approximations via expansions.**

**T.2.a)**

Find the expansion of $x/(9 - x)$ in powers of $x$.

**Answer:**

Look at:

$In[1]:=$ `Series[x/(9 - x), {x, 0, 7}]`
$Out[1]=$
$$\frac{x}{9} + \frac{x^2}{81} + \frac{x^3}{729} + \frac{x^4}{6561} + \frac{x^5}{59049} + \frac{x^6}{531441} + \frac{x^7}{4782969} + O[x]^8$$

The expansion of $x/(9 - x)$ in powers of $x$ is

$$x/9 + x^2/9^2 + x^3/9^3 + \ldots + x^k/9^k + \ldots$$

Check it out:

$In[2]:=$ `Sum[x^k/9^k, {k, 1, 7}]`
$Out[2]=$
$$\frac{x}{9} + \frac{x^2}{81} + \frac{x^3}{729} + \frac{x^4}{6561} + \frac{x^5}{59049} + \frac{x^6}{531441} + \frac{x^7}{4782969}$$

$In[3]:=$ `Normal[Series[x/(9 - x), {x, 0, 7}]]`
$Out[3]=$
$$\frac{x}{9} + \frac{x^2}{81} + \frac{x^3}{729} + \frac{x^4}{6561} + \frac{x^5}{59049} + \frac{x^6}{531441} + \frac{x^7}{4782969}$$

It checks out.

**T.2.b.i)**

Plot $x/(9 - x)$ and the sum of the of the terms in its expansion through the $x^3$ term on the same axes for $-2 \le x \le 2$.

Describe what you see.

**Answer:**

$In[1]:=$ `expan = Normal[Series[x/(9 - x), {x, 0, 3}]]`
$Out[1]=$
$$\frac{x}{9} + \frac{x^2}{81} + \frac{x^3}{729}$$

Here comes the plot. The thicker curve is the plot of the partial expansion.

$In[2]:=$ `Plot[{x/(9 - x), expan}, {x, -2, 2}]`

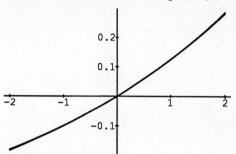

Sharing ink all the way.

**T.2.b.ii)**

Plot $x/(9 - x)$ and the sum of the of the terms in its expansion through the $x^3$ term on the same axes for $-5 \le x \le 5$.

Describe what you see.

**Answer:**

Calculate again:

$In[1]:=$ `expan = Normal[Series[x/(9 - x), {x, 0, 3}]]`
$Out[1]=$
$$\frac{x}{9} + \frac{x^2}{81} + \frac{x^3}{729}$$

Here comes the plot. The thicker curve is the partial expansion.

$In[2]:=$ `Plot[{x/(9 - x), expan}, {x, -5, 5},`
`    PlotStyle->{{Thickness[.004]},`
`    {Thickness[.008], RGBColor[0, 0, 1]}}]`

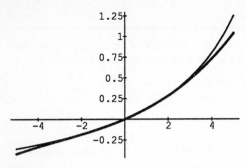

Sharing lots of ink for $-3 \leq x \leq 3$, but nasty split ends are developing at the ends.

### T.2.b.iii)

Plot $x/(9-x)$ and the sum of the of the terms in its expansion through the $x^5$ term on the same axes for $-5 \leq x \leq 5$.

Describe what you see.

---

**Answer:**

```
In[1]:=   expan = Normal[Series[x/(9 - x),{x,0,5}]]
Out[1]=         2    3     4      5
          x   x    x     x      x
          - + -- + --- + ---- + -----
          9   81   729   6561   59049
```

Here comes the plot. The thicker curve is the partial expansion.

```
In[2]:=   Plot[{x/(9 - x), expan}, {x,-5,5},
          PlotStyle->{{Thickness[.004],
          GrayLevel[0.00]},
          {Thickness[.008],RGBColor[0,0,1]}}]
```

The split ends are almost gone. Apparently it is true that using more of the expansion results in a better approximation.

### T.2.b.iv)

Plot $x/(9-x)$ and the sum of the of the terms in its expansion through the $x^9$ term on the same axes for $-5 \leq x \leq 5$.

Describe what you see.

---

**Answer:**

```
In[1]:=   expan = Normal[Series[x/(9 - x),{x,0,9}]]
Out[1]=         2    3     4      5       6
          x   x    x     x      x       x
          - + -- + --- + ---- + ----- + ------ +
          9   81   729   6561   59049   531441

               7         8         9
              x         x         x
          ------- + --------- + ---------
          4782969   43046721   387420489
```

Here comes the plot. The thicker curve is the partial expansion.

```
In[2]:=   Plot[{x/(9 - x), expan}, {x,-5,5},
          PlotStyle->{{Thickness[.004],
          GrayLevel[0.00]},
          {Thickness[.008],RGBColor[0,0,1]}}]
```

The split ends are gone.

### T.2.b.v)

Plot $x/(9-x)$ and the sum of the of the terms in its expansion through the $x^9$ term on the same axes for the larger interval $-8 \leq x \leq 8$.

Describe what you see.

---

**Answer:**

```
In[1]:=   expan = Normal[Series[x/(9 - x),{x,0,9}]]
Out[1]=         2    3     4      5       6
          x   x    x     x      x       x
          - + -- + --- + ---- + ----- + ------ +
          9   81   729   6561   59049   531441

               7         8         9
              x         x         x
          ------- + --------- + ---------
          4782969   43046721   387420489
```

Here comes the plot. The thicker curve is the partial expansion.

```
In[2]:=   Plot[{x/(9 - x), expan}, {x,-8,8},
          PlotStyle->{{Thickness[.004],
          GrayLevel[0.00]},
          {Thickness[.008],RGBColor[0,0,1]}}]
```

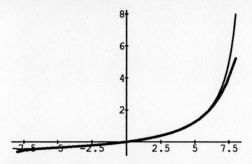

The split ends have reappeared at the ends of the larger interval.

**T.2.b.vi)**

Try to increase the quality of the approximation by plotting $x/(9 - x)$ and the sum of the of the terms in its expansion through the $x^{20}$ term on the same axes for the larger interval $-8 \leq x \leq 8$.

Describe what you see.

**Answer:**

Expand and show:

```
In[1]:=    expan = Normal[Series[x/(9 - x),{x,0,20}]]

           Plot[{x/(9 - x), expan}, {x,-8,8},
           PlotStyle->{{Thickness[.004],
           GrayLevel[0.00]},
           {Thickness[.008],RGBColor[0,0,1]}}]
```

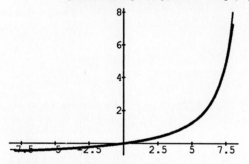

Sharing ink all the way.

**T.2.c)**

Interpret your plots.

**Answer:**

The cohabitation interval grows as we use more of the expansion.

## Literacy Sheet

1. What is the expansion of $1/(1-x)$ in powers of $x$?

.

.

2. How do you use the expansion of $1/(1-x)$ in powers of $x$ to get the expansion of $1/(1+x)$ in powers of $x$?

.

.

3. How do you use the expansion of $1/(1-x)$ in powers of $x$ to get the expansion of $1/(1+x^2)$ in powers of $x$?

.

.

4. How do you use the expansion of $1/(1-x)$ in powers of $x$ to get the expansion of $1/(1-x^3)$ in powers of $x$?

.

.

5. How do you use the expansion of $1/(1-x)$ in powers of $x$ to get the expansion of $1/(1-x^9)$ in powers of $x$?

.

.

6. How do you use the expansion of $1/(1-x)$ in powers of $x$ to get the expansion of $x^2/(1+x^5)$ in powers of $x$?

.

7. Take polynomials $P[x]$ and $Q[x]$ with $Q[0] \neq 0$. Suppose the expansion of $P[x]/Q[x]$ in powers of $x$ is

$$a[0] + a[1]x + a[2]x^2 + a[3]x^3 + ... + a[k]x^k + ...$$

What happens to the graphs of $P[x]/Q[x]$ and

$$a[0] + a[1]x + a[2]x^2 + a[3]x^3 + ... + a[n]x^n$$

as $n$ is increased?

.

.

8. Remembering that the derivative of $1/(1-x)$ is $1/(1-x)^2$, how can you get the expansion of $1/(1-x)^2$ in powers of $x$ directly from the expansion of $1/(1-x)$ in powers of $x$?

.

.

9. In what way are expansions in powers of $x$ like decimal expansions of numbers?

.

# Approximations by expansions: Integration

## Guide

Fueled by our success with finding expansions of quotients of polynomials, we move to the main functions in calculus:

$$e^x$$

$$\sin[x]$$

$$\cos[x].$$

In this section, you will see how to obtain expansions of these functions and many other related functions.

As an added bonus, you will also see how the imaginary number $I = \sqrt{-1}$ can be used to relate $e^x$, $\cos[x]$ and $\sin[x]$ via Euler's formula

$$e^{Ix} = \cos[x] + I\sin[x].$$

## Basics

■ **B.1) Geometric series and expansions of the Logarithm and Arc-Tangent.**

The expansion of $1/(1-x)$ was discovered by Mercator (of map fame) in 1668 . It is of signal importance in finding expansions of the building block calculus functions. It is so important that it even has its own name; it is called the **geometric series.** To refresh the memory, recall:

*The expansion of $1/(1-x)$ in powers of $x$ is*

$$1 + x + x^2 + x^3 + x^4 + x^5 + x^6 + \ldots + x^k + \ldots$$

As we saw in the last lesson, this underlying expansion can be used as the basic building block for a variety of other expansions.

**B.1.a.i)**

Look at:

```
In[1]:=    Integrate[- 1/ (1 - t),{t,0,x}]
Out[1]=    Log[1 - x]
```

This tells us that

$$\log[1-x] = -\int_0^x 1/(1-t)\,dt.$$

As a result, a natural way to get the expansion of $\log[1-x]$ in powers of $x$ is to take the expansion of $1/(1-t)$ in powers of $t$, integrate it from 0 to $x$ and then multiply the result by $-1$. See what happens when you do this.

**Answer:**

First let's take a look:

```
In[2]:=    Series[ Log[1 - x],{x,0,8}]
Out[2]=         2   3   4   5   6   7   8
              x   x   x   x   x   x   x
        -x - -- - -- - -- - -- - -- - -- - -- +
              2   3   4   5   6   7   8

              9
          0[x]
```

The expansion of $\log[1-x]$ in powers of $x$ is

$$-x - x^2/2 - x^3/3 - x^4/4 - \ldots - x^k/k - \ldots$$

Now let's see whether the procedure of taking the expansion of $1/(1-t)$ in powers of $t$, integrating it from 0 to $x$ and then multipling by $-1$ outlined above will explain this formula for the expansion.

Take the expansion of $1/(1-t)$ in powers of $t$:

$$1 + t + t^2 + t^3 + t^4 + t^5 + \ldots + t^k + \ldots$$

Integrate it from 0 to $x$ to get

$$\int_0^x 1\,dt + \int_0^x t\,dt + \int_0^x t^2\,dt + \int_0^x t^3\,dt + \ldots + \int_0^x t^k\,dt + \ldots$$

$$= x + x^2/2 + x^3/3 + x^4/4 + \ldots + x^{k+1}/(k+1) + \ldots$$

Multiply the result by $-1$ to get

$$-x - x^2/2 - x^3/3 - x^4/4 - \ldots - x^{k+1}/(k+1) + \ldots$$

This is in harmony with the formula we got above.

Sweet and easy.

**B.1.a.ii)**

Try out the expansion

$$-x - x^2/2 - x^3/3 - x^4/4 - \ldots - x^k/k - \ldots$$

of $\log[1-x]$ in powers of $x$ by plotting $\log[1-x]$ and a few of its partial expansions on the same axes for $-1 \le x \le 1$. Assess the results.

**Answer:**

```
In[1]:=    expan4 =
           Normal[Series[Log[1 - x],{x,0,4}]]
```

```
Out[1]=          2    3    4
                x    x    x
          -x - -- - -- - --
                2    3    4
```

The plot of the partial expansion is the thicker of the two plots.

```
In[2]:=    Plot[{Log[1 - x],expan4},{x,-1,1},
           PlotRange->{-4,2},PlotStyle->
           {{Thickness[.004],GrayLevel[.0]},
           {Thickness[.008],RGBColor[0,0,1]}}];
```

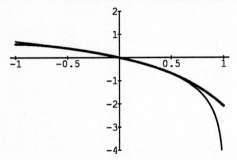

Pretty darn good, but let's do even better:

```
In[3]:=    expan8 =
           Normal[Series[Log[1 - x],{x,0,8}]]

Out[3]=          2    3    4    5    6    7    8
                x    x    x    x    x    x    x
          -x - -- - -- - -- - -- - -- - -- - --
                2    3    4    5    6    7    8
```

(The plot of the partial expansion is the thicker of the two plots.)

```
In[4]:=    Plot[{Log[1 - x],expan8},{x,-1,1},
           PlotRange->{-4,2},PlotStyle->
           {{Thickness[.004],GrayLevel[0.00]},
           {Thickness[.008],RGBColor[0,0,1]}}];
```

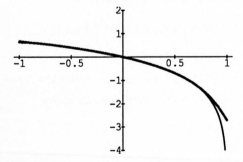

Even better; now let's go for the moon:

```
In[5]:=    expan20 =
           Normal[Series[Log[1 - x],{x,0,20}]]

           Plot[{Log[1 - x],expan20},{x,-1,1},
           PlotRange->{-4,2},
           PlotStyle->
           {{Thickness[.004],GrayLevel[0.00]},
           {Thickness[.008],RGBColor[0,0,1]}}];
```

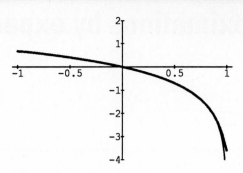

Mighty satisfying.

### B.1.a.iii)

Use the expansion

$$-x - x^2/2 - x^3/3 - x^4/4 - \ldots - x^k/k - \ldots$$

of $\log[1-x]$ in powers of $x$ to get the expansion of $\log[1+x]$ in powers of $x$.

### Answer:

This is pretty easy. To get $\log[1 + x]$, we just take $\log[1 - x]$ and change $x$ to $-x$:

```
In[1]:=    Log[1 - x]/.x->-x

Out[1]=    Log[1 + x]
```

So to get the expansion of $\log[1 + x]$ in powers of $x$, we just take

$$-x - x^2/2 - x^3/3 - x^4/4 - \ldots - x^k/k - \ldots$$

and change $x$ to $-x$ to get:

$$x - x^2/2 + x^3/3 - x^4/4 - x^5/5 \ldots + (-1)^k x^k/k + \ldots$$

Let's run this through *Mathematica* :

```
In[2]:=    Normal[Series[Log[1 - x],{x,0,7}]]

Out[2]=          2    3    4    5    6    7
                x    x    x    x    x    x
          -x - -- - -- - -- - -- - -- - --
                2    3    4    5    6    7
```

```
In[3]:=    Normal[Series[
           Log[1 - x],{x,0,7}]]/.x->-x

Out[3]=          2    3    4    5    6    7
                x    x    x    x    x    x
          x - -- + -- - -- + -- - -- + --
               2    3    4    5    6    7
```

Of course this agrees with:

```
In[4]:=    Normal[Series[Log[1 + x],{x,0,7}]]

Out[4]=          2    3    4    5    6    7
                x    x    x    x    x    x
          x - -- + -- - -- + -- - -- + --
               2    3    4    5    6    7
```

### B.1.b.i)

Look at:

```
In[1]:=    Integrate[1/(1 + t^2),{t,0,x}]
Out[1]=    ArcTan[x]
```

This tells us that

$$\arctan[x] = \int_0^x 1/(1 + t^2)\,dt.$$

As a result, a natural way to get the expansion of $\arctan[x]$ in powers of $x$ is to take the expansion of $1/(1 + t^2)$ in powers of $t$ and integrate it from $0$ to $x$. See what happens when you do this.

**Answer:**

First let's take a look:

```
In[2]:=    Series[ArcTan[x],{x,0,11}]
Out[2]=         3    5    7    9    11
           x    x    x    x    x            12
       x - -- + -- - -- + -- - --- + O[x]
           3    5    7    9    11
```

The expansion of $\arctan[x]$ in powers of $x$ is

$$x - x^3/3 + x^5/5 - x^7/7 + \ldots + (-1)^k x^{2k+1}/(2k+1)\ldots$$

Now let's see whether the procedure of taking the expansion of $1/(1 + t^2)$ in powers of $t$ and integrating it from $0$ to $x$ outlined above will explain this formula.

To get the expansion of $1/(1 + t^2)$ in powers of $t$, take the expansion of $1/(1 - t)$ in powers of $t$

$$1 + t + t^2 + t^3 + t^4 + t^5 + \ldots + t^k + \ldots$$

and change $t$ to $-t^2$:

$$1 - t^2 + (-t^2)^2 + (-t^2)^3 + (-t^2)^4 + \ldots + (-t^2)^k + \ldots$$
$$= 1 - t^2 + t^4 - t^6 + t^8 - t^{10} + \ldots + (-1)^k t^{2k} + \ldots$$

Integrate it from $0$ to $x$ to get

$$\int_0^x 1\,dt - \int_0^x t^2\,dt + \int_0^x t^4\,dt - \int_0^x t^6\,dt + \ldots + (-1)^k \int_0^x t^{2k}\,dt + \ldots$$
$$= x - x^3/3 + x^5/5 - x^7/7 + \ldots + (-1)^k x^{2k+1}/(2k+1) + \ldots$$

This is in harmony with the formula we got above:

```
In[3]:=    Series[ArcTan[x],{x,0,11}]
Out[3]=         3    5    7    9    11
           x    x    x    x    x            12
       x - -- + -- - -- + -- - --- + O[x]
           3    5    7    9    11
```

Neat.

### ■ B.2) The expansion of $e^x$ in powers of $x$.

Having had so much success with integration in B.1), we now get a little cocky and try to combine integration and the method of undetermined coefficients to get one of the most important expansions of them all.

### B.2.a)

Use the method of undetermined coefficients and the facts that

$$e^x - 1 = \int_0^x e^t\,dt$$

to get the expansion of $e^x$ in powers of $x$.

**Answer:**

Fix a positive integer $n$:

```
In[1]:=    n = 9
Out[1]=    9
```

Let

$$a[0] + a[1]x + a[2]x^2 + a[3]x^3 + \ldots + a[k]x^k + \ldots$$

stand for the expansion of $e^x$ in powers of $x$. Our object is to determine what the constants

$$a[0], a[1], a[2], \ldots a[k], \ldots$$

must be.

```
In[2]:=    partexpan[x_] = Sum[a[k] x^k ,{k,0,n}]
Out[2]=                          2        3
           a[0] + x a[1] + x  a[2] + x  a[3] +

              4        5        6
           x  a[4] + x  a[5] + x  a[6] +

              7        8        9
           x  a[7] + x  a[8] + x  a[9]
```

Recall

$$e^x - 1 = \int_0^x e^t\,dt.$$

We'll treat $partexpan[x]$ as if it were equal to $e^x$ by replacing $e^x$ by $partexpan[x]$ on both sides and then solve for the coefficients $a[0], a[1], a[2], \ldots a[n]$.

Replace $e^x - 1$ on the left by $partexpan[x] - 1$ plus the error term:

```
In[3]:=    left = partexpan[x] - 1 + O[x]^(n+1)
Out[3]=                                        2
           (-1 + a[0]) + a[1] x + a[2] x  +

                3         4         5
           a[3] x  + a[4] x  + a[5] x  +

                6         7         8
           a[6] x  + a[7] x  + a[8] x  +

                9          10
           a[9] x  + O[x]
```

Replace $\int_0^x e^t\,dt$ on the right by $\int_0^x partexpan[t]\,dt$:

```
In[4]:=    right = Integrate[partexpan[t],{t,0,x}]
```

*Out[4]=*
$$x\, a[0] + \frac{x^2\, a[1]}{2} + \frac{x^3\, a[2]}{3} + \frac{x^4\, a[3]}{4} +$$
$$\frac{x^5\, a[4]}{5} + \frac{x^6\, a[5]}{6} + \frac{x^7\, a[6]}{7} +$$
$$\frac{x^8\, a[7]}{8} + \frac{x^9\, a[8]}{9} + \frac{x^{10}\, a[9]}{10}$$

Now find the equations that result from equating the coefficients of like powers of $x$ on the left and right sides of the equation $left == right$.

You can spot a few of them:

$$-1 + a[0] == 0,\ a[1] == a[0],$$
$$a[2] == a[1]/2, ...$$

Now let *Mathematica* set up these equations for you:

*In[5]:=*  `coeffeqns = LogicalExpand[left == right]`
*Out[5]=*  $-1 + a[0] == 0$ && $-a[0] + a[1] == 0$ &&

$$\frac{-a[1]}{2} + a[2] == 0\ \&\&$$
$$\frac{-a[2]}{3} + a[3] == 0\ \&\&$$
$$\frac{-a[3]}{4} + a[4] == 0\ \&\&$$
$$\frac{-a[4]}{5} + a[5] == 0\ \&\&$$
$$\frac{-a[5]}{6} + a[6] == 0\ \&\&$$
$$\frac{-a[6]}{7} + a[7] == 0\ \&\&$$
$$\frac{-a[7]}{8} + a[8] == 0\ \&\&\ \frac{-a[8]}{9} + a[9] == 0$$

Solve for the coefficients.

*In[6]:=*  `coeffs = Solve[coeffeqns]`
*Out[6]=*  $\{\{a[0] \to 1,\ a[1] \to 1,\ a[2] \to \frac{1}{2},$

$$a[3] \to \frac{1}{6},\ a[4] \to \frac{1}{24},\ a[5] \to \frac{1}{120},$$
$$a[6] \to \frac{1}{720},\ a[7] \to \frac{1}{5040},$$
$$a[8] \to \frac{1}{40320},\ a[9] \to \frac{1}{362880}\}\}$$

Put these coefficients into the partial expansion:

*In[7]:=*  `partexpan[x]/.coeffs[[1]]`
*Out[7]=*
$$1 + x + \frac{x^2}{2} + \frac{x^3}{6} + \frac{x^4}{24} + \frac{x^5}{120} + \frac{x^6}{720} +$$
$$\frac{x^7}{5040} + \frac{x^8}{40320} + \frac{x^9}{362880}$$

Let's rerun for a larger $n$:

*In[8]:=*  `n = 12`
*Out[8]=*  12

*In[9]:=*
```
partexpan[x_] =
Sum[a[k] x^k ,{k,0,n}]
left =
partexpan[x] - 1 + O[x]^(n+1)
right =
Integrate[partexpan[t],{t,0,x}]
coeffeqns =
LogicalExpand[left == right]
```
*Out[9]=*  $-1 + a[0] == 0$ && $-a[0] + a[1] == 0$ &&

$$\frac{-a[1]}{2} + a[2] == 0\ \&\&$$
$$\frac{-a[2]}{3} + a[3] == 0\ \&\&$$
$$\frac{-a[3]}{4} + a[4] == 0\ \&\&$$
$$\frac{-a[4]}{5} + a[5] == 0\ \&\&$$
$$\frac{-a[5]}{6} + a[6] == 0\ \&\&$$
$$\frac{-a[6]}{7} + a[7] == 0\ \&\&$$
$$\frac{-a[7]}{8} + a[8] == 0\ \&\&$$
$$\frac{-a[8]}{9} + a[9] == 0\ \&\&$$
$$\frac{-a[9]}{10} + a[10] == 0\ \&\&$$
$$\frac{-a[10]}{11} + a[11] == 0\ \&\&$$
$$\frac{-a[11]}{12} + a[12] == 0$$

*In[10]:=*
```
coeffs = Solve[coeffeqns]
partexpan[x]/.coeffs[[1]]
```

```
Out[10]=
                  2    3    4     5      6
                 x    x    x     x      x
          1 + x + -- + -- + -- + --- + --- +
                  2    6    24   120    720

                 7      8       9        10
                x      x       x        x
             ---- + ----- + ------ + ------- +
             5040   40320   362880   3628800

                 11          12
                x           x
             --------- + ----------
             39916800    479001600
```

Rerun this for several larger values of $n$ and study the coefficient equations to learn that the expansion of $e^x$ in powers of $x$ is

$$1 + x + x^2/2! + x^3/3! + x^4/4! + x^5/5! + ... + x^k/k! + ...$$

This is the most famous expansion of them all! It ranks right up there with the expansion of $1/(1 - x)$ in importance. Learn it!

```
In[11]:=   Series[E^x,{x,0,2}]
Out[11]=
                  2
                 x         3
          1 + x + -- + O[x]
                 2
```

```
In[12]:=   Series[E^x,{x,0,5}]
Out[12]=
                  2    3    4     5
                 x    x    x     x        6
          1 + x + -- + -- + -- + --- + O[x]
                  2    6    24   120
```

```
In[13]:=   Series[E^x,{x,0,9}]
Out[13]=
                  2    3    4     5      6
                 x    x    x     x      x
          1 + x + -- + -- + -- + --- + --- +
                  2    6    24   120    720

                 7      8       9
                x      x       x         10
             ---- + ----- + ------ + O[x]
             5040   40320   362880
```

■ **B.3) The expansions of sin[x] and cos[x] in powers of x.**

**B.3.a)**

Use the method of undetermined coefficients and the facts that

$$\sin[x] = \int_0^x \cos[t]dt$$

and

$$1 - \cos[x] = \int_0^x \sin[t]dt$$

to get the expansions of $\sin[x]$ and $\cos[x]$ in powers of $x$.

**Answer:**

We'll use the same approach we used in B.2). In fact, the whole procedure is nothing but a copy, paste and edit job on the answer to B.2)

Fix a positive integer $n$:

```
In[1]:=    n = 9
Out[1]=    9
```

Let

$$a[0] + a[1]x + a[2]x^2 + a[3]x^3 + ... + a[k]x^k + ...$$

stand for the expansion of $\cos[x]$ in powers of $x$.

And let

$$b[0] + b[1]x + b[2]x^2 + b[3]x^3 + ... + b[k]x^k + ...$$

stand for the expansion of $\sin[x]$ in powers of $x$.

Our object is to determine what the constants

$$a[0], a[1], a[2], ...a[k], ...$$
$$b[0], b[1], b[2], ...b[k], ...$$

must be.

```
In[2]:=    cospartexpan[x_] = Sum[a[k] x^k ,{k,0,n}]
Out[2]=
                       2        3
         a[0] + x a[1] + x  a[2] + x  a[3] +

            4        5        6
           x  a[4] + x  a[5] + x  a[6] +

            7        8        9
           x  a[7] + x  a[8] + x  a[9]
```

```
In[3]:=    sinpartexpan[x_] = Sum[b[k] x^k ,{k,0,n}]
Out[3]=
                       2        3
         b[0] + x b[1] + x  b[2] + x  b[3] +

            4        5        6
           x  b[4] + x  b[5] + x  b[6] +

            7        8        9
           x  b[7] + x  b[8] + x  b[9]
```

Recall

$$\sin[x] = \int_0^x \cos[t]dt$$

and

$$1 - \cos[x] = \int_0^x \sin[t]dt$$

We'll treat *cospartexpan[x]* as if it were equal to $\cos[x]$ and we'll treat *sinpartexpan[x]* as if it were equal to $\sin[x]$ by replacing $\cos[x]$ by *cospartexpan[x]* and by replacing $\sin[x]$ by *sinpartexpan[x]* on both sides of both equations. Then we will solve for the coefficients $a[0], a[1], a[2], ...a[n], b[0], b[1], b[2], ..., b[n]$.

The Big Oh error term, $O[x]^{(n+1)}$, is included to assure equality.

```
In[4]:=    left1 = sinpartexpan[x] + O[x]^(n + 1)
```

*Out[4]=*
$$b[0] + b[1] \, x + b[2] \, x^2 + b[3] \, x^3 +$$
$$b[4] \, x^4 + b[5] \, x^5 + b[6] \, x^6 +$$
$$b[7] \, x^7 + b[8] \, x^8 + b[9] \, x^9 + O[x]^{10}$$

*In[5]:=*  `right1 = Integrate[cospartexpan[t], {t, 0, x}]`

*Out[5]=*
$$x \, a[0] + \frac{x^2 \, a[1]}{2} + \frac{x^3 \, a[2]}{3} + \frac{x^4 \, a[3]}{4} +$$
$$\frac{x^5 \, a[4]}{5} + \frac{x^6 \, a[5]}{6} + \frac{x^7 \, a[6]}{7} +$$
$$\frac{x^8 \, a[7]}{8} + \frac{x^9 \, a[8]}{9} + \frac{x^{10} \, a[9]}{10}$$

*In[6]:=*  `left2 = 1 - cospartexpan[x] + O[x]^(n + 1)`

*Out[6]=*
$$(1 - a[0]) - a[1] \, x - a[2] \, x^2 -$$
$$a[3] \, x^3 - a[4] \, x^4 - a[5] \, x^5 -$$
$$a[6] \, x^6 - a[7] \, x^7 - a[8] \, x^8 -$$
$$a[9] \, x^9 + O[x]^{10}$$

*In[7]:=*  `right2 = Integrate[sinpartexpan[t], {t, 0, x}]`

*Out[7]=*
$$x \, b[0] + \frac{x^2 \, b[1]}{2} + \frac{x^3 \, b[2]}{3} + \frac{x^4 \, b[3]}{4} +$$
$$\frac{x^5 \, b[4]}{5} + \frac{x^6 \, b[5]}{6} + \frac{x^7 \, b[6]}{7} +$$
$$\frac{x^8 \, b[7]}{8} + \frac{x^9 \, b[8]}{9} + \frac{x^{10} \, b[9]}{10}$$

Now find the equations that result from equating the co-efficients of like powers of $x$ on the left and right sides of the equation *left1 == right1*:

*In[8]:=*  `coeffeqns1 = LogicalExpand[left1 == right1]`

*Out[8]=*   $b[0] == 0$ && $-a[0] + b[1] == 0$ &&
$$\frac{-a[1]}{2} + b[2] == 0 \text{ \&\& }$$
$$\frac{-a[2]}{3} + b[3] == 0 \text{ \&\& }$$
$$\frac{-a[3]}{5} + b[4] == 0 \text{ \&\& }$$

$$\frac{-a[4]}{5} + b[5] == 0 \text{ \&\& }$$
$$\frac{-a[5]}{6} + b[6] == 0 \text{ \&\& }$$
$$\frac{-a[6]}{7} + b[7] == 0 \text{ \&\& }$$
$$\frac{-a[7]}{8} + b[8] == 0 \text{ \&\& } \quad \frac{-a[8]}{9} + b[9] == 0$$

And find the equations that result from equating the coefficients of like powers of $x$ on the left and right sides of the equation *left2 == right2*:

*In[9]:=*  `coeffeqns2 = LogicalExpand[left2 == right2]`

*Out[9]=*   $1 - a[0] == 0$ && $-a[1] - b[0] == 0$ &&
$$-a[2] - \frac{b[1]}{2} == 0 \text{ \&\& }$$
$$-a[3] - \frac{b[2]}{3} == 0 \text{ \&\& }$$
$$-a[4] - \frac{b[3]}{4} == 0 \text{ \&\& }$$
$$-a[5] - \frac{b[4]}{5} == 0 \text{ \&\& }$$
$$-a[6] - \frac{b[5]}{6} == 0 \text{ \&\& }$$
$$-a[7] - \frac{b[6]}{7} == 0 \text{ \&\& }$$
$$-a[8] - \frac{b[7]}{8} == 0 \text{ \&\& } \quad -a[9] - \frac{b[8]}{9} == 0$$

Solve for the coefficients:

*In[10]:=*  `coeffs = Solve[{coeffeqns1, coeffeqns2}]`

*Out[10]=*   $\{\{b[0] \to 0,\ a[0] \to 1,\ b[1] \to 1,$
$$a[1] \to 0,\ b[2] \to 0,\ a[2] \to -\left(\frac{1}{2}\right),$$
$$b[3] \to -\left(\frac{1}{6}\right),\ a[3] \to 0,\ b[4] \to 0,$$
$$a[4] \to \frac{1}{24},\ b[5] \to \frac{1}{120},\ a[5] \to 0,$$
$$b[6] \to 0,\ a[6] \to -\left(\frac{1}{720}\right),$$
$$1$$

```
        b[7] -> -(----), a[7] -> 0,
                  5040

        b[8] -> 0, a[8] -> -----,
                            40320

              1
        b[9] -> ------, a[9] -> 0}}
              362880
```

Here is the expansion of $\cos[x]$ through the $x^n$ term:

```
In[11]:=   cospartexpan[x]/.coeffs[[1]]
Out[11]=        2    4    6    8
               x    x    x    x
           1 - -- + -- - --- + -----
               2    24   720   40320
```

Check:

```
In[12]:=   Normal[Series[Cos[x],{x,0,n}]]
Out[12]=        2    4    6    8
               x    x    x    x
           1 - -- + -- - --- + -----
               2    24   720   40320
```

Cool.

Here is the expansion of $\sin[x]$ through the $x^n$ term:

```
In[13]:=   sinpartexpan[x]/.coeffs[[1]]
Out[13]=       3    5    7     9
              x    x    x     x
           x - -- + --- - ---- + ------
              6    120  5040   362880
```

Check:

```
In[14]:=   Normal[Series[Sin[x],{x,0,n}]]
Out[14]=       3    5    7     9
              x    x    x     x
           x - -- + --- - ---- + ------
              6    120  5040   362880
```

Just as cool.

Let's rerun for a larger $n$:

```
In[15]:=   n = 15
Out[15]=   15
```

```
In[16]:=   cospartexpan[x_] =
           Sum[a[k] x^k ,{k,0,n}]
           sinpartexpan[x_] =
           Sum[b[k] x^k ,{k,0,n}]
           left1 =
           sinpartexpan[x] + O[x]^(n + 1)
           right1 =
           Integrate[cospartexpan[t],{t,0,x}]
           left2 =
           1 - cospartexpan[x] + O[x]^(n + 1)
           right2 =
           Integrate[sinpartexpan[t],{t,0,x}]
           coeffeqns1 =
           LogicalExpand[left1 == right1]
           coeffeqns2 =
           LogicalExpand[left2 == right2]
```

```
           coeffs =
           Solve[{coeffeqns1,coeffeqns2}];
```

Here is the expansion of $\cos[x]$ through the $x^n$ term:

```
In[17]:=   cospartexpan[x]/.coeffs[[1]]
Out[17]=        2    4    6    8      10
               x    x    x    x      x
           1 - -- + -- - --- + ----- - -------- +
               2    24   720  40320   3628800

               12         14
              x          x
           --------- - -----------
           479001600   87178291200
```

This is in harmony with:

```
In[18]:=   Normal[Series[Cos[x],{x,0,n}]]
Out[18]=        2    4    6    8      10
               x    x    x    x      x
           1 - -- + -- - --- + ----- - -------- +
               2    24   720  40320   3628800

               12         14
              x          x
           --------- - -----------
           479001600   87178291200
```

Here is the expansion of $\sin[x]$ through the $x^n$ term:

```
In[19]:=   sinpartexpan[x]/.coeffs[[1]]
Out[19]=       3    5    7     9
              x    x    x     x
           x - -- + --- - ---- + ------ -
              6    120  5040   362880

               11          13            15
              x           x             x
           -------- + ----------- - --------------
           39916800   6227020800   1307674368000
```

This is in harmony with:

```
In[20]:=   Normal[Series[Sin[x],{x,0,n}]]
Out[20]=       3    5    7     9
              x    x    x     x
           x - -- + --- - ---- + ------ -
              6    120  5040   362880

               11          13            15
              x           x             x
           -------- + ----------- - --------------
           39916800   6227020800   1307674368000
```

Rerun this for several other values of $n$ and study the coefficient equations to learn that the expansion of $\sin[x]$ in powers of $x$ is

$$x - x^3/3! + x^5/5! - x^7/7!.. + (-1)^k x^{2k+1}/(2k+1)! + ...$$

And the expansion of $\cos[x]$ in powers of $x$ is

$$1 - x^2/2! + x^4/4! - x^6/6!.. + (-1)^k x^{2k}/(2k)! + ...$$

where the usual convention $0! = 1$ is understood.

■ **B.4) The complex exponential function.**

**B.4.a)**

The imaginary number $I = \sqrt{-1}$ can do a lot to explain the similarity of the expansions

$e^x : 1 + x + x^2/2! + x^3/3! + x^4/4! + x^5/5! + \ldots + x^k/k! + \ldots$

$\sin[x] : x - x^3/3! + x^5/5! - x^7/7! + \ldots + (-1)^k x^{2k+1}/(2k+1)! + \ldots$

$\cos[x] : 1 - x^2/2! + x^4/4! - x^6/6! \ldots + (-1)^n x^{2k}/(2k)! + \ldots$ Look at:

```
In[1]:=    Normal[Series[
           E^x,{x,0,6}]]/.x->I x
Out[1]=
                 2         4          6
                x    -I 3 x     I 5  x
           1 + I x - -- + -- x + -- x + --- x - ---
                 2    6     24    120    720
```

```
In[2]:=    Normal[Series[Cos[x],{x,0,6}]] +
           I Normal[Series[Sin[x],{x,0,6}]]
Out[2]=
                 2   4    6          3   5
                x   x    x          x   x
           1 - -- + -- - --- + I (x - -- + ---)
                2   24   720         6   120
```

Try to determine the relationship.

---

**Answer:**

Try a few:

```
In[3]:=    Normal[Series[
           E^x,{x,0,4}]]/.x->I x
Out[3]=
                 2         4
                x    -I 3 x
           1 + I x - -- + -- x + --
                 2    6     24
```

```
In[4]:=    Expand[Normal[
           Series[Cos[x],{x,0,4}]] +
           I Normal[Series[Sin[x],{x,0,4}]]]
Out[4]=
                 2         4
                x    -I 3 x
           1 + I x - -- + -- x + --
                 2    6     24
```

Golly-oskee; both lines came out the same! Let's try again:

```
In[5]:=    Normal[Series[
           E^x,{x,0,12}]]/.x->I x
Out[5]=
                 2         4
                x    -I 3 x     I 5
           1 + I x - -- + -- x + --- x -
                 2    6     24    120

                 6         8
                x    -I 7 x     I    9
           --- + ---- x + ----- + ------ x -
           720   5040      40320   362880

                10          12
                x    -I 11 x     x
           ------- + -------- x + ---------
           3628800   39916800     479001600
```

```
In[6]:=    Expand[Normal[
           Series[Cos[x],{x,0,12}]] +
```

```
           I Normal[Series[Sin[x],{x,0,8}]]]
Out[6]=
                 2         4
                x    -I 3 x     I 5
           1 + I x - -- + -- x + --- x -
                 2    6     24    120

                 6         8          10
                x    -I 7 x     x          x
           --- + ---- x + ----- - ------- +
           720   5040      40320   3628800

                12
                x
           ---------
           479001600
```

```
In[7]:=
```

Again both lines came out the same. Evidently, if we take the expansion of $e^x$ in powers of $x$ and replace $x$ by $I$ times $x$, then we get the expansion of $\cos[x]$ plus $I$ times the expansion of $\sin[x]$. For this reason, $e^{Ix}$ is defined to be $\cos[x] + I\sin[x]$.

This relationship between the exponential function $e^x$ and the trigonometric functions $\cos[x]$ and $\sin[x]$ is called *Euler's formula*.

**B.4.b.i)**

---

For a real number $x$, we are pretty confident about calculating $e^x$.

But if $z = x + Iy$ is a complex number, then how can we make sense of $e^z$?

---

**Answer:**

Just write $z = x + Iy$ and
$$e^z = e^{x+Iy} = e^x e^{Iy} = e^x(\cos[y] + I\sin[y]).$$
So
$$e^{x+Iy} = e^x\cos[y] + Ie^x\sin[y].$$

Not much to it.

**B.4.b.ii)**

---

How do complex numbers allow us to take the natural logarithm of a negative real number?

---

**Answer:**

Use the usual agreement that $y = \log[x]$ if $e^y = x$. Under this agreement, if we want $\log[-2]$, for instance, we find a complex number $y = a + Ib$ with $e^y = -2$. The resulting complex number $y$ will work for $\log[-2]$.

So we go after a complex number $y = a + Ib$ with
$$e^y = e^{a+Ib} = e^a(\cos[b] + I\sin[b]) = -2.$$

The first observation is that Sin[b] must be zeroed out: This gives a couple of possibilities:

```
In[1]:=    E^a (Cos[b] + I Sin[b])/.b->0
Out[1]=      a
           E
```

```
In[2]:=    E^a (Cos[b] + I Sin[b])/.b->Pi
Out[2]=       a
           -E
```

The first of these cannot work because there is no real number $a$ that makes $e^a = -2$. So we go with the second.

This is easy because now all we have to do is to solve $-e^a = -2$ for $a$.

This is the same as $e^a = 2$. This sets $a = \log[2]$.

Let's try it out:

```
In[3]:=    E^a (Cos[b] + I Sin[b])/.
           {b->Pi,a->Log[2]}
Out[3]=    -2
```

It works!

This means
$$\log[-2] = \log[2] + I\pi.$$

Similarly, we can see that if $x$ is any positive real number, then
$$\log[-x] = \log[x] + I\pi.$$

Take a look:

```
In[4]:=    N[{Log[-2],Log[2] + I Pi},8]
Out[4]=    {0.69314718 + 3.1415927 I,

             0.69314718 + 3.1415927 I}
```

```
In[5]:=    N[{Log[-E],Log[E] + I Pi},8]
Out[5]=    {1. + 3.1415927 I, 1. + 3.1415927 I}
```

```
In[6]:=    N[{Log[-Pi],Log[Pi] + I Pi},8]
Out[6]=    {1.1447299 + 3.1415927 I,

             1.1447299 + 3.1415927 I}
```

```
In[7]:=    N[{Log[-6],Log[6] + I Pi},8]
Out[7]=    {1.7917595 + 3.1415927 I,

             1.7917595 + 3.1415927 I}
```

```
In[8]:=    N[{Log[-17],Log[17] + I Pi},8]
Out[8]=    {2.8332133 + 3.1415927 I,

             2.8332133 + 3.1415927 I}
```

Nifty as hell.

Shame on that bureaucratic math teacher who once told you that you cannot take the log of a negative number.

Still this doesn't tell you how to take the log of zero.

This is going to be impossible because logs of negative numbers near zero all carry the term $I\pi$; while the logs of positive numbers near zero do not.

**B.4.b.iii)**

If $z = x + Iy$ is a complex number, then how can we make sense of sin[z] and cos[z]?

**Answer:**

Hmmmm.
$$e^{Ix} = \cos[x] + I\sin[x]$$
$$e^{-Ix} = \cos[-x] + I\sin[-x] = \cos[x] - I\sin[x].$$

Solve for sin[x] and cos[x] in terms of $e^{Ix}$ and $e^{-Ix}$:

```
In[1]:=    Solve[{E^(I x) == Cos[x] + I Sin[x],
           E^(-I x) == Cos[x] - I Sin[x]},
           {Sin[x],Cos[x]}]
Out[1]=                  I  -I x    -I  I x
           {{Sin[x] -> - E      + -- E    ,
                       2         2

                          -I x    I x
                         E       E
             Cos[x] -> ----- + ----}}
                         2       2
```

So for a real number $x$, we see
$$\sin[x] = I(e^{-Ix} - e^{Ix})/2$$
and
$$\cos[x] = (e^{Ix} + e^{-Ix})/2.$$

So to get sin[z] and cos[z] for a complex $z = x + Iy$, we just replace $x$ by $z$ in the formulas above to get:
$$\sin[z] = I(e^{-Iz} - e^{Iz})/2$$
and
$$\cos[z] = (e^{Iz} + e^{-Iz})/2.$$

Try it:

```
In[2]:=    N[{Cos[z],(E^(I z) + E^(-I z))/2}/.
           z->1 + 2 I]
Out[2]=    {2.03272 - 3.0519 I, 2.03272 - 3.0519 I}
```

```
In[3]:=    N[{Cos[z],(E^(I z) + E^(-I z))/2}/.
           z->Pi/2]
Out[3]=    {0., 0.}
```

```
In[4]:=    N[{Sin[z],I(E^(-I z) - E^(I z))/2}/.
           z->1 + 2 I]
Out[4]=    {3.16578 + 1.9596 I, 3.16578 + 1.9596 I}
```

```
In[5]:=    N[{Sin[z],I(E^(-I z) - E^(I z))/2}/.
           z->Pi/2]
```

`Out[5]=     {1., 1.}`

# Tutorial

### ■ T.1) Expansions by substitution.

#### T.1.a)

Give the expansion of $e^{x^2}$ in powers of $x$.

### Answer:

The expansion of $e^x$ in powers of $x$ is

$$1 + x + x^2/2! + x^3/3! + x^4/4! + x^5/5! + \ldots + x^k/k! + \ldots$$

So to get the expansion of $e^{x^2}$, just replace $x$ by $x^2$ throughout; this gives the expansion

$$1 + x^2 + x^4/2! + x^6/3! + x^8/4! + x^{10}/5! + \ldots + x^{2n}/n! + \ldots$$

of $e^{x^2}$ in powers of $x$.

Let's run this through *Mathematica* :

```
In[1]:=     Normal[Series[E^x,{x,0,8}]]
```
```
Out[1]=                  2    3    4     5     6
                         x    x    x     x     x
              1 + x + -- + -- + -- + --- + --- +
                         2    6    24   120   720

                  7      8
                 x      x
              ---- + -----
              5040   40320
```

```
In[2]:=     Normal[Series[E^x,{x,0,8}]]/.
            x->x^2
```
```
Out[2]=                  4    6    8     10    12
                 2       x    x    x     x     x
              1 + x + -- + -- + -- + --- + --- +
                         2    6    24   120   720

                  14     16
                 x      x
              ---- + -----
              5040   40320
```

```
In[3]:=     Normal[Series[E^(x^2),{x,0,16}]]
```
```
Out[3]=                  4    6    8     10    12
                 2       x    x    x     x     x
              1 + x + -- + -- + -- + --- + --- +
                         2    6    24   120   720

                  14     16
                 x      x
              ---- + -----
              5040   40320
```

Right on the money.

#### T.1.b)

Give the expansion of $\sin[x^3]$ in powers of $x$.

### Answer:

The expansion of $\sin[x]$ in powers of $x$ is

$$x - x^3/3! + x^5/5! - x^7/7!.. + (-1)^k x^{2k+1}/(2k+1)! + \ldots$$

So to get the expansion of $\sin[x^3]$ , just replace $x$ by $x^3$ throughout; this gives the expansion

$$x^3 - x^9/3! + x^{15}/5! - x^{21}/7!.. + (-1)^k x^{6k+3}/(2k+1)! + \ldots$$

of $\sin[x^3]$ in powers of $x$.

Check:

```
In[1]:=     Normal[Series[Sin[x^3],{x,0,39}]]
```
```
Out[1]=              9    15    21      27
             3      x    x     x       x
            x  - -- + --- - ---- + ------ -
                  6    120   5040   362880

                33         39
               x          x
            --------- + -----------
            39916800    6227020800
```

O. K.

### ■ T.2) Tricks of the trade.

#### T.2.a)

Explain the output from:

```
In[1]:=     Series[E^x Cos[x],{x,0,8}]
```
```
Out[1]=              3    4    5     7      8
                     x    x    x     x      x
            1 + x - -- - -- - -- + --- + ---- +
                     3    6    30   630   2520

                  9
             0[x]
```

### Answer:

This must come from the expansion of $e^x$ multiplied by the expansion of $\cos[x]$. Let's check this out:

```
In[2]:=     Eexpansion = Series[E^x,{x,0,8}]
```
```
Out[2]=                  2    3    4     5     6
                         x    x    x     x     x
              1 + x + -- + -- + -- + --- + --- +
                         2    6    24   120   720

                  7      8
                 x      x                9
              ---- + ----- + 0[x]
              5040   40320
```

```
In[3]:=     Cosexpansion =
            Series[Cos[x],{x,0,8}]
```
```
Out[3]=              2    4     6      8
                     x    x     x      x             9
            1 - -- + -- - --- + ----- + 0[x]
                  2   24   720   40320
```

Their product is:

*In[4]:=*     **Eexpansion Cosexpansion**

*Out[4]=*
$$1 + x - \frac{x^3}{3} - \frac{x^4}{6} - \frac{x^5}{30} + \frac{x^7}{630} + \frac{x^8}{2520} +$$
$$O[x]^9$$

Compare with:

*In[5]:=*     **Series[E^x Cos[x],{x,0,8}]**

*Out[5]=*
$$1 + x - \frac{x^3}{3} - \frac{x^4}{6} - \frac{x^5}{30} + \frac{x^7}{630} + \frac{x^8}{2520} +$$
$$O[x]^9$$

On the mark!

**T.3.b)**

---

Explain the output from:

*In[1]:=*     **Series[(E^x)/Cos[x],{x,0,8}]**

*Out[1]=*
$$1 + x + x^2 + \frac{2 x^3}{3} + \frac{x^4}{2} + \frac{3 x^5}{10} + \frac{19 x^6}{90} +$$
$$\frac{13 x^7}{105} + \frac{31 x^8}{360} + O[x]^9$$

---

**Answer:**

This must come from the expansion of $e^x$ divided by the expansion of cos[$x$]. Let's check this out:

*In[2]:=*     **Eexpansion = Series[E^x,{x,0,8}]**

*Out[2]=*
$$1 + x + \frac{x^2}{2} + \frac{x^3}{6} + \frac{x^4}{24} + \frac{x^5}{120} + \frac{x^6}{720} +$$
$$\frac{x^7}{5040} + \frac{x^8}{40320} + O[x]^9$$

*In[3]:=*     **Cosexpansion =**
              **Series[Cos[x],{x,0,8}]**

*Out[3]=*
$$1 - \frac{x^2}{2} + \frac{x^4}{24} - \frac{x^6}{720} + \frac{x^8}{40320} + O[x]^9$$

Their quotient is:

*In[4]:=*     **Eexpansion/Cosexpansion**

*Out[4]=*
$$1 + x + x^2 + \frac{2 x^3}{3} + \frac{x^4}{2} + \frac{3 x^5}{10} + \frac{19 x^6}{90}$$

---

$$1 + x + x^2 + \frac{2 x^3}{3} + \frac{x^4}{2} + \frac{3 x^5}{10} + \frac{19 x^6}{90} +$$
$$\frac{13 x^7}{105} + \frac{31 x^8}{360} + O[x]^9$$

Compare with:

*In[5]:=*     **Series[(E^x)/Cos[x],{x,0,8}]**

*Out[5]=*
$$1 + x + x^2 + \frac{2 x^3}{3} + \frac{x^4}{2} + \frac{3 x^5}{10} + \frac{19 x^6}{90} +$$
$$\frac{13 x^7}{105} + \frac{31 x^8}{360} + O[x]^9$$

On the mark!

**T.2.c)**

---

Explain the output from:

*In[1]:=*     **Series[E^Sin[x],{x,0,8}]**

*Out[1]=*
$$1 + x + \frac{x^2}{2} - \frac{x^4}{8} - \frac{x^5}{15} - \frac{x^6}{240} + \frac{x^7}{90} +$$
$$\frac{31 x^8}{5760} + O[x]^9$$

---

**Answer:**

This must come about by taking the expansion for $e^x$ and replacing $x$ by the expansion of sin[$x$]. Let's check this out:

*In[2]:=*     **Eexpansion = Series[E^x,{x,0,8}]**

*Out[2]=*
$$1 + x + \frac{x^2}{2} + \frac{x^3}{6} + \frac{x^4}{24} + \frac{x^5}{120} + \frac{x^6}{720} +$$
$$\frac{x^7}{5040} + \frac{x^8}{40320} + O[x]^9$$

*In[3]:=*     **Sexpansion =**
              **Series[Sin[x],{x,0,8}]**

*Out[3]=*
$$x - \frac{x^3}{6} + \frac{x^5}{120} - \frac{x^7}{5040} + O[x]^9$$

Now replace $x$ by the *Sexpansion*:

*In[4]:=*     **Expand[Eexpansion/.x->Sexpansion]**

*Out[4]=*
$$x^2 \quad x^4 \quad x^5 \quad x^6 \quad x^7$$

```
             x    x    x     x      x
1 + x +  -- - -- - -- - --- + -- +
             2    8    15   240    90

        8
   31 x              9
   ----- + O[x]
    5760
```

Compare with:

```
In[5]:=    Series[(E^Sin[x]),{x,0,8}]
```
```
Out[5]=              2    4    5    6    7
                    x    x    x    x    x
          1 + x +  -- - -- - -- - --- + -- +
                    2    8    15   240   90

              8
         31 x              9
         ----- + O[x]
          5760
```

Nothing could be finer.

### ■ T.3) Binomial Series

#### T.3.a)

---

Explain how to get the expansion of $\sqrt{1+x}$ in powers of $x$.
$\sqrt{1+x}$ and the sum of the terms of its expansion through
the $x^6$ term on the same axes for $-1 \le x \le 1$. Then plot
$\sqrt{1+x}$ and the sum of the terms of its expansion through
the $x^{12}$ term on the same axes for $-1 \le x \le 1$.

---

**Answer:**

Fix a positive integer $n$:

```
In[1]:=    n = 6
Out[1]=    6
```

Let

$$a[0] + a[1]x + a[2]x^2 + a[3]x^3 + ... + a[k]x^k + ...$$

stand for the expansion of $\sqrt{1+x}$ in powers of $x$. Our
object is to determine what the constants

$$a[0], a[1], a[2], ...a[k], ...$$

must be.

```
In[2]:=    partexpan[x_] = Sum[a[k] x^k ,{k,0,n}]
```
```
Out[2]=                      2            3
          a[0] + x a[1] + x  a[2] + x  a[3] +

              4         5         6
          x  a[4] + x  a[5] + x  a[6]
```

Recall

$$(\sqrt{1+x})^2 = 1 + x$$

Replace $(\sqrt{1+x})^2$ on the left by $partexpan[x]^2$ plus the
error term, and just put $1+x$ on the right. Then set up the
equations that result from requiring that $left == right$,
and solve them:

```
In[3]:=    left = partexpan[x]^2+ O[x]^(n+1)
           right = 1 + x
           coeffeqns =
           LogicalExpand[left == right]
           coeffs = Solve[coeffeqns]
```
```
Out[3]=              21              7
          {{a[6] -> -(----), a[5] -> ---,
                    1024            256

               5              1
          a[4] -> -(---), a[3] -> --,
                   128            16

               1              1
          a[2] -> -(-), a[1] -> -, a[0] -> 1},
                   8            2

               21             7
          {a[6] -> ----, a[5] -> -(---),
                  1024           256

               5              1
          a[4] -> ---, a[3] -> -(--),
                  128            16

               1              1
          a[2] -> -, a[1] -> -(-), a[0] -> -1}}
                  8            2
```

We get two sets of solutions. One is (naturally) the negative of the other. One expansion is:

```
In[4]:=    pos6 = partexpan[x]/.coeffs[[1]]
```
```
Out[4]=               2    3      4      5        6
               x    x    x    5 x    7 x    21 x
          1 + - - -- + -- - ---- + ---- - -----
               2    8   16   128    256    1024
```

and the other is:

```
In[5]:=    neg6 =partexpan[x]/.coeffs[[2]]
```
```
Out[5]=               2    3      4      5        6
                x    x    x    5 x    7 x    21 x
          -1 - - + -- - -- + ---- - ---- + -----
                2    8   16   128    256    1024
```

By custom we take the expansion **pos6** for Sqrt[1 + x].

```
In[6]:=    pos6
```
```
Out[6]=               2    3      4      5        6
               x    x    x    5 x    7 x    21 x
          1 + - - -- + -- - ---- + ---- - -----
               2    8   16   128    256    1024
```

Check:

```
In[7]:=    Normal[Series[Sqrt[1 + x],{x,0,6}]]
```
```
Out[7]=               2    3      4      5        6
               x    x    x    5 x    7 x    21 x
          1 + - - -- + -- - ---- + ---- - -----
               2    8   16   128    256    1024
```

Good.

Here is a plot of the expansion through the $x^6$ term and

$\sqrt{1+x}$ on the same axes (the plot of the partial expansion
is the thicker of the two):

In[8]:=    Plot[{Sqrt[1 + x],pos6},{x,-1,1},
           PlotStyle->{{Thickness[.004],
           GrayLevel[.0]},
           {Thickness[.008],RGBColor[0,0,1]}}];

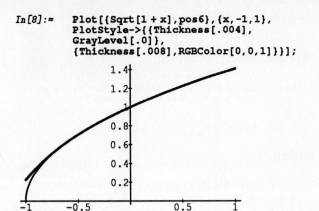

Not bad.

Let's see what happens when we take the expansion through the $x^{12}$ term:

In[9]:=    expan12 =
           Normal[Series[Sqrt[1 + x],{x,0,12}]]

Out[9]=
                  2    3     4      5       6
           x    x    x    5 x    7 x    21 x
       1 + - -  -- + -- - ---- + ---- - ----- +
           2    8    16   128    256    1024

             7        8       9        10
         33 x    429 x    715 x    2431 x
         ----- - ------ + ------ - --------- +
         2048    32768    65536    262144

            11        12
       4199 x    29393 x
       --------- - ----------
        524288    4194304


In[10]:=   Plot[{Sqrt[1 + x],expan12},{x,-1,1},
           PlotStyle->{{Thickness[.004],
           GrayLevel[.0]},
           {Thickness[.008],RGBColor[0,0,1]}}];

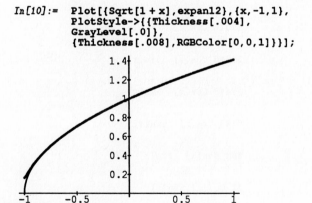

Sharing lots of ink.

### T.3.b)

---

Explain why $\sqrt{x}$ has no expansion in powers of $x$.

---

### Answer:

Let's pretend that $\sqrt{x}$ does have a partial expansion make our usual attempt at finding out what the expansion must be. Fix a positive integer $n$:

In[1]:=    n = 2
Out[1]=    2

Let

$$a[0] + a[1]x + a[2]x^2 + a[3]x^3 + ... + a[k]x^k + ...$$

stand for the expansion of $\sqrt{x}$ in powers of $x$. We will attempt to determine the constants

$$a[0], a[1], a[2], ...a[k], ...$$

In[2]:=    partexpan[x_] = Sum[a[k] x^k ,{k,0,n}]

Out[2]=
                                2
           a[0] + x a[1] + x  a[2]

Recall

$$(\sqrt{x})^2 = x.$$

Replace $(\sqrt{x})^2$ on the left by $partexpan[x]^2$ plus the error term.

In[3]:=    left = partexpan[x]^2+ O[x]^(n+1)

Out[3]=
                 2
           a[0]  + 2 a[0] a[1] x +

                       2
                2   a[1]    2 a[2]   2      3
           a[0]  (----- + ------) x  + O[x]
                        2    a[0]
                   a[0]

And put $x$ on the right.

In[4]:=    right = x
Out[4]=    x

Now find the equations that result from $left == right$.

In[5]:=    coeffeqns =
           LogicalExpand[left == right]

Out[5]=
                 2
           a[0]   == 0 && -1 + 2 a[0] a[1] == 0 &&

                        2
                2   a[1]    2 a[2]
           a[0]  (----- + ------) == 0
                        2    a[0]
                   a[0]

and solve for the coefficients:

In[6]:=    coeffs = Solve[coeffeqns]
Out[6]=    {}

This is *Mathematica*'s way of telling us that there are no solutions for the coefficients. Take a close look.

The first equation gives

$$a[0] = 0.$$

No problem with this one. But the second equation

$$2a[0]a[1] = 1$$

is not consistent with the first equation (unless you are comfortable with having 0 = 1).

Now we see why the coefficient equations have no solutions. They are inconsistent! And this is the reason $\sqrt{x}$ has no expansion in (integral) powers of $x$.

■ **T.4) Using the complex exponential to generate trigonometric identities.**

**T.4.a)**

Why is it true that if m is a positive integer, then

$$\cos[mx] + I\sin[mx] = (\cos[x] + I\sin[x])^m$$

and

$$\cos[mx] - I\sin[mx] = (\cos[x] - I\sin[x])^m?$$

**Answer:**

Well, $\cos[x] + I\sin[x] = e^{Ix}$. Raising both sides to the $m$th power gives

$$(\cos[x] + I\sin[x])^m = e^{mIx} = \cos[mx] + I\sin[mx].$$

On the other hand,

$$\cos[x] - I\sin[x] = \cos[-x] + I\sin[-x] = e^{-Ix}.$$

Raising both sides to the $m$th power gives

$$(\cos[x] - I\sin[x])^m = e^{-mIx}$$

$$= \cos[-mx] + I\sin[-mx] = \cos[mx] - I\sin[mx].$$

These identities were first observed by DeMoivre, a French mathematician banished to England as a result of the French revolution, who supported himself by supplying information on games of chance to gamblers.

**T.4.b)**

What good are the identities in part a above?

**Answer:**

The identities

$$\cos[mx] + I\sin[mx] = (\cos[x] + I\sin[x])^m$$

and

$$\cos[mx] - I\sin[mx] = (\cos[x] - I\sin[x])^m$$

let us slam out many of the identies that were such a pain in trigonometry classes.

Adding the two identities gives:

$$2\cos[mx] = (\cos[x] + I\sin[x])^m + (\cos[x] - I\sin[x])^m.$$

So we can get the usual formula for cos[2x] by activating:

```
In[1]:=   Expand[(1/2)((Cos[x] + I Sin[x])^m +
          (Cos[x] - I Sin[x])^m)/.m->2]
```

```
Out[1]=         2            2
          Cos[x]  - Sin[x]
```

Here is a formula for cos[3x]:

```
In[2]:=   Expand[(1/2)((Cos[x] + I Sin[x])^m +
          (Cos[x] - I Sin[x])^m)/.m->3]
```

```
Out[2]=         3                 2
          Cos[x]  - 3 Cos[x] Sin[x]
```

cos[4x]:

```
In[3]:=   Expand[(1/2)((Cos[x] + I Sin[x])^m +
          (Cos[x] - I Sin[x])^m)/.m->4]
```

```
Out[3]=         4              2      2          4
          Cos[x]  - 6 Cos[x]  Sin[x]  + Sin[x]
```

cos[5x]:

```
In[4]:=   Expand[(1/2)((Cos[x] + I Sin[x])^m +
          (Cos[x] - I Sin[x])^m)/.m->5]
```

```
Out[4]=         5               3      2
          Cos[x]  - 10 Cos[x]  Sin[x]  +

                         4
          5 Cos[x] Sin[x]
```

cos[17x]:

```
In[5]:=   Expand[(1/2)((Cos[x] + I Sin[x])^m +
          (Cos[x] - I Sin[x])^m)/.m->17]
```

```
Out[5]=         17                 15      2
          Cos[x]   - 136 Cos[x]   Sin[x]  +

                          13      4
          2380 Cos[x]   Sin[x]  -

                           11      6
          12376 Cos[x]   Sin[x]  +

                           9      8
          24310 Cos[x]   Sin[x]  -

                           7      10
          19448 Cos[x]   Sin[x]   +

                          5      12
          6188 Cos[x]   Sin[x]   -

                         3      14
          680 Cos[x]   Sin[x]   +

                       16
          17 Cos[x] Sin[x]
```

Very few mortals have ever seen this formula. Maybe we are getting carried away with this.

# Literacy Sheet

**1.** What is the expansion of $e^x$ in powers of $x$ ?

**2.** What is the expansion of $\sin[x]$ in powers of $x$ ?

**3.** What is the expansion of $\cos[x]$ in powers of $x$ ?

**4.** What is the expansion of $1/(1-x)$ in powers of $x$ ?

**5.** What is the expansion of $e^{-x}$ in powers of $x$ ?

**6.** What is the expansion of $e^{x^2}$ in powers of $x$ ?

**7.** What is the expansion of $e^{-x^2}$ in powers of $x$ ?

**8.** What is the expansion of $\sin[x^2]$ in powers of $x$ ?

**9.** What is the expansion of $\cos[x^2]$ in powers of $x$ ?

**10.** What happens to the graphs of $e^x$ and $1 + x + x^2/2 + x^3/3! + x^4/4! + ... + x^n/n!$ as $n$ is increased?

**11.** If you had a calculator that only adds, subtracts, multiplies and divides, then what numbers would you enter to calculate a reasonably accurate value for $e^{1/2}$ ?

**12.** Obtain the first three non-zero terms of the expansion of $e^x \sin[x]$ by multiplying together the first three non-zero terms of the expansion of $e^x$ in powers of $x$ times the first three non-zero terms of the expansion of $\sin[x]$ in powers of $x$ .

**13.** Give the identity that relates $e^{ix}, \cos[x]$ and $\sin[x]$ .

**1 4.** What familiar number wears the disguise of $e^{I\pi}$ ?

**15.** What is the value of $\log[-e^4]$ ?

# Convergence: the explanation of our observations.

## Guide

Richard Feynman once said:

"With more knowledge comes a deeper, more wonderful mystery, luring one on to penetrate deeper still."

In this section you will see that complex numbers unlock all the mystery of why expansions in powers of $x$ behave the way you have already observed. Maybe you too will want "to penetrate deeper still" by continuing your study of mathematical phenomena beyond calculus.

## Basics

■ **B.1) Complex numbers and convergence of expansions.**

Let's attempt to consolidate our position.

We have seen:

**(1)** Most of the functions we have looked at have expansions in powers of $x$.

**(2)** Some functions (like $\sqrt{x}$ and $x^{1/3}$) do not have expansions in powers of $x$.

**(3)** When a function has an expansion in powers of $x$, then the plots of the function and the sum of the early terms of its expansion seem to run together for a while and then brake sharply away from each other.

These three observations have a common mathematical explanation. Instead of looking at an $f[x]$ as a function of a real number $x$, we try to see how

$$f[x], f'[x], f''[x]$$

and the higher derivatives look when we replace the real numbers $x$ by complex numbers $z = x + Iy$.

We find the distance $R$ from $0$ to the closest *complex* singularity (blow up) of $f[z]$ or of any derivative $f^n[z]$.

If there are no complex singularities, then we agree $R = \infty$.

For instance if

$$f[x] = 1/(x^4 + x^3 + 9x^2 + x + 28),$$

then we look at:

```
In[1]:=    1/(x^4 + x^3 + 9 x^2 + x + 28)/.x->z
Out[1]=                    1
           -------------------------
                        2    3    4
           28 + z + 9 z  + z  + z
```

```
In[2]:=    D[1/(x^4 + x^3 + 9 x^2 + x + 28),
           {x,1}]/.x->z
Out[2]=                          2      3
               1 + 18 z + 3 z  + 4 z
           -(------------------------)
                            2    3    4 2
              (28 + z + 9 z  + z  + z )
```

```
In[3]:=    Together[D[
           1/(x^4 + x^3 + 9 x^2 + x + 28),
           {x,2}]/.x->z]
Out[3]=                             2         3
           (-502 - 114 z + 156 z  + 148 z  +

                 4        5        6
           174 z  + 30 z  + 20 z  ) /

                         2    3    4 3
           (28 + z + 9 z  + z  + z )
```

```
In[4]:=    Together[
           D[1/(x^4 + x^3 + 9 x^2 + x + 28),
           {x,3}]/.x->z]
Out[4]=                                    2
           (-1686 + 36072 z + 21924 z  +

                 3         4          5
           20808 z  + 540 z  - 2160 z  -

                 6          7         8          9
           2412 z  - 1512 z  - 270 z  - 120 z  )

                        2    3    4 4
           / (28 + z + 9 z  + z  + z )
```

It becomes clear that the only singularities (blow ups) happen at solutions of

$$z^4 + z^3 + 9z^2 + z + 28 = 0.$$

Find where these happen:

```
In[5]:=    Solve[z^4 + z^3 + 9 z^2 + z + 28 == 0]
Out[5]=               1 + Sqrt[-15]
           {{z -> -------------},
                        2

                   1 - Sqrt[-15]
            {z -> -------------},
                        2

                   -2 + 2 Sqrt[-6]
            {z -> ---------------},
                         2

                   -2 - 2 Sqrt[-6]
            {z -> ---------------}}
                         2
```

We want to find distance $R$ from $0$ to the closest **complex** singularity. The distance of a complex number $a + Ib$ from zero is defined to be $\sqrt{a^2 + b^2}$.

So we look at:

```
In[6]:=    Sqrt[(1/2)^2+(Sqrt[15]/2)^2]
Out[6]=    2
```

```
In[7]:=    Sqrt[(1/2)^2+(-Sqrt[15]/2)^2]
Out[7]=    2
```

```
In[8]:=    Sqrt[(-2/2)^2+(2 Sqrt[6]/2)^2]
Out[8]=    Sqrt[7]
```

```
In[9]:=    Sqrt[(-2/2)^2+(-2 Sqrt[6]/2)^2]
Out[9]=    Sqrt[7]
```

The smallest of these numbers is 2. So in this case, we set $R = 2$.

Now back to expansions:

(a) If $R = 0$, then $f[x]$ has no expansion in powers of $x$.

(b) If $R > 0$, then $f[x]$ has an expansion in powers of $x$. Furthermore if $0 \leq r < R$, then the expansion **converges** to $f[x]$ on the interval $[-r, r]$ in the following sense:

Suppose the expansion of $f[x]$ in powers of $x$ is

$$a[0] + a[1]x + a[2]x^2 + a[3]x^3 + ... + a[k]x^k + ...$$

Then the largest difference on $[-r, r]$ between $f[x]$ and

$$a[0] + a[1]x + a[2]x^2 + a[3]x^3 + ...... + a[n]x^n$$

tends to 0 as $n$ tends to $\infty$.

This notion of convergence means that once you know $R$, then you are guaranteed of being able to match the plot of $f[x]$ with the plot of a partial expansion on any interval $[-r, r]$ provided $0 \leq r < R$.

In the case we looked at above, we found $R = 2$. In this case the expansion of

$$f[x] = 1/(x^4 + x^3 + 9x^2 + x + 28)$$

converges in the sense described above to $f[x]$ on any interval $[-r, r]$ as long as $0 \leq r < 2$.

We are going to leave the justification of this basic procedure a mystery in hopes of luring you to "penetrate deeper still" by continuing your mathematical studies to the point at which you too can see why this these facts are perfectly natural. In the meantime use these facts with vigor and joy because they do no more or less than confirm what you have already observed.

■ **B.2) Intervals of convergence of expansions - polynomials in the denominator.**

Give the intervals of convergence of the expansion in powers of $x$ for each of the following functions; then plot to see the expansions converge:

**B.2.a)**

$$1/(1 - x)$$

(The expansion of this function is the most famous of them all; it is called the geometric series.)

**Answer:**

Let's look at some of the higher derivatives.

```
In[1]:=    D[1/(1 - x),{x,1}]/.x->z
Out[1]=        -2
           (1 - z)
```

```
In[2]:=    D[1/(1 - x),{x,2}]/.x->z
Out[2]=         2
           --------
                 3
           (1 - z)
```

```
In[3]:=    D[1/(1 - x),{x,3}]/.x->z
Out[3]=         6
           --------
                 4
           (1 - z)
```

```
In[4]:=    D[1/(1 - x),{x,7}]/.x->z
Out[4]=       5040
           --------
                 8
           (1 - z)
```

```
In[5]:=    D[1/(1 - x),{x,12}]/.x->z
Out[5]=    479001600
           ---------
                 13
           (1 - z)
```

It is evident that the only complex singularity (blow up) is at $z = 1$. This z is 1 unit from 0. Consequently $R = 1$ and the expansion of $1/(1 - x)$ in powers of $x$ converges to $1/(1 - x)$ on any interval of the form $[-r, r]$ as long as $0 \leq r < 1$.

Watch the expansion converge to $1/(1 - x)$ on intervals $[-r, r]$ as long as $0 < r < R = 1$:

In each plot, the expansion is the thicker of the two curves.

```
In[6]:=    R = 1
Out[6]=    1
```

```
In[7]:=    expansion8 = Normal[Series[1/(1 - x),
           {x,0,8}]]
Out[7]=                 2    3    4    5    6    7    8
           1 + x + x  + x  + x  + x  + x  + x  + x
```

```
In[8]:=    Plot[{1/(1 - x),expansion8},{x,-R,R},
```

```
     PlotRange->{0,10},
     PlotStyle->{{Thickness[.003]},
     {Thickness[.006],RGBColor[0,0,1]}}]
```

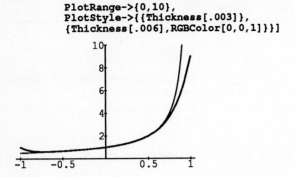

```
In[9]:=    expansion14 = Normal[Series[1/(1 - x),
           {x,0,14}]]
```

$$Out[9]= 1 + x + x^2 + x^3 + x^4 + x^5 + x^6 + x^7 +$$
$$x^8 + x^9 + x^{10} + x^{11} + x^{12} + x^{13} + x^{14}$$

```
In[10]:=   Plot[{1/(1 - x),expansion14},{x,-R,R},
           PlotRange->{0,10},
           PlotStyle->{{Thickness[.003]},
           {Thickness[.006],RGBColor[0,0,1]}}]
```

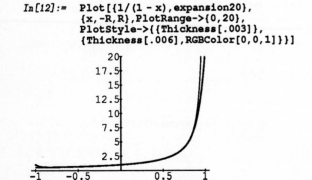

```
In[11]:=   expansion20 = Normal[Series[1/(1 - x),
           {x,0,20}]]
```

$$Out[11]= 1 + x + x^2 + x^3 + x^4 + x^5 + x^6 + x^7 +$$
$$x^8 + x^9 + x^{10} + x^{11} + x^{12} + x^{13} +$$
$$x^{14} + x^{15} + x^{16} + x^{17} + x^{18} + x^{19} +$$
$$x^{20}$$

```
In[12]:=   Plot[{1/(1 - x),expansion20},
           {x,-R,R},PlotRange->{0,20},
           PlotStyle->{{Thickness[.003]},
           {Thickness[.006],RGBColor[0,0,1]}}]
```

The more terms of the expansion you use, the better co-

---

habitation you get.  But everything breaks down at the endpoints. This is why we say we have convergence on intervals $[-r, r]$ provided $0 \leq r < R = 1$. We do **not** have convergence on the full interval $[-1, 1]$.

### B.2.b)

---

$$1/(4 + x^2)$$

---

**Answer:**

Let's look at some of the higher derivatives.

```
In[1]:=    D[1/(4 + x^2),{x,1}]/.x->z
```

$$Out[1]= \frac{-2 z}{(4 + z^2)^2}$$

```
In[2]:=    Together[
           D[1/(4 + x^2),{x,4}]/.x->z]
```

$$Out[2]= \frac{384 - 960 z^2 + 120 z^4}{(4 + z^2)^5}$$

```
In[3]:=    Together[
           D[1/(4 + x^2),{x,8}]/.x->z]
```

$$Out[3]= (10321920 - 92897280 z^2 + 81285120 z^4 -$$
$$13547520 z^6 + 362880 z^8) / (4 + z^2)^9$$

```
In[4]:=    Together[
           D[1/(4 + x^2),{x,12}]/.x->z]
```

$$Out[4]= (1961990553600 - 38258815795200 z^2 +$$
$$87676452864000 z^4 -$$
$$52605871718400 z^6 +$$
$$9863600947200 z^8 -$$
$$547977830400 z^{10} + 6227020800 z^{12})$$
$$/ (4 + z^2)^{13}$$

It is evident that the only complex singularities (blow ups) happen when the denominator is $0$ :

```
In[5]:=    Solve[4 + z^2 == 0]
```

$$Out[5]= \{\{z \rightarrow 2 I\}, \{z \rightarrow -2 I\}\}$$

To find $R$, look at:

*In[6]:=*     `Sqrt[0^2 + 2^2]`

*Out[6]=*     2

*In[7]:=*     `Sqrt[0^2 + (-2)^2]`

*Out[7]=*     2

Consequently $R = 2$ and expansion of $1/(4+x^2)$ in powers of $x$ converges to $1/(4 + x^2)$ on any interval of the form $[-r, r]$ as long as $0 < r < 2$.

Watch the expansion converge to $1/(4 + x^2)$ on intervals $[-r, r]$ as long as $0 < r < R = 2$:

In each plot, the expansion is the thicker of the two curves.

*In[8]:=*     `R = 2`

*Out[8]=*     2

*In[9]:=*     `expansion6 =`
             `Normal[Series[1/(4 + x^2),{x,0,6}]]`

*Out[9]=*
```
      2    4     6
 1   x    x     x
 - - -- + -- - ---
 4   16   64   256
```

*In[10]:=*    `Plot[{1/(4 + x^2),expansion6},`
             `{x,-R,R},PlotRange->{0,.3},`
             `PlotStyle->{{Thickness[.003]},`
             `{Thickness[.006],RGBColor[0,0,1]}}]`

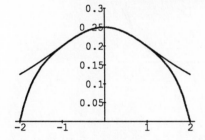

*In[11]:=*    `expansion12 =`
             `Normal[Series[1/(4 + x^2),{x,0,12}]]`

*Out[11]=*
```
      2    4     6     8      10      12
 1   x    x     x     x       x       x
 - - -- + -- - --- + ---- - ---- + -----
 4   16   64   256   1024   4096   16384
```

*In[12]:=*    `Plot[{1/(4 + x^2),expansion12},`
             `{x,-R,R},PlotRange->{0,.3},`
             `PlotStyle->{{Thickness[.003]},`
             `{Thickness[.006],RGBColor[0,0,1]}}]`

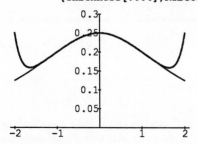

*In[13]:=*    `expansion18 =`
             `Normal[Series[1/(4 + x^2),{x,0,18}]]`

*Out[13]=*
```
      2    4    6      8      10
 1   x    x    x      x       x
 - - -- + -- - --- + ---- - ---- +
 4   16   64   256   1024   4096

    12      14      16       18
   x       x       x        x
 ----- - ----- + ------ - -------
 16384   65536   262144   1048576
```

*In[14]:=*    `Plot[{1/(4 + x^2),expansion18},`
             `{x,-R,R},PlotRange->{0,.3},`
             `PlotStyle->{{Thickness[.003]},`
             `{Thickness[.006],RGBColor[0,0,1]}}]`

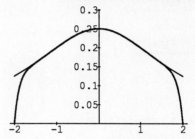

The more terms of the expansion you use, the better co-habitation you get. But everything breaks down at the endpoints. This is why we say we have convergence on intervals $[-r, r]$ provided $0 \leq r < R = 2$. We do **not** have convergence on the full interval $[-2, 2]$.

**B.2.c)**

$$x^5/(3 - x + x^2)$$

**Answer:**

Let's look at some of the higher derivatives.

*In[1]:=*     `Together[`
             `D[x^5/(3 - x + x^2),{x,1}]/.x->z]`

*Out[1]=*
```
      4     5      6
 15 z  - 4 z  + 3 z
 -------------------
            2 2
     (3 - z + z )
```

*In[2]:=*     `Together[`
             `D[x^5/(3 - x + x^2),{x,2}]/.x->z]`

*Out[2]=*
```
      3      4      5       6      7
 180 z  - 90 z  + 66 z  - 16 z  + 6 z
 ------------------------------------
                   2 3
            (3 - z + z )
```

*In[3]:=*     `Short[Together[`
             `D[x^5/(3 - x + x^2),{x,5}]/.x->z]]`

*Out[3]=*
```
                        6
 29160 + <<4>> - 120 z
 ----------------------
          2 6
```

```
                    (3 - z + z )
```

```
In[4]:=    Short[Together[
           D[x^5/(3 - x + x^2),{x,8}]/.x->z]]
```

```
Out[4]=    -48988800 + <<8>>
           -----------------
                        2 9
              (3 - z + z )
```

It is evident that the only complex singularities (blow ups) happen when the denominator is 0:

```
In[5]:=    Solve[3 - z + z^2 == 0]
```

```
Out[5]=         1 + Sqrt[-11]
           {{z -> -------------},
                        2

                1 - Sqrt[-11]
            {z -> -------------}}
                        2
```

To find R, look at:

```
In[6]:=    Sqrt[(1/2)^2 + (Sqrt[11]/2)^2]
```

```
Out[6]=    Sqrt[3]
```

```
In[7]:=    Sqrt[(1/2)^2 + (-Sqrt[11]/2)^2]
```

```
Out[7]=    Sqrt[3]
```

Consequently $R = \sqrt{3}$ and expansion of $x^5/(3 - x + x^2)$ in powers of $x$ converges to $x^5/(3 - x + x^2)$ on any interval of the form $[-r, r]$ as long as $0 < r < \sqrt{3}$. Watch the expansion converge to $x^5/(3 - x + x^2)$ on intervals $[-r, r]$ as long as $0 < r < \sqrt{3}$:

```
In[8]:=    R = Sqrt[3]
```

```
Out[8]=    Sqrt[3]
```

```
In[9]:=    expansion5 =
           Normal[Series[x^5/(3 - x + x^2),
           {x,0,5}]]
```

```
Out[9]=     5
            x
           --
            3
```

```
In[10]:=   Plot[{x^5/(3 - x + x^2),expansion5},
           {x,-R,R},PlotRange->{-4,4},
           PlotStyle->{{Thickness[.003]},
           {Thickness[.006],RGBColor[0,0,1]}}]
```

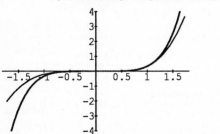

```
In[11]:=   expansion9 =
           Normal[Series[x^5/(3 - x + x^2),
           {x,0,9}]]
```

```
Out[11]=    5    6     7      8     9
            x    x   2 x    5 x    x
           -- + -- - ---- - ---- + ---
            3    9    27     81    243
```

```
In[12]:=   Plot[{x^5/(3 - x + x^2),expansion9},
           {x,-R,R},PlotRange->{-4,4},
           PlotStyle->{{Thickness[.003]},
           {Thickness[.006],RGBColor[0,0,1]}}]
```

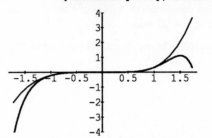

```
In[13]:=   expansion20 =
           Normal[Series[x^5/(3 - x + x^2),
           {x,0,20}]];
```

```
In[14]:=   Plot[{x^5/(3 - x + x^2),expansion20},
           {x,-R,R},PlotRange->{-4,4},
           PlotStyle->{{Thickness[.003]},
           {Thickness[.006],RGBColor[0,0,1]}}]
```

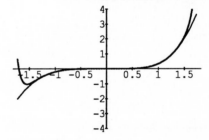

The more terms of the expansion you use, the better cohabitation you get. But everything breaks down at the endpoints. This is why we say we have convergence on intervals $[-r, r]$ provided $0 \le r < R = \sqrt{3}$.

■ **B.3) Intervals of convergence of expansions-** $e^x$, $\sin[x]$ **and** $\cos[x]$.

Give the intervals of convergence of the expansion in powers of $x$ for each of the following functions; then plot to watch the expansions converge:

**B.3.a)**

---

$$e^x$$

---

**Answer:**

All the derivatives of $e^x$ are the same- they are all $e^x$ itself. So all we have to do is to examine $e^z$ for complex singularities (blow ups).

Remember that for $z = x + Iy$,

$$e^z = e^x \cos[y] + I e^x \sin[y].$$

Because none of the functions $e^x$, $\cos[y]$ or $\sin[y]$ have any singularities (blow ups), we conclude $e^z$ has no complex singularities. Therefore $R = \infty$. Consequently the expansion of $e^x$ in powers of $x$ converges to $e^x$ on any interval $[-r, r]$ as long as $0 \le r < \infty$.

Watch the expansion converge to $e^x$ on $[-3, 3]$:

```
In[1]:=    expansion4 = Normal[Series[E^x,{x,0,4}]]
```
```
Out[1]=              2    3    4
                    x    x    x
            1 + x + -- + -- + --
                    2    6    24
```

```
In[2]:=    Plot[{E^x,expansion4},{x,-3,3},
             PlotStyle->{{Thickness[.003]},
             {Thickness[.006],RGBColor[0,0,1]}}]
```

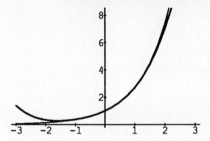

```
In[3]:=    expansion8 = Normal[Series[E^x,{x,0,8}]]
```
```
Out[3]=              2    3    4    5     6
                    x    x    x    x     x
            1 + x + -- + -- + -- + --- + --- +
                    2    6    24   120   720

                7      8
                x      x
               ---- + -----
               5040   40320
```

```
In[4]:=    Plot[{E^x,expansion8},{x,-3,3},
             PlotStyle->{{Thickness[.003]},
             {Thickness[.006],RGBColor[0,0,1]}}]
```

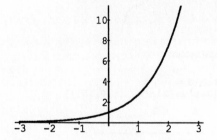

Sharing the same ink very nicely.

Let's see what's happening on $[-7, 7]$:

```
In[5]:=    expansion4 = Normal[Series[E^x,{x,0,4}]]
```
```
Out[5]=              2    3    4
                    x    x    x
            1 + x + -- + -- + --
                    2    6    24
```

```
In[6]:=    Plot[{E^x,expansion4},{x,-7,7},
             PlotStyle->{{Thickness[.003]},
             {Thickness[.006],RGBColor[0,0,1]}}]
```

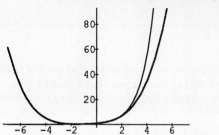

```
In[7]:=    expansion8 = Normal[Series[E^x,{x,0,8}]]
```
```
Out[7]=              2    3    4    5     6
                    x    x    x    x     x
            1 + x + -- + -- + -- + --- + --- +
                    2    6    24   120   720

                7      8
                x      x
               ---- + -----
               5040   40320
```

```
In[8]:=    Plot[{E^x,expansion8},{x,-7,7},
             PlotRange->All,
             PlotStyle->{{Thickness[.003]},
             {Thickness[.006],RGBColor[0,0,1]}}]
```

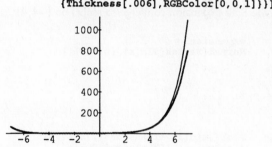

```
In[9]:=    expansion13 =
             Normal[Series[E^x,{x,0,13}]];
```

```
In[10]:=   Plot[{E^x,expansion13},{x,-7,7},
             PlotRange->All,
             PlotStyle->{{Thickness[.003]},
             {Thickness[.006],RGBColor[0,0,1]}}]
```

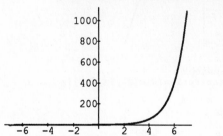

On the large interval $[-7, 7]$, the convergence is a little slower than the convergence on the short interval $[-3, 3]$. We say this because we needed more of the expansion to get cohabitation on $[-7, 7]$ than we needed to get cohabitation on $[-3, 3]$. If we use more and more of the expansion, we can get good results on larger and larger intervals.

## B.3.b)

sin[$x$] and cos[$x$]

**Answer:**

All the higher derivatives of sin[$x$] and cos[$x$] are just multiples of sin[$x$] or cos[$x$] themselves. Let's change $x$ to $z$ and check for any complex singularities of these functions.

Recall

$$\sin[z] = I(e^{-Iz} - e^{Iz})/2$$

and

$$\cos[z] = (e^{Iz} + e^{-Iz})/2.$$

Because neither $e^{Iz}$ nor $e^{-Iz}$ have any complex singularities (blow ups), we conclude that neither sin[$z$] nor cos[$z$] have any complex singularities. This means none of the higher derivatives has any complex singularities. Consequently $R = \infty$. As a result, the expansion of sin[$x$] in powers of $x$ converges to sin[$x$] on any interval $[-r, r]$ as long as $0 \le r < \infty$. And the expansion of cos[$x$] in powers of $x$ converges to cos[$x$] on any interval $[-r, r]$ as long as $0 \le r < \infty$.

Watch the Sine expansion converge to sin[$x$] on $[-4, 4]$:

```
In[1]:=   expansion3 =
          Normal[Series[Sin[x],{x,0,3}]]
```

```
Out[1]=        3
               x
          x - ---
               6
```

```
In[2]:=   Plot[{Sin[x],expansion3},{x,-4,4},
            PlotStyle->{{Thickness[.003]},
            {Thickness[.006],RGBColor[0,0,1]}}]
```

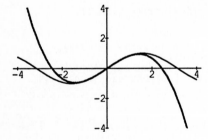

```
In[3]:=   expansion7 =
          Normal[Series[Sin[x],{x,0,7}]]
```

```
Out[3]=        3     5     7
               x     x     x
          x - --- + --- - ----
               6    120   5040
```

```
In[4]:=   Plot[{Sin[x],expansion7},{x,-4,4},
            PlotStyle->{{Thickness[.003]},
            {Thickness[.006],RGBColor[0,0,1]}}]
```

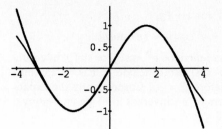

```
In[5]:=   expansion11 =
          Normal[Series[Sin[x],{x,0,11}]]
```

```
Out[5]=        3     5     7      9        11
               x     x     x      x        x
          x - --- + --- - ---- + ------ - --------
               6    120   5040   362880   39916800
```

```
In[6]:=   Plot[{Sin[x],expansion11},{x,-4,4},
            PlotStyle->{{Thickness[.003]},
            {Thickness[.006],RGBColor[0,0,1]}}]
```

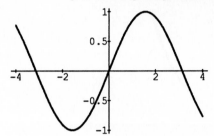

Sharing ink all the way on $[-4, 4]$.

Let's see what's happening on $[-6, 6]$:

```
In[7]:=   Plot[{Sin[x],expansion11},{x,-6,6},
            PlotStyle->{{Thickness[.003]},
            {Thickness[.006],RGBColor[0,0,1]}}]
```

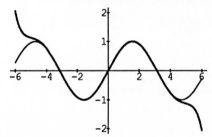

Use more of the expansion:

```
In[8]:=   expansion15 =
          Normal[Series[Sin[x],{x,0,15}]]
```

```
Out[8]=        3     5     7      9
               x     x     x      x
          x - --- + --- - ---- + ------ -
               6    120   5040   362880

               11          13              15
               x           x               x
          -------- + ----------- - --------------
          39916800   6227020800   1307674368000
```

```
In[9]:=   Plot[{Sin[x],expansion15},{x,-6,6},
            PlotStyle->{{Thickness[.003]},
            {Thickness[.006],RGBColor[0,0,1]}}]
```

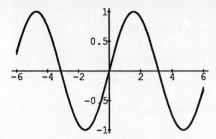

Just a hint of split ends.

We can get rid of them by using more of the expansion:

```
In[10]:=   expansion17 =
           Normal[Series[Sin[x],{x,0,17}]]
```
```
                   3     5      7        9
                  x     x      x        x
Out[10]=     x - -- + --- - ---- + ------ -
                  6    120   5040   362880

                  11          13
                 x           x
            --------- + ----------- -
            39916800   6227020800

                   15             17
                  x              x
            -------------- + ----------------
            1307674368000   355687428096000
```

```
In[11]:=   Plot[{Sin[x],expansion17},{x,-6,6},
           PlotStyle->{{Thickness[.003]},
           {Thickness[.006],RGBColor[0,0,1]}}]
```

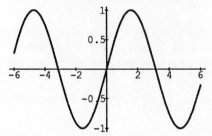

Good.

If we use more and more of the expansion, then we can get good results on larger and larger intervals.

## Tutorial

### ■ T.1) Convergence intervals and plots.

For each of the following functions $f[x]$, give the sum of first eight non-zero terms of the expansion in powers of $x$ and a description of the intervals on which the expansion converges to the function. Plot the function and the sum of its expansion through the $x^8$ term on the same axes on $[-R, R]$ where $R$ is the distance from 0 to the closest complex singularity of $f[z]$ or of any derivative $f^n[z]$.

### T.1.a)

$$f[x] = e^x/(5 - 3x^2 - x^4)$$

**Answer:**

```
In[1]:=   firsteight =
          Normal[Series[(E^x)/
          (5 - 3 x^2 -x^4), {x,0,8}]]
```
```
                    2       3        4
          1   x   11 x   23 x    541 x
Out[1]=   - + - + ----- + ----- + ------ +
          5   5    50     150     3000

               5          6          7
          401 x    13723 x    69871 x
          ------ + -------- + -------- +
           3000     90000      630000

                 8
          3214469 x
          -----------
          25200000
```

To get the convergence intervals, let's look at some of the higher derivatives.

```
In[2]:=   Together[
          D[(E^x)/(5 - 3 x^2 - x^4),{x,1}]/.
          x->z]
```
```
                 z       z        z  2      z  3
Out[2]=   -((-5 E  - 6 E  z + 3 E  z  - 4 E  z  +

             z  4       2     4 2
            E  z ) / (-5 + 3 z  + z ) )
```

```
In[3]:=   Together[
          D[(E^x)/(5 - 3 x^2 - x^4),{x,2}]/.
          x->z]
```
```
                z        z        z  2
Out[3]=   (-55 E  - 60 E  z - 84 E  z  -

                z  3       z  4       z  5
            4 E  z  - 53 E  z  + 36 E  z  -

                 z  6      z  7    z  8
            26 E  z  + 8 E  z  - E  z ) /

                    2     4 3
            (-5 + 3 z  + z )
```

```
In[4]:=   Short[Together[
          D[(E^x)/(5 - 3 x^2 - x^4),{x,5}]/.x->z],4]
```
```
                      z          z
Out[4]=   -((-250625 E  - 1464750 E  z -

                    z  2
            1874625 E  z  + <<16>> -

                z  19    z  20
            20 E  z   + E  z  ) /

                    2     4 6
            (-5 + 3 z  + z ) )
```

Because $e^z$ has no complex singularties, it is clear that the only complex singularities are at the complex solutions $z$ of

$$5 - 3z^2 - z^4 = 0.$$

```
In[5]:=   sings = Solve[5 - 3 z^2 - z^4 == 0,z]
```
```
                Sqrt[-3 + Sqrt[29]]
Out[5]=   {{z -> -------------------},
                         2
```

```
              Sqrt[2]

           Sqrt[-3 + Sqrt[29]]
  {z -> -(-------------------)},
               Sqrt[2]

          Sqrt[-3 - Sqrt[29]]
  {z -> -------------------},
              Sqrt[2]

           Sqrt[-3 - Sqrt[29]]
  {z -> -(-------------------)}}
               Sqrt[2]
```

To see how these line up, look at:

```
In[6]:=   N[sings]

Out[6]=   {{z -> 1.09205}, {z -> -1.09205},

          {z -> 2.04758 I}, {z -> -2.04758 I}}
```

The closest is

$$\sqrt{-3 + \sqrt{29}}/\sqrt{2}$$

units from 0. So

$$R = \sqrt{-3 + \sqrt{29}}/\sqrt{2}$$

and the expansion converges to $e^x/(5 - 3x^2 - x^4)$ on each interval $[-r, r]$ as long as

$$0 < r < R = \sqrt{-3 + \sqrt{29}}/\sqrt{2}.$$

Here is the plot:

```
In[7]:=   R = N[Sqrt[-3 +Sqrt[29]]/Sqrt[2]]

Out[7]=   1.09205
```

```
In[8]:=   Plot[{(E^x)/(5 - 3 x^2 - x^4),
          firsteight}, {x, -R, R},
          PlotStyle->{{Thickness[.003]},
          {Thickness[.006],RGBColor[0,0,1]}}]
```

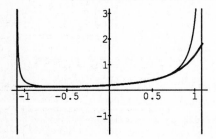

The crazy action at the end points comes from the singularities of $e^x/(5 - 3x^2 - x^4)$ at the endpoints.

**T.1.b)**

---

$$f[x] = e^x \sin[x]$$

---

**Answer:**

```
In[1]:=   firsteight =
          Normal[Series[(E^x) Sin[x], {x, 0, 8}]]

Out[1]=           3    5    6    7
              2  x    x    x    x
          x + x  + -- - -- - -- - ---
                   3    30   90   630
```

To get the convergence intervals, let's look at some of the higher derivatives.

```
In[2]:=   D[(E^x) Sin[x], {x,1}]/.x->z

Out[2]=    z           z
          E  Cos[z] + E  Sin[z]
```

```
In[3]:=   D[(E^x) Sin[x], {x,2}]/.x->z

Out[3]=      z
          2 E  Cos[z]
```

```
In[4]:=   D[(E^x) Sin[x], {x,3}]/.x->z

Out[4]=      z           z
          2 E  Cos[z] - 2 E  Sin[z]
```

```
In[5]:=   D[(E^x) Sin[x], {x,6}]/.x->z

Out[5]=       z
          -8 E  Cos[z]
```

```
In[6]:=   D[(E^x) Sin[x], {x,8}]/.x->z

Out[6]=       z
          16 E  Sin[z]
```

It's now clear that none of the higher derivatives has any complex singularities at all. Therefore $R = \infty$. The infinite series expansion converges to $e^x \sin[x]$ on every interval $[-r, r]$ as long as $0 < r < \infty$.

We cannot plot on $(-\infty, \infty)$, so we plot on a reasonable substitute:

```
In[7]:=   Plot[{(E^x) Sin[x], firsteight},
          {x, -5, 5},
          PlotStyle->{{Thickness[.003]},
          {Thickness[.006],RGBColor[0,0,1]}}]
```

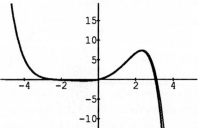

■ **T.2) Find the infinite sum.**

**T.2.a)**

---

Find the infinite sum and illustrate the convergence of

$$1 + (1/2) + (1/2)^2 + (1/2)^3 + (1/2)^4 + \ldots + (1/2)^k + \ldots$$

---

**Answer:**

Recall from B.1.a)

$$1/(1 - x) = 1 + x + x^2 + x^3 + x^4 + x^5 + \ldots + x^k + \ldots$$

provided $-r \le x \le r$ and $0 < r < 1$.

Because $x = 1/2$ is in many such intervals, we know

$$1 + (1/2) + (1/2)^2 + (1/2)^3 + (1/2)^4 + \ldots + (1/2)^k + \ldots$$

is given by:

```
In[1]:=    1/(1 - x)/.x->1/2
Out[1]=    2
```

So

$$2 = 1 + (1/2) + (1/2)^2 + (1/2)^3 + (1/2)^4 + \ldots + (1/2)^k + \ldots$$

Neat.

Watch it converge:

```
In[2]:=    N[Sum[(1/2)^k,{k,0,10}],12]
Out[2]=    1.9990234375
```

```
In[3]:=    N[Sum[(1/2)^k,{k,0,20}],12]
Out[3]=    1.99999904633
```

```
In[4]:=    N[Sum[(1/2)^k,{k,0,30}],12]
Out[4]=    1.99999999907
```

```
In[5]:=    N[Sum[(1/2)^k,{k,0,100}],40]
Out[5]=    1.99999999999999999999999999992111390
           95
```

## T.2.b)

Find the infinite sum and illustrate the convergence of

$$(1/5) - (1/5)^2 + (1/5)^3 - (1/5)^4 + \ldots (-1)^k/5^{k+1} + \ldots$$

**Answer:**

Recall

$$1/(1 - x) = 1 + x + x^2 + x^3 + x^4 + x^5 + \ldots + x^k + \ldots$$

provided $-r \le x \le r$ and $0 < r < 1$.

So

$$-x/(1 - x) = -x - x^2 - x^3 - x^4 - x^5 - \ldots - x^{k+1} - \ldots$$

provided $-r \le x \le r$ and $0 < r < 1$.

Replacing $x$ by $-x$ gives

$$x/(1 + x) = x - x^2 + x^3 - x^4 + x^5 + \ldots + (-1)^k x^{k+1} + \ldots$$

provided $-r \le x \le r$ and $0 < r < 1$.

Because $x = 1/5$ is in many such intervals, we know

$$(1/5) - (1/5)^2 + (1/5)^3 - (1/5)^4 + \ldots + (-1)^k (1/5)^{k+1} + \ldots$$

is given by:

```
In[1]:=    x/(1 + x)/.x->1/5
Out[1]=    1
           -
           6
```

So

$$1/6 = (1/5) - (1/5)^2 + (1/5)^3 - (1/5)^4 + \ldots + (-1)^{k+2}(1/5)^{k+1} + \ldots$$

Watch it converge:

```
In[2]:=    N[{Sum[(-1)^(k+2)
           (1/5)^(k+1),{k,0,5}],1/6},20]
Out[2]=    {0.166656, 0.16666666666666666667}
```

```
In[3]:=    N[{Sum[(-1)^(k+2)
           (1/5)^(k+1),{k,0,10}],1/6},20]
Out[3]=    {0.16666667008, 0.16666666666666666667}
```

```
In[4]:=    N[{Sum[(-1)^(k+2)
           (1/5)^(k+1),{k,0,20}],1/6},20]
Out[4]=    {0.16666666666666701619,
           
           0.16666666666666666667}
```

```
In[5]:=    N[{Sum[(-1)^(k+2)
           (1/5)^(k+1),{k,0,30}],30]
Out[5]=    {0.166666666666666666666702458061,
           
           0.166666666666666666666666666667}
```

```
In[6]:=    N[{Sum[(-1)^(k+2)
           (1/5)^(k+1),{k,0,40}],40]
Out[6]=    {0.16666666666666666666666666666670331705
           4259, 0.166666666666666666666666666666666
           666666666667}
```

That's convergence!

## T.2.c)

Find the infinite sum and illustrate the convergence of

$$1 - 1/2! + 1/4! - 1/6! + 1/8! + \ldots + (-1)^k 1/(2k)! + \ldots$$

**Answer:**

$$\cos[x] = 1 - x^2/2! + x^4/4! - x^6/6!$$
$$+ x^8/8! + \ldots + (-1)^k x^k/(2k)! + \ldots$$

on any interval $[-r, r]$ provided $0 \le r < \infty$.

$x = 1$ is in many such intervals; so

$$\cos[1] = 1 - 1/2! + 1/4! - 1/6!$$
$$+ 1/8! + \ldots + (-1)^k 1/(2k)! + \ldots$$

Watch it converge:

```
In[1]:=   N[{Sum[((-1)^k)/(2 k)!,{k,0,5}],
          Cos[1]},12]
Out[1]=   {0.540302303792, 0.540302305868}

In[2]:=   N[{Sum[((-1)^k)/(2 k)!,{k,0,8}],
          Cos[1]},20]
Out[2]=   {0.5403023058681397318,

            0.5403023058681397174}

In[3]:=   N[{Sum[((-1)^k)/(2 k)!,{k,0,12}],
          Cos[1]},20]
Out[3]=   {0.5403023058681397174,

            0.5403023058681397174}

In[4]:=   N[{Sum[((-1)^k)/(2 k)!,{k,0,30}],
          Cos[1]},30]
Out[4]=   {0.540302305868139717400936607443,

            0.540302305868139717400936607443}

In[5]:=   N[{Sum[((-1)^k)/(2 k)!,{k,0,40}],
          Cos[1]},40]
Out[5]=   {0.5403023058681397174009366074429766037
            7323, 0.5403023058681397174009366074
            42976603732}
```

Wow!

■ **T.3) Drugs and the geometric series.**

### T.3.a)

A dose of a certain drug results in an immediate blood-stream concentration of $C$ units per cubic centimeter of blood. It is known that $t$ hours after one dose, the concentration of the drug resulting from that single dose is

$$Ce^{-at}$$

where a is a parameter that depends on the individual drug and not on the dose.

Suppose the drug is taken every $p$ hours. Give a formula for the concentration in the blood immediately before the 8th dose.

**Answer:**

The concentration in the blood immediately before the 8th dose is

$$Ce^{-7ap} + Ce^{-6ap} + Ce^{-5ap} + Ce^{-4ap} +$$
$$Ce^{-3ap} + Ce^{-2ap} + Ce^{-ap}.$$

### T.3.b)

Assume this drug is taken every $p$ hours on a regular basis for a very long time. What is the approximate concentration of the drug before each new dose?

How is this information used in setting doses of drugs?

**Answer:**

Recall

$$1/(1-x) = 1 + x + x^2 + x^3 + x^4 + x^5 + \ldots + x^k + \ldots$$

provided $-r \le x \le r$ and $0 < r < 1$.

The approximate concentration of the drug before each new dose is

$$Ce^{-ap} + Ce^{-2ap} + Ce^{-3ap} + Ce^{-4ap} +$$
$$Ce^{-5ap} + \ldots + Ce^{-kap} + \ldots$$

$$= Ce^{-ap}(1 + e^{-ap} + (e^{-ap})^2 + (e^{-ap})^3 + (e^{-ap})^4 +$$
$$(e^{-ap})^5 + \ldots + (e^{-ap})^k + \ldots)$$

$$= Ce^{-ap}(1/(1 - e^{-ap}))$$

(because $0 < e^{-ap} < 1$ and so $e^{-ap}$ is in an interval of the form $[-r, r]$ with $0 < r < 1$).

The formula for the total concentration is:

```
In[1]:=   Simplify[C E^(- a p)/(1 - E^(- a p))]
Out[1]=        C
          ---------
                a p
          -1 + E
```

In practice the pharacologist knows the given the number a and the maximum safe concentration $S$ of the drug in the bloodstream. Then the dose size $C$ and the dose spacing $p$ are set such that

$$C/(-1 + e^{ap}) < S.$$

This ensures that the concentration never exceeds the maximum safe concentration $S$.

■ **T.4) Intervals of convergence of expansions.**

Give the intervals of convergence of the expansion in powers of $x$ of each of the following functions:

### T.4.a)

$$\sqrt{x}$$

**Answer:**

Look at some of the higher derivatives:

```
In[1]:=   D[Sqrt[x],{x,1}]/.x->z
```

$Out[1]=$

$$\frac{1}{2 \; Sqrt[z]}$$

$In[2]:=$  `D[Sqrt[x],{x,2}]/.x->z`

$Out[2]=$

$$\frac{-1}{4 \; z^{3/2}}$$

$In[3]:=$  `D[Sqrt[x],{x,3}]/.x->z`

$Out[3]=$

$$\frac{3}{8 \; z^{5/2}}$$

All of these derivatives blow up at $z = 0$. Consequently the singularity closest to 0 is 0 itself. So $R = 0$ and there is no expansion of $\sqrt{x}$ in powers of $x$.

**T.4.b)**

$$\sqrt{3+x}$$

**Answer:**

Look at some of the higher derivatives:

$In[1]:=$  `D[Sqrt[3+x],{x,1}]/.x->z`

$Out[1]=$

$$\frac{1}{2 \; Sqrt[3+z]}$$

$In[2]:=$  `D[Sqrt[3+x],{x,2}]/.x->z`

$Out[2]=$

$$\frac{-1}{4 \; (3+z)^{3/2}}$$

$In[3]:=$  `D[Sqrt[3+x],{x,3}]/.x->z`

$Out[3]=$

$$\frac{3}{8 \; (3+z)^{5/2}}$$

$In[4]:=$  `D[Sqrt[3+x],{x,4}]/.x->z`

$Out[4]=$

$$\frac{-15}{16 \; (3+z)^{7/2}}$$

$In[5]:=$  `D[Sqrt[3+x],{x,7}]/.x->z`

$Out[5]=$

$$\frac{10395}{128 \; (3+z)^{13/2}}$$

$In[6]:=$  `D[Sqrt[3+x],{x,11}]/.x->z`

$Out[6]=$

$$\frac{654729075}{2048 \; (3+z)^{21/2}}$$

The higher derivatives of $\sqrt{3+x}$ all blow up at $x = -3$ and have no other complex singulaities any closer to 0 than this one. Therefore $R = 3$. Consequently the expansion of $\sqrt{3+x}$ converges to $\sqrt{3+x}$ on any interval $[-r, r]$ as long as $0 \le r < 3$.

**T.4.c)**

$$\log[7+x]$$

**Answer:**

Look at some of the higher derivatives.

$In[1]:=$  `D[Log[7+x],{x,1}]/.x->z`

$Out[1]=$

$$\frac{1}{7+z}$$

$In[2]:=$  `D[Log[7+x],{x,2}]/.x->z`

$Out[2]=$

$$-(7+z)^{-2}$$

$In[3]:=$  `D[Log[7+x],{x,3}]/.x->z`

$Out[3]=$

$$\frac{2}{(7+z)^3}$$

$In[4]:=$  `D[Log[7+x],{x,4}]/.x->z`

$Out[4]=$

$$\frac{-6}{(7+z)^4}$$

$In[5]:=$  `D[Log[7+x],{x,7}]/.x->z`

$Out[5]=$

$$\frac{720}{(7+z)^7}$$

$In[6]:=$  `D[Log[7+x],{x,11}]/.x->z`

$Out[6]=$

$$\frac{3628800}{(7+z)^{11}}$$

The derivatives of $\log[7+x]$ all blow up at $-7$ and have no other complex singulaities any closer to 0 than this one. Therefore $R = 7$. Consequently the expansion of $\log[7+x]$ in powers of $x$ converges to $\log[7+x]$ on any interval $[-r, r]$ as long as $0 \le r < 7$.

**T.4.d)**

---

$$\text{arctan}[x]$$

---

**Answer:**

Let's look at some of the higher derivatives.

```
In[1]:=    D[ArcTan[x],{x,1}]/.x->z
Out[1]=       1
           ------
                2
           1 + z
```

```
In[2]:=    D[ArcTan[x],{x,2}]/.x->z
Out[2]=      -2 z
           ---------
                 2 2
           (1 + z )
```

```
In[3]:=    Together[
           D[ArcTan[x],{x,3}]/.x->z]
Out[3]=            2
           -2 + 6 z
           ---------
                 2 3
           (1 + z )
```

```
In[4]:=    Together[
           D[ArcTan[x],{x,4}]/.x->z]
Out[4]=               3
           24 z - 24 z
           ------------
                 2 4
            (1 + z )
```

```
In[5]:=    Together[
           D[ArcTan[x],{x,7}]/.x->z]
Out[5]=                2          4          6
           -720 + 15120 z  - 25200 z  + 5040 z
           -----------------------------------
                          2 7
                    (1 + z )
```

```
In[6]:=    Short[Together[
           D[ArcTan[x],{x,11}]/.x->z]]
Out[6]=    -3628800 + <<5>>
           -----------------
                   2 11
             (1 + z )
```

To find the singularities, look at:

```
In[7]:=    Solve[1 + z^2 == 0]
Out[7]=    {{z -> I}, {z -> -I}}
```

To find $R$, look at:

```
In[8]:=    Sqrt[0^2 + 1^2]
```

```
Out[8]=    1
```

```
In[9]:=    Sqrt[0^2 + (-1)^2]
Out[9]=    1
```

Consequently $R = 1$ and expansion of arctan$[x]$ in powers of $x$ converges to arctan$[x]$ on any interval of the form $[-r, r]$ as long as $0 \le r < 1$.

# Literacy Sheet

**1.** What is the expansion in powers of $x$ of $e^x$ ? On what intervals $[-r, r]$ are you guaranteed that this expansion converges to $e^x$ ?

.

.

**2.** What is the expansion in powers of $x$ of $e^{-x}$ ? On what intervals $[-r, r]$ are you guaranteed that this expansion converges to $e^{-x}$ ?

.

.

**3.** What is the expansion in powers of $x$ of $\sin[x]$ ? On what intervals $[-r, r]$ are you guaranteed that this expansion converges to $\sin[x]$ ?

.

.

.

**4.** What is the expansion in powers of $x$ of $\sin[x^2]$ ? On what intervals $[-r, r]$ are you guaranteed that this expansion converges to $\sin[x^2]$ ?

.

.

**5.** What is the expansion in powers of $x$ of $\cos[x]$ ? On what intervals $[-r, r]$ are you guaranteed that this expansion converges to $\cos[x]$ ?

.

.

.

**6.** What is the expansion in powers of $x$ of $\cos[x^3]$ ? On what intervals $[-r, r]$ are you guaranteed that this expansion converges to $\cos[x^3]$ ?

.

.

.

**7.** What is the expansion in powers of $x$ of $1/(1-x)$ ? On what intervals $[-r, r]$ are you guaranteed that this expansion converges to $1/(1-x)$ ?

.

.

.

**8.** What is the expansion in powers of $x$ of $1/(1+x^2)$ ? On what intervals $[-r, r]$ are you guaranteed that this expansion converges to $1/(1+x^2)$ ?

.

.

.

**9.** On what intervals $[-r, r]$ are you guaranteed that the expansion of $\sqrt{1+x}$ in powers of $x$ converges to $\sqrt{1+x}$ ?

.

.

.

**10.** Use the quadratic formula to find the intervals $[-r, r]$ on which you are guaranteed convergence of the expansion of $(x^2 + 1)/(1 - x - x^2)$ in powers of $x$ to $(x^2 + 1)/(1 - x - x^2)$ .

.

.

**11.** Give a common everyday function that has no expansion in powers of $x$ .

.

.

**12.** Why is it probably a good idea to approximate $e^{.5}$ by

$$1 + (.5) + (.5)^2/2! + (.5)^3/3! + (.5)^4/4! + (.5)^5/5! ?$$

And why is it probably a bad idea to approximate $e^{20}$ by

$$1 + (20) + (20)^2/2! + (20)^3/3! + (20)^4/4! + (20)^5/5! ?$$

.

.

**13.** Give the value of the infinite sum

$$1 + 1/2 + 1/2^2 + 1/2^3 + 1/2^4 + \ldots + 1/2^n + \ldots$$

.

.

**14.** Give the value of the infinite sum

$$1 - 1/2 + 1/4! - 1/6! + 1/8! + \ldots + (-1)^n/(2n)! + \ldots$$

.

.

**15.** Give the value of the infinite sum

$$1 + 1 + 1/2 + 1/3! + 1/4! + 1/5! + \ldots + 1/n! + \ldots$$

.

.

.

**16.** Give the value of the infinite sum

$$1 - 1 + 1/2 - 1/3! + 1/4! - 1/5! + \ldots + (-1)^n/n! + \ldots$$

.

.

.

**17.** Find a constant $a$ such that the plots of the curves $y = e^{3x} - 1$ and $y = \sin[ax]$ share a lot of ink on small intervals centered on $0$.

.

.

.

# Error estimates

## Guide

Some functions are hard to integrate, but expansions (as polynomials) are easy to integrate. If you can get a good approximation of a function by a partial expansion, then you can integrate the partial expansion to get a good estimate of the integral.

In this lesson, you are going to see how to control the error in these approximations of functions and how to control the error in the resulting estimates of their integrals.

## Basics

■ **B.1) Precise polynomial error estimates.**

### B.1.a.i)

Find a polynomial that runs within $10^{-9}$ of sin[$x$] for $-1 \leq x \leq 1$.

**Answer:**

It seems natural to try something like:

```
In[1]:=   guess1 = Normal[Series[Sin[x],{x,0,9}]]
```
```
Out[1]=        3    5     7      9
           x   x    x     x      x
         x - -- + --- - ---- + ------
           6   120  5040  362880
```

To test this, look at:

```
In[2]:=   Plot[{Abs[Sin[x] - guess1],10^(-9)},
           {x,-1,1},PlotRange->All,
           PlotStyle->{{Thickness[.003]},
           {Thickness[.006],RGBColor[1,0,0]}}}]
```

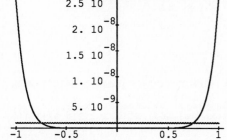

Close but no cigar. Let's try:

```
In[3]:=   guess2 =
           Normal[Series[Sin[x],{x,0,11}]]
```

```
Out[3]=         3    5     7      9        11
             x    x     x      x         x
           x - -- + --- - ---- + ------ - --------
             6   120  5040  362880  39916800
```

To test this, look at:

```
In[4]:=   Plot[{Abs[Sin[x] - guess2],10^(-9)},
           {x,-1,1},PlotRange->All,
           PlotStyle->{{Thickness[.003]},
           {Thickness[.006],RGBColor[1,0,0]}}}]
```

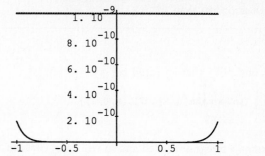

Great: **guess2** is running well within $10^{-9}$ of sin[$x$] for $-1 \leq x \leq 1$.

### B.1.a.ii)

Find a polynomial that runs within $10^{-9}$ of sin[$x^3$] for $-1 \leq x \leq 1$.

**Answer:**

From part a.i) immediately above, we know that the polynomial **guess2** runs within $10^{-9}$ of sin[$x$] for $-1 \leq x \leq 1$. So the following polynomial runs within $10^{-9}$ of sin[$x^3$] for $-1 \leq x \leq 1$.

```
In[1]:=   goodpoly = guess2/.x->x^3
```

```
Out[1]=         9    15    21     27       33
             3   x     x     x      x        x
           x - -- + --- - ---- + ------ - --------
             6   120  5040  362880  39916800
```

If you are skeptical, then look at:

```
In[2]:=   Plot[{Abs[Sin[x^3] - goodpoly],10^(-9)},
           {x,-1,1},PlotRange->All,
           PlotStyle->{{Thickness[.003]},
           {Thickness[.006],RGBColor[1,0,0]}}}]
```

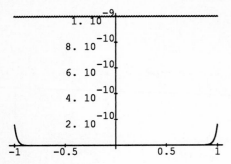

It does a great job.

**B.1.b)**

Estimate the integral

$$\int_0^1 \sin[x^3]dx$$

within an error less than $10^{-9}$.

**Answer:**

A quick and dirty attempt could be:

```
In[1]:=    Integrate[Sin[x^3],{x,0,1}]
Out[1]=    0.23385
```

*Mathematica* 's integrator failed and so will every other integrator fail on this. But here's a thought: We already know that the following polynomial is running within $10^{-9}$ of $\sin[x^3]$ for $0 \le x \le 1$ :

```
In[2]:=    p[x_] = goodpoly
```
$$Out[2]= \quad x^3 - \frac{x^9}{6} + \frac{x^{15}}{120} - \frac{x^{21}}{5040} + \frac{x^{27}}{362880} - \frac{x^{33}}{39916800}$$

Therefore

$$\int_0^1 \sin[x^3]dx - \int_0^1 p[x]dx$$

$$\le \int_0^1 \sin[x^3]dx - p[x]dx < \int_0^1 10^{-9}dx = 10^{-9}.$$

This means $\int_0^1 p[x]dx$ is within $10^{-9}$ of $\int_0^1 \sin[x^3]dx$. So our guaranteed estimate of $\int_0^1 \sin[x^3]dx$ within an error of $10^{-9}$ is:

```
In[3]:=    estimate = Integrate[p[x],{x,0,1}]
```
$$Out[3]= \quad \frac{555394057}{2375049600}$$

or in decimal form:

```
In[4]:=    N[estimate,10]
```

```
Out[4]=    0.2338452456
```

This gives us extra **accurate** decimals that *Mathematica* 's instruction **NIntegrate** did not deliver.

## Tutorial

■ **T.1) Precise evaluation of some integrals.**

**T.1.a)**

Estimate the integral

$$\int_0^{1/2} e^{x^3}dx$$

within an error less than $10^{-8}$.

**Answer:**

The integrator won't touch this one:

```
In[1]:=    Integrate[E^(x^3),{x,0,1/2}]
```
$$Out[1]= \quad Integrate[E^{x^3}, \{x, 0, \frac{1}{2}\}]$$

But all is not lost. If we can find a polynomial $p[x]$ such that $|p[x] - e|^{x^3} < 2.0 10^{-8}$ for $0 \le x \le 1/2$ , then we'll know (here $*$ means "times")

$$\int_0^{1/2} e^{x^3}dx - \int_0^{1/2} p[x]dx \le \int_0^{1/2} |e^{x^3} - p[x]|dx$$

$$< \int_0^{1/2} 2.0 * 10^{-8}dx = 2.0 * 10^{-8}(1/2 - 0) = 10^{-8}.$$

Thus if we can find such a $p[x]$, we'll integrate it with absolute assurance that $\int_0^{1/2} p[x]dx$ estimates $\int_0^{1/2} e^{x^3}dx$ within a guaranteed error of less than $10^{-8}$.

Let's hunt for $p[x]$ :

```
In[2]:=    guess1 =
           Normal[Series[E^(x^3),{x,0,9}]]
```
$$Out[2]= \quad 1 + x^3 + \frac{x^6}{2} + \frac{x^9}{6}$$

To test this guess, look at:

```
In[3]:=    Plot[{Abs[E^(x^3) - guess1],2*10^(-8)},
           {x,0,1/2},PlotRange->All,
           PlotStyle->{{Thickness[.003]},
           {Thickness[.006],RGBColor[1,0,0]}}]
```

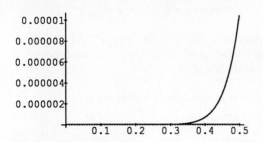

Not good enough; so we make another guess:

```
In[4]:=   guess2 =
          Normal[Series[E^(x^3),{x,0,15}]]
Out[4]=              6    9    12    15
                3   x    x    x     x
          1 + x  + -- + -- + --- + ---
                   2    6    24    120
```

And try it out:

```
In[5]:=   Plot[{Abs[E^(x^3) - guess2],2*10^(-8)},
          {x,0,1/2},PlotRange->All,
          PlotStyle->{{Thickness[.003]},
          {Thickness[.006],RGBColor[1,0,0]}}]
```

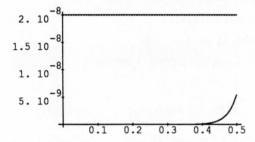

Good. The plot shows that:

```
In[6]:=   p[x_] = guess2
Out[6]=              6    9    12    15
                3   x    x    x     x
          1 + x  + -- + -- + --- + ---
                   2    6    24    120
```

satisfies $|p[x] - e|^{x^3}| < 2.010^{-8}$ for $0 \le x \le 1/2$.

So we can announce to the world that an estimate of

$$\int_0^{1/2} e^{x^3} dx$$

within a guaranteed error of less that $10^{-8}$ is:

```
In[7]:=   estimate =
          Integrate[p[x],{x,0,1/2}]
Out[7]=   5910718939
          -----------
          11450449920
```

Or in decimal form:

```
In[8]:=   N[estimate,9]
```

```
Out[8]=   0.516199711
```

### T.1.b)

Estimate the integral

$$\int_0^{1/3} \sqrt{1 + x^6} dx$$

within an error less than $10^{-10}$.

**Answer:**

Again, the integrator won't touch this one:

```
In[1]:=   Integrate[Sqrt[1 + x^6],{x,0,1/3}]
Out[1]=                     6     1
          Integrate[Sqrt[1 + x ], {x, 0, -}]
                                          3
```

If we can find a polynomial $p[x]$ such that

$$|p[x] - \sqrt{1 + x^6}| < 3.0 * 10^{-10}$$

for $0 \le x \le 1/3$, then we'll know

$$\int_0^{1/3} \sqrt{1 + x^6} dx - \int_0^{1/3} p[x] dx$$

$$\le \int_0^{1/3} |\sqrt{1 + x^6} - p[x]| dx$$

$$< \int_0^{1/3} 3.0 * 10^{-7} dx = 3.0 * 10^{-10}(1/3 - 0) = 10^{-10}.$$

Thus if we can find such a $p[x]$, we'll integrate it with absolute assurance that $\int_0^{1/3} p[x] dx$ estimates $\int_0^{1/3} \sqrt{1 + x^6} dx$ within a guaranteed error of less than $10^{-10}$.

Let's hunt for $p[x]$:

```
In[2]:=   guess1 =
          Normal[Series[Sqrt[1 + x^6],{x,0,24}]]
Out[2]=           6    12    18     24
                 x    x     x     5 x
          1 + -- - -- + --- - -----
                 2    8     16    128
```

To test this guess, look at:

```
In[3]:=   Plot[{Abs[Sqrt[1 + x^6] - guess1],
          3*10^(-10)},
          {x,0,1/2},PlotRange->All,
          PlotStyle->{{Thickness[.003]},
          {Thickness[.006],RGBColor[1,0,0]}}]
```

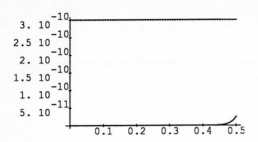

Cool. The plot shows that:

```
In[4]:=    p[x_] = guess1

Out[4]=          6     12    18     24
                x     x     x     5 x
           1 + -- - --- + --- - -----
                2     8    16    128
```

satisfies $|p[x] - \sqrt{1 + x^6}| < 3.0 * 10^{-10}$ for $0 \le x \le 1/3$. So we can announce with authority that an estimate of the integral

$$\int_0^{1/3} \sqrt{1 + x^6}\,dx$$

within a guaranteed error of less that $10^{-10}$ is:

```
In[5]:=    estimate =
           Integrate[p[x],{x,0,1/3}]

Out[5]=    312555843951985511
           ------------------
           937575683665246080
```

Or in decimal form:

```
In[6]:=    N[estimate,11]

Out[6]=    0.33336598783
```

## ■ T.2) Elliptic Integrals.

The problem set here seems easy enough. We want to find the length of the ellipse

$$(x/3)^2 + (y/2)^2 = 1.$$

To this end:

```
In[7]:=    Solve[(x/3)^2 + (y/2)^2 == 1,y]

Out[7]=                   2
                  Sqrt[36 - 4 x ]
           {{y -> ---------------},
                        3

                   2
                  -Sqrt[36 - 4 x ]
           {y -> ----------------}}
                        3
```

Set:

```
In[8]:=    yprime = D[Sqrt[36 - 4 x^2]/3,x]

Out[8]=        -4 x
           -----------------
                   2
```

```
3 Sqrt[36 - 4 x ]
```

The arc length of the ellipse $(x/3)^2 + (y/2)^2 = 1$ should be given by the following calculation, but *Mathematica* won't be able to handle this one:

```
In[9]:=    4 Integrate[Sqrt[1 + yprime^2],{x,0,3}]

Out[9]=    $Interrupted
```

Don't forget that Command-Period will allow you to abort. The fact that *Mathematica* does not do this integral is no shortcoming of *Mathematica*; this type of integral is impossible to solve by usual means.

Looks like we'll have to estimate this nasty integral:

```
In[10]:=   4 NIntegrate[Sqrt[1 + yprime^2],{x,0,3}]

Power::infy:

1

Infinite expression -- encountered.

0.

NIntegrate::notnum:

Integrand ComplexInfinity

is not numerical at 3..
Out[10]=                                          2
                4 NIntegrate[Sqrt[1 + yprime ],

                  {x, 0, 3}]
```

Still out of luck. The trouble is that

```
In[11]:=   Sqrt[1 + yprime^2]

Out[11]=                    2
                        16 x
           Sqrt[1 + -------------]
                           2
                    9 (36 - 4 x )
```

blows up at $x = 3$.

## T.2.a)

How can we get the estimate?

**Answer:**

Substitute in the integral.

Look at:

```
In[1]:=    integrand = Sqrt[1 + yprime^2]

Out[1]=                     2
                        16 x
           Sqrt[1 + -------------]
                           2
                    9 (36 - 4 x )
```

The most obnoxious part of this integral is the term

$$\sqrt{36 - 4x^2}.$$

We squash it by setting $x = 3\sin[t]$. This substitution gives the pairings:

$$\int_0^3 \rightarrow\rightarrow\rightarrow\rightarrow\rightarrow \int_0^{\pi/2}$$

```
In[2]:=    subs = x->3 Sin[t]
Out[2]=    x -> 3 Sin[t]

In[3]:=    tintegrand =
           (integrand/.subs) D[x/.subs,t]
Out[3]=
                              2
                      16 Sin[t]
           3 Cos[t] Sqrt[1 + ---------------]
                                        2
                             36 - 36 Sin[t]
```

Let's do a little hand work on **tintegrand** :

Because

$$36 - 36\sin[t]^2 = 36\cos[t]^2,$$

we see that

$$3\cos[t]\sqrt{1 + 16\sin[t]^2/(36 - 36\sin[t]^2)}$$

simplifies to

$$(3\cos[t]/(6\cos[t]))\sqrt{36\cos[t]^2 + 16\sin[t]^2}$$

$$= (1/2)\sqrt{36(1 - \sin[t])^2 + 16\sin[t]^2}$$

$$= (1/2)\sqrt{36 - 20\sin[t]^2} = (1/2)\sqrt{36(1 - (20/36)\sin[t]^2}$$

$$= (6/2)\sqrt{1 - (20/36)\sin[t]^2} = 3\sqrt{1 - (5/9)\sin[t]^2}$$

Now we can try:

```
In[4]:=    4 NIntegrate[3 Sqrt[1 - (5/9)
           Sin[t]^2],{t,0,Pi/2}]
Out[4]=    15.8654
```

This gives us the estimate: Length = 15.8654.

### T.2.b)

---

What can we do to get an estimate with guaranteed accuracy within $10^{-7}$?

---

**Answer:**

Acording to the calculations in T.2.a), the exact length is

$$4\int_0^{\pi/2} 3\sqrt{1 - (5/9)\sin[t]^2}\,dt$$

At first glance, we might want to try to look at the expansion of

$$3\sqrt{1 - (5/9)\sin[t]^2}$$

in powers of $t$ , but on account of the presence of the sine term, we decide to sneak up on this.

As $t$ advances from $0$ to $\pi/2$ , the function

$$(5/9)\sin[t]^2$$

advances from $0$ to $5/9$.

If we can find a polynomial $p[x]$ such that

$$|p[x] - 3\sqrt{1 - x}| < (1/4)(2/\pi)10^{-7}$$

for $0 \le x \le 5/9$ , then setting $x = (5/9)\sin[t]^2$ we'll know

$$|p[(5/9)\sin[t]^2] - 3\sqrt{1 - (5/9)\sin[t]^2}|$$

$$< (1/4)(2/\pi)10^{-7},$$

for $0 \le t \le \pi/2$.

As a result, we'll know for sure that

$$|4\int_0^{\pi/2} p[(5/9)\sin[t]^2]dt - 4\int_0^{\pi/2} 3\sqrt{1 - (5/9)\sin[t]^2}dt|$$

$$\le 4\int_0^{\pi/2} |p[(5/9)\sin[t]^2] - 3\sqrt{1 - (5/9)\sin[t]^2}|dt$$

$$< 4\int_0^{\pi/2} (1/4)(2/\pi)10^{-7}dt$$

$$= 4(1/4)(2/\pi)10^{-7}(\pi/2) = 10^{-7}.$$

This means

$$4\int_0^{\pi/2} p[(5/9)\sin[t]^2]dt$$

$$= 4\int_0^{\pi/2} 3\sqrt{1 - (5/9)\sin[t]^2}dt$$

within a guaranteed error less than $10^{-7}$. Accordingly, if we can find such a $p[x]$ with

$$|p[x] - 3\sqrt{1 - x}| < (1/4)(2/\pi)10^{-7}$$

for $0 \le x \le 5/9$ , then we can announce with absolute certainty that the length of the ellipse is

$$\int_0^{\pi/2} p[(5/9)\sin[t]^2]dt$$

within a guaranteed error smaller than $10^{-7}$. Let's go a-hunting.

```
In[1]:=    guess1 =
           Normal[Series[3 Sqrt[1 - x],{x,0,23}]];

In[2]:=    allowederror = (1/4) (2/Pi) 10^(-7)
```

*Out[2]=*

$$\frac{1}{20000000 \; Pi}$$

*In[3]:=*   `Plot[{Abs[3 Sqrt[1 - x] - guess1],`
                 `allowederror},`
                 `{x,0,5/9},PlotRange->All,`
                 `PlotStyle->{{Thickness[.003]},`
                 `{Thickness[.006],RGBColor[1,0,0]}}}]`

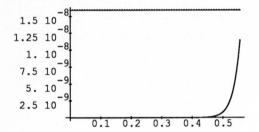

That took a while, but we hit it on the first try. The plot shows that:

*In[4]:=*   `p[x_] = guess1`

*Out[4]=*

$$3 - \frac{3x}{2} - \frac{3x^2}{8} - \frac{3x^3}{16} - \frac{15x^4}{128} - \frac{21x^5}{256} -$$

$$\frac{63x^6}{1024} - \frac{99x^7}{2048} - \frac{1287x^8}{32768} - \frac{2145x^9}{65536} -$$

$$\frac{7293x^{10}}{262144} - \frac{12597x^{11}}{524288} - \frac{88179x^{12}}{4194304} -$$

$$\frac{156009x^{13}}{8388608} - \frac{557175x^{14}}{33554432} -$$

$$\frac{1002915x^{15}}{67108864} - \frac{29084535x^{16}}{2147483648} -$$

$$\frac{53036505x^{17}}{4294967296} - \frac{194467185x^{18}}{17179869184} -$$

$$\frac{358229025x^{19}}{34359738368} - \frac{2650894785x^{20}}{274877906944} -$$

$$\frac{4923090315x^{21}}{549755813888} - \frac{18349700265x^{22}}{2199023255552} -$$

$$\frac{34305961365x^{23}}{4398046511104}$$

satisfies

$$|p[x] - 3\sqrt{1-x}| < (1/4)(2/\pi)10^{-7}$$

for $0 \le x \le 5/9$. So we can announce with authority that within an guaranteed accuracy of than $10^{-7}$ the length of the ellipse is the integral from $0 \, to \, \pi/2$ of

$$4 \int_0^{\pi/2} p[(5/9)\sin[t]^2]dt.$$

Here is the estimate (it is going to take more than a few seconds to complete):

*In[5]:=*   `estimate =`
              `4 Integrate[p[ (5/9) Sin[t]^2],{t,0,Pi/2}]`

*Out[5]=*   (178140768239242316723595863607567356711
            93702699 Pi) /

            35274517631258074813986133320361022270
            47399424

That took a long time; old *Mathematica* had to work its ears off on that integral. Here is a decimal accurate within $10^{-7}$ :

*In[6]:=*   `N[estimate,12]`

*Out[6]=*   `15.8654395974`

This is more accurate than:

*In[7]:=*   `4 NIntegrate[3 Sqrt[1 - (5/9)`
              `Sin[t]^2],{t,0,Pi/2}]`

*Out[7]=*   `15.8654`

We just did some very good mathematics.

■ **T.3) Pade's idea: Approximation by quotients of polynomials.**

Pade had the idea that we can improve the quality of approximation by using quotients of polynomials instead of polynomials themselves. One of his reasons is that polynomial approximation of a bounded function like sin[x] is bound to fail on large intervals because the polynomial eventually goes into global scale and then it must pull way from the Sine curve. So it might be a good idea to approximate sin[x] by a function that cannot grow like a high degree polynomial.

**T.3.a.i)**

What kind of function should we look for?

**Answer:**

Because sin[x] is an odd function, we should try to approximate sin[x] by an odd function g[x]. One easy way to do this is to take

$$g[x] = xh[x]$$

where $h[x]$ is an even function. Because we don't wnat $g[x]$ to grow too fast, we should take $h[x]$ bounded in the global scale. A simple form for this kind of function is

$$h[x] = (a + bx^2 + cx^4)/(1 + dx^2 + ex^4)$$

where $a, b, c, d$ and $e$ are to be determined.

**T.3.a.ii)**

Find $a, b, c, d$ and $e$ that make the expansions of

$$f[x] = \sin[x]$$

and

$$g[x] = x(a + bx^2 + cx^4)/(1 + dx^2 + ex^4)$$

in powers of $x$ coincide up to as high a power of $x$ as possible. Plot the two functions on the same axes on an interval chosen to show off most of their similarities and a little of their differences.

**Answer:**

Set:

```
In[1]:=    prelimg =
           x ( a + b x^2 + c x^4)/(1 + d x^2 + e x^4)
```

```
Out[1]=           2     4
           x (a + b x  + c x )
           -------------------
                  2     4
           1 + d x  + e x
```

There are five constants to determine, so we want to run the expansion of $g[x]$ out to generate five equations when we set the result equal to the corresponding partial expansion of $\sin[x]$.

```
In[2]:=    left = Series[prelimg,{x,0,9}]
```

```
Out[2]=                    3
           a x + (b - a d) x  +

                           2       5
           (c - b d + a (d  - e)) x  +

                      2
           (-(c d) + b (d  - e) -

                 2
             a (d (d  - e) - d e)) x  +

              2           2
           (c (d  - e) - b (d (d  - e) - d e) +

                 2
             a (-((d  - e) e) +

                     2                  9
                d (d (d  - e) - d e))) x  +

            10
           O[x]
```

```
In[3]:=    right = Series[Sin[x],{x,0,9}]
```

```
Out[3]=         3     5     7       9
               x     x     x       x        10
           x - -- + --- - ---- + ------ + O[x]
               6    120   5040   362880
```

Set the two equal and then find the equations that result from equating the coefficients of like powers of $x$:

```
In[4]:=    coeffequations =
           LogicalExpand[left == right]
```

```
Out[4]=                         1
           -1 + a == 0 && - + b - a d == 0 &&
                          6

             1
           -(---) + c - b d + a (d  - e) == 0 &&
            120

             1                 2
           ---- - c d + b (d  - e) -
           5040

                2
           a (d (d  - e) - d e) == 0 &&

              1            2
           -(------) + c (d  - e) -
            362880

                2
           b (d (d  - e) - d e) +

                2
           a (-((d  - e) e) +

                   2
              d (d (d  - e) - d e)) == 0
```

Solve these equations for $a, b, c, d$ and $e$:

```
In[5]:=    coeffs = Solve[coeffequations]
```

```
Out[5]=             551            5             53
           {{c -> ------,   e -> -----,   b -> -(---),
                  166320        11088           396

                   13
              d -> ---,   a -> 1}}
                   396
```

Substitute these values of $a, b, c, d$ and $e$ into **prelimg** to get the $g[x]$ that we were after:

```
In[6]:=    g[x_] = prelimg/.coeffs[[1]]
```

```
Out[6]=              2        4
                53 x     551 x
           x (1 - ----- + ------)
                  396     166320
           ----------------------
                 2       4
              13 x     5 x
           1 + ----- + -----
              396     11088
```

Here come some plots:

```
In[7]:=    Plot[{Sin[x],g[x]},{x,-Pi,Pi}]
```

Beautiful fit on $[-\pi, \pi]$.

*In[8]:=*　Plot[{Sin[x],g[x]},{x,-6,6}]

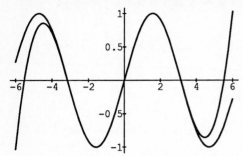

The quality of the fit breaks down on the larger interval $[-6,6]$.

### T.3.a.iii)

---

We got the $g[x]$ above by matching the beginning of its expansion with the beginning of the Sine expansion through the $x^9$ term. Which gives the more satisfactory fit of of $\sin[x]$ on $[-7,7]$: the beginning of the Sine expansion through the $x^9$ term or the $g[x]$ we found above?

---

**Answer:**

Look at the plots:

The thickest is the plot of $g[x]$; the thinnest is the plot of $\sin[x]$.

*In[1]:=*　exp9 = Normal[Series[Sin[x],{x,0,9}]]

*Out[1]=*
$$x - \frac{x^3}{6} + \frac{x^5}{120} - \frac{x^7}{5040} + \frac{x^9}{362880}$$

*In[2]:=*　Plot[{g[x],exp9,Sin[x]},
　　　{x,-6,6},
　　　PlotStyle->{{Thickness[.003],
　　　RGBColor[1,0,0]},{Thickness[.01],
　　　GrayLevel[0.6]},{Thickness[.006]}}];

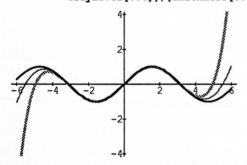

The thick gray plot of $g[x]$ is sharing more ink with the thin $\sin[x]$ plot than is the heavy black plot of **exp9**. This is amazing because we used only the data from **exp9** to produce $g[x]$. Pade had a good idea.

### ■ T.4) Fourier's idea: Approximation by Sine and Cosine waves.

Fourier (1768-1830) had the idea that all functions defined on $[-\pi,\pi]$ can be built from Sine and Cosine waves the same way that all music can be built from basic harmonics.

Here is one way to try out Fourier's idea:

### T.4.a)

---

Find the choices of $a, b, c, d$ and $e$ that make the expansions of
$$f[x] = (x^2 + x)e^{-x^2}$$
and
$$g[x] = a + b\cos[x] + c\sin[x] + d\cos[2x] + e\sin[2x]$$
in powers of $x$ coincide up to as high a power of $x$ as possible.

Plot the two functions on the same axes on an interval chosen to show off most of their similarities and a little of their differences.

---

**Answer:**

*In[1]:=*　f[x_] := (x^2 + x) E^(-x^2)

*In[2]:=*　prelimg =
　　　a + b Cos[x] + c Sin[x] +
　　　d Cos[2 x] + e Sin[2 x]

*Out[2]=*　a + b Cos[x] + d Cos[2 x] + c Sin[x] +

　　　e Sin[2 x]

There are five constants to determine, so we want to run the expansion of $g[x]$ out to allow us to generate five equations.

*In[3]:=*　gexpan =Series[prelimg,{x,0,4}]

*Out[3]=*　(a + b + d) + (c + 2 e) x +

$$\left(\frac{-b}{2} - 2d\right) x^2 + \left(\frac{-c}{6} - \frac{4e}{3}\right) x^3 +$$

$$\left(\frac{b}{24} + \frac{2d}{3}\right) x^4 + O[x]^5$$

Match this against:

*In[4]:=*　fexpan = Series[f[x],{x,0,4}]

*Out[4]=*　$x + x^2 - x^3 - x^4 + O[x]^5$

Now find the equations that result from equating the coefficients of like powers of $x$ on the left and right sides of the equation **gexpan == fexpan**. You can spot some of them:
$$a + b + d == 0, c + 2e == 1,$$
etc.

Now let *Mathematica* set up these equations for you:

```
In[5]:=    coeffeqns =
           LogicalExpand[fexpan == gexpan]
```

```
Out[5]=    -(a + b + d) == 0 &&

                 1 - (c + 2 e) == 0 &&

                      -b
                 1 - (-- - 2 d) == 0 &&
                       2

                      -c    4 e
                -1 - (-- - ---) == 0 &&
                       6     3

                      b     2 d
                -1 - (-- + ---) == 0
                      24     3
```

Solve for the coefficients.

```
In[6]:=    coeffs = Solve[coeffeqns]
```

```
Out[6]=           7          16         11
           {{a -> -(-),  b -> --,  d -> -(--),
                   2          3          6

                      2         5
                c -> -(-),  e -> -}}
                      3         6
```

Put these coefficients into the partial expansion.

```
In[7]:=    g[x_] = prelimg/.coeffs[[1]]
```

```
Out[7]=        7    16 Cos[x]   11 Cos[2 x]
           -(-) + --------- - ----------- -
            2         3            6

             2 Sin[x]   5 Sin[2 x]
             -------- + ----------
                3           6
```

Now plot:

```
In[8]:=    Plot[{f[x],g[x]},{x,-1,1}]
```

Not half bad; maybe Fourier knew what he was talking about.

**T.4.b)**

What could you do to get a better approximation of

$$f[x] = (x^2 + x)e^{-x^2}$$

by Sine and Cosine waves?

**Answer:**

To try for more accuracy, we'll use additional waves; like $\cos(3x)$, $\sin(3x)$, $\cos(4x)$, $\sin(4x)$, $\cos(5x)$, $\sin(5x)$, $\cos(6x)$, and $\sin(6x)$.

```
In[1]:=    f[x_] = (x^2 + x) E^(-x^2)
```

```
Out[1]=        2
           x + x
           ------
             2
            x
           E
```

```
In[2]:=    prelimg =
           c + Sum[a[k] Cos[k x] +
           b[k] Sin[k x],{k,1,6}]
```

```
Out[2]=    c + Cos[x] a[1] + Cos[2 x] a[2] +

           Cos[3 x] a[3] + Cos[4 x] a[4] +

           Cos[5 x] a[5] + Cos[6 x] a[6] +

           Sin[x] b[1] + Sin[2 x] b[2] +

           Sin[3 x] b[3] + Sin[4 x] b[4] +

           Sin[5 x] b[5] + Sin[6 x] b[6]
```

There are 13 constants to determine; so we want to run the expansion of $g[x]$ out far enough to give us 13 equations.

```
In[3]:=    gexpan = Series[prelimg,{x,0,12}]
```

```
Out[3]=    (c + a[1] + a[2] + a[3] + a[4] + a[5] +

             a[6]) + (b[1] + 2 b[2] + 3 b[3] +

             4 b[4] + 5 b[5] + 6 b[6]) x +

            -a[1]               9 a[3]
           (----- - 2 a[2] - ------ - 8 a[4] -
              2                  2

             25 a[5]              2
             ------- - 18 a[6]) x  +
                2

            -b[1]   4 b[2]   9 b[3]   32 b[4]
           (----- - ------ - ------ - ------- -
              6       3        2         3

             125 b[5]            3
             -------- - 36 b[6]) x  +
                6

            a[1]   2 a[2]   27 a[3]   32 a[4]
           (---- + ------ + ------- + ------- +
             24      3         8         3

             625 a[5]             4
             -------- + 54 a[6]) x  +
                24

            b[1]   4 b[2]   81 b[3]   128 b[4]
           (---- + ------ + ------- + -------- +
            120     15        40         15

             625 b[5]   324 b[6]    5
             -------- + --------) x  +
                24          5

            -a[1]   4 a[2]   81 a[3]
           (----- - ------ - ------- -
             720     45        80

             256 a[4]   3125 a[5]   324 a[6]   6
```

```
     --------- - ---------- - --------) x
        45          144         5

      -b[1]    8 b[2]   243 b[3]
    + (----- - ------ - --------
       5040     315      560

      1024 b[4]   15625 b[5]   1944 b[6]
      --------- - ---------- - ---------)
         315        1008         35

     7   a[1]     2 a[2]   729 a[3]
    x + (----- + ------ + -------- +
         40320    315      4480

      512 a[4]   78125 a[5]   1458 a[6]
      -------- + ---------- + --------)
         315       8064         35

     8   b[1]      4 b[2]   243 b[3]
    x + (------ + ------ + -------- +
         362880    2835     4480

      2048 b[4]   390625 b[5]   972 b[6]
      --------- + ---------- + --------)
         2835        72576        35

     9    -a[1]      4 a[2]
    x + (------- - ------ -
         3628800    14175

      729 a[3]   4096 a[4]
      -------- - --------- -
        44800      14175

      390625 a[5]   2916 a[6]    10
      ----------- - ---------) x   +
        145152         175

      -b[1]        8 b[2]   2187 b[3]
    (-------- - -------- - --------- -
     39916800    155925     492800

      16384 b[4]   1953125 b[5]
      ---------- - ----------- -
        155925      1596672

      17496 b[6]    11
      ----------) x   +
         1925

      a[1]        4 a[2]    2187 a[3]
    (--------- + ------ + --------- +
     479001600    467775    1971200

      16384 a[4]   9765625 a[5]
      ---------- + ------------ +
        467775      19160064

      8748 a[6]    12       13
      ---------) x   + O[x]
         1925
```

Kowa-Bonga!

Match this against:

```
In[4]:=    fexpan = Series[f[x],{x,0,12}]
Out[4]=
                          5    6    7    8
          2    3    4    x    x    x    x
     x + x  - x  - x  + -- + -- - -- - -- +
                        2    2    6    6

      9    10    11    12
     x    x     x     x           13
     -- + --- - --- - --- + O[x]
     24   24    120   120
```

Now repeat the procedure of part a) above:

```
In[5]:=    coeffeqns =
           LogicalExpand[fexpan == gexpan]
           coeffs = Solve[coeffeqns]
           g[x_] = prelimg/.coeffs[[1]]
Out[5]=    1367      43 Cos[x]    1303 Cos[2 x]
           ----- + ---------- - ------------- -
           21600       175           6720

           337 Cos[3 x]    2627 Cos[4 x]
           ------------ - ------------- -
              5670            50400

           83 Cos[5 x]    11461 Cos[6 x]
           ----------- - -------------- +
              34650          9979200

           2701 Sin[x]    4327 Sin[2 x]
           ----------- + ------------- +
              12600          20160

           47 Sin[3 x]    1753 Sin[4 x]
           ----------- + ------------- +
              560            75600

           65 Sin[5 x]    127 Sin[6 x]
           ----------- + ------------
              33264          369600
```

That's quite a function. Let's see how much shared ink we get on $[-1,1]$:

```
In[6]:=    Plot[{f[x],g[x]},{x,-1,1}]
```

Super! Old Fourier **did** know what he was talking about. Let's check the precision of the approximation on $[-1/4,1/4]$:

```
In[7]:=    errorplot =
           Plot[Abs[f[x] - g[x]],
           {x,-1/4,1/4},PlotRange->All]
```

Damn good.

T.4.c)

Use approximation by Sine and Cosine waves to estimate

$$\int_0^{1/4} (x^2 + x)e^{-x^2}\,dx$$

within a guaranteed error of less that $10^{-12}$.

---

**Answer:**

Let's check out the $g[x]$ we found in the last part: To see how well $g[x]$ approximates

$$f[x] = (x^2 + x)e^{-x^2}$$

for $0 \le x \le 1/4$, take another look at the plot of $|f[x] - g[x]|$ for $0 \le x \le 1/4$. The plot shows that $|f[x] - g[x]| < 10^{-12}$ for $0 \le x \le 1/4$.

Therefore

$$\int_0^{1/4} f[x]\,dx - \int_0^{1/4} g[x]\,dx$$

$$= \int_0^{1/4} (f[x] - g[x])\,dx \le \int_0^{1/4} |f[x] - g[x]|\,dx$$

$$< \int_0^{1/4} 10^{-12}\,dx = 10^{-12}/4 < 10^{-12}.$$

Consequently

$$\int_0^{1/4} g[x]\,dx$$

estimates

$$\int_0^{1/4} f[x]\,dx$$

within a guaranteed error less than $10^{-12}$. Our estimate is:

```
In[1]:=   estimate =
          Integrate[g[x],{x,0,1/4}]
```

```
Out[1]=
                      1            1
               2701 Cos[-]  4327 Cos[-]
        32117          4            2
        ----- - ----------- - ----------- -
        86400      12600         40320

                3                      5
          47 Cos[-]              13 Cos[-]
                4    1753 Cos[1]        4
        --------- - ----------- - --------- -
           1680       302400        33264

                3           1
          127 Cos[-]  43 Sin[-]
                2           4
        ---------- + --------- -
         2217600       175

                1           3
          1303 Sin[-]  337 Sin[-]
                2           4
        ----------- - ---------- -
           13440       17010

                      5            3
                83 Sin[-]  11461 Sin[-]
                      4            2
        2627 Sin[1]
```

---

```
        ----------- - ---------- - -----------
          201600       173250       59875200
```

Or in decimal form:

```
In[2]:=   N[estimate,12]
Out[2]=   0.0353107793317
```

Actually *Mathematica* can do this integral in exact form. This gives us the chance to check our estimate

```
In[3]:=   exact = Integrate[f[x],{x,0,1/4}]
Out[3]=
                                    1
                           Sqrt[Pi] Erf[-]
            1        5                  4
          - - - ------- + ---------------
            2     1/16          4
                 8 E
```

```
In[4]:=   N[exact,12]
Out[4]=   0.0353107793317
```

Our estimate holds up within $10^{-12}$ accuracy. Can not argue with success.

■ **T.5) Inappropriate attempts at approximation by expansion.**

In some situations, approximations by expansions are theoretically possible, but are useless from a practical point of view.

**T.5.a.i)**

---

Try to use expansions in powers of $x$ to get a polynomial that runs within $10^{-1}$ of $1/(1 + x^2)$ for $-2 \le x \le 2$.

---

**Answer:**

At first it seems that the following polynomial ought to work:

```
In[1]:=   guess1 = Normal[Series[1/(1+x^2),{x,0,20}]]
Out[1]=
                2    4    6    8    10    12
            1 - x  + x  - x  + x  - x   + x   -

            14    16    18    20
            x   + x   - x   + x
```

To test this, look at:

```
In[2]:=   Plot[{Abs[1/(1 + x^2) -
          guess1],10^(-1)},
          {x,-2,2},PlotRange->All]
```

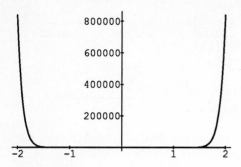

Something is not working the way we might have expected.

### T.5.a.ii)

What's going wrong?

### Answer:

The obstacle is the complex singularity of $1/(1 + z^2)$:

```
In[1]:=    Solve[1 + z^2 == 0]
Out[1]=    {{z -> I}, {z -> -I}}
```

Both singularities are one unit from 0. Therefore we are guaranteed convergence of the expansion only on intervals $[-r, r]$ as long as $0 \le r < 1$.

The interval $[-2, 2]$ we were working on above is **not** one of these intervals. As a result we cannot expect convergence on $[-2, 2]$. This is the reason that the approximation did not work.

### T.5.a.iii)

What advice can be derived from this?

### Answer:

If you want to approximate a function on an interval $[a, b]$ by means of expansions in powers of $x$, then you should at the very least know that $[a, b]$ is inside an interval on which the expansions in powers of $x$ converge to the function. But even this might not be enough as the next part shows.

### T.5.b.i)

Try to use expansions in powers of x to get a polynomial that runs within $10^{-1}$ of $e^x$ for $19 \le x \le 20$.

### Answer:

The expansion of $e^x$ in powers of $x$ converges to $e^x$ on any interval $[-r, r]$ provided $0 \le r < \infty$. The interval $[19, 20]$ in question here is inside many of these convergence intervals $[-r, r]$. So it might seem natural to try something like:

```
In[1]:=    guess1 = Normal[Series[E^x,{x,0,20}]]
           Plot[{Abs[E^x - guess1],
           10^(-1)},{x,19,20},PlotRange->All]
```

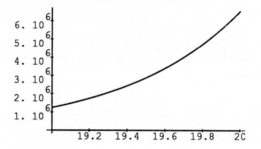

Not even close. Let's try:

```
In[2]:=    guess2 =
           Normal[Series[E^x,{x,0,30}]]
           Plot[{Abs[E^x - guess2],
           10^(-1)},{x,19,20,
           PlotRange->All]
```

Still far off.

In this problem, theory and practice do not completely agree: The **theory** tells us that if we go for a tremendously high degree expansion of $e^x$ in powers of $x$, then we can get the approximation that we want.

The **practicalities** of the situation prevent us from using expansions of very high degree because the equipment we are running on cannot deal with expansions of tremendously high degree.

### T.5.b.ii)

What advice can be derived from this?

### Answer:

If you are using expansions in powers of $x$ for precise approximations on intervals $-a \le x \le a$, then smaller $a's$ will give you good luck and larger $a's$ will give you problems.

You will never have good luck on an interval $[-a, a]$ in a situation in which the expansions do not converge to the function on $[-a, a]$.

**T.5.c)**

Speculate on why when you use expansions of a function $f[x]$ in powers of $x$ for precise approximations, then it seems that you get better results for small $x$'s than for large $x$'s?

**Answer:**

Look at the beginning of a typical expansion:

```
In[1]:=    typical = Normal[Series[Sin[x],
           {x,0,11}]]
```

```
Out[1]=        3    5     7       9        11
               x    x     x       x        x
          x - --- + --- - ---- + ------ - --------
               6   120   5040   362880   39916800
```

```
In[2]:=    short =
           Normal[Series[Sin[x],{x,0,5}]]
```

```
Out[2]=        3    5
               x    x
          x - --- + ---
               6   120
```

Look at the plot of $\sin[x]$, **typical** and **short** for $-1 \le x \le 1$. The plot of $\sin[x]$ is the thinnest and the plot of **typical** is the thickest.

```
In[3]:=    Plot[{Sin[x],short,typical},{x,-1,1},
           PlotStyle->{{Thickness[.003]},
           {Thickness[.006],RGBColor[0,0,1]},
           {Thickness[.009],RGBColor[1,0,0]}}];
```

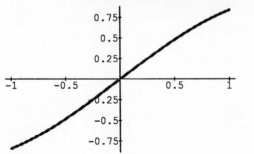

When $|x|$ is smaller than 1, then the early terms are the most influential and the high degree terms make little difference, but when $|x|$ is larger than 1, then the high degree terms start taking over:

```
In[4]:=    Plot[{Sin[x],short,typical},{x,-4,4},
           PlotStyle->{{Thickness[.003]},
           {Thickness[.006],RGBColor[0,0,1]},
           {Thickness[.009],RGBColor[1,0,0]}}];
```

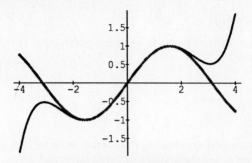

When $|x|$ is large, the highest degree terms in **typical** and **short** dominate and pull each into its own global scale:

```
In[5]:=    Plot[{Sin[x],short,typical},{x,-10,10},
           PlotStyle->{{Thickness[.003]},
           {Thickness[.006],RGBColor[0,0,1]},
           {Thickness[.009],RGBColor[1,0,0]}}];
```

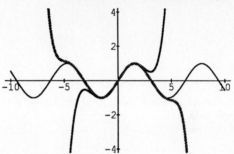

This is one reason that approximations by expansions in powers of $x$ are obliged to give poor results on very large intervals.

# Literacy Sheet

**1.** What some ways of trying to get good polynomial approximations?

.

.

.

**2.** Does the method we used in this lesson work to find good polynomial approximations on large intervals?

.

.

.

**3.** What is the basic idea we used in finding approximations by Sine and Cosine waves?

.

.

.

**4.** What powers of $x$ appear in the expansion of an odd function in powers of $x$? How about an even function?

.

.

.

**5.** If $f[x]$ approximates $g[x]$ within $1/3000$ for $10 \leq x \leq 12$, then $\int_{10}^{12} f[x]dx$ approximates $\int_{10}^{12} g[x]dx$ to how many accurate decimals?

.

.

.

**6.** Given a function $f[x]$ and a positive integer $k$, tell how to set a number $c$ such that if $p[x]$ is another function with $|p[x] - f[x]| < c$ for $0 \leq x \leq 1/2$, then

$$\int_0^{1/2} f[x]dx = \int_0^{1/2} p[x]dx$$

with an error of less than $10^{-k}$. Why is this of interest?

.

.

.

**7.** Why is it probably a good idea to approximate $\sin[.5]$ by

$$.5 - (.5)^3/3! + (.5)^5/5! - (.5)^7/7!?$$

.

.

.

And why is it probably a bad idea to approximate $\sin[50]$ by

$$50 - (50)^3/3! + (50)^5/5! - (50)^7/7!?$$

.

.

.

Is it a good idea to try to approximate $\sin[50]$ by any polynomial arising from the expansion of $\sin[x]$ in powers of $x$? Why?

.

.

**8.** Give some examples of situations in which approximation by expansions are likely to give good results

.

.

.

**9.** Give some examples of situations in which approximation by expansions are likely to give poor results

.

.

# Power Series

## Guide

Up to thus point we have had a lot of fun expanding functions, plotting functions and their expansions and approximating functions by their expansions.

In this section we alter the viewpoint but not the fundamental ideas. Instead of looking at infinite sums that arise from expansions of functions, we are going to look at functions that arise from infinite sums.

## Basics

■ **B.1) What is a power series and where do power series come from?**

**B.1.a)**

What is a power series?

**Answer:**

A power series is any infinite sum that has the form

$$a[0] + a[1]x + a[2]x^2 + a[3]x^3 + ... + a[k]x^k + ...$$

where the numbers $a[0], a[1], a[2], ..., a[k], ...$ are constants. It is called a power series because of the powers of $x$ evident in each term of the series.

**B.1.b)**

Give three concrete examples of power series.

**Answer:**

Any expansion in powers of $x$ is a power series.

*Example 1:* An archetypical example is:

$$1 + x + x^2/2! + x^3/3! + x^4/4! + x^5/5! + ... + x^k/k! + ...$$

This power series is the expansion of $e^x$ in powers of $x$; it converges to $e^x$ on every interval $[-r, r]$ as long as $0 \le r < \infty$.

*Example 2:* A second and *important* example is the **Geometric Series** :

$$1 + x + x^2 + x^3 + x^4 + ... + x^k + ...$$

This power series is the expansion of $1/(1-x)$ in powers of $x$; it converges to $1/(1-x)$ on every interval $[-r, r]$ as long as $0 \le r < 1$.

*Example 3:* A third and slightly mysterious example is

$$1 + x + x^2/2^2 + x^3/3^2 + x^4/4^2 + x^5/5^2 + ... + x^k/k^2 + ...$$

At this point, it is not clear to us what function (if any) has this power series as its expansion in powers of $x$.

■ **B.2) The Basic Convergence Principle: The straight scoop on convergence of power series.**

**B.2.a)**

How can we try to determine a Basic Convergence Principle to find the intervals on which a power series

$$a[0] + a[1]x + a[2]x^2 + a[3]x^3 + ... + a[k]x^k + ...$$

converges?

**Answer:**

If the power series

$$a[0] + a[1]x + a[2]x^2 + a[3]x^3 + ... + a[k]x^k + ...$$

comes into into our hands as an expansion of a certain familiar function $f[x]$, we can look at the complex singularities of $f[z], f'[z], f''[z], ...$ to determine its intervals of convergence.

If, on the other hand, we cannot get our hands on the nature of $f[x]$, then we are forced to deal with the series itself. Here is the starting point:

**Basic Convergence Principle :**

Given a power series

$$a[0] + a[1]x + a[2]x^2 + a[3]x^3 + ... + a[k]x^k + ...$$

If for some number $x = R$, the (infinite) list of individual terms

$$\{a[0], a[1]R, a[2]R^2, a[3]R^3, ..., a[k]R^k, ...\}$$

stays bounded (i.e. does not blow up to $\infty$ or down to $-\infty$ ), then the power series

$$a[0] + a[1]x + a[2]x^2 + a[3]x^3 + ... + a[k]x^k + ...$$

converges on any interval $[-r, r]$ as long as $0 \le r < |R|$.

Here is a detailed discussion about why this is true: Take any $r$ with $0 < r < |R|$. Saying that the (infinite) list of individual terms

$$\{a[0], a[1]R, a[2]R^2, a[3]R^3, ..., a[k]R^k, ...\}$$

stays bounded is the same as saying that we can get our hands on a number $M$ such that $M$ is larger than all members of the list of terms

$$\{|a[0]|, |a[1]R|, |a[2]R|^2, |a[3]R|^3, ..., |a[k]R|^k, ...\}$$

Notice that for $|x| \leq r$, we have

$$|a[k]x|^k \leq |a[k]r|^k = |a[k]R|^k r^k / |R|^k$$

$$= |a[k]R|^k (r/|R|)^k = |a[k]R|^k t^k \leq Mt^k$$

where $t = r/|R|$.

Make careful note of the fact that $0 < t < 1$. This ensures that the geometric series

$$1 + t + t^2 + t^3 + t^4 + t^5 + ... + t^k + ...$$

is convergent to $1/(1-t)$.

Therefore the series

$$M + Mt + Mt^2 + Mt^3 + Mt^4 + ... + Mt^k + ...$$

is convergent to $M/(1-t)$. Here is the situation:

*For $|x| \leq r$, the term $|a[k]x|^k$ is under the corresponding term $Mt^k$ of the convergent geometric series. This means that for $|x| \leq r$, the series*

$$a[0] + a[1]x + a[2]x^2 + a[3]x^3 + ... + a[k]x^k + ...$$

*has no choice but to converge at least as fast as the convergent geometric series*

$$M + Mt + Mt^2 + Mt^3 + Mt^4 + ... + Mt^k + ...$$

This shows why

$$a[0] + a[1]x + a[2]x^2 + a[3]x^3 + ... + a[k]x^k + ...$$

is convergent for $|x| \leq r$ provided $r < |R|$.

The salient point of the detailed dicussion is that on the advertised interval $[-r, r]$, *the power series converges faster than a multiple of the geometric series.*

This is why most folks regard geometric series as the keystone of power series.

### B.2.b.i)

---

Determine some convergence intervals for the power series

$$1 + 2x + 2^2 x^2 / 2! + 2^3 x^3 / 3! + ... + 2^k x^k / k! + ...$$

---

**Answer:**

This looks familiar! Recall

$$1 + x + x^2 / 2! + x^3 / 3! + x^4 / 4! + ... + x^k / k! + ...$$

is the expansion of $e^x$ in powers of $x$. It converges to $e^x$ on every interval $[-r, r]$ as long as $0 \leq r < \infty$.

Replacing $x$ by $2x$ above shows that the series in question is the expansion of $e^{2x}$ in powers of $x$. It converges to $e^{2x}$ on every interval $[-r, r]$ as long as $0 \leq r < \infty$.

Consequently, the power series

$$1 + 2x + 2^2 x^2 / 2! + 2^3 x^3 / 3! + ... + 2^k x^k / k! + ...$$

converges on every interval $[-r, r]$ as long as $0 \leq r < \infty$.

### B.2.b.ii)

---

Determine some convergence intervals for the power series

$$1 + x + x^2 / 2^2 + x^3 / 3^2 + x^4 / 4^2 + ... + x^n / n^2 + ...$$

Illustrate with some plots.

---

**Answer:**

For $x = R = 1$, the (infinite) list of individual terms

$$\{1, 1, 1/2^2, 1/3^2, 1/4^2, 1/5^2, ..., 1/n^2, ...\}$$

stays bounded because the individual terms get closer and closer to 0. As a result, this power series converges on any interval $[-r, r]$ provided $0 \leq r < 1 = |R|$.

Lets look a some plots:

```
In[1]:=   sum5 = 1 + Sum[x^n/n^2, {n,1,5}]
                      2    3    4    5
Out[1]=              x    x    x    x
          1 + x + -- + -- + -- + --
                   4    9    16   25
```

```
In[2]:=   sum6 = 1 + Sum[x^n/n^2, {n,1,6}]
                      2    3    4    5    6
Out[2]=              x    x    x    x    x
          1 + x + -- + -- + -- + -- + --
                   4    9    16   25   36
```

```
In[3]:=   Plot[{sum5, sum6}, {x,-1.5,1.5}];
```

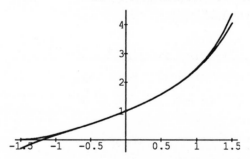

Convergence on subintervals of $[-1, 1]$ is suggested by lots of shared ink.

```
In[4]:=   sum10 = 1 + Sum[x^n/n^2, {n,1,10}]
                      2    3    4    5    6    7
Out[4]=              x    x    x    x    x    x
          1 + x + -- + -- + -- + -- + -- + -- +
                   4    9    16   25   36   49

                   8    9    10
                  x    x    x
                  -- + -- + ---
                  64   81   100
```

In[5]:=    sum16 = 1 + Sum[x^n/n^2, {n,1,16}]

Out[5]=
$$1 + x + \frac{x^2}{4} + \frac{x^3}{9} + \frac{x^4}{16} + \frac{x^5}{25} + \frac{x^6}{36} + \frac{x^7}{49} +$$

$$\frac{x^8}{64} + \frac{x^9}{81} + \frac{x^{10}}{100} + \frac{x^{11}}{121} + \frac{x^{12}}{144} + \frac{x^{13}}{169} +$$

$$\frac{x^{14}}{196} + \frac{x^{15}}{225} + \frac{x^{16}}{256}$$

In[6]:=    Plot[{sum10, sum16}, {x,-1.3,1.3}];

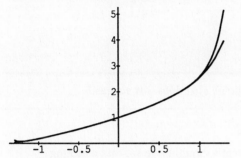

Again convergence on subintervals of $[-1,1]$ is suggested by lots of shared ink.

### ■ B.3) Differentiating and integrating power series.

The following two facts should come as no surprise:

### B.3.a)

The integral of the power series is the power series of the integrals.

**Answer:**

Suppose $R$ is a positive number and the power series

$$a[0] + a[1]x + a[2]x^2 + a[3]x^3 + ... + a[k]x^k + ...$$

converges on $[-r, r]$ as long as $0 \le r < R$.
Then

$$a[0] + a[1]t^2/2 + a[2]t^3/3 + a[3]t^4/4 +$$

$$... + a[n]t^{k+1}/(k+1) + ...$$

converges to

$$\int_0^t (a[0] + a[1]x + a[2]x^2 + a[3]x^3 + ... + a[k]x^k + ...)dx$$

for $-r \le t \le r$.

*Argument:*

Let $f[x]$ stand for the infinite sum

$$a[0] + a[1]x + a[2]x^2 + a[3]x^3 + ... + a[k]x^k + ...$$

Remember that to say

$$a[0] + a[1]x + a[2]x^2 + a[3]x^3 + ... + a[k]x^k + ...$$

converges on $[-r, r]$ is to say that the largest difference between the polynomial

$$a[0] + a[1]x + a[2]x^2 + a[3]x^3 + ... + a[k]x^k$$

and

$$f[x]$$

on $[-r, r]$ tends to $0$ as $k$ grows without bound. As a result, as $t$ varies in $[-r, r]$ the largest difference between the polynomial

$$\int_0^t (a[0] + a[1]x + a[2]x^2 + a[3]x^3 + ... + a[k]x^k)dx$$

and

$$\int_0^t f[x]dx$$

tends to $0$ as $k$ grows without bound.

This is the same as saying

$$a[0] + a[1]t^2/2 + a[2]t^3/3 + a[3]t^4/4 + ...$$

$$... + a[k]t^{k+1}/(k+1) + ...$$

converges to

$$\int_0^t (a[0] + a[1]x + a[2]x^2 + a[3]x^3 + ... + a[k]x^k + ...)dx$$

on the interval $-r \le t \le r$.

### B.3.b)

The derivative of the power series is the power series of the derivatives.

**Answer:**

Suppose $R$ is a positive number and the power series

$$a[0] + a[1]x + a[2]x^2 + a[3]x^3 + ... + a[k]x^k + ...$$

converges on $[-r, r]$ as long as $0 \le r < R$. Then

$$a[1] + 2a[2]x + 3a[3]x^2 + ... + ka[k]x^{k-1} + ...$$

converges to

$$D[a[0] + a[1]x + a[2]x^2 + a[3]x^3 + ... + a[k]x^k + ..., x]$$

for $-r \le x \le r$.

Argument:

The underlying reason is the fundamental theorem of calculus which says that

$$\int_0^t f'[x]dx = f[t] - f[0].$$

Leaving out some of the technical details, we can see why this is true by putting

$$f[x] = a[0] + a[1]x + a[2]x^2 + a[3]x^3 + ... + a[k]x^k + ...$$

and noting that

$$\int_0^t (a[1] + 2a[2]x + 3a[3]x^2 + ... + ka[k]x^{k-1} + ...)dx$$

$$= a[1]t + a[2]t^2 + a[3]t^3 + ... + a[k]t^k + ...$$

$$= f[t] - f[0]$$

because $f[0] = a[0]$.

Consequently

$$f'[x] = a[1] + 2a[2]x + 3a[3]x^2 + ... + ka[k]x^{k-1} + ...$$

as advertised.

## B.3.c)

In B.2.b.ii) above, we saw that the power series

$$1 + x + x^2/2^2 + x^3/3^2 + x^4/4^2 + ... + x^n/n^2 + ...$$

converges on every interval $[-r, r]$ as long as $0 \le r < 1$. For $x$ in such an interval, put

$$f[x] = 1 + x + x^2/2^2 + x^3/3^2 + x^4/4^2 + ... + x^n/n^2 + ...$$

Give a power series that converges to $f'[x]$ and give another power series that converges to $\int_0^x f[t]dt$ on the intervals specified above.

**Answer:**

To get a power series expansion of $f'[x]$, just take each term of

$$f[x] = 1 + x + x^2/2^2 + x^3/3^2 + x^4/4^2 + ... + x^n/n^2 + ...,$$

differentiate and add them up to get:

$$f'[x] = 1 + x/2 + x^2/3 + x^3/4 + x^4/5 + ... + x^{n-1}/n + ...$$

Convergence is guaranteed on each interval $[-r, r]$ as long as $0 \le r < 1$.

To get a power series expansion of $\int_0^x f[t]dt$ just take each term of

$$f[t] = 1 + t + t^2/2^2 + t^3/3^2 + t^4/4^2 + ... + t^n/n^2 + ...$$

and integrate from $0$ to $x$ and add them up to get:

$$\int_0^x f[t]dt = x + x^2/2 + x^3/(3*2^2) + x^4/(4*3^2)+$$

$$x^5/(5*4^2) + ... + x^{n+1}/((n+1)n^2) + ...$$

Convergence is guaranteed on each interval $[-r, r]$ as long as $0 \le r < 1$.

Very easy.

■ **B.4) Relating the coefficients of the expansion to the higher derivatives of the function: Taylor's formula.**

Take an unspecified function $f[x]$ and execute:

```
In[1]:=     n = 7
            Series[f[x], {x, 0, n}]
```

```
Out[1]=
                                            2
                                    f''[0] x
            f[0] + f'[0] x + --------- +
                                 2

                (3)   3        (4)   4
              f   [0] x      f   [0] x
              ---------- + ---------- +
                  6             24

                (5)   5        (6)   6
              f   [0] x      f   [0] x
              ---------- + ---------- +
                 120            720

                (7)   7
              f   [0] x                8
              ---------- + O[x]
                 5040
```

## B.4.a.i)

What formula for $a[n]$ does this suggest?

**Answer:**

After a moment's thought, we see that matching

$$f[x] = a[0] + a[1]x + a[2]x^2 + a[3]x^3 + ... + a[n]x^n + ...$$

against the output above, then we arrive at the neat formula

$$a[n] = f^{(n)}[0]/n!$$

for $n = 0, 1, 2, 3, ...$ where $f^{(n)}[0]$ stands for the $n$ th derivative of $f[x]$ evaluated at $0$. This is called *Taylor's formula.*

We agree that $0! = 1$, so for $n = 0$ we have

$$f^{(0)}[0]/0! = f^{(0)}[0]/1 = f[0]/1 = f[0].$$

## B.4.a.ii)

Check out Taylor's formula for the coefficients of the expansion of $f[x] = e^x$.

**Answer:**

The expansion of $e^x$ in powers of $x$ starts out with:

```
In[1]:=     n = 6
            Normal[Series[E^x, {x, 0, n}]]
```

```
Out[1]=
                       2    3    4     5     6
                       x    x    x     x     x
            1 + x + -- + -- + -- + --- + ---
                       2    6    24   120   720
```

The corresponding coefficients predicted by Taylor's formula are:

```
In[2]:=     Table[(D[E^x, {x, k}]/.x->0)/k!, {k, 0, n}]
```

$$Out[2]= \{1, 1, \frac{1}{2}, \frac{1}{6}, \frac{1}{24}, \frac{1}{120}, \frac{1}{720}\}$$

Right on the money. Taylor's formula seems to work well.

### B.4.b.i)

Where does Taylor's formula come from?

**Answer:**

Suppose $R$ is fixed and we know

$$f[x] = a[0] + a[1]x + a[2]x^2 + a[3]x^3 + ... + a[n]x^n + ...$$

on intervals $[-r, r]$ for $0 \leq r < R$. Taking $x = 0$ on both sides gives:

$$f[0] = a[0] + 0 + 0 + 0 + ... + 0 + ...$$

Consequently

$$a[0] = f[0] = f^{(0)}[0]/0!.$$

Also

$$f'[x] = a[1] + 2a[2]x + 3a[3]x^2 + 4a[4]x^3 + ...$$

$$... + na[n]x^{n-1} + ...$$

Taking $x = 0$ on both sides gives:

$$f'[0] = a[1] + 0 + 0 + 0 + ... + 0 + ...$$

Consequently

$$a[1] = f'[0] = f'[0]/1!.$$

Also

$$f''[x] = 2a[2] + 3*2a[3]x + 4*3a[4]x^2 + ...$$

$$... + n(n-1)a[n]x^{n-2} + ...$$

Taking $x = 0$ on both sides gives:

$$f''[0] = 2a[2] + 0 + 0 + 0 + ... + 0 + ...$$

Consequently

$$a[2] = f''[0]/2 = f''[0]/2!.$$

Also

$$f'''[x] = 3*2a[3] + 4*3*2a[4]x + 5*4*3x^2...$$

$$... + n(n-1)(n-2)a[n]x^{n-3} + ...$$

Taking $x = 0$ on both sides gives:

$$f'''[0] = 3*2a[3] + 0 + 0 + 0 + ... + 0 + ...$$

Consequently

$$a[3] = f'''[0]/(3*2) = f'''[0]/3!.$$

We are convinced: we can grind out the formulas for rest of the coefficients $a[n]$ the same way. It will be rather boring, though.

### B.4.b.ii)

What use is Taylor's formula?

**Answer:**

Many people want to jump on this formula to obtain all expansions. This is not a good idea even if you are as fast as *Mathematica*. As a matter of fact, this formula is usually the **least efficient** way to obtain an expansion of a given function. Just think of the misery involved in calculating many derivatives and plugging in.

Although Taylor's formula is not very useful for down-to-earth computations, it does have quite a bit of theoretical value. And since theory feeds new computations, it has some practical value as well.

You'll see it again in the Tutorial.

## Tutorial

### ■ T.1) The Ratio Test and convergence intervals.

Given a power series

$$a[0] + a[1]x + a[2]x^2 + a[3]x^3 + ... + a[n]x^n + ...$$

we can attempt to find some of its convergence intervals by using the Basic Convergence Principle or by using

*The Ratio Test:*

If for some positive number $R$, we can ascertain that

$$|a[n+1]R^{n+1}/(a[n]R^n)| \leq 1$$

for all large $n's$, then

$$a[0] + a[1]x + a[2]x^2 + a[3]x^3 + ... + a[n]x^n + ...$$

converges on any interval $[-r, r]$ as long as $0 \leq r < R$.

Notice that the Ratio test does not apply in the case that infinitely many of the coefficients $a[n]$ are zero.

### T.1.a)

Explain why the ratio test works.

**Answer:**

Recall the **Basic Convergence Principle:**

Given a power series

$$a[0] + a[1]x + a[2]x^2 + a[3]x^3 + ... + a[k]x^k + ...$$

If for some number $x = R$, the (infinite) list of individual terms

$$\{a[0], a[1]R, a[2]R^2, a[3]R^3, ..., a[k]R^k, ...\}$$

stays bounded (i.e. does not blow up to $\infty$ or down to $-\infty$), then the power series

$$a[0] + a[1]x + a[2]x^2 + a[3]x^3 + ... + a[k]x^k + ...$$

converges on any interval $[-r, r]$ as long as $0 \le r < |R|$.

To see why the Ratio Test works, suppose we have our hands on a number $R$ such that

$$|a[n+1]R^{n+1}/(a[n]R^n)| \le 1$$

for all large $n's$. This means

$$|a[n+1]R|^{n+1} \le |a[n]R|^n$$

for all large $n's$. Thus for all large $n's$, the terms $|(a[n]R^n)|$ are getting smaller as $n$ grows toward $\infty$. As a result, the (infinite) list of individual terms

$$\{a[0], a[1]R, a[2]R^2, a[3]R^3, ..., a[n]R^n, ...\}$$

cannot blow up or down and therefore must stay bounded.

And now the Basic Convergence Principle guarantees convergence on any interval $[-r, r]$ as long as $0 \le r < R$.

Find convergence intervals of the following power series.

**T.1.b.i)**

---

$$1 - x + x^2 - x^3 + x^4 - x^5 + ... + (-1)^n x^n + ...$$

---

**Answer:**

**The easiest way:** This is the expansion in powers of $x$ for

$$f[x] = 1/(1 + x).$$

On account of the lone complex singularity at $x = -1$, the power series converges on $[-r, r]$ provided $0 \le r < 1$.

**By the Basic Convergence Principle:** The power series is

$$1 - x + x^2 - x^3 + x^4 - x^5 + ... + (-1)^n x^n + ...$$

For $R = 1$, the list of terms

$$\{1, -R, R, -R^3, R^4, ..., (-1)^n R^n, ...\}$$

$$= \{1, -1, 1, -1, 1, -1, ..., (-1)^n, ...\}$$

stays bounded. Therefore the power series converges on intervals $[-r, r]$ provided $0 \le r < R = 1$.

**By the ratio test:** The power series is

$$1 - x + x^2 - x^3 + x^4 - x^5 + ... + (-1)^n x^n + ...$$

Note that $|a[n]| = 1$ for all $n's$. For a positive number $R$, look at

$$|a[n+1]R^{n+1}/(a[n]R^n)| = R^{n+1}/R^n = R$$

Clearly these are $\le 1$ for $R = 1$. Therefore the series converges on intervals $[-r, r]$ provided $0 \le r < R = 1$.

Of course *each* of these is a complete answer: one good argument is all you need to explain true facts (but explaining lies takes many long sentences.)

**T.1.b.ii)**

---

$$x^2/(2*3^2) + x^3/(3*3^3) + x^4/(4*3^4) + ... + x^n/(n*3^n) + ...$$

---

**Answer:**

**By the ratio test:** For $n \ge 2$, read off:

```
In[1]:=    Clear[n]
           a[n_] =1/(n 3^n)

Out[1]=      1
           ----
             n
           3  n
```

For a positive number $R$, look at:

```
In[2]:=    ratio =
           Simplify[a[n+1] R^(n + 1)/(a[n] R^n)]

Out[2]=      R n
           -------
           3 + 3 n
```

Look at the ratios for $R = 3$:

```
In[3]:=    ratio/.R->3

Out[3]=      3 n
           -------
           3 + 3 n
```

Evidently these ratios are $\le 1$ for $R = 3$. Therefore the series converges on intervals $[-r, r]$ provided $0 \le r < R = 3$.

Just for kicks, let us give a second argument using the **Basic Convergence Principle:** The power series is:

$$x^2/(2*3^2) + x^3/(3*3^3) + x^4/(4*3^4) +$$

$$... + x^n/(n*3^n) + ...$$

For $R = 3$, the list of terms

$$\{R^2/(2*3^2), R^3/(3*3^3), R^4/(4*3^4), ..., R^n/(n*3^n), ...\}$$

$$= \{3^2/(2*3^2), 3^3/(3*3^3), 3^4/(4*3^4), ..., 3^n/(n*3^n), ...\}$$

$$= \{1/2, 1/3, 1/4, 1/5, ..., 1/n, ...\}$$

stays bounded. Therefore the series converges on intervals $[-r, r]$ provided $0 \le r < R = 3$.

Let's look at some plots. The power series is

$$x^2/(2*3^2) + x^3/(3*3^3) + x^4/(4*3^4) +$$

$$... + x^n/(n*3^n) + ...$$

$In[4]:=$    `sum12 = Sum[x^n/(n 3^n),{n,2,12}]`

$Out[4]=$

$$\frac{x^2}{18} + \frac{x^3}{81} + \frac{x^4}{324} + \frac{x^5}{1215} + \frac{x^6}{4374} + \frac{x^7}{15309} +$$

$$\frac{x^8}{52488} + \frac{x^9}{177147} + \frac{x^{10}}{590490} + \frac{x^{11}}{1948617} +$$

$$\frac{x^{12}}{6377292}$$

$In[5]:=$    `sum15 = Sum[x^n/(n 3^n),{n,2,15}];`

$In[6]:=$    `Plot[{sum12,sum15},{x,-3.5,3.5},`
            `AxesLabel->{"x","y"},`
            `PlotRange->{0,3}];`

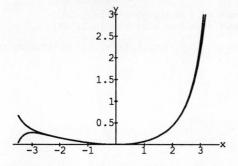

Convergence on subintervals of $[-3,3]$ is suggested by lots of shared ink.

**T.1.b.iii)**

---

$$1 + x + x^2/2^2 + x^3/3^3 + x^4/4^4 + ... + x^n/n^n + ...$$

---

**Answer:**

The Basic Convergence Principle works like a charm: For any positive number $x = R$, look at the list of terms

$$\{1, R, R^2/2^2, R^3/3^3, R^4/4^4, ..., R^n/n^n, ...\}.$$

Notice that at least by the point at which $n$ becomes larger than $R$, then the terms get smaller and smaller.

Take a look in the case that $R = 9$:

$In[1]:=$    `ListPlot[Table[{n, (9^n)/n^n},{n,1,25}],`
            `PlotStyle->`
            `{PointSize[0.02],RGBColor[1,0,0]}];`

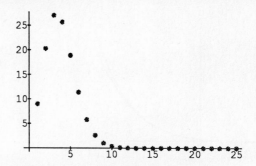

Take a look in the case that $R = 30$:

$In[2]:=$    `ListPlot[Table[{n, (30^n)/n^n},{n,1,50}],`
            `PlotStyle->`
            `{PointSize[0.02],RGBColor[1,0,0]}];`

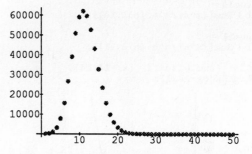

As a result, no matter what positive $R$ we take, then the list of terms

$$\{1, R, R^2/2^2, R^3/3^3, R^4/4^4, ..., R^n/n^n, ...\}$$

cannot blow up or down; so no matter what positive R we take, then these terms stay bounded.

The Basic Convergence Principle steps in to tell us that no matter what positive $R$ we take, then this power series converges on intervals $[-r, r]$ provided $0 \le r < R$. In short this power series converges on every interval $[-r, r]$ as long as $0 \le r < \infty$.

Lets look at some plots. The series is

$$1 + x + x^2/2^2 + x^3/3^3 + x^4/4^4 + ... + x^n/n^n + ...$$

$In[3]:=$    `sum4 = 1 + Sum[ (x^n)/n^n, {n,1,4}]`

$Out[3]=$

$$1 + x + \frac{x^2}{4} + \frac{x^3}{27} + \frac{x^4}{256}$$

$In[4]:=$    `sum5 = 1 + Sum[ (x^n)/n^n, {n,1,5}]`

$Out[4]=$

$$1 + x + \frac{x^2}{4} + \frac{x^3}{27} + \frac{x^4}{256} + \frac{x^5}{3125}$$

$In[5]:=$    `Plot[{sum4,sum5},{x,-8,8},`
            `PlotRange->{-10,80}];`

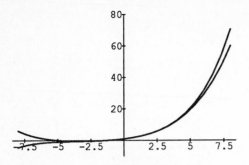

Convergence on $[-8, 8]$ is suggested by lots of shared ink.

Let us try more terms and a longer interval (we type a ;
to suppress the long outputs):

```
In[6]:=    sum10 =
           1 + Sum[ (x^n) /n^n, {n,1,10}]

           sum11 =
           1 + Sum[ (x^n) /n^n, {n,1,11}];

           Plot[{sum10, sum11}, {x,-14,14},
           PlotRange->All];
```

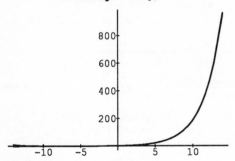

Here convergence on the large interval $[-11, 11]$ is suggested by lots of shared ink.

**T.1.b.iv)**

$$1 - 2^2 x + 3^2 x^2 - 4^2 x^3 + ... + (-1)^n (n+1)^2 x^n + ...$$

**Answer:**

The Ratio Test works nicely here.

```
In[1]:=    a[n_] = (-1)^n (n + 1)^2
```
$$Out[1]=\quad (-1)^n (1 + n)^2$$

```
In[2]:=    firstlook =
           Cancel[a[n + 1] R^(n + 1)/(a[n] R^n)]
```
$$Out[2]=\quad -\left(\frac{R (2 + n)^2}{(1 + n)^2}\right)$$

```
In[3]:=    secondlook =
           Together[ExpandAll[firstlook]]
```

$$Out[3]= \quad \frac{-4R - 4Rn - Rn^2}{1 + 2n + n^2}$$

So
$$\lim |_{n \to \infty} a[n+1]R^{n+1}/(a[n]R^n)| = R.$$

As a result if $R$ as any number with $0 \leq R < 1$, then
$$|a[n+1]R^{n+1}/(a[n]R^n)| < 1$$

for all large $n's$. The Ratio Test tells us that if $R$ is any number with $0 \leq R < 1$, then this power series converges on intervals $[-r, r]$ as long as $0 < r < R$. In short:

*this power series converges on intervals* $[-r, r]$ *as long as* $0 < r < 1$.

Lets look at some plots.

```
In[4]:=    sum20 = 1 + Sum[a[n] x^n, {n,1,15}]
           sum21 = 1 + Sum[a[n] x^n, {n,1,16}];

           Plot[{sum20, sum21}, {x,-1,1}, PlotRange->{-
           30,30}];
```

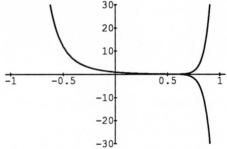

Convergence on subintervals of $[-1, 1]$ is suggested by lots of shared ink.

**T.1.b.v)**

$$x + (2/2^2)x^2 + (3!/3^3)x^3 + (4!/4^4)x^4 + ...$$
$$... + (n!/n^n)x^n + ...$$

**Answer:**

The Ratio Test is the right way to do this one. Read off for $n \geq 1$
$$a[n] = n!/n^n$$

For a positive number $R$, look at
$$|a[n+1]R^{n+1}/(a[n]R^n)|$$

$$= ((n+1)! R^{n+1}/(n+1)^{n+1})/(n! R^n/n^n)$$

$$= ((n+1)!/n!)(n^n/(n+1)^{n+1})(R^{n+1}/R^n)$$

$$= (n+1)(n^n/(n+1)^n)(1/(n+1)R$$

$$= (n+1)(n^n/(n+1)^n)(1/(n+1)R = (n^n/(n+1)^n)R$$

Clearly these are $\leq 1$ for $R = 1$. Therefore this power series converges on intervals $[-r, r]$ provided $0 < r < R = 1$. But we can do better. Note that

$$(n+1)^n/n^n = ((n+1)/n)^n = (1+1/n)^n.$$

Therefore

$$\lim_{n\to\infty} (n+1)^n/n^n = \lim_{n\to\infty} (1+1/n)^n = e.$$

Thus for large $n$, the ratios

$$|a[n+1]R^{n+1}/(a[n]R^n)| = (n^n/(n+1)^n)R$$

are close to $R/e$. Accordingly if $R < e$, then for large $n's$, the ratios

$$|a[n+1]R^{n+1}/(a[n]R^n)| = (n^n/(n+1)^n)R$$

are all less than 1.

Therefore this power series converges on intervals $[-r, r]$ provided $0 < r < R < e$. In short, this power series converges on intervals $[-r, r]$ provided $0 < r < e$. Let's look at some plots.

```
In[1]:=   sum9 = Sum[ (n!/n^n) x^n,{n,1,9}]
          sum10 = Sum[ (n!/n^n) x^n,{n,1,10}];

          plot1 =
          Plot[{sum9,sum10},{x,-E-.5, E+.5},
          PlotRange->{-10,40}];
```

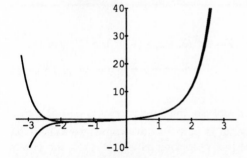

Because $e$ is approximately 2.7, convergence on subintervals of $[-e, e]$ is suggested by lots of shared ink.

Lets push out further and take a look:

```
In[2]:=   sum14 = Sum[ (n!/n^n) x^n,{n,1,14}]
          sum15 = Sum[ (n!/n^n) x^n,{n,1,15}];

In[3]:=   plot2 =
          Plot[{sum14,sum15},{x,-E-.5, E+.5},
          PlotRange->{-20,50}];
```

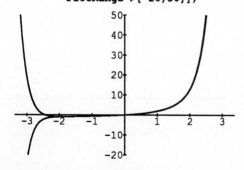

Again convergence on subintervals of $[-e, e]$ is suggested by lots of shared ink.

To get a another look at the convergence, show the combined plot:

```
In[4]:=   Show[plot1,plot2];
```

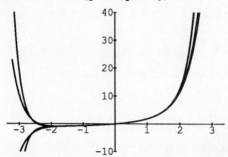

Look at those curves trying to glue themselves together for $-e \leq x \leq e$.

**T.1.b.vi)**

$$1 + 2x - 3x^2 + 4x^3 - 5x^4 + \ldots + (-1)^{n+1}(n+1)x^n + \ldots$$

**Answer:**

Say this one looks like the derivative of a familiar Geometric Series.

```
In[1]:=   Series[D[1/(1 + x),x],{x,0,7}]
```
```
Out[1]=
                      2       3       4       5
          -1 + 2 x - 3 x  + 4 x  - 5 x  + 6 x  -

                6      7        8
             7 x  + 8 x  + 0[x]
```

Looks good except for the first term. Let's adjust it:

```
In[2]:=   Series[2 + D[1/(1 + x),x],{x,0,7}]
```
```
Out[2]=
                     2       3       4       5
          1 + 2 x - 3 x  + 4 x  - 5 x  + 6 x  -

                6      7        8
             7 x  + 8 x  + 0[x]
```

Looks good. Now we are confident that the power series

$$1 + 2x - 3x^2 + 4x^3 - 5x^4 + \ldots + (-1)^{n+1}(n+1)x^n + \ldots$$

is the expansion in powers of $x$ of:

```
In[3]:=   2 + D[1/(1 + x),x]
```
```
Out[3]=
                          -2
          2 - (1 + x)
```

Because of the lone singularity at $x = -1$, we conclude that

$$1 + 2x - 3x^2 + 4x^3 - 5x^4 + \ldots + (-1)^{n+1}(n+1)x^n + \ldots$$

converges to $2 - 1/(1 + x)^2$ on intervals $[-r, r]$ as long as $0 \le r < 1$. Watch it converge on subintervals of $[-1, 1]$: here are a few terms,

```
In[4]:=    expan6 = 1 + Sum[ (-1)^(n + 1)
           (n + 1) x^n, {n, 1, 6}]
Out[4]=
                    2      3      4      5
           1 + 2 x - 3 x  + 4 x  - 5 x  + 6 x  -

               6
           7 x
```

Note that the first term is not given by the formula for the general term.

```
In[5]:=    expan7 = 1 + Sum[ (-1)^(n + 1)
           (n + 1) x^n, {n, 1, 7}];

           Plot[{2 - 1/ (1 + x)^2,
           expan6, expan7}, {x, -1, 1},
           PlotRange->{-20, 2},
           AxesLabel->{"x", "y"}];
```

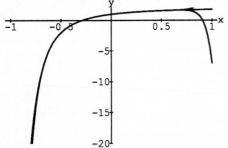

Let's go further out in the power series with the expectation of seeing more cohabitation:

```
In[6]:=    expan15 = 1 + Sum[ (-1)^(n + 1)
           (n + 1) x^n, {n, 1, 15}]

           expan16 = 1 + Sum[ (-1)^(n + 1)
           (n + 1) x^n, {n, 1, 16}]

           Plot[{2 - 1/ (1 +x)^2, expan15, expan16},
           {x, -1, 1}, PlotRange->{-20, 2},
           AxesLabel->{"x", "y"}];
```

Yes! More cohabitation on sunbintervals of $[-1, 1]$.

■ **T.2) Recognizing some power series as expansions**

**T.2.a)**

Some power series are quickly recognizable as expansions of familiar functions. Try your hand at recognizing the following power series. In each case, specify the intervals on which the power series converges.

**T.2.a.i)**

$$1 + x + x^2/2 + x^3/3! + \ldots + x^n/n! + \ldots$$

**Answer:**

This is just the expansion in powers of $x$ of the function $e^x$. It converges to $e^x$ on every interval $[-r, r]$ as long as $0 \le r < \infty$.

**T.2.a.ii)**

$$1 - x^3 + x^6/2 - x^9/3! + \ldots + (-1)^n x^{3n}/n! + \ldots$$

**Answer:**

This is the power series in part i) above with $x$ changed to $-x^3$ It converges to $e^{-x^3}$ on every interval $[-r, r]$ as long as $0 \le r < \infty$.

**T.2.a.iii)**

$$-3x^2 + 6x^5/2 - 9x^8/3! + \ldots + (-1)^n 3n x^{3n-1}/n! + \ldots$$

**Answer:**

This is the derivative of the power series in part ii) above. Because the series in part ii) above converges to $e^{-x^3}$ on every interval $[-r, r]$ as long as $0 \le r < \infty$, the power series

$$-3x^2 + 6x^5/2 - 9x^8/3! + \ldots + (-1)^n 3n x^{3n-1}/n! + \ldots$$

converges to:

```
In[1]:=    D[E^(-x^3), x]

Out[1]=
                2
           -3 x
           -----
               3
              x
             E
```

The convergence takes place on every interval $[-r, r]$ as long as $0 \le r < \infty$.

**T.2.a.iv)**

$$1 + 7x + 4x^2 + 9x^3 + 16x^4 + \ldots + n^2 x^n + \ldots$$

**Answer:**

This is a big bite to chew on. Set

$$f[x] = 1 + 7x + 4x^2 + 9x^3 + 16x^4 + \ldots + n^2 x^n + \ldots$$

Note that

$$(f[x] - 1 - 7x)/x = 4x + 9x^2 + 16x^3 + \ldots + n^2 x^{n-1} + \ldots$$

Call

$$g[x] = 4x + 9x^2 + 16x^3 + \ldots + n^2 x^{n-1} + \ldots$$

and note

$$(f[x] - 1 - 7x)/x = g[x]$$

Now if we can find what a formula for $g[x]$, then finding a formula for $f[x]$ will be duck soup.

Hmmmmmmm.

Well, $g[x] = h'[x]$ where

$$h[x] = 2x^2 + 3x^3 + 4x^4 + \ldots + nx^n + \ldots$$

And

$$h[x]/x = 2x + 3x^2 + 4x^3 + \ldots + nx^n - 1 + \ldots$$

So $(h[x]/x) + 1 =$

$$1 + 2x + 3x^2 + 4x^3 + \ldots + nx^n - 1 + \ldots$$

This means $(h[x]/x) + 1$ is the derivative of the **geometric series**

$$1 + x + x^2 + x^3 + x^4 + \ldots + x^n + \ldots$$

which we recognize (finally) as $1/(1-x)$ on every interval $[-r, r]$ such that $0 \le r < 1$. Thus

$$(h[x]/x) + 1 = D[1/(1-x), x]$$

on every interval $[-r, r]$ such that $0 \le r < 1$. Let's see what $h[x]$ is:

```
In[1]:=   hsolved = Solve[(h[x]/x) + 1 ==
          D[1/(1 - x),x],h[x]]
```
```
Out[1]=                    2    3
                        -2 x  + x
          {{h[x] -> -(------------)}}
                           2
                      1 - 2 x + x
```

Here is the formula for $h[x]$:

```
In[2]:=   h[x_] = hsolved[[1,1,2]]
```
```
Out[2]=              2    3
                  -2 x  + x
          -(------------)
                   2
           1 - 2 x + x
```

Remember $g[x] = h'[x]$:

```
In[3]:=   g[x_] = D[h[x],x]
```

```
Out[3]=                   2
                    -4 x + 3 x
          -(------------) +
                        2
              1 - 2 x + x

                              2     3
              (-2 + 2 x) (-2 x  + x )
          ------------------------
                          2 2
               (1 - 2 x + x )
```

Now remember that $(f[x] - 1 - 7x)/x = g[x]$

```
In[4]:=   fsolved = Solve[
          (f[x] - 1 - 7 x)/x == g[x],f[x]]
```
```
Out[4]=   {{f[x] ->

                            2      3      4
              1 + 4 x - 14 x  + 17 x  - 6 x
          -(-------------------------------)}}
                            2     3
              -1 + 3 x - 3 x  + x
```

Here is $f[x]$:

```
In[5]:=   f[x_] = fsolved[[1,1,2]]
```
```
Out[5]=                     2      3      4
              1 + 4 x - 14 x  + 17 x  - 6 x
          -(-------------------------------)
                            2     3
              -1 + 3 x - 3 x  + x
```

This tells us that

$$1 + 7x + 4x^2 + 9x^3 + 16x^4 + \ldots + n^2 x^n + \ldots$$

converges to this $f[x]$ on every interval $[-r, r]$ with $0 \le r < 1$. Let's check this out with a plot:

```
In[6]:=   sum13 = 1 + 7 x + Sum[n^2 x^n, {n,2,13}]
          Plot[{f[x],sum13}, {x,-1,1}];
```

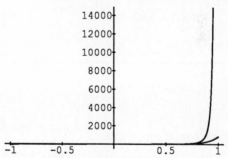

This doesn't tell us much; let's limit the plot range:

```
In[7]:=   Plot[{f[x],sum13}, {x,-1,1},
          PlotRange->{-10,100}];
```

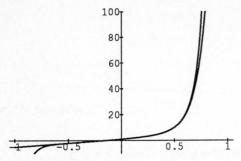

Nice ink sharing on the subinterval $[-0.6, 0.6]$ of $[-1, 1]$. You can get more sharing of ink by using more of the power series. Try it.

### T.2.b)

Every expansion of a function in powers of $x$ is a power series. If a power series converges on an interval $[-r, r]$ with $r > 0$, then must it be the expansion in powers of $x$ of a function?

**Answer:**

Yes.

If you have a power series that converges on an interval $[-r, r]$ for some $r > 0$, then define a function $f[x]$ to be what the power series converges to. The power series is the expansion of this function $f[x]$.

■ **T.3) Limits at 0; L'Hospital's rule.**

### T.3.a.i)

Use expansions and power series to find
$$\lim_{x \to 0} (1 - \cos[x])/x^2.$$
Illustrate with a plot.

**Answer:**

Look at:

```
In[1]:=    Series[Cos[x],{x,0,12}]
Out[1]=        2    4    6      8       10
           x    x    x    x      x      x
       1 - -- + -- - --- + ----- - ------- +
           2    24   720   40320   3628800

             12
           x             13
       --------- + 0[x]
       479001600
```

On $[-1, 1]$, we know
$$\cos[x] = 1 - x^2/2! + x^4/4! - x^6/6! + \ldots$$

So
$$(1 - \cos[x]) = 1 - (1 - x^2/2! + x^4/4! - x^6/6! + \ldots)$$
$$= x^2/2! - x^4/4! + x^6/6! - x^8/8! + \ldots.$$

Consequently $(1 - \cos[x])/x^2 =$
$$1/2! - x^2/4! + x^4/6! - x^6/8! + \ldots.$$

As $x$ closes in on 0, we see that $(1 - \cos[x])/x^2$ closes in on
$$1/2! - 0 + 0 - 0 + \ldots = 1/2.$$

So we report
$$\lim_{x \to 0} (1 - \cos[x])/x^2 = 1/2.$$

Lets check this out with a plot:

```
In[2]:=    Plot[{(1 - Cos[x])/x^2,1/2},
           {x,-.1,.1},AxesLabel->{"x","y"},
           Axes->{0,0.4999},PlotStyle->
           {{Thickness[0.004]},
           {Thickness[0.008],RGBColor[1,0,0]}}];
```

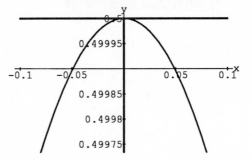

Looks fine.

### T.3.a.ii)

How can we calculate
$$\lim_{x \to 0} (1 - \cos[x])/x^2$$
without all the writing we did above?

**Answer:**

Replace the numerator and the denominator by the early parts of their series expansions plus remainders; then divide.

```
In[1]:=    numerator = Series[1 - Cos[x],{x,0,4}]
Out[1]=     2    4
           x    x         5
           -- - -- + 0[x]
           2    24
```

```
In[2]:=    denominator = Series[x^2,{x,0,4}]
Out[2]=     2         5
           x  + 0[x]
```

Dividing this out gives the first few terms of the power series converging to $(1 - \cos[x])/x^2$:

```
In[3]:=    numerator/denominator
```

*Out[3]=*
$$\frac{1}{2} - \frac{x^2}{24} + O[x]^3$$

The leading term (which in this case is 1/2) is the limit.

### T.3.a.iii)

Use expansions and power series to calculate
$$\lim_{x\to 0}(x - \sin[x])/\sin[7x^3].$$
Illustrate with a plot.

### Answer:

Replace the numerator and the denominator by the early parts of their series expansions plus remainders; then divide.

*In[1]:=*    `numerator =`
`Series[x - Sin[x],{x,0,5}]`

*Out[1]=*
$$\frac{x^3}{6} - \frac{x^5}{120} + O[x]^6$$

*In[2]:=*    `denominator =`
`Series[Sin[7 x^3],{x,0,5}]`

*Out[2]=*
$$7 x^3 + O[x]^6$$

*In[3]:=*    `numerator/denominator`

*Out[3]=*
$$\frac{1}{42} - \frac{x^2}{840} + O[x]^3$$

Sending $x$ to 0 gives
$$\lim_{x\to 0}(x - \sin[x])/\sin[7x^3] = 1/42.$$

Let's check this out with a plot:

*In[4]:=*    `Plot[{(x - Sin[x])/Sin[7 x^3],1/42},`
`{x,-.1,.1},AxesLabel->{"x","y"},`
`PlotStyle->{{Thickness[0.004]},`
`{Thickness[0.008],RGBColor[1,0,0]}}];`

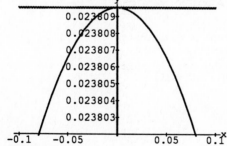

Looks fine.

### T.3.a.iv)

Use expansions and power series to find
$$\lim_{x\to 0}\log[1 + \sin[x^3]]/(\sqrt{(1+x)/(1-x)} - e^x).$$
Illustrate with a plot.

### Answer:

Replace the numerator and the denominator by the early parts of their series expansions plus remainders; then divide.

*In[1]:=*    `numerator = Series[Log[1 + Sin[x^3]],{x,0,5}]`

*Out[1]=*
$$x^3 + O[x]^6$$

*In[2]:=*    `denominator = Series[Sqrt[`
`(1 + x)/(1 - x)] - E^x,{x,0,5}]`

*Out[2]=*
$$\frac{x^3}{3} + \frac{x^4}{3} + \frac{11 x^5}{30} + O[x]^6$$

*In[3]:=*    `numerator/denominator`

*Out[3]=*
$$3 - 3 x - \frac{3 x^2}{10} + O[x]^3$$

Sending $x$ to 0 gives
$$\lim_{x\to 0}\log[1 + \sin[x^3]]/(\sqrt{(1+x)/(1-x)} - e^x) = 3.$$

Here's a plot:

*In[4]:=*    `Plot[{ Log[1 + Sin[x^3]]/`
`(Sqrt[(1 + x)/(1 - x)] -`
`E^x),3},{x,-.1,.1},Axes->{0,2.8},`
`AxesLabel->{"x","y"},`
`PlotStyle->{{Thickness[0.004]},`
`{Thickness[0.008],RGBColor[1,0,0]}}];`

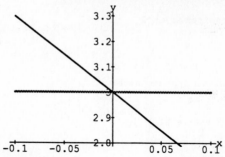

Looks fine.

### T.3.a.v)

Do you need expansions and power series to find
$$\lim_{x\to\infty}\log[x]/x?$$

**Answer:**

This is a ringer. Everyone knows that $\log[x]$ grows much more slowly than any positive power of $x$. The limit is 0.

### T.3.a.vi)

Use expansions and power series to find
$$\lim_{x\to\infty} x\log[(x-1)/(x+1)]$$
Illustrate with a plot.

**Answer:**

Recall that $\lim_{x\to\infty} f[x] = \lim_{t\to 0} f[1/t^2]$. So we change the variable from $x$ to $1/t^2$ and expand as a function of $t$ in hopes of finding what happens as $t\to 0$.

```
In[1]:=   yx = x Log[(x - 1)/(1 + x)]
Out[1]=
                 -1 + x
          x Log[------]
                 1 + x
```

```
In[2]:=   yt = yx/.x->1/t^2
Out[2]=
                    -2
              -1 + t
          Log[--------]
                    -2
              1 + t
          --------------
                 2
                t
```

```
In[3]:=   numerator =
          Series[Numerator[yt],{t,0,6}]
Out[3]=
              2    2 t
          -2 t  - ---- + O[t]
                   3
                            7
```

```
In[4]:=   denominator = t^2
Out[4]=
              2
             t
```

```
In[5]:=   numerator/denominator
Out[5]=
                  4
              2 t        5
          -2 - ---- + O[t]
                3
```

As $t\to 0$, the result is $-2$.

Therefore
$$\lim_{x\to\infty} x\log[(x-1)/(x+1)] = -2.$$

Let's check this out with a plot:

```
In[6]:=   Plot[{x Log[(x - 1)/(x + 1)],-2},
          {x,2,400},PlotRange->{-2.012,-1.998},
          Axes->{0,-2.002},PlotStyle->
```

```
{{Thickness[0.004]},{Thickness[0.008],
RGBColor[1,0,0]}}];
```

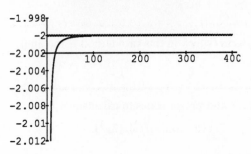

Looks just right.

### T.3.b): L'Hospital's Rule.

Suppose $f[x]$ and $g[x]$ are functions with $f[0] = g[0] = 0$, but $g'[0] \neq 0$. Explain the formula
$$\lim_{x\to 0} f[x]/g[x] = f'[0]/g'[0].$$

Some folks like to call this formula by the name of L'Hospital's Rule. Some old-fashioned calculus courses like to make a big deal about this formula and make it mysterious as heck. You will see that it is no more mysterious than the Wizard of Oz.

**Answer:**

Replace the numerator and the denominator by the early parts of their series expansions plus remainders; then divide.

```
In[1]:=   Clear[f]
          numerator =
          Series[f[x],{x,0,2}]/.f[0]->0
Out[1]=
                      2
               f''[0] x        3
          f'[0] x + --------- + O[x]
                        2
```

```
In[2]:=   Clear[g]
          denominator =
          Series[g[x],{x,0,2}]/.g[0]->0
Out[2]=
                      2
               g''[0] x        3
          g'[0] x + --------- + O[x]
                        2
```

```
In[3]:=   numerator/denominator
Out[3]=   f'[0]    f''[0]    f'[0] g''[0]
          ----- + (------- - -------------) x +
          g'[0]    2 g'[0]             2
                                2 g'[0]

              2
          O[x]
```

Sending $x$ to 0 gives
$$\lim_{x\to 0} f[x]/g[x] = f'[0]/g'[0].$$

Really easy.

■ **T.4) More infinite sums of numbers.**

**T.4.a.i)**

Find the value of the infinite sum of numbers

$$\pi - \pi^3/3! + \pi^5/5! - \pi^7/7! + \ldots + (-1)^n \pi^{2n+1}/(2n+1)! + \ldots$$

**Answer:**

This looks like an old friend.

```
In[1]:=  Normal[Series[Sin[x],{x,0,12}]]/.
         x->Pi
```
```
Out[1]=      3    5    7     9
            Pi   Pi   Pi    Pi
      Pi - --- + --- - ---- + ------ -
           6    120  5040   362880

             11
           Pi
          --------
          39916800
```

```
In[2]:=  Sum[(-1)^n (Pi^(2 n + 1))/
         (2 n + 1)!,{n,0,5}]
```
```
Out[2]=      3    5    7     9
            Pi   Pi   Pi    Pi
      Pi - --- + --- - ---- + ------ -
           6    120  5040   362880

             11
           Pi
          --------
          39916800
```

The infinite sum

$$\pi - \pi^3/3! + \pi^5/5! + \ldots + (-1)^n \pi^{2n+1}/(2n+1)! + \ldots$$

is the expansion of sin[x] in powers of x evaluated at x = π. It converges to sin[π] = 0. Watch it converge to 0.

```
In[3]:=  N[Sum[((-1)^n Pi^(2 n + 1))/
         (2 n + 1)!, {n,0,5}]]
```
```
Out[3]=  -0.00044516
```

```
In[4]:=  N[Sum[((-1)^n Pi^(2 n + 1))/
         (2 n + 1)!, {n,0,10}],20]
```
```
Out[4]=            -11
         1.0347853 10
```

```
In[5]:=  N[Sum[((-1)^n Pi^(2 n + 1))/
         (2 n + 1)!, {n,0,20}],40]
```
```
Out[5]=          -32
         3.927693 10
```

```
In[6]:=  N[Sum[((-1)^n Pi^(2 n + 1))/
         (2 n + 1)!, {n,0,30}],60]
```
```
Out[6]=       -56
         1.05 10
```

```
In[7]:=  N[Sum[((-1)^n Pi^(2 n + 1))/
         (2 n + 1)!, {n,0,40}],100]
```

```
Out[7]=                  -84
         4.6423170759635 10
```

Damn right.

**T.4.a.ii)**

Find the value of the infinite sum of numbers

$$1/3 + 2/3^2 + 3/3^3 + 4/3^4 + \ldots + n/3^n \ldots$$

**Answer:**

This hints of the derivative of the geometric series.

```
In[1]:=  geom = Series[1/(1 - x),{x,0,10}]
```
```
Out[1]=      2   3   4   5   6   7
      1 + x + x + x + x + x + x + x +

       8   9   10      11
      x + x + x  + O[x]
```

Take the derivative:

```
In[2]:=  dergeom = D[geom,x]
```
```
Out[2]=          2     3      4      5
      1 + 2 x + 3 x + 4 x + 5 x + 6 x +

          6     7     8      9       10
       7 x + 8 x + 9 x + 10 x  + O[x]
```

Take x = 1/3:

```
In[3]:=  dergeom/.x->1/3
```
```
Out[3]=    2   3   4    5    6     7
      1 + - + - + -- + -- + --- + --- +
          3   9   27   81   243   729

       8     9    10         1 10
      ---- + ---- + ----- + O[-]
      2187  6561   19683       3
```

Multipying (1/3) times **dergeom/.x->1/3** gives:

$$1/3 + 2/3^2 + 3/3^3 + 4/3^4 + \ldots + n/3^n \ldots$$

which is the series we are studying. This means

$$1/3 + 2/3^2 + 3/3^3 + 4/3^4 + \ldots + n/3^n \ldots$$

converges to

```
In[4]:=  (1/3)D[1/(1 - x),x]/.x->1/3
```
```
Out[4]=  3
         -
         4
```

Watch it converge.

```
In[5]:=  {N[Sum[n/3^n,{n,1,20}],10],
         N[3/4,12]}
```
```
Out[5]=  {0.7499999969, 0.75}
```

```
In[6]:=    {N[Sum[n/3^n,{n,1,30}],20],
            N[3/4,20]}

Out[6]=    {0.74999999999992350326, 0.75}


In[7]:=    {N[Sum[n/3^n,{n,1,50}],30],
            N[3/4,20]}

Out[7]=    {0.7499999999999999999999964131394, 0.75}
```

Beautiful.

**T.4.b)**

---

Explain why the following infinite sum of numbers is convergent. Try to estimate the sum.

$$-2+7/8-5/24+13/384-1/240+...+(-1)^n(3n+1)/(2^n n!)+...$$

---

**Answer:**

Look at the power series

$$-2x+(7/8)x^2-(5/24)x^3+...+(-1)^n(3n+1)/(2^n n!))x^n+...$$

If we can determine that this power series convergences on an interval including $x = 1$, then we'll have explained why the sum of numbers in question converges. To this end, we try to determine some convergence intervals of

$$-2x+(7/8)x^2-(5/24)x^3+...+(-1)^n(3n+1)/(2^n n!))x^n+...$$

The Ratio Test is a good way to do this one.

```
In[1]:=    a[n_] = (-1)^n ( 3 n + 1)/(2^n n!)

Out[1]=           n
            (-1)  (1 + 3 n)
           ----------------
                 n
              2  n!
```

Take a positive $R$ and look at $a[n+1]R^{n+1}/(a[n]R^n)$:

```
In[2]:=    Simplify[a[n + 1] R^(n+ 1)/(a[n] R^n)]

Out[2]=     -(R (4 + 3 n) n!)
           ---------------------
           2 (1 + 3 n) (1 + n)!
```

So $|a[n+1]R^{n+1}/(a[n]R^n)|$ is:

```
In[3]:=    firstshot = ExpandAll[R (4 + 3 n)/
            (2 (1 + 3 n) (n + 1))]

Out[3]=          4 R                3 R n
           ---------------  +  ---------------
                      2                   2
           2 + 8 n + 6 n      2 + 8 n + 6 n


In[4]:=    bettershot = Together[firstshot]

Out[4]=      4 R + 3 R n
           ---------------
                      2
           2 + 8 n + 6 n
```

The deonominator of $|a[n+1]R^{n+1}/(a[n]R^n)|$ is a quadratic function of $n$ and the numerator is a linear function of $n$.

So, no matter what R is,

$$\lim_{n \to \infty} |a[n+1]R^{n+1}/(a[n]R^n)| = 0.$$

Thus, no matter what $R$ is, $|a[n+1]R^{n+1}/(a[n]R^n)| \leq 1$ for all large $n's$.

This means that the power series

$$-2x+(7/8)x^2-(5/24)x^3+...+(-1)^n(3n+1)/(2^n n!))x^n+...$$

converges on every interval $[-r, r]$ as long as $0 \leq r < \infty$.

Because $x = 1$ is in many such intervals, the infinite sum

$$-2 + (7/8) - (5/24) + ... + (-1)^n(3n+1)/(2^n n!)) + ...$$

is convergent. Watch it converge:

```
In[5]:=    N[Sum[a[n],{n,1,1}],12]

Out[5]=    -2.


In[6]:=    N[Sum[a[n],{n,1,7}],12]

Out[6]=    -1.30326760913


In[7]:=    N[Sum[a[n],{n,1,10}],12]

Out[7]=    -1.30326532946
```

This one seems to be converging like the proverbial bat out of the bad place. No wonder; look at all the power in the denominators!

```
In[8]:=    N[Sum[a[n],{n,1,15}],18]

Out[8]=    -1.30326532985631675


In[9]:=    N[Sum[a[n],{n,1,20}],30]

Out[9]=    -1.303265329856316711801899918407
```

Settling down beautifully.

```
In[10]:=   N[Sum[a[n],{n,1,30}],30]

Out[10]=   -1.303265329856316711801899767675


In[11]:=   N[Sum[a[n],{n,1,50}],50]

Out[11]=   -1.3032653298563167118018997674955900226
           7209590677436
```

A **very** confident educated guess it that

$$-4/2+7/8-5/24+13/384-1/240+$$

$$...+(-1)^n(3n+1)/(2^n n!)+...=-1.303265329856$$

to 12 accurate decimals.

**T.4.c): The harmonic series**

---

Try to explain why the infinite sum

$$1+1/2+1/3+1/4+...+1/n+...$$

is **not** convergent.

## Answer.

Note:

$$x + x^2/2 + x^3/3 + x^4/4 + ... + x^n/n + ...$$

$$= \int_0^x (1 + t + t^2 + t^3 + ... + t^{n-1} + ...)dt$$

$$= \int_0^x 1/(1-t)dt$$

provided $0 \le x < 1$.

```
In[1]:=    Integrate[1/(1 - t), {t,0,x}]
Out[1]=    -Log[1 - x]
```

This tells us that for $0 \le x < 1$,

$$x + x^2/2 + x^3/3 + x^4/4 + ... + x^n/n + ...$$
$$= -\log[1-x].$$

Now

$$1 + 1/2 + 1/3 + 1/4 + ... + 1/n + ...$$

$$= \lim_{x \to 1} (x + x^2/2 + x^3/3 + x^4/4 + ... + x^n/n + ...)$$

$$= \lim_{x \to 1} -\log[1-x] = \infty.$$

This tells us that the infinite sum

$$1 + 1/2 + 1/3 + 1/4 + ... + 1/n + ...$$

blows up and cannot converge. Let's see what happens when we take some partial sums:

```
In[2]:=    N[Sum[1/n, {n,1,9}]]
Out[2]=    2.82897
```

```
In[3]:=    N[Sum[1/n, {n,1,20}]]
Out[3]=    3.59774
```

```
In[4]:=    N[Sum[1/n, {n,1,50}]]
Out[4]=    4.49921
```

```
In[5]:=    N[Sum[1/n, {n,1,100}]]
Out[5]=    5.18738
```

```
In[6]:=    N[Sum[1/n, {n,1,150}]]
Out[6]=    5.59118
```

```
In[7]:=    N[Sum[1/n, {n,1,400}]]
Out[7]=    6.56993
```

The partial sums are not setting down. They are going up about just a little faster than $\log[x]$ goes up:

```
In[8]:=    N[{Sum[1/n, {n,1,9}], Log[9]}]
Out[8]=    {2.82897, 2.19722}
```

```
In[9]:=    N[{Sum[1/n, {n,1,20}], Log[20]}]
Out[9]=    {3.59774, 2.99573}
```

```
In[10]:=   N[{Sum[1/n, {n,1,50}], Log[50]}]
Out[10]=   {4.49921, 3.91202}
```

```
In[11]:=   N[{Sum[1/n, {n,1,100}], Log[100]}]
Out[11]=   {5.18738, 4.60517}
```

```
In[12]:=   N[{Sum[1/n, {n,1,150}], Log[150]}]
Out[12]=   {5.59118, 5.01064}
```

```
In[13]:=   N[{Sum[1/n, {n,1,400}], Log[400]}]
Out[13]=   {6.56993, 5.99146}
```

You might be able to see why these sums behave this way. Think about it. If you want to see another intriguing reason why the infinite sum

$$1 + 1/2 + 1/3 + 1/4 + 1/5 + ... + 1/n + ...$$

does not converge, then read on. Some people love this argument.

We'll assume it does converge and then try for a contradiction. To this end suppose it converges to a definite value $s$. Thus

$$s = 1 + 1/2 + 1/3 + 1/4 + 1/5 + ... + 1/n + ...$$

Divide through by 2 to get

$$s/2 = 1/2 + 1/4 + 1/6 + 1/8 + 1/10 + ... + 1/(2n) + ...$$

Subtract the expression for $s/2$ from the expression for $s$. This gives

$$s/2 = s - s/2$$

$$= 1 + 1/3 + 1/5 + 1/7 + ... + 1/(2n+1) + ...$$

We get **two** expressions for $s/2$ :

$$s/2 = 1 + 1/3 + 1/5 + 1/7 + 1/9 + ...$$

$$s/2 = 1/2 + 1/4 + 1/6 + 1/8 + 1/10 + ...$$

Note that each term in the first expression for $s/2$ is strictly greater than the corresponding term in the second expression. The inescapable conclusion is that

$$s/2 > s/2.$$

But there is no number $s$ with this property. This means that there is no number s with

$$s = 1 + 1/2 + 1/3 + 1/4 + 1/5 + ... + 1/n + ...$$

and we have another reason why the harmonic series

$$1 + 1/2 + 1/3 + 1/4 + 1/5 + ... + 1/n + ...$$

cannot converge.

■ **T.5) Taylor's formula.**

Taylor's formula in its raw form is not a good calculational device. For example, look at:

```
In[14]:=   Series[E^(3 x) Cos[5 x],{x,0,9}]
```

```
Out[14]=
                         2        3    161 x
         1 + 3 x - 8 x  - 33 x  - ------ +
                                     6

              5          6          7
         239 x     2444 x     5713 x
         ------ + ------- + ------- -
           10        45        210

              8           9
         31679 x    17831 x            10
         -------- - -------- + O[x]
           2520        840
```

To get the coefficients Taylor's formula, you look at:

```
In[15]:=   coeffs = Table[(D[E^(3 x)
           Cos[5 x]/k!,{x,k}])/.x->0,
           {k,0,9}]
```

```
Out[15]=                     161   239  2444
         {1, 3, -8, -33, -(---), ---, ----,
                            6     10    45

          5713     31679      17831
          ----, -(-----), -(-----)}
           210     2520       840
```

*Mathematica* takes longer to calculate the list of derivatives than it took to give the expansion directly. The reason is that *Mathematica* did not use Taylor's formula. Instead it took:

```
In[16]:=   factor1 = Series[E^(3 x),{x,0,9}]
```

```
Out[16]=
                     2     3      4      5
                   9 x   9 x   27 x   81 x
         1 + 3 x + ---- + ---- + ----- + ----- +
                    2     2      8      40

            6       7       8       9
          81 x   243 x   729 x   243 x
          ----- + ------ + ------ + ------ +
           80     560    4480    4480

          10
         O[x]
```

```
In[17]:=   factor2 = Series[Cos[5 x],{x,0,9}]
```

```
Out[17]=
                2        4        6
             25 x    625 x   3125 x
         1 - ----- + ------ - ------- +
              2        24       144

               8
         78125 x            10
         -------- + O[x]
           8064
```

And then it multiplied the two factors together dropping in the product all the $x^n$ terms for $n > 9$:

```
In[18]:=   product =
           ExpandAll[factor1 factor2]
```

```
Out[18]=
                         2        3    161 x
         1 + 3 x - 8 x  - 33 x  - ------ +
                                     6

              5          6          7
         239 x     2444 x     5713 x
         ------ + ------- + ------- -
           10        45        210

              8           9
         31679 x    17831 x            10
         -------- - -------- + O[x]
           2520        840
```

This agrees with:

```
In[19]:=   Series[E^(3 x) Cos[5 x],
           {x,0,9}]
```

```
Out[19]=
                         2        3    161 x
         1 + 3 x - 8 x  - 33 x  - ------ +
                                     6

              5          6          7
         239 x     2444 x     5713 x
         ------ + ------- + ------- -
           10        45        210

              8           9
         31679 x    17831 x            10
         -------- - -------- + O[x]
           2520        840
```

This method is quick, easy and is the method you should use in hand calculations.

**T.5.a)**

How does Taylor's formula guarantee that it is O.K. to get expansions by devices other than Taylor's formula?

**Answer:**

Taylor's formula tells us that there is a definite formula for expansions.

```
In[1]:=    Clear[f]
           Series[f[x],{x,0,7}]
```

```
Out[1]=
                                 2
                            f''[0] x
         f[0] + f'[0] x + --------- +
                              2

          (3)   3    (4)   4
         f   [0] x  f   [0] x
         ---------- + ---------- +
             6            24

          (5)   5    (6)   6
         f   [0] x  f   [0] x
         ---------- + ---------- +
            120          720

          (7)   7
         f   [0] x            8
         ---------- + O[x]
            5040
```

The fact there is such a formula means that there is only

one possibility for an expansion in powers of $x$. As a result, you always get the same expansion for a given function no matter how you go about it.

So get it any way you can!

### T.5.b): Taylor's formula in reverse.

Here is the expansion in powers of $x$ of

$f[x] = 1/(1 + x - x^2)$ through the $x^{12}$ term:

```
In[1]:=    f[x_] = 1/(1 - x - x^2)
           expan12 =
           Normal[Series[f[x],{x,0,12}]]
```
```
Out[1]=               2     3       4       5
           1 + x + 2 x + 3 x + 5 x + 8 x +

                   6      7       8       9
              13 x + 21 x + 34 x + 55 x +

                   10        11        12
              89 x   + 144 x   + 233 x
```

Use this to give a table of the values of the derivatives

$$\{f'[0], f''[0], f'''[0], ..., f^{(11)}[0], f^{(12)}[0]\}$$

### Answer:

Taylor's formula tells us that the table is:

```
In[2]:=    Table[k! Coefficient[expan12,x^k],
           {k,1,12}]
```
```
Out[2]=    {1, 4, 18, 120, 960, 9360, 105840,

            1370880, 19958400, 322963200,

            5748019200, 111607372800}
```

Check:

```
In[3]:=    Table[D[f[x],{x,k}]/.x->0,{k,1,12}]
```
```
Out[3]=    {1, 4, 18, 120, 960, 9360, 105840,

            1370880, 19958400, 322963200,

            5748019200, 111607372800}
```

Note how *Mathematica* grinds on the last one because along the way it has to get some hairy things.

# Literacy Sheet

**1.** Give an example of a power series.

.

.

**2.** Give convergence intervals fof the following power series:

$1 + x + x^2 + x^3 + x^4 + x^5 + \ldots + x^n + \ldots$

.

.

$1 + 2x + 4x^2 + 8x^3 + 16x^4 + 32x^5 + \ldots + 2^n x^n + \ldots$

.

.

$1 + x + x^2/5^2 + x^3/5^3 + x^4/5^4 + \ldots + x^n/5^n + \ldots$

.

.

$1 + x + x^2/2 + x^3/3 + x^4/4 + x^5/5 + \ldots + x^n/n + \ldots$

.

.

$1 + x + x^2/2! + x^3/3! + x^4/4! + x^5/5! + \ldots + x^n/n! + \ldots$

.

.

Calculate the limits:

**3.** $\lim_{x \to 0} \sin[x]/x$

.

.

**4.** $\lim_{x \to 0} \sin[3x^2]/x^2$

.

.

**5.** $\lim_{x \to 0} \sin[3x^4]/\sin[2x]^4$

.

.

**6.** $\lim_{x \to 0} (1 + x - e^x)/x^2$

.

.

**7.** $\lim_{x \to 0} (1 - x - e^{-x})/(1 - \cos[x])$

.

.

**8.** What is Taylor's formula?

.

.

**9.** If $f[0] = 0$ and $f'[0] = 7$, then what are $\lim_{x \to 0} \sin[x]/f[x]$ and $\lim_{x \to 0} \sin[x^5]/f[x^5]$?

.

.

**10.** Give the functions with the following power series:

$1 + x + x^2 + x^3 + x^4 + x^5 + \ldots + x^n + \ldots$

.

$1 - x + x^2 - x^3 + x^4 - x^5 + \ldots + (-1)^n x^n + \ldots$

.

$1 - x^2 + x^4 - x^6 + x^8 - x^{10} + \ldots + (-1)^n x^{2n} + \ldots$

.

$1 + x + x^2/2! + x^3/3! + x^4/4! + \ldots + x^n/n! + \ldots$

.

$1 - x^2/2! + x/4! - x^6/6! + \ldots + (-1)^n x^{2n}/n! + \ldots$

.

$1 - x^2 + x^4/2! - x^6/3! + \ldots + (-1)^n x^{2n}/n! + \ldots$

.

$x - x^3/3! + x^5/5! - x^7/7! + \ldots + (-1)^n x^{2n+1}/(2n+1)! + \ldots$

.

$1 + 2x + 3x^2 + 4x^3 + 5x^4 + 6x^5 + \ldots + (n+1)x^n + \ldots$

.

$x - x^2/2 + x^3/3 + x^4/4 - x^5/5 + \ldots + (-1)^n x^n/n + \ldots$

.

**11.** Give the value of the following infinite sums:

$\sum_0^\infty 1/2^n$

.

.

$\sum_0^\infty 1/n!$

.

.

$\sum_0^\infty (-1)^n \pi^{2n}/(2n)!$

$\sum_0^\infty n/2^n$

**12.** Why would it be a real sign of calculus illiteracy if you try to find

$$\lim_{x \to \infty} f[x]/g[x]$$

by expanding both $f[x]$ and $g[x]$ in powers of $x$ and then trying to analyze the quotient of the expansions?

What should you do before you try to use expansions?

**13.** Suppose

$$f[x] = e + e^2 x + e^2 x^2 + e^3 x^3 \ldots + e^k x^k + \ldots.$$

What is the value of $f[0]$?

What is the value of $f'[0]$?

What is the value of $f''[0]$?

What is a formula for $f^{(k)}[0]$?

What is the equation of the tangent through $\{0, f[0]\}$?

**14.** Write out the expansions in powers of $x$ of $e^{-x^2}$ and $\cos[3x]$ through the $x^4$ terms.

Multiply the two partial expansions to obtain the expansion in powers of $x$ of $e^{-x^2} \cos[3x]$ through the $x^4$ terms.

Use Taylor's formula to obtain the expansion in powers of $x$ of $e^{-x^2} \cos[3x]$ through the $x^4$ term.

**15.** Is the infinite sum

$$1 + 1/2 + 1/3 + 1/4 + 1/5 + 1/6 + \ldots + 1/n + \ldots$$

convergent?

**16.** Discuss:

**a.)** The role of the geometric series in the explanation of the Basic Convergence Principle.

**b.)** The role of the Basic Convergence Principle in the explanation of the Ratio Test.

# Trustworthy plots

## Guide

A lot of engineering activity is in the direction of finding a differential equation that models a physical process at hand and then using mathematics to solve the resulting differential equation. Often it is not necessary to have the precise form of the solution and in these cases sometimes engineers and scientists are quite content with approximations coming from the sum of the first terms of the expansion of the true solution.

But the question comes up:

For what $x's$ should this work?

This section attempts to deal with this question by showing how to get partial expansions of solutions of a variety of differential equations and then trying to use the partial expansions to find empirically trustworthy plots of the true solutions.

## Basics

### ■ B.1) Getting reasonably trustworthy plots of power series.

The Basic Convergence Principle tells us that the power series

$$1 + x + x^2/2^2 + x^3/3^3 + x^4/4^4 + ... + x^n/n^n + ...$$

converges on every interval $[-r, r]$ provided $0 \le r < \infty$.

This is very satisfying, but it ignores the issue of how the function $f[x]$ to which

$$1 + x + x^2/2^2 + x^3/3^3 + x^4/4^4 + ... + x^n/n^n + ...$$

converges looks. To get an idea of how $f[x]$ looks, let's plot some partial expansions on the same axes on as large an interval as we can find on which all three plots share the same ink:

```
In[1]:=   Clear[x,n]
          sum8 = 1 + Sum[x^n/n^n, {n,1,8}]
```
```
Out[1]=            2    3    4     5      6
                  x    x    x     x      x
          1 + x + -- + -- + --- + ---- + ----- +
                  4    27   256   3125   46656

                 7       8
                x       x
          ------- + --------
          823543    16777216
```

```
In[2]:=   sum11 = 1 + Sum[x^n/n^n, {n,1,11}]
```
```
Out[2]=            2    3    4     5      6
```

```
                  x    x    x     x      x
          1 + x + -- + -- + --- + ---- + ----- +
                  4    27   256   3125   46656

                 7       8          9
                x       x          x
          ------- + -------- + --------- +
          823543    16777216   387420489

                 10            11
                x             x
          ------------ + -------------
          10000000000    285311670611
```

```
In[3]:=   sum13 = 1 + Sum[x^n/n^n, {n,1,13}]
```
```
Out[3]=            2    3    4     5      6
                  x    x    x     x      x
          1 + x + -- + -- + --- + ---- + ----- +
                  4    27   256   3125   46656

                 7       8          9
                x       x          x
          ------- + -------- + --------- +
          823543    16777216   387420489

                 10            11
                x             x
          ------------ + ------------- +
          10000000000    285311670611

                 12                 13
                x                  x
          -------------- + ----------------
          8916100448256    302875106592253
```

```
In[4]:=   Plot[{sum8,sum11,sum13},{x,-4,4}];
```

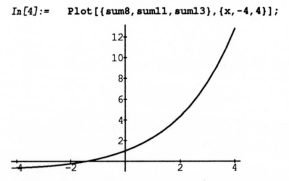

Sharing ink all the way. Let's go for a longer interval:

```
In[5]:=   plot = Plot[{sum8,sum11,sum13},{x,-12,12},
          PlotRange->All];
```

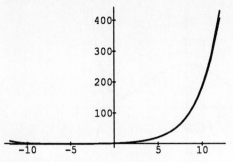

Split ends. Let's cut back the interval; say to $-7 \le x \le 7$:

```
In[6]:=    Show[plot,PlotRange->{{-7,7},{-10,50}}];
```

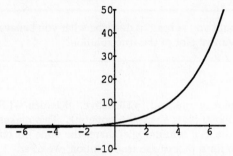

This is a reasonably trustworthy plot of $f[x]$.

### B.1.a)

Conventional wisdom often says that to get a good plot of $f[x]$ over any interval, just choose a large number like 10 and plot the partial expansion with that many terms. Why is this not always reliable?

**Answer:**

Look at:

```
In[1]:=    sum10=Sum[x^n/n^n, {n,1,10}]
Out[1]=
            2    3    4     5      6
           x    x    x     x      x
       x + -- + -- + --- + ---- + ----- +
           4    27   256   3125   46656

            7          8          9
           x          x          x
       ------ + --------- + --------- +
       823543   16777216   387420489

          10
         x
       -----------
       10000000000
```

This is a tenth degree polynomial and the $x^{10}$ term at the end exerts its influence for larger $x's$ and pulls the graph of **sum10** away from the graph of $f[x]$.

The trick is to plot several partial sums together. Where they share ink will tell you what $f[x]$ looks like.

■ **B.2) Plotting the solution of a differential equation.**

### B.2.a.i)

Find the expansions in powers of $x$ through the eighth and tenth degree terms of the solution of

$$y'[x] = -xy[x]$$

with $y[0] = 1$. Plot them on the same axes.

**Answer:**

Set:

```
In[1]:=    Clear[a,n,j,x]
           n = 10
Out[1]=    10
```

Here is the partial expansion for $y$ :

```
In[2]:=    yexpan10 =
           (Sum[a[j] x^j,{j,0,n}]+O[x]^(n+1))/.
           a[0]->1
Out[2]=
                               2          3
           1 + a[1] x + a[2] x  + a[3] x  +

                 4          5          6
           a[4] x  + a[5] x  + a[6] x  +

                 7          8          9
           a[7] x  + a[8] x  + a[9] x  +

                  10       11
           a[10] x   + O[x]
```

The reason we take $a[0] = 1$ is that by Taylor's Formula $1 = y[0] = a[0]$. Our differential equation is $y'[x] = -xy[x]$:

```
In[3]:=    left = D[yexpan10,x]
Out[3]=
                                        2
           a[1] + 2 a[2] x + 3 a[3] x  +

                   3          4          5
           4 a[4] x  + 5 a[5] x  + 6 a[6] x  +

                   6          7          8
           7 a[7] x  + 8 a[8] x  + 9 a[9] x  +

                    9         10
           10 a[10] x  + O[x]
```

```
In[4]:=    right = - x yexpan10
Out[4]=
                         2          3          4
           -x - a[1] x  - a[2] x  - a[3] x  -

                 5          6          7
           a[4] x  - a[5] x  - a[6] x  -

                 8          9          10
           a[7] x  - a[8] x  - a[9] x   -

                  11        12
           a[10] x   + O[x]
```

Set the left side equal to the right side and equate the coefficients of like powers of $x$ : (Sound familiar?)

```
In[5]:=    coeffeqns = LogicalExpand[left == right]
```

```
Out[5]=    a[1] == 0 && 1 + 2 a[2] == 0 &&

           a[1] + 3 a[3] == 0 &&

           a[2] + 4 a[4] == 0 &&

           a[3] + 5 a[5] == 0 &&

           a[4] + 6 a[6] == 0 &&

           a[5] + 7 a[7] == 0 &&

           a[6] + 8 a[8] == 0 &&

           a[7] + 9 a[9] == 0 &&

           a[8] + 10 a[10] == 0
```

Solve these equations for the coefficients:

```
In[6]:=    coeffsolved = Solve[coeffeqns]
Out[6]=                              1
           {{a[1] -> 0, a[2] -> -(-), a[3] -> 0,
                              2

              1                      1
           a[4] -> -, a[5] -> 0, a[6] -> -(--),
              8                      48

                         1
           a[7] -> 0, a[8] -> ---, a[9] -> 0,
                        384

              1
           a[10] -> -(----)}}
              3840
```

Substitute these into the expansion of for $y$ :

```
In[7]:=    yexpan10/.coeffsolved[[1]]
Out[7]=       2    4    6     8     10
           x    x    x    x    x            11
       1 - -- + -- - -- + --- - ---- + O[x]
           2    8    48   384   3840
```

Chop the error term to get the expansion through the tenth degree term:

```
In[8]:=    y10 = Normal[yexpan10/.coeffsolved[[1]]]
Out[8]=       2    4    6     8     10
           x    x    x    x    x
       1 - -- + -- - -- + --- - ----
           2    8    48   384   3840
```

To get the eighth degree approximate solution just chop:

```
In[9]:=    y8 = Normal[Series[y10,{x,0,8}]]
Out[9]=       2    4    6    8
           x    x    x    x
       1 - -- + -- - -- + ---
           2    8    48   384
```

Now plot **y8** and **y10** on the same axes.

```
In[10]:=   plot = Plot[{y8,y10},{x,-3,3}];
```

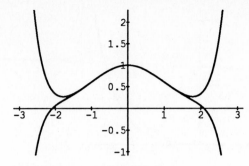

Lots of love and lots of hate.

### B.2.a.ii)

What plot would you submit to describe what you believe to be a trustworthy plot of the true solution?

### Answer:

Look at the plot of **y8** and **y10** above. Between −1.5 and 1.5, the above plots share the same ink. This makes us think that the true solution $y[x]$ shares the same ink on $[-1.5, 1.5]$. So for a plot of the true solution, we offer:

```
In[1]:=    Plot[y10,{x,-1.5,1.5},Axes->{0,0},
           PlotRange->{{-2.5,2.5},{-.2,1.2}},
           PlotLabel->
           "Trustworthy plot"];
```

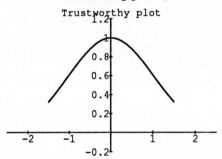

### B.2.a.iii)

One special feature of this problem is that we can find the actual solution. All we do is write down the differential equation

$$y'[x] = -xy[x]$$

with $y[0] = 1$ , separate the variables

$$y'[x]/y[x] = -x$$

and integrate

$$\int_0^x y'[x]/y[x]dx = \int_0^x -xdx.$$

.This gives

$$\log[y[x]] - \log[y[0]] = -x^2/2$$

$$\log[y[x]] - \log[1] = -x^2/2$$

$$\log[y[x]] = -x^2/2$$

$$y[x] = e^{-x^2/2}$$

as the true solution of the differential equation $y'[x] = -xy[x]$ with $y[0] = 1$.

Plot this true solution with your trustworthy plot over an interval in which you have confidence in your trustworthy plot.

---

**Answer:**

*In[1]:=*  `Plot[{E^(-(x^2)/2),y10},{x,-1.5,1.5},`
           `PlotRange->{{-2.5,2.5},{-.2,1.2}}];`

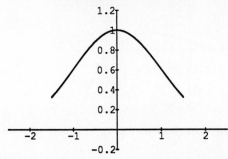

Sharing ink all the way! The method works.

■ **B.3) Plotting the solution of a second order differential equation.**

**B.3.a.i)**

---

Give a reasonably trustworthy plot on an interval centered on 0 of the solution of

$$y'' - xy' + 2y = e^{-x}$$

with $y[0] = 1$ and $y'[0] = 2$.

---

**Answer:**

First choose $n$ :

*In[1]:=*  `Clear[a,n,j,x,y,coeffs,coeffsolved]`
           `n = 12`
*Out[1]=*  `12`

And then write the expansion for $y$ :

*In[2]:=*  `yexpan12 =`
           `(Sum[a[j] x^j,{j,0,n}] +`
           `O[x]^(n+1))/.{a[0]->1,a[1]->2}`
*Out[2]=*
```
             2        3        4
1 + 2 x + a[2] x + a[3] x + a[4] x +

     5        6        7
a[5] x + a[6] x + a[7] x +

     8        9         10
a[8] x + a[9] x + a[10] x +

     11       12       13
```

```
a[11] x    + a[12] x    + O[x]
```

The reason we take $a[0] = 1$ and $a[1] = 2$ is that Taylor's Formula tells us that

$$1 = y[0] = a[0]$$

and

$$2 = y'[0] = a[1].$$

The differential equation is $y'' - xy' + 2y = e^{-x}$.

*In[3]:=*  `left =`
           `D[yexpan12,{x,2}]-x D[yexpan12,x] +`
           `2 yexpan12`
*Out[3]=*  `(2 + 2 a[2]) + (2 + 6 a[3]) x +`

```
                2                       3
    12 a[4] x + (-a[3] + 20 a[5]) x +

                        4
    (-2 a[4] + 30 a[6]) x +

                        5
    (-3 a[5] + 42 a[7]) x +

                        6
    (-4 a[6] + 56 a[8]) x +

                        7
    (-5 a[7] + 72 a[9]) x +

                         8
    (-6 a[8] + 90 a[10]) x +

                          9
    (-7 a[9] + 110 a[11]) x +

                            10        11
    (-8 a[10] + 132 a[12]) x   + O[x]
```

*In[4]:=*  `right = Series[E^(-x),{x,0,n}]`
*Out[4]=*
```
         2   3    4     5     6
        x   x    x     x     x
1 - x + -- - -- + -- - --- + --- -
        2   6    24   120   720

   7      8       9       10
  x      x       x       x
---- + ----- - ------ + ------- -
5040   40320   362880   3628800

    11         12
   x          x          13
-------- + --------- + O[x]
39916800   479001600
```

Set the left side equal to the right side and equate the coefficients of like powers of $x$ and solve these equations for the coefficients: :

*In[5]:=*  `coeffeqns =`
           `LogicalExpand[left == right]`
           `coeffsolved = Solve[coeffeqns]`
*Out[5]=*
```
              1                1
{{a[2] -> -(-), a[3] -> -(-),
              2                2

                1               1
     a[4] -> --, a[5] -> -(--),
             24            30
```

$$a[6] \to \frac{1}{240}, \quad a[7] \to -\left(\frac{13}{5040}\right),$$

$$a[8] \to \frac{13}{40320}, \quad a[9] \to -\left(\frac{11}{60480}\right),$$

$$a[10] \to \frac{79}{3628800},$$

$$a[11] \to -\left(\frac{463}{39916800}\right),$$

$$a[12] \to \frac{211}{159667200}\}\}$$

Substitute these into the expression for **yexpan12** :

```
In[6]:=   y12 =
          Normal[yexpan12/.coeffsolved[[1]]]
```

```
Out[6]=                2    3    4    5    6
                       x    x    x    x    x
          1 + 2 x - -- - -- + -- - -- + --- -
                     2    2    24   30   240

              7        8        9       10
          13 x     13 x     11 x     79 x
          ----- + ----- - ------ + ------- -
          5040     40320    60480   3628800

               11           12
          463 x        211 x
          --------- + ----------
          39916800    159667200
```

Notice that if we use just the expansion through the $x^{11}$ term then the this partial expansion will end in a negative term and will pull abruptly away from the plot of **y12** .

```
In[7]:=   y11 = Normal[Series[y12,{x,0,11}]]
```

```
Out[7]=                2    3    4    5    6
                       x    x    x    x    x
          1 + 2 x - -- - -- + -- - -- + --- -
                     2    2    24   30   240

              7        8        9       10
          13 x     13 x     11 x     79 x
          ----- + ----- - ------ + ------- -
          5040     40320    60480   3628800

               11
          463 x
          ---------
          39916800
```

Now plot both on the same axes:

```
In[8]:=   Plot[{y11,y12},{x,-4,4},
          PlotRange->{-40,40}];
```

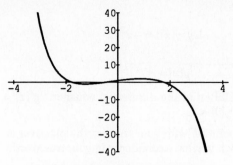

A reasonably trustworthy plot seems to be:

```
In[9]:=   Plot[y12,{x,-2.5,2.5},PlotRange->All];
```

Let's put an additional check on our approximate solution **y12** .

The true solution $y[x]$ satisfies

$$y[x]'' - xy[x]' + 2y[x] = e^{-x}.$$

To see how good **y12** is on this interval, we'll plot

$$(y12)'' - x(y12)' + 2 y12$$

and $e^{-x}$ on this interval:

```
In[10]:=  test =
          Expand[D[y12,{x,2}]
          - x D[y12,x] + 2 y12]
```

```
Out[10]=               2    3    4     5     6
                       x    x    x     x     x
          1 - x + -- - -- + -- - --- + --- -
                   2    6    24   120   720

             7      8       9        10
            x      x       x        x
          ---- + ----- - ------ + ------- +
          5040   40320   362880   3628800

                11          12
           463 x       211 x
          ------- - ---------
          4435200   15966720
```

```
In[11]:=  Plot[{test,E^(-x)},{x,-3,3},
          PlotRange->All];
```

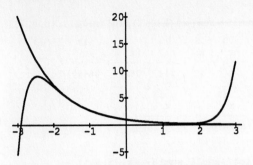

This means **y12** does a pretty good job for $-2 \le x \le 2$. So maybe a more conservative plot of the true solution is:

*In[12]:=* **Plot[y12,{x,-2,2}];**

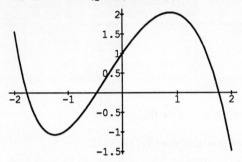

### B.3.a.ii)

Give an estimate of the the largest value of the solution on the interval $[0, 2]$.

**Answer:**

The plot above indicates that the maximizer seems to be close to $x = 0.9$.

*In[1]:=* **deriv = D[y12,x]**

*Out[1]=*
$$2 + x + \frac{x^2}{2} + \frac{x^3}{2} + \frac{x^4}{12} + \frac{13 x^5}{120} + \frac{x^6}{80} +$$
$$\frac{79 x^7}{5040} + \frac{31 x^8}{20160} + \frac{211 x^9}{120960} + \frac{557 x^{10}}{3628800} +$$
$$\frac{6331 x^{11}}{39916800}$$

*In[2]:=* **FindRoot[deriv,{x,0.9}]**

*Out[2]=* **{x -> -1.36545}**

We estimate the that the solution is largest at $x = 0.874$. The corresponding maximum value is about:

*In[3]:=* **y12/.x->0.874**

*Out[3]=* **3.33216**

### ◼ B.4) A perspective.

Many differential equations texts make a big deal of finding the general term of the series expansion of a solution to a differential equation and then using the ratio test to find intervals of convergence. In fact advanced theory tells us that in all three of the above problems, the series expansion converges to the true solution of every interval $[-r, r]$ provided $0 \le r < \infty$. This is fact is comforting in theory. In practice, it does not mean much because *expansions in powers of x usually converge too slowly to give us any real feeling for the behavior of true solution except on small intervals centered at 0.*

For studying behavior in the large, many advanced methods are available. When you take a course in differential equations, you should learn about some of these methods; if you don't see them in your course, demand that the instructor include some of them.

For good independent reading, look at the book, *Numerical Methods and Software* , by David Kahaner, Cleve Moler and Stephen Nash (Prentice Hall, 1989).

## Tutorial

### ◼ T.1) Empirical plots of power series.

### T.1.a)

If the signs of the terms alternate, you can get a pretty good empirical plot of a function defined by a power series. Give a plot you believe to be trustworthy of

$$f[x] = \sum_{0}^{\infty} (-1)^n x^{2n+1} / (n!(n+1)!).$$

**Answer:**

*In[1]:=* **Clear[x,n]**
**sum13 =**
**Sum[((-1)^(n)) x^(2n+1)/(n!(n+1)!),**
**{n,0,6}]**

*Out[1]=*
$$x - \frac{x^3}{2} + \frac{x^5}{12} - \frac{x^7}{144} + \frac{x^9}{2880} - \frac{x^{11}}{86400} +$$
$$\frac{x^{13}}{3628800}$$

*In[2]:=* **sum15 =**
**Sum[((-1)^(n)) x^(2n +1)/(n!(n+1)!),**
**{n,0,7}]**

*Out[2]=*
$$x - \frac{x^3}{2} + \frac{x^5}{12} - \frac{x^7}{144} + \frac{x^9}{2880} - \frac{x^{11}}{86400} +$$
$$\frac{x^{13}}{} \qquad \frac{x^{15}}{}$$

```
------- - ---------
3628800    203212800
```

The plus sign at the end of **sum13** and the minus sign at the end of **sum15** will help to show off the differences in the two plots.

*In[3]:=*    `Plot[{sum13, sum15}, {x, -3, 3}];`

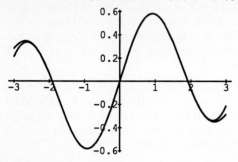

The high degree terms are taking over near the ends. A reasonably trustworthy plot of $f[x]$ seems to be:

*In[4]:=*    `Plot[sum13, {x, -2.4, 2.4}];`

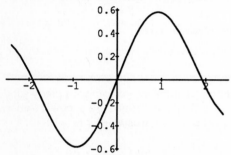

The plot looks something like the plot of $\sin[x]$.

### T.1.b)

If the signs of the terms do not alternate, you can still get a pretty good empirical plot of a function defined by a power series.

Give a plot you believe to be trustworthy of

$$f[x] = \sum_{0}^{\infty} x^{2n+1}/(n!(n+1)!).$$

### Answer:

Because all the signs are plus signs, we won't use consecutive sums.

*In[1]:=*    `Clear[x,n]`
             `sum11 = Sum[ x^(2 n+1)/(n! (n+1)!), {n, 0, 5}]`

*Out[1]=*
$$x + \frac{x^3}{2} + \frac{x^5}{12} + \frac{x^7}{144} + \frac{x^9}{2880} + \frac{x^{11}}{86400}$$

*In[2]:=*    `sum15 = Sum[ x^(2 n+1)/(n! (n+1)!), {n, 0, 7}]`

*Out[2]=*
$$x + \frac{x^3}{2} + \frac{x^5}{12} + \frac{x^7}{144} + \frac{x^9}{2880} + \frac{x^{11}}{86400} +$$
$$\frac{x^{13}}{3628800} + \frac{x^{15}}{203212800}$$

*In[3]:=*    `Plot[{sum11, sum15}, {x, -5, 5}];`

Cut the interval to eliminate the split ends.

*In[4]:=*    `Plot[{sum11, sum15}, {x, -3, 3}];`

Good. A reasonably trustworthy plot of $f[x]$ appears to be:

*In[5]:=*    `Plot[sum15, {x, -3, 3}];`

Up, up and away.

### ■ T.2) Two more differential equations.

### T.2.a.i)

Give a reasonably trustworthy plot on an interval centered

on 0 of the solution $y$ of

$$y'' - xy' + 2\sin[x]y = e^x$$

with $y[0] = 1$ and $y'[0] = -1$.

---

**Answer:**

```
In[1]:=    Clear[a,x,y]
           n = 12
Out[1]=    12
```

And then write the expansion for $y$:

```
In[2]:=    yexpan12 =
           (Sum[a[j] x^j,{j,0,n}] +
           O[x]^(n + 1))/.{a[0]->0,a[1] ->-1}
```

$$Out[2]= \quad -x + a[2]\ x^2 + a[3]\ x^3 + a[4]\ x^4 +$$

$$a[5]\ x^5 + a[6]\ x^6 + a[7]\ x^7 +$$

$$a[8]\ x^8 + a[9]\ x^9 + a[10]\ x^{10} +$$

$$a[11]\ x^{11} + a[12]\ x^{12} + O[x]^{13}$$

The reason we take $a[0] = 1$ and $a[1] = -1$ is that Taylor's formula tells us that

$$1 = y[0] = a[0]$$

and

$$-1 = y'[0] = a[1].$$

The differential equation is

$$y'' - xy' + 2\sin[x]y = e^x.$$

```
In[3]:=    left =
           D[yexpan12,{x,2}] - x D[yexpan12,x]
           +2*yexpan12*Series[Sin[x],{x,0,n}]
           right = Series[E^x,{x,0,n}]
           coeffeqns = LogicalExpand[left == right]
           coeffsolved = Solve[coeffeqns]
```

$$Out[3]= \quad \{\{a[2] \to \frac{1}{2},\ a[3] \to 0,\ a[4] \to \frac{1}{8},$$

$$a[5] \to \frac{1}{120},\ a[6] \to \frac{13}{720},$$

$$a[7] \to \frac{1}{840},\ a[8] \to \frac{79}{40320},$$

$$a[9] \to \frac{43}{362880},\ a[10] \to \frac{211}{1209600},$$

$$a[11] \to \frac{97}{9979200},\ a[12] \to \frac{6331}{479001600}\}$$

$$\}$$

Substitute these into the expression for **yexpan12** .

```
In[4]:=    y12 = Normal[yexpan12/.coeffsolved[[1]]]
```

$$Out[4]= \quad -x + \frac{x^2}{2} + \frac{x^4}{8} + \frac{x^5}{120} + \frac{13\ x^6}{720} + \frac{x^7}{840} +$$

$$\frac{79\ x^8}{40320} + \frac{43\ x^9}{362880} + \frac{211\ x^{10}}{1209600} + \frac{97\ x^{11}}{9979200} +$$

$$\frac{6331\ x^{12}}{479001600}$$

Now chop to get the expansion through the $x^{11}$ term: We like the fact that the $x^{11}$ term carries a minus sign while the $x^{12}$ term carries a plus sign.

```
In[5]:=    y11 = Normal[Series[y12,{x,0,11}]]
```

$$Out[5]= \quad -x + \frac{x^2}{2} + \frac{x^4}{8} + \frac{x^5}{120} + \frac{13\ x^6}{720} + \frac{x^7}{840} +$$

$$\frac{79\ x^8}{40320} + \frac{43\ x^9}{362880} + \frac{211\ x^{10}}{1209600} + \frac{97\ x^{11}}{9979200}$$

Now plot on the same axes.

```
In[6]:=    Plot[{y11,y12},{x,-3,3}];
```

It looks like a reasonably trustworthy plot of true solution is:

```
In[7]:=    Plot[y12,{x,-2,2}];
```

Let's put an additional check on our approximate solution **y12** . The true solution y satisfies

$$y'' - xy' + 2\sin[x]y = e^x.$$

To see how good **y12** is on this interval, we'll plot

$$(y12)'' - x(y12)' + 2\sin[x]y$$

and $e^x$ on this interval:

```
In[8]:=  test =
         D[y12,{x,2}] - x D[y12,x] + 2 Sin[x] y12
         Plot[{test,E^x},{x,-2,2},PlotRange->All];
```

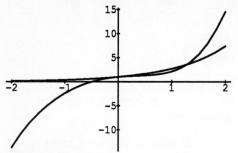

This means **y12** does a pretty good job of solving the differential equation for $-1.4 \leq x \leq 1.4$. So maybe a more conservative plot of the true solution is:

```
In[9]:=  Plot[y12,{x,-1.4,1.4}];
```

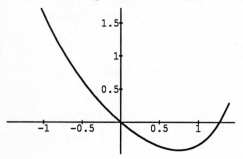

And we're out of here.

### T.2.b)

---

Give a reasonably trustworthy plot on an interval centered on 0 of the solution $y$ of

$$y' + e^y x = y\sin[x]$$

with $y[0] = 0$.

---

**Answer:**

```
In[1]:=  Clear[a,x,y]
```

```
In[2]:=  n = 10
Out[2]=  10
```

And write the expansion for $y$:

```
In[3]:=  yexpan10 =
         (Sum[a[j] x^j,{j,0,n}] +O[x]^(n+1))/.a[0] -
>0
```

```
Out[3]=       2       3       4
         a[1] x + a[2] x + a[3] x + a[4] x +

              5       6       7
         a[5] x + a[6] x + a[7] x +

              8       9        10        11
         a[8] x + a[9] x + a[10] x  + O[x]
```

The reason we take $a[0] = 0$ is that Taylor's formula tells us that $0 = y[0] = a[0]$. The differential equation is $y' + e^y x = y\sin[x]$.

```
In[4]:=  Clear[t]
         left =
         D[yexpan10,x] + x Series[E^t,{t,0,n}]/.
         t->yexpan10
         right = yexpan10 Series[Sin[x],{x,0,n}]
         coeffeqns = LogicalExpand[left == right]
         coeffsolved = Solve[coeffeqns]
```

```
Out[4]=               583
         {{a[10] -> -(-------), a[9] -> 0,
                     1209600

                    1
          a[8] -> ---, a[7] -> 0,
                  480

                     1
          a[6] -> -(---), a[5] -> 0,
                   144

                                         1
          a[4] -> 0, a[3] -> 0, a[2] -> -(-),
                                         2

          a[1] -> 0}}
```

Substitute these into the expression for **yexpan10** .

```
In[5]:=  y10 = Normal[yexpan10/.coeffsolved[[1]]]
Out[5]=    2     6     8        10
         -x     x     x     583 x
         --- - --- + --- - -------
          2    144   480   1209600
```

Now chop to get the expansion through the $x^8$ term:

```
In[6]:=  y8 = Normal[Series[y10,{x,0,8}]]
Out[6]=    2     6     8
         -x     x     x
         --- - --- + ---
          2    144   480
```

Now plot on the same axes.

```
In[7]:=  Plot[{y8,y10},{x,-2,2}];
```

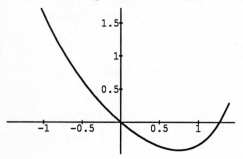

It looks like a reasonably trustworthy plot of the true solution is:

*In[8]:=*    `Plot[y10,{x,-1.5,1.5}];`

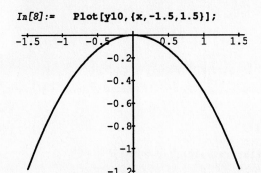

Let's put an additional check on our approximate solution
**y10** . The true solution $y$ satisfies $y' + e^y x = y \sin[x]$. To see
how good **y10** is on this interval, we'll plot $(y10)' + e^{y10}x$
and $(y10) \sin[x]$

on this interval:

*In[9]:=*    `test = D[y10,x] + x E^(y10)`
             `Plot[{test,y10 Sin[x]},{x,-2,2},`
             `PlotRange->All];`

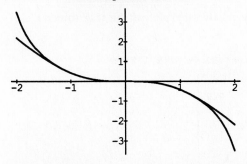

This means **y10** does a pretty good job of solving the differential equation for $-1.4 \le x \le 1.4$. So maybe a more
conservative plot of the true solution is:

*In[10]:=*    `Plot[y10,{x,-1.4,1.4}];`

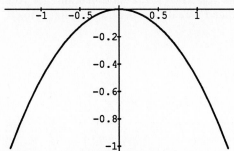

Not bad for beginners.

■ **T.3) Simultaneous differential equations.**

**T.3.a)**

Give reasonably trustworthy plots on an interval centered

on 0 of the solutions $y[x]$ and $z[x]$ of of the system
$$z'[x] = y[x] - z[x]$$
and
$$y'[x] = x + z[x]^2$$
with $y[0] = 0$ and $z[0] = 1$.

---

**Answer:**

*In[1]:=*    `Clear[x,y,z,a,b]`
             `n = 10`
             `yexpan10 =`
             `(Sum[a[j] x^j,{j,0,n}]`
             `+ O[x]^(n+1))/.a[0]->0`

             `zexpan10 =`
             `(Sum[b[j] x^j,{j,0,n}]`
             `+ O[x]^(n+1))/.b[0]->1`

             `left1 = D[zexpan10,x]`
             `right1 = yexpan10 - zexpan10`

             `left2 = D[yexpan10,x]`
             `right2 = x + zexpan10^2`

             `coeffeqns1 =`
             `LogicalExpand[left1 == right1]`
             `coeffeqns2 =`
             `LogicalExpand[left2 == right2]`

             `coeffsolved =`
             `Solve[{coeffeqns1,coeffeqns2}]`

*Out[1]=*
```
                16967              929
{{a[10] -> -(------), b[10] -> -----,
                302400             80640

       2837            2587
a[9] -> -----, b[9] -> -(------),
       30240            120960

        49            529
a[8] -> -(---), b[8] -> -----,
        320           13440

       41            17
a[7] -> ---, b[7] -> -(---),
       168           240

        11            31
a[6] -> -(--), b[6] -> ---,
        30            240

       11            9
a[5] -> --, b[5] -> -(--),
       20            40

        3           3
a[4] -> -(-), b[4] -> -, a[3] -> 1,
        4           8

        1            1
b[3] -> -(-), a[2] -> -(-),
        2            2

b[2] -> 1, a[1] -> 1, b[1] -> -1}}
```

Substitute these into the expressions for *yexpan*10 and
*zexpan*10.

*In[2]:=*    `y10 = Normal[yexpan10/.coeffsolved[[1]]]`

*Out[2]=*
```
     2           4      5      6
    x      3   3 x   11 x   11 x
x - -- + x  - ---- + ----- - ----- +
    2          4      20      11
```

```
     2          4       20      30
    7         8         9              10
  41 x      49 x     2837 x     16967 x
  -----  -  -----  +  -------  -  ---------
   168      320       30240       302400
```

For checking accuracy, we'll bounce this off:

```
In[3]:=    y9 = Normal[Series[y10,{x,0,9}]]

Out[3]=       2          4       5       6
          x         3   3 x    11 x    11 x
      x - -- + x  - ---- + ----- - ----- +
          2          4      20      30

             7       8        9
          41 x    49 x    2837 x
          -----  ----- + -------
           168    320     30240
```

And now for $z[x]$:

```
In[4]:=    z10 = Normal[zexpan10/.coeffsolved[[1]]]

Out[4]=              3      4      5       6
                2   x    3 x    9 x    31 x
      1 - x + x  - -- + ---- - ---- + ----- -
                   2     8     40     240

             7       8        9       10
          17 x    529 x    2587 x    929 x
          ----- + ------ - ------- + -----
           240    13440    120960    80640
```

We'll bounce this off:

```
In[5]:=    z9 = Normal[Series[z10,{x,0,9}]]

Out[5]=              3      4      5       6
                2   x    3 x    9 x    31 x
      1 - x + x  - -- + ---- - ---- + ----- -
                   2     8     40     240

             7       8        9
          17 x    529 x    2587 x
          ----- + ------ - -------
           240    13440    120960
```

Now check for accuracy of the plots:

```
In[6]:=    Plot[{y9,y10},{x,-2,2}];
```

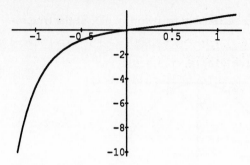

A reasonably trustworthy plot of the true solution $y[x]$ seems to be:

```
In[7]:=    Plot[y10,{x,-1.2,1.2}];
```

Now for $z[x]$:

```
In[8]:=    Plot[{z9,z10},{x,-2,2}];
```

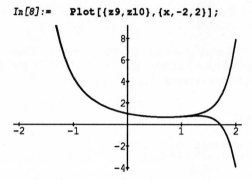

A reasonably trustworthy plot of the true solution $z[x]$ seems to be:

```
In[9]:=    Plot[z10,{x,-1.2,1.2}];
```

Aren't you glad you have *Mathematica* working for you? Doing this by hand would have been a long, long project.

# Literacy Sheet

**1.** Describe the emiprical technique for getting trustworthy plots of functions specified via power series.

.

.

**2.** Can plotting in this way be expected to give accurate results on large intervals? Why?

.

.

**3.** Draw hand sketches of $y = 1 + x$ and

$$y = 1 + x + x^{10} + x^{12} + x^{100} + x^{102}$$

for $-1.2 \leq x \leq 1.2$ on the same axes.

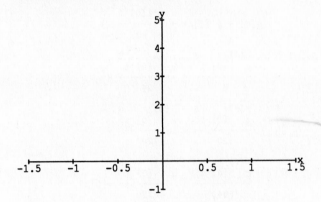

**4.** Describe the emiprical technique for trying to get trustworthy plots of differential equations.

.

.

**5.** When you are trying to use power series to try to get trustworthy plots of solutions of differential equations, why do you find Taylor's formula to be often handy?

.

.

**6.** Use expansions to solve by hand the differential equation $y''[x] = 1 - x$ with $y[0] = 1$ and $y'[0] = 2$. Then check yourself by solving it by performing two integrations.

.

.

.

.

.

# Expansions in powers of (x - a)

## Guide

By now we know how to expand almost any function $f[x]$ we can get our hands on into power series

$$a[0] + a[1]x + a[2]x^2 + a[3]x^3 + ... + a[n]x^n + ...$$

We usually know the intervals on which the series converges to the function and have some idea of how to use the early terms of the series for precise work near zero. But sometimes we have to do precise work near points other than zero.

This is not a big problem because by a simple substitution you can use what you already know to expand functions in an infinite sum

$$a[0] + a[1](x - a) + a[2](x - a)^2$$
$$+a[3](x - a)^3 + ... + a[n](x - a)^n + ...$$

for numbers $a$ other than 0.

## Basics

■ **B.1) Expansions in powers of (x - a).**

Expanding in powers of $(x - a)$ is just a small variation on the theme of expanding in powers of $x$.

**B.1.a)**

Fix a constant $a$ and expand

$$f[x] = 1/(1 - x)$$

in powers of $(x - a)$.

**Answer:**

Here's the strategy:

→ i) Substitute $x = t + a$.

→ ii) Expand in powers of $t$.

→ iii) Replace $t$ by $x - a$.

```
In[1]:=    Clear[x,t,a,f,g]
           f[x_] = 1/(1 - x)
Out[1]=        1
            -----
            1 - x
```

```
In[2]:=    g[t] = f[x]/.x->(t + a)
```

```
Out[2]=        1
            -----------
            1 - (a + t)
```

Expand in powers of $t$:

```
In[3]:=    tExpansion = Series[g[t],{t,0,8}]
```

$$Out[3]= \ g[0] + g'[0]\,t + \frac{g''[0]\,t^2}{2} +$$
$$\frac{g^{(3)}[0]\,t^3}{6} + \frac{g^{(4)}[0]\,t^4}{24} +$$
$$\frac{g^{(5)}[0]\,t^5}{120} + \frac{g^{(6)}[0]\,t^6}{720} +$$
$$\frac{g^{(7)}[0]\,t^7}{5040} + \frac{g^{(8)}[0]\,t^8}{40320} + O[t]^9$$

Replace $t$ by the equivalent expression $x - a$:

```
In[4]:=    xMinusaExpan = tExpansion/.t->(x - a)
```

$$Out[4]= \ \frac{1}{1 - a} + \frac{-a + x}{(1 - a)^2} + \frac{(-a + x)^2}{(1 - a)^3} +$$
$$\frac{(-a + x)^3}{(1 - a)^4} + \frac{(-a + x)^4}{(1 - a)^5} + \frac{(-a + x)^5}{(1 - a)^6} +$$
$$\frac{(-a + x)^6}{(1 - a)^7} + \frac{(-a + x)^7}{(1 - a)^8} + \frac{(-a + x)^8}{(1 - a)^9} +$$
$$O[-a + x]^9$$

This is the beginning of the expansion of $f[x] = 1/(1 - x)$ in powers of $(x - a)$. We can accomplish the same thing with a single *Mathematica* command:

```
In[5]:=    n=8
           Series[f[x],{x,a,n}]
```

$$Out[5]= \ \frac{1}{1 - a} + \frac{-a + x}{(1 - a)^2} + \frac{(-a + x)^2}{} +$$

$$1 - a \qquad (1 - a)^2 \qquad (1 - a)^3$$

$$\frac{(-a + x)^3}{(1 - a)^4} + \frac{(-a + x)^4}{(1 - a)^5} + \frac{(-a + x)^5}{(1 - a)^6} +$$

$$\frac{(-a + x)^6}{(1 - a)^7} + \frac{(-a + x)^7}{(1 - a)^8} + \frac{(-a + x)^8}{(1 - a)^9} +$$

$$O[-a + x]^9$$

The expansion of $f[x] = 1/(1 - x)$ in powers of $(x - a)$ is

$$1/(1 - a) + (x - a)/(1 - a)^2 + (x - a)^2/(1 - a)^3 +$$

$$... + (x - a)^k/(1 - a)^{k+1} + ...$$

This is **not** worth committing to memory.

Note that the expansion $1/(1 - x)$ in powers of $(x - a)$ does not work for $a = 1$.

Why? You guessed it! The expansion of $1/(1 - x)$ in powers of $(x - 1)$ fails because $1/(1 - x)$ has a big fat singularity (blow up) at $x = 1$.

### B.1.b)

Fix a constant $a$ and expand $e^x$ in powers of $(x - a)$.

### Answer:

Here's the strategy:

→ i) Substitute $x = t + a$.

→ ii) Expand in powers of $t$.

→ iii) Replace $t$ by $x - a$.

```
In[1]:=    Clear[x,t,a,g,f]
           f[x_] = E^x
           g[t] = f[x]/.x->(t + a)

Out[1]=    a + t
           E
```

This is the same as $e^a e^t$. Expand in powers of $t$:

```
In[2]:=    tExpansion = Series[g[t],{t,0,6}]
```

$$Out[2]= \quad E^a + E^a\, t + \frac{E^a\, t^2}{2} + \frac{E^a\, t^3}{6} + \frac{E^a\, t^4}{24} +$$

$$\frac{E^a\, t^5}{120} + \frac{E^a\, t^6}{720} + O[t]^7$$

Looks good and feels good. Replace $t$ by the equivalent expression $x - a$:

```
In[3]:=    xMinusaExpan = tExpansion/.t->(x - a)
```

$$Out[3]= \quad E^a + E^a\,(-a + x) + \frac{E^a\,(-a + x)^2}{2} +$$

$$\frac{E^a\,(-a + x)^3}{6} + \frac{E^a\,(-a + x)^4}{24} +$$

$$\frac{E^a\,(-a + x)^5}{120} + \frac{E^a\,(-a + x)^6}{720} +$$

$$O[-a + x]^7$$

This is the beginning of the expansion of $f[x] = e^x$ in powers of $(x - a)$. We can accomplish the same thing with a single *Mathematica* command:

```
In[4]:=    Series[E^x,{x,a,6}]
```

$$Out[4]= \quad E^a + E^a\,(-a + x) + \frac{E^a\,(-a + x)^2}{2} +$$

$$\frac{E^a\,(-a + x)^3}{6} + \frac{E^a\,(-a + x)^4}{24} +$$

$$\frac{E^a\,(-a + x)^5}{120} + \frac{E^a\,(-a + x)^6}{720} +$$

$$O[-a + x]^7$$

The expansion of $e^x$ in powers of $(x - a)$ is

$$e^a + e^a(x - a) + e^a(x - a)^2/2 + e^a(x - a)^3/3! +$$

$$... + e^a(x - a)^n/n! + ...$$

This expansion has a comfortable feeling because it looks so familiar. This is **might** worth committing to memory but only if you are really into this sort of activity.

### B.1.c)

Fix a constant $a$ and expand $\sin[x]$ in powers of $(x - a)$.

### Answer:

Here's the strategy:

→ i) Substitute $x = t + a$.

→ ii) Expand in powers of $t$.

→ iii) Replace $t$ by $x - a$.

```
In[1]:=    Clear[x,t,a,f,g]
           f[x] = Sin[x]
```

*Out[1]=* Sin[x]

Substitute $x = t + a$:

*In[2]:=* `g[t_] = f[x]/.x->(t + a)`

*Out[2]=* Sin[a + t]

An old trig identity tells us that

$$g[t] = \sin[a + t] = \sin[a]\cos[t] + \cos[a]\sin[t].$$

So $g[t]$ is the same as $\sin[a]\cos[t] + \cos[a]\sin[t]$. Expand in powers of $t$:

*In[3]:=* `n = 7`
`tExpansion =`
`Series[g[t],{t,0,n}]`

*Out[3]=*

$$\text{Sin}[a] + \text{Cos}[a]\ t - \frac{\text{Sin}[a]\ t^2}{2} -$$

$$\frac{\text{Cos}[a]\ t^3}{6} + \frac{\text{Sin}[a]\ t^4}{24} + \frac{\text{Cos}[a]\ t^5}{120} -$$

$$\frac{\text{Sin}[a]\ t^6}{720} - \frac{\text{Cos}[a]\ t^7}{5040} + O[t]^8$$

Intriguing. Replace $t$ by the equivalent expression $x - a$:

*In[4]:=* `xMinusaExpan = tExpansion/.t->(x - a)`

*Out[4]=* Sin[a] + Cos[a] (-a + x) -

$$\frac{\text{Sin}[a]\ (-a + x)^2}{2} - \frac{\text{Cos}[a]\ (-a + x)^3}{6} +$$

$$\frac{\text{Sin}[a]\ (-a + x)^4}{24} + \frac{\text{Cos}[a]\ (-a + x)^5}{120} -$$

$$\frac{\text{Sin}[a]\ (-a + x)^6}{720} - \frac{\text{Cos}[a]\ (-a + x)^7}{5040} +$$

$$O[-a + x]^8$$

This is the beginning of the expansion of $\sin[x]$ in powers of $(x - a)$. We can accomplish the same thing with a single *Mathematica* command:

*In[5]:=* `n=7`
`Series[Sin[x],{x,a,n}]`

*Out[5]=* Sin[a] + Cos[a] (-a + x) -

$$\frac{\text{Sin}[a]\ (-a + x)^2}{2} - \frac{\text{Cos}[a]\ (-a + x)^3}{6} +$$

$$\frac{\text{Sin}[a]\ (-a + x)^4}{24} + \frac{\text{Cos}[a]\ (-a + x)^5}{120} -$$

$$\frac{\text{Sin}[a]\ (-a + x)^6}{720} - \frac{\text{Cos}[a]\ (-a + x)^7}{5040} +$$

$$O[-a + x]^8$$

The expansion of $\sin[x]$ in powers of $(x - a)$ is the sum of two familiar-looking expansions

$$\sin[a](1 - (x - a)^2/2 + (x - a)^4/4! - (x - a)^6/6! + \dots$$

$$+ (-1)^k (x - a)^{2k}/(2k)! + \dots$$

and

$$\cos[a]((x - a) - (x - a)^3/3! + (x - a)^5/5! - (x - a)^7/7! + \dots$$

$$+ (-1)^k (x - a)^{2k+1}/(2k + 1)! + \dots$$

This expansion has a comfortable feeling because has a familiar ring. This expansion **is not** worth committing to memory.

### ■ B.2) Taylor's formula for expansions in powers of (x - a).

**B.2.a)**

In each of the expansions in powers $(x - a)$ that we did in B.1) above, we followed the general outline:

→ i) Substitute $x = t + a$.

→ ii) Expand in powers of $t$.

→ iii) Replace $t$ by $x - a$.

Follow the same outline to obtain Taylor's Formula for the expansion of any $f[x]$ in powers of $(x - a)$.

**Answer:**

Take $f[x]$ and substitute $x = t + a$

*In[1]:=* `Clear[f,g,x,t,a]`
`g[t_] = f[t + a]`

*Out[1]=* f[a + t]

Expand in powers of $t$ using the version of Taylor's formula for expansions in powers of $t$:

*In[2]:=* `expant = Series[g[t],{t,0,6}]`

*Out[2]=*

$$f[a] + f'[a]\ t + \frac{f''[a]\ t^2}{2} +$$

$$\frac{f^{(3)}[a]\ t^3}{6} + \frac{f^{(4)}[a]\ t^4}{24} +$$

```
      (5)   5      (6)   6
    f   [a] t    f   [a] t             7
    ---------- + ---------- + O[t]
       120          720
```

Replace $t$ by $(x-a)$ to get the expansion of $f[x]$ in powers of $(x-a)$:

```
In[3]:=    expanxMinusa = expant/.t->(x- a)

Out[3]=    f[a] + f'[a] (-a + x) +

                        2
              f''[a] (-a + x)
              ---------------- +
                     2

               (3)          3
              f   [a] (-a + x)
              ---------------- +
                     6

               (4)          4
              f   [a] (-a + x)
              ---------------- +
                     24

               (5)          5
              f   [a] (-a + x)
              ---------------- +
                     120

               (6)          6
              f   [a] (-a + x)                7
              ---------------- + O[-a + x]
                     720
```

After a moment's thought, we see that matching

$$f[x] = a[0] + a[1](x-a) + a[2](x-a)^2$$

$$+a[3](x-a)^3 + ... + a[n](x-a)^n + ...$$

against the output above, then we arrive at the neat formula

$$a[n] = f^{(n)}[a]/n!$$

for $n = 0,1,2,3,...$ where $f^{(n)}[a]$ stands for the $n$ th derivative of $f[x]$ evaluated at $a$ and we agree that $0! = 1$ ; so that

$$f^{(0)}[a]/0! = f^{(0)}[a]/1 = f[a]/1 = f[a].$$

This is Taylor's formula for expansions in powers of $(x-a)$. It agrees with:

```
In[4]:=    Series[f[x],{x,a,6}]

Out[4]=    f[a] + f'[a] (-a + x) +

                        2
              f''[a] (-a + x)
              ---------------- +
                     2

               (3)          3
              f   [a] (-a + x)
              ---------------- +
                     6

               (4)          4
              f   [a] (-a + x)
              ---------------- +
```

```
                     24
               (5)          5
              f   [a] (-a + x)
              ---------------- +
                     120

               (6)          6
              f   [a] (-a + x)                7
              ---------------- + O[-a + x]
                     720
```

It's a sweet life.

### ■ B.3 A perspective.

If you have examined B.1) and B.2) then you have seen a consistent procedure in dealing with expansions in powers of (x - a):

→ i) Substitute $x = t + a$.

→ ii) Expand in powers of $t$.

→ iii) Replace $t$ by $x - a$.

Does this mean that expansions in powers of $(x-a)$ is really a new inquiry? Or does it mean that that expansions in powers of $(x-a)$ is just a footnote within the theory of expansions in powers of $x$?

---

**Answer:**

It looks as if you'll have to agree that expansions in powers of $(x-a)$ is just a footnote within the theory of expansions in powers of $x$.

### ■ B.4) Convergence of expansions in powers of ( x - a).

### B.4.a)

---

How do we determine the intervals of convergence for expansions in powers of $(x-a)$?

---

**Answer:**

Instead of looking at $f[x]$ as a function of a real number $x$, we try to see how $f[x], f'[x], f''[x]$ and the higher derivatives look when we replace the real numbers $x$ by complex numbers $z = x + Iy$.

We find the distance $R$ from $a$ to the closest *complex* singularity (blow up) of $f[z]$ or of any derivative $f^n[z]$. If there are no complex singularities, then we agree $R = \infty$.

For instance if

$$f[x] = 1/(x^4 + x^3 + 9x^2 + x + 28),$$

then we set

```
In[1]:=    f[x_]=1/(x^4 + x^3 + 9 x^2 + x + 28)

Out[1]=                 1
           -----------------------
                      2    3    4
             28 + x + 9 x  + x  + x
```

and we look at:

*In[2]:=*     **f[z]**

*Out[2]=*
```
                 1
        -------------------------
                     2     3    4
        28 + z + 9 z  + z  + z
```

*In[3]:=*     **f'[z]**

*Out[3]=*
```
                            2      3
              1 + 18 z + 3 z  + 4 z
        -(-------------------------)
                     2     3    4 2
           (28 + z + 9 z  + z  + z )
```

*In[4]:=*     **f''[z]**

*Out[4]=*
```
                            2      3 2
        2 (1 + 18 z + 3 z  + 4 z )
        -------------------------- -
                     2     3    4 3
           (28 + z + 9 z  + z  + z )

                             2
             18 + 6 z + 12 z
        --------------------------
                     2     3    4 2
           (28 + z + 9 z  + z  + z )
```

or better yet:

*In[5]:=*     **Short[Together[f''[z]],2]**

*Out[5]=*
```
                     2                6
        -502 - 114 z + 156 z  + <<3>> + 20 z
        ------------------------------------
                     2     3    4 3
           (28 + z + 9 z  + z  + z )
```

*In[6]:=*     **Short[Together[f'''[z]],2]**

*Out[6]=*
```
                                        9
        -1686 + 36072 z + <<7>> - 120 z
        -------------------------------
                    2     3    4 4
           (28 + z + 9 z  + z  + z )
```

It becomes clear that the only singularities (blow ups) happen at solutions of

$$z^4 + z^3 + 9z^2 + z + 28 = 0.$$

Find where these happen:

*In[7]:=*     **Solve[z^4 + z^3 + 9 z^2 + z + 28 == 0]**

*Out[7]=*
```
              1 + Sqrt[-15]
        {{z -> -------------},
                    2

              1 - Sqrt[-15]
         {z -> -------------},
                    2

              -2 + 2 Sqrt[-6]
         {z -> ---------------},
                     2

              -2 - 2 Sqrt[-6]
         {z -> ---------------}}
                     2
```

We want to find distance $R$ from $a$ to the closest **complex singularity**. The distance of a complex number $c + Id$ from $a$ is defined to be $\sqrt{(c-a)^2 + d^2}$. So we look at:

*In[8]:=*     **Sqrt[(1/2 - a)^2 +(Sqrt[15]/2)^2]**

*Out[8]=*
```
              15    1       2
        Sqrt[-- + (- - a) ]
              4     2
```

*In[9]:=*     **Sqrt[(1/2 - a)^2 +(-Sqrt[15]/2)^2]**

*Out[9]=*
```
              15    1       2
        Sqrt[-- + (- - a) ]
              4     2
```

*In[10]:=*     **Sqrt[(-2/2 - a)^2 +(2 Sqrt[6]/2)^2]**

*Out[10]=*
```
                        2
        Sqrt[6 + (-1 - a) ]
```

*In[11]:=*     **Sqrt[(-2/2 - a)^2 +(-2 Sqrt[6]/2)^2]**

*Out[11]=*
```
                        2
        Sqrt[6 + (-1 - a) ]
```

We look at the smallest of these numbers and set it equal to $R$.

(i) If $R = 0$, then $f[x]$ has no expansion in powers of $x - a$.

(ii) If $R > 0$, then $f[x]$ has an expansion in powers of $x$. Furthermore if $0 \le r < R$, then the expansion in powers of $(x - a)$ **converges** to $f[x]$ on the interval $[a - r, a + r]$ in the following sense:

Suppose the expansion of $f[x]$ in powers of $x$ is

$$a[0] + a[1](x - a) + a[2](x - a)^2 + a[3](x - a)^3$$

$$+ \ldots + a[k](x - a)^k + \ldots$$

Then the largest difference on $[a - r, a + r]$ between $f[x]$ and

$$a[0] + a[1](x - a) + a[2](x - a)^2$$

$$+ a[3](x - a)^3 + \ldots + a[n](x - a)^n$$

tends to $0$ as $n$ tends to $\infty$.

*Note that the intervals of convergence of expansions in powers of $(x - a)$ are centered at the number $a$ and are not centered at $0$ unless $a = 0$.*

**B.4.b.i)**

On what intervals does the expansion of $1/(4 + x^2)$ in powers of $(x - 3)$ converge to $1/(4 + x^2)$? Illustrate with some plots.

**Answer:**

First look at a few of the higher derivatives:

```
In[1]:=   Clear[x, f]
          f[x_] = 1/( 4 + x^2)
```
```
Out[1]=       1
          ------
               2
          4 + x
```

```
In[2]:=   Together[f[z]]
```
```
Out[2]=       1
          ------
               2
          4 + z
```

```
In[3]:=   Together[f'[z]]
```
```
Out[3]=      -2 z
          ---------
               2 2
          (4 + z )
```

```
In[4]:=   Together[f''[z]]
```
```
Out[4]=         2
          -8 + 6 z
          ---------
               2 3
          (4 + z )
```

```
In[5]:=   Together[f'''[z]]
```
```
Out[5]=               3
          96 z - 24 z
          -------------
               2 4
          (4 + z )
```

```
In[6]:=   Together[f'''''[z]]
```
```
Out[6]=                    3        5
          -11520 z + 9600 z  - 720 z
          --------------------------
                   2 6
               (4 + z )
```

```
In[7]:=   Together[f'''''''''[z]]
```
```
Out[7]=                                            3
          (-928972800 z + 2786918400 z  -

                       5              7
          1463132160 z  + 174182400 z  -

                    9        2 10
          3628800 z ) / (4 + z )
```

The only complex singularies (blow ups) are solutions of $4 + z^2 = 0$:

```
In[8]:=   Clear[z]
          Solve[4 + z^2 == 0]
```
```
Out[8]=   {{z -> 2 I}, {z -> -2 I}}
```

Because we are going to expand $f[x]$ in powers of $(x - 3)$, we look at the distances from $+3$:

```
In[9]:=   Sqrt[(0 - 3)^2 + 2^2]
```
```
Out[9]=   Sqrt[13]
```

```
In[10]:=  Sqrt[(0 - 3)^2 + (-2)^2]
```
```
Out[10]=  Sqrt[13]
```

Read off $R = \sqrt{13}$. The expansion of $f[x] = 1/(4 + x^2)$ in powers of $(x - 3)$ converges to $1/(4 + x^2)$ on intervals $[3 - r, 3 + r]$ as long as $0 \leq r < R = \sqrt{13}$. Let's see some plots:

```
In[11]:=  R = Sqrt[13]
          exp5 = Normal[Series[f[x], {x, 3, 5}]]
```
```
Out[11]=                                       2
           1     6 (-3 + x)   23 (-3 + x)
           -- - ----------- + ------------- -
           13       169           2197

                       3              4
             60 (-3 + x)    61 (-3 + x)
             ------------ + ------------- +
                28561          371293

                         5
             414 (-3 + x)
             --------------
                4826809
```

The plot of $f[x] = 1/(4 + x^2)$ is the thicker of the two:

```
In[12]:=  Plot[{f[x], exp5}, {x, 3 - R, 3 + R},
               AxesLabel->{"x", "y"},
               PlotStyle->{{Thickness[0.01]},
               {RGBColor[0, 0, 1]}}];
```

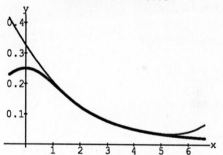

Let's see what happens when we use more of the expansion: Again the plot of $f[x] = 1/(4 + x^2)$ is the thicker of the two.

```
In[13]:=  exp12 = Normal[Series[f[x], {x, 3, 12}]]
          Plot[{f[x], exp12}, {x, 3 - R, 3 + R},
               AxesLabel->{"x", "y"},
               PlotStyle->{{Thickness[0.01]},
               {RGBColor[0, 0, 1]}}];
```

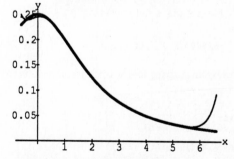

Convergence on subintervals of $[3 - R, 3 + R]$ is indicated by all the shared ink.

**B.4.b.ii)**

On what intervals does the expansion of $1/(4 + x^2)$ in powers of $(x + 5)$ converge to $1/(4 + x^2)$? Illustrate with some plots.

**Answer:**

First look at a few of the higher derivatives:

```
In[1]:=    Clear[x,f]
           f[x_] = 1/( 4 + x^2)
Out[1]=       1
           ------
              2
           4 + x
```

```
In[2]:=    Together[f[z]]
Out[2]=       1
           ------
              2
           4 + z
```

```
In[3]:=    Together[f'[z]]
Out[3]=      -2 z
           ---------
               2 2
           (4 + z )
```

```
In[4]:=    Together[f''[z]]
Out[4]=          2
           -8 + 6 z
           ---------
               2 3
           (4 + z )
```

```
In[5]:=    Together[f'''''[z]]
Out[5]=                  3        5
           -11520 z + 9600 z  - 720 z
           ---------------------------
                        2 6
                   (4 + z )
```

```
In[6]:=    Together[f'''''''''[z]]
Out[6]=                            3
           (-928972800 z + 2786918400 z  -

                    5               7
           1463132160 z  + 174182400 z  -

                    9        2 10
           3628800 z ) / (4 + z )
```

The only complex singularies (blow ups) are solutions of $4 + z^2 = 0$.

```
In[7]:=    Clear[z]
           Solve[4 + z^2 == 0]
Out[7]=    {{z -> 2 I}, {z -> -2 I}}
```

Because we are going to expand $f[x]$ in powers of $(x + 5)$,

we look at the distances from $-5$:

```
In[8]:=    Sqrt[(0 - (-5))^2 + 2^2]
Out[8]=    Sqrt[29]
```

```
In[9]:=    Sqrt[(0 - (-5))^2 + (-2)^2]
Out[9]=    Sqrt[29]
```

Read off $R = \sqrt{29}$. The expansion of $f[x] = 1/(4 + x^2)$ in powers of $(x + 5)$ converges to $1/(4 + x^2)$ on intervals $[-5 - r, -5 + r]$ as long as $0 \le r < R = \sqrt{29}$.

Let's see some plots:

```
In[10]:=   R = Sqrt[29]
           exp6 = Normal[Series[f[x],{x,-5,6}]];
```

The plot of $f[x] = 1/(4 + x^2)$ is the thicker of the two:

```
In[11]:=   Plot[{f[x],exp6},{x,-5 - R, -5 + R},
              AxesLabel->{"x","y"},
              PlotStyle->{{Thickness[0.01]},
              {RGBColor[0,0,1]}}];
```

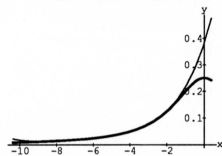

Let's see what happens when we use more of the expansion. Again the plot of $f[x] = 1/(4 + x^2)$ is the thicker of the two.

```
In[12]:=   exp12 = Normal[Series[f[x],{x,-5,12}]]
           Plot[{f[x],exp12},{x,-5 - R, -5 + R},
              AxesLabel->{"x","y"},
              PlotStyle->{{Thickness[0.01]},
              {RGBColor[0,0,1]}}];
```

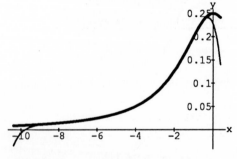

Convergence on subintervals of $[-5 - R, -5 + R]$ is indicated by all the shared ink.

**B.4.c)**

On what intervals does the expansion of $\cos[x]$ in powers

of $(x - \pi/4)$ converge to $\cos[x]$? Illustrate with some plots.

**Answer:**

All the higher derivatives of $\cos[x]$ are multiples of either $\sin[x]$ or $\cos[x]$. Neither $\sin[z]$ nor $\cos[z]$ have any complex singulaities.

Consequently $R = \infty$, and the expansion of $\cos[x]$ in powers of $(x - \pi/4)$ converges to $\cos[x]$ on any interval $[\pi/4 - r, \pi/4 + r]$ as long as $0 \leq r < \infty$.

Here come some plots: (The plot of $\cos[x]$ is the thicker of the two)

```
In[1]:=    Clear[x]
           exp6 = Normal[Series[Cos[x],{x,Pi/4,6}]]
           Plot[{Cos[x],exp6},{x,Pi/4 - 4, Pi/4 + 4},
           AxesLabel->{"x","y"},
           PlotStyle->{{Thickness[0.01]},
           {RGBColor[0,0,1]}}];
```

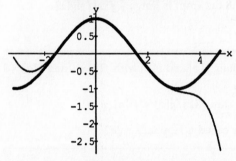

Let's see what happens when we use more of the expansion (again the plot of $\cos[x]$ is the thicker of the two):

```
In[2]:=    exp10 = Normal[Series[Cos[x],{x,Pi/4,10}]]

           Plot[{Cos[x],exp10},{x,Pi/4 - 4, Pi/4 + 4},
           AxesLabel->{"x","y"},
           PlotStyle->{{Thickness[0.01]},
           {RGBColor[0,0,1]}}];
```

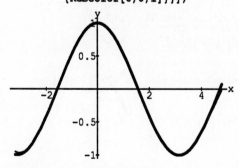

Just about what we expected.

# Tutorial

■ **T.1) Limits.**

**T.1.a)**

Find $\lim_{x \to 1} \sin[\pi x]/(e^x - e)$.

**Answer:**

Look at the early parts of the expansions in powers of $(x - 1)$ of the numerator and the denominator:

```
In[1]:=    numerator = Series[Sin[Pi x],{x,1,3}]
```
$$Out[1]= \quad -(Pi\ (-1 + x)) + \frac{Pi^3\ (-1 + x)^3}{6} + O[-1 + x]^4$$

```
In[2]:=    denominator = Series[E^x - E,{x,1,3}]
```
$$Out[2]= \quad E\ (-1 + x) + \frac{E\ (-1 + x)^2}{2} + \frac{E\ (-1 + x)^3}{6} + O[-1 + x]^4$$

You can spot the limit; it is $-\pi/e$. If you don't see this, then divide the expansions:

```
In[3]:=    numerator/denominator
```
$$Out[3]= \quad -\left(\frac{Pi}{E}\right) + \frac{Pi\ (-1 + x)}{2\ E} + \left(\frac{-Pi}{12\ E} + \frac{Pi}{6\ E}\right)(-1 + x)^2 + O[-1 + x]^3$$

As $x$ closes in on $1$, this closes in on
$$-\pi/e + 0 + 0 + 0 + ...$$

As a result,
$$\lim_{x \to 1} \sin[\pi x]/(e^x - e) = -\pi/e.$$

Note that this is the same value you get by dividing the first term of the denominator into the first term of the numerator.

**T.1.b)**

Find $\lim_{x \to 7}(2 - \sqrt{x - 3})/(x^2 - 49)$.

**Answer:**

Look at the early parts of the expansions in powers of $(x - 7)$ of the numerator and the denominator:

```
In[1]:=    numerator =
           Series[2 - Sqrt[x - 3],{x,7,3}]
```
$$Out[1]= \quad -\frac{(-7 + x)}{4} + \frac{(-7 + x)^2}{64} - \frac{(-7 + x)^3}{512} +$$

```
             O[-7 + x]
```

```
In[2]:=    denominator = Series[x^2 - 49,{x,7,3}]
Out[2]=
                        2              4
           14 (-7 + x) + (-7 + x)  + O[-7 + x]
```

You can spot the limit; it is −1/56. If you don't see this, then divide the expansions:

```
In[3]:=    numerator/denominator
Out[3]=
             1      15 (-7 + x)    109 (-7 + x)
          -(--)  + ----------- -  ------------- +
            56         6272           351232

                        3
             O[-7 + x]
```

As $x$ closes in on 7, this closes in on $-1/56 + 0 + 0 + 0 + \dots$ As a result,

$$\lim_{x \to 7} (2 - \sqrt{x - 3})/(x^2 - 49) = -1/56.$$

Note that this is the same value you get by dividing the first term of the denominator into the first term of the numerator.

### T.1.c)

All you know about a certain function $f[x]$ is that $f[\pi] = 0$ and $f'[\pi] = 8$. Find $\lim_{x \to \pi} f[x]/\sin[x]$.

### Answer:

Taylor's formula tells us that the expansion of $f[x]$ in powers of $(x - \pi)$ opens with:

```
In[1]:=    Clear[f,x]
           Series[f[x],{x,Pi,2}]
Out[1]=    f[Pi] + f'[Pi] (-Pi + x) +

                           2
           f''[Pi] (-Pi + x)          3
           ------------------ + O[-Pi + x]
                   2
```

Because $f[\pi] = 0$ and $f[\pi] = 8$, the expansion of $f[x]$ in powers of $(x - \pi)$ opens with

$$8(x - \pi) + f''[\pi](x - \pi)^2/2 + O[x - \pi]^3.$$

The expansion of $\sin[x]$ in powers of $(x - \pi)$ opens with

$$\sin[\pi] + \cos[\pi](x - \pi) - \sin[\pi](x - \pi)^2/2$$

$$+O[x - \pi]^3 = -(x - \pi) + O[x - \pi]^3.$$

Dividing the first terms gives

$$\lim_{x \to \pi} f[x]/\sin[x] = -8.$$

You should be able to do this problem without computer support, but it does lend itself nicely to automation:

```
In[2]:=    numerator =
           Series[f[x],{x,Pi,2}]/.
           {f[Pi]->0,f'[Pi]->8}
Out[2]=
                                            2
                        f''[Pi] (-Pi + x)
           8 (-Pi + x) + ------------------ +
                                2

                      3
           O[-Pi + x]
```

```
In[3]:=    denominator = Series[Sin[x],{x,Pi,2}]
Out[3]=
                                   3
           -(-Pi + x) + O[-Pi + x]
```

```
In[4]:=    numerator/denominator
Out[4]=
                f''[Pi] (-Pi + x)                2
           -8 - ----------------- + O[-Pi + x]
                        2
```

That big fat −8 out front is $\lim_{x \to \pi} f[x]/\sin[x]$.

### T.1.d)

For two functions $f[x]$ and $g[x]$ with $f[a] = 0$ and $g[a] = 0$ and $g'[a] \neq 0$,

$$\lim_{x \to a} f[x]/g[x] = f'[a]/g'[a].$$

Why? (This is called L'Hopital's rule.)

### Answer:

Easy: write the expansions of $f[x]$ and of $g[x]$ and divide:

```
In[1]:=    Clear[f,g]
           numer=Series[f[x],{x,a,2}]/.f[a]->0
Out[1]=
                                        2
                        f''[a] (-a + x)
           f'[a] (-a + x) + --------------- +
                                   2

                    3
           O[-a + x]
```

```
In[2]:=    denom=Series[g[x],{x,a,2}]/.g[a]->0
Out[2]=
                                        2
                        g''[a] (-a + x)
           g'[a] (-a + x) + --------------- +
                                   2

                    3
           O[-a + x]
```

```
In[3]:=    numer/denom
Out[3]=    f'[a]   f''[a]    f'[a] g''[a]
           -----  + (------- - ------------)
           g'[a]    2 g'[a]            2
                                2 g'[a]

                        2
           (-a + x) + O[-a + x]
```

Then

$$f'[x]/g'[x] = f'[a]/g'[a] + ((f''[a]/2g'[a])$$
$$-(f'[a]g''[a]/2g'[a]^2))(x - a) + O[x - a]^2$$

and this clearly closes in on $f'[a]/g'[a]$. We can think on more exciting things to do.

# Literacy Sheet

1. How are expansions in powers of $(x - a)$ derived from expansions in powers of $x$?

2. Take a pencil and use the expansion of $e^x$ in powers of $x$ to derive the expansion of $e^x$ in powers of $(x - 1)$.

3. Why is it that if you know everything about expansions in powers of $x$, then you are in a position to know everything about expansions in powers of $(x - a)$?

4. At what point on the $x$–axis are the intervals of convergence of the expansions in powers of $(x - a)$ centered?

At what point on the $x$–axis are the intervals of convergence of the expansions in powers of $(x + a)$ centered?

5. Give intervals of convergence of the expansions of the following functions $f[x]$ in powers of $(x - a)$ and choices of $a$:

a) $f[x] = 1/(1 - x)$ with $a = 4$

b) $f[x] = 1/(1 + x)$ with $a = 4$

c) $f[x] = e^x$ with $a = 1$

d) $f[x] = \sin[x]$ with $a = 5\pi$

e) $f[x] = 1/(1 + x^2)$ with $a = 2$.

6. Neither $\sqrt{x}$ nor $\log[x]$ have an expansion in powers of $x$. But if $a > 0$,, then both functions have expansions in powers of $(x - a)$. How do you account for this?

On what intervals $[a - r, a + r]$ do you expect convergence of the expansions in powers of $(x - a)$?

7. Which of the following do you think is the best overall approximation on $[0, 1]$ of $f[x] = e^x$:

$approx1[x] = \textbf{Normal[Series[} E^x \textbf{, } x \textbf{, 0, 8]]}$

$approx2[x] = \textbf{Normal[Series[} E^x \textbf{, } x \textbf{, 1/2, 8]]}$

$approx3[x] = \textbf{Normal[Series[} E^x \textbf{, } x \textbf{, 1, 8]]}$

How would you use your selection to estimate the value of $\int_0^1 e^{-x^4} dx$?

8. What is Taylor's formula for the coefficients in expansions in powers of $(x - a)$?

9. What is L'Hospital's rule? How does L'Hospital's rule come from Taylor's formula? When can you use L'Hospital's rule?

Why is L'Hospital's rule a big deal if you don't know Taylor's formula but not such a big deal if you do know Taylor's formula?

Why would it be a sign of calculus illiteracy if you find $\lim_{x \to 0} \sin[x]/x$ by using L'Hospital's rule?